S0-BRY-206

Fundamentals of Refrigeration

Fundamentals of Refrigeration

Billy C. Langley

I(T)P An International Thomson Publishing Company

Albany • Bonn • Boston • Cincinnati • Detroit • London • Madrid • Melbourne
Mexico City • New York • Pacific Grove • Paris • San Francisco • Singapore • Tokyo
Toronto • Washington

NOTICE TO THE READER

Publisher does not warrant or guarantee any of the products described herein or perform any independent analysis in connection with any of the product information contained herein. Publisher does not assume, and expressly disclaims, any obligation to obtain and include information other than that provided to it by the manufacturer.

The reader is expressly warned to consider and adopt all safety precautions that might be indicated by the activities herein and to avoid all potential hazards. By following the instructions contained herein, the reader willingly assumes all risks in connections with such instructions.

The publisher makes no representation or warranties of any kind, including but not limited to, the warranties of fitness for particular purpose or merchantability, nor are any such representations implied with respect to the material set forth herein, and the publisher takes no responsibility with respect to such material. The publisher shall not be liable for any special, consequential, or exemplary damage resulting, in whole or part, from the readers' use of, or reliance upon, this material.

Cover photo courtesy of: Henry Valve Company
Delmar Staff
Administrative Editor: Vernon Anthony
Project Editor: Eleanor Isenhart
Production Coordinator: Karen Smith
Art Coordinator: Cheri Plasse

Printed in the United States of America

For more information, contact:

Delmar Publishers
3 Columbia Circle, Box 15015
Albany, New York 12212–5015

International Thomson Publishing Europe
Berkshire House 168–173
High Holborn
London WC1V 7 AA
England

Thomas Nelson Australia
102 Dodds Street
South Melbourne, 3205
Victoria, Australia

Nelson Canada
1120 Birchmount Road
Scarborough, Ontario
Canada M1K 5G4

International Thomson Editores
Campos Eliseos 385, Piso 7
Col Polanco
11560 Mexico D F Mexico

International Thomson Publishing GmbH
Königswinterer Strasse 418
53227 Bonn
Germany

International Thomson Publishing Asia
221 Henderson Road
#05–10 Henderson Building
Singapore 0315

International Thomson Publishing - Japan
Hirakawacho Kyowa Building, 3F
2–2–1 Hirakawacho
Chiyoda-ku, Tokyo 102
Japan

1 2 3 4 5 6 7 8 9 10 XXX 01 00 99 98 97 96 95

Library of Congress Cataloging-in-Publication Data

Langley, Billy C., 1931–
Fundamentals of refrigeration / Billy C. Langley.
p. cm.
Includes index.
ISBN 0–8273–6529–2
1. Refrigeration and refrigerating machinery. I. Title.
TP492.L353 1995
621.56--dc20 94–42864
CIP

BRIEF CONTENTS

Preface		xiii
Chapter 1	Refrigeration Fundamentals	1
Chapter 2	Heat	11
Chapter 3	Temperature	35
Chapter 4	Pressure	43
Chapter 5	Vapor Compression Refrigeration Systems	55
Chapter 6	Refrigeration Hand Tools and Test Instruments	67
Chapter 7	Refrigeration Materials	95
Chapter 8	Compressors and Lubrication	117
Chapter 9	Condensers and Receivers	149
Chapter 10	Evaporators	171
Chapter 11	Flow-Control Devices	183
Chapter 12	Accessories	217
Chapter 13	Refrigerants	239
Chapter 14	Refrigerant Recovery, Recycling, and Reclaim	277
Chapter 15	Introduction to Electricity	293
Chapter 16	Electric Motors and Controls	345
Glossary		377
Index		409

CONTENTS

CHAPTER 1 REFRIGERATION FUNDAMENTALS 1

1–1 Fundamentals 2
1–2 Refrigeration 3
1–3 Common Elements 5
1–4 Atoms, Molecules, Chemical Compounds, and Molecular Motion Atoms 6
1–5 Solids, Liquids, Gases, and Change of State 8

CHAPTER 2 HEAT 11

2–1 Heat Movement 12
2–2 Heat Flow 14
2–3 Specific Heat 17
2–4 Sensible Heat 18
2–5 Latent Heat 18
2–6 Relationship of Heat and Temperature 20
2–7 Heat of Compression 21
2–8 Superheat 22
2–9 Heat Calculation 22
2–10 Enthalpy 24
2–11 Mechanical Equivalent of Heat 25
2–12 Cooling 25
2–13 Subcooling 27
2–14 Humidity 27
2–15 Dehydration and Noncondensables 29
2–16 Energy, Work, and Power 31

CHAPTER 3 TEMPERATURE 35

3–1 Temperature Scales 36
3–2 Critical Temperature 38
3–3 Saturation Temperature 38
3–4 Dry-Bulb Temperature 39
3–5 Wet-Bulb Temperature 39
3–6 Wet-Bulb Depression 41
3–7 Dew-Point Temperature 41

CHAPTER 4 PRESSURE 43

4–1 Atmospheric Pressure 44
4–2 Gauge Pressure 45
4–3 Absolute Pressure 46
4–4 Critical Pressure 46
4–5 Pressure Measurement 47
4–6 Laws Affecting Pressure 49
4–7 Pressure-Temperature Relationships 53

CHAPTER 5 VAPOR COMPRESSION REFRIGERATION SYSTEMS 55

5–1 Refrigeration by Evaporation 56
5–2 Compression Refrigeration System Principles 57
5–3 The Basic Refrigeration Cycle 59
5–4 Types of Flooded Systems 62

CHAPTER 6 REFRIGERATION HAND TOOLS AND TEST INSTRUMENTS 67

6–1 Hand Tools 68
6–2 Welding Unit 80
6–3 Test Instruments 85
6–4 Leak Detectors 93

CHAPTER 7 REFRIGERATION MATERIALS 95

7–1 Piping and Tubing 96
7–2 Sweat Fittings 100
7–3 Flare Fittings 107
7–4 Compression, Hose, and "O"-Ring Fittings 109
7–5 Bending and Changing Tube Size 111
7–6 Equivalent Lengths of Pipe (Tubing) 112
7–7 Materials 113

CHAPTER 8 COMPRESSORS AND LUBRICATION 117

8–1 Compressor Types 118
8–2 Compressor Designs 131
8–3 Compressor Valves 134
8–4 Compressor Lubrication 139
8–5 Compression Ratio 141
8–6 Clearance Volume 143
8–7 Compressor Cooling 144
8–8 Factors Controlling Compressor Output 146

CHAPTER 9	**CONDENSERS AND RECEIVERS**	**149**
	9–1 Purpose of the Condenser	150
	9–2 Air-Cooled Condensers	151
	9–3 Water-Cooled Condensers	154
	9–4 Evaporative Condensers	159
	9–5 Counterflow of Cooling Water	160
	9–6 Condenser Capacity	161
	9–7 Condensing Temperature	162
	9–8 Noncondensable Gases	163
	9–9 Cleaning the Condenser	164
	9–10 Condenser Location	165
	9–11 Water-Flow Control Valves	166
	9–12 Liquid Receivers	167
CHAPTER 10	**EVAPORATORS**	**171**
	10–1 Types of Evaporators	172
	10–2 Evaporator Styles	173
	10–3 Heat Transfer in Evaporators	175
	10–4 Calculation of Heat Transferred	176
	10–5 Evaporator Design Factors	177
	10–6 Temperature Difference and Dehumidification	178
	10–7 Evaporator Frosting	179
	10–8 Oil Circulation	180
CHAPTER 11	**FLOW-CONTROL DEVICES**	**183**
	11–1 Purpose	184
	11–2 Types of Flow-Control Devices	184
	11–3 Automatic Expansion Valves (AXVS)	187
	11–4 Thermostatic Expansion Valves (AXVS)	192
	11–5 Capillary Tubes	206
CHAPTER 12	**ACCESSORIES**	**217**
	12–1 Accumulators	218
	12–2 Filter-Driers	220
	12–3 Strainers	225
	12–4 Moisture-Liquid Indicators	226
	12–5 Oil Separator	227
	12–6 Vibration Eliminators	230
	12–7 Discharge Mufflers	232
	12–8 Crankcase Heaters	233
	12–9 Compressor Service Valves	235
	12–10 Check Valves	236
	12–11 Water-Regulating Valves	237

CHAPTER 13 REFRIGERANTS 239

13–1 Refrigerant Characteristics 240
13–2 Effects of Pressure on the Boiling Point 241
13–3 Critical Temperature 242
13–4 Standard Conditions 243
13–5 Condensing Pressure 243
13–6 Vaporizing Pressure 246
13–7 Latent Heat of Vaporization 247
13–8 Types of Refrigerants 249
13–9 Refrigerant-Oil Relationships 259
13–10 Refrigerant Tables 260
13–11 Pocket Pressure-Temperature Charts 262
13–12 The P-H Diagram 262
13–13 Handling of Refrigerant Cylinders 276

CHAPTER 14 REFRIGERANT RECOVERY, RECYCLING, AND RECLAIM 277

14–1 Stratospheric Ozone 278
14–2 The Clean Air Act 280
14–3 National Recycling and Emission-Reduction Program 282
14–4 Federal Tax 283
14–5 Section 608 of the Clean Air Act 284
14–6 Recovery and Recycling Equipment 289

CHAPTER 15 INTRODUCTION TO ELECTRICITY 293

15–1 The Nature of Matter 294
15–2 Structure of an Atom 297
15–3 Law of Charges—Static Electricity 301
15–4 Electrostatic Charges 303
15–5 Electrostatic Fields 305
15–6 Electron Orbits 306
15–7 Electrical Conductors 311
15–8 Electrical Insulators 312
15–9 Electric Current 313
15–10 Voltage (EMF—Electromotive Force) 315
15–11 Current (Amperage) 316
15–12 Resistance 316
15–13 Ohm's Law 317
15–14 Electrical Circuit 318
15–15 Power and Energy 326
15–16 Conductors 327
15–17 Resistors 330
15–18 Magnetism 331
15–19 Electromagnets 334
15–20 Inductors 336
15–21 Capacitance 338

CHAPTER 16 ELECTRIC MOTORS AND CONTROLS 345

16–1 Electric Motor Theory 346
16–2 Capacitors 350
16–3 Split-Phase Motors 353
16–4 Capacitor-Start (CSR) Motors 355
16–5 Permanent-Split Capacitor (PSC) Motors 355
16–6 Capacitor-Start/Capacitor-Run (CSCR) Motors 356
16–7 Shaded-Pole Motors 357
16–8 Two-Speed Motors 359
16–9 Centrifugal Switch 360
16–10 Single-Phase Motor Protectors 361
16–11 Testing Single-Phase Motors 365
16–12 Replacing Capacitors 368
16–13 Starting Relays 369
16–14 Starters and Contactors 375

GLOSSARY 377

INDEX 409

PREFACE

Fundamentals of Refrigeration is a basic text written to aid the student in learning the underlying fundamentals of this industry. It should also be helpful to the instructor in planning the class schedule and in making reading assignments. The text will be of benefit to those who are currently working in the industry at the beginning level, as well as those that are studying to learn this industry.

Fundamentals of Refrigeration was written in a manner that allows it to be used as an instructor's curriculum guide, as a textbook for classroom study, or for independent study. The text presents the practical fundamentals and service procedures and safety instructions with which the learner should become familiar. The material in this text is organized to be used in short survey courses or in comprehensive course study in high schools, two-year technical programs, and trade schools.

ORGANIZATION

Fundamentals of Refrigeration presents the material in chapters that deal with specific topics. Each chapter is divided into numbered sections that present smaller parts of the topic covered in the chapter. Every chapter begins with objectives indicating the things that the reader should gain from that study and an introduction. At the end of each numbered topic are a summary and review questions. The review questions usually require thought-provoking answers. There are safety procedures introduced at the appropriate point of study. There are also troubleshooting and service procedures introduced in areas that allow this type of activity. The language used is as close as possible to that typically used in the field. The text is supported by the generous use of photos, line drawings, tables, and charts. There are examples used to aid in the understanding of the material presented.

Chapter Organization

Chapter 1 introduces the reader to the fundamentals of refrigeration which must be learned before success is possible. Chapter 2 presents heat and is divided into sixteen separate units. Temperature and its importance to refrigeration and air conditioning work are discussed in Chapter 3. Pressure and its various aspects are covered in Chapter 4. Vapor compression refrigeration systems are presented in Chapter 5. Chapter 6 discusses tools and test instruments that are commonly used in refrigeration and air conditioning work. Refrigeration materials which include piping, fittings, working with tubing, and how to determine the equivalent length of pipe are presented in Chapter 7. Chapter 8 presents compressors and lubricants and how they are used in the system. Chapter 9 discusses condensers and receivers and their importance to the operation of the system. Evaporator styles, heat transfer, and temperature difference are discussed in Chapter 10. The various flow control valves and capillary tubes that are popular for use on refrigeration systems are the topics of Chapter 11. Accessories and how they fit into the overall system are presented in Chapter 12. Chapter 13 discusses both old and new refrigerants and their purpose and safe working procedures in overall system operation. Refrigerant recovery, recycling, and reclaim procedures are presented in Chapter 14. Chapter 15 is an introduction to electricity, how it works, and how it is used in refrigeration and air conditioning systems. Electric motors, their controls, and purpose in refrigeration and air conditioning systems are presented in Chapter 16.

TO THE READER

Study this material with assurance that you are being exposed to a very comprehensive presentation. I wish you the very best in your endeavors in this exciting industry.

Billy C. Langley

CHAPTER

1

REFRIGERATION FUNDAMENTALS

OBJECTIVES

Upon completion of this chapter, you should be able to:

- Understand the fundamentals of refrigeration
- Define refrigeration
- Describe the theories of molecular motion
- Explain how chemical compounds exist
- Describe the three states of matter

INTRODUCTION

Cooling of some kind has been used since the beginning of time. In the early years, water and ice were used to provide cooling. It is only in the last century that mechanical refrigeration has been used on a wide basis. One of the earliest methods of providing cooling was with the ice box. Movie houses and department stores were the only places that used air conditioning in the early years.

The modern way of living requires that refrigeration equipment be used for the proper preservation of food and for human comfort. Many of the modern industrial and commercial processes, such as food storage, textile manufacturing, and printing, depend on the proper operation of air conditioning and refrigeration equipment for efficient, economical operation of the manufacturing processes and equipment.

There are many fields included in the refrigeration and air conditioning industry, such as physics and chemistry. Having good mechanical ability is very important to those who design, install, or service these systems. Customer relations is a very important aspect that must be practiced faithfully for success in this industry.

For those who are interested in selling, designing, manufacturing, servicing, and installing air conditioning and refrigeration equipment there are many opportunities available. Anyone who becomes proficient in any of the areas encompassed by this industry will be in great demand. However, before one can become proficient in this industry, it is necessary to thoroughly understand the basic principles of refrigeration.

1–1 FUNDAMENTALS

The fundamentals of refrigeration must be understood before any success can be achieved in this industry. There is a great desire by many beginners to bypass learning and understanding the underlying fundamentals necessary. When this step is taken, the learner is not going to have the foundation upon which to build a secure future.

Brief History of Refrigeration

As stated before, since the beginning of time people have used some means of food preservation. In the beginning, the food was lowered into a well or was stored in caves that were cooler than the surrounding temperatures. Then, natural ice was used. The ice was cut from rivers and lakes during the winter and stored until it was needed during the warmer weather.

When the harvesting of natural ice became efficient and plentiful, the ice box was more widely used. However, transporting the ice from the cooler to the warmer climates was a problem. Because of this, natural ice was considered a luxury. In some instances, ice frozen from dirty water and containing germs could not be safely used.

More than a century ago, an English scientist used pressure and a lower temperature to successfully change ammonia gas to a liquid. This was done by increasing the pressure and reducing the temperature. When the pressure was released, the ammonia liquid boiled off very rapidly and changed back to a gas. When the liquid changed to a gas, heat was absorbed from the objects surrounding the ammonia. This was an important discovery that eventually led to the development of the refrigeration equipment used today.

The first commercial ice machine was used around 1825. The ice machine produced ice that was purer than that made by nature. Also, the ice-making process did not depend on atmospheric conditions.

As time passed, the demand for more cooling was increased. The need for refrigeration became more apparent. This led to a varied and interesting industry. It was discovered that when foods were stored below 50°F and above 32°F they lasted longer. The spoilage resulting from microbic growth or freezing was prevented. Today, this temperature range is known as the food safety zone, and is commonly known as safety zone refrigeration.

SUMMARY 1–1

- The fundamentals of refrigeration must be understood before success in the refrigeration industry can be achieved.
- Since the beginning of time people have used some means of food preservation.
- Natural ice was used for cooling foods but it sometimes contained germs and could not be safely used. Natural ice was considered to be a luxury.
- An experiment proved that adding pressure and lowering the temperature of ammonia gas could change it to a liquid.
- Safety zone refrigeration is between 50°F and 32°F, which is the most satisfactory storage temperature for food.

REVIEW QUESTIONS 1–1

1. Name two things that are very important for success in the refrigeration industry.
2. What are some of the early uses for air conditioning?
3. What is the major use of refrigeration in modern times?
4. What were the first methods used for cooling purposes?
5. Why does textile manufacturing depend on air conditioning?
6. For what health reason was natural ice not used?
7. What happens to a gas when its temperature is lowered and its pressure is raised?
8. What is the correct storage temperature range for foods?
9. What foundation is necessary before success in the refrigeration and air conditioning industry can be achieved?
10. What happens to a liquid when the pressure applied to it is released?

1–2 REFRIGERATION

Refrigeration is commonly defined as the process of removing heat from a space or material and maintaining that space or material at a temperature lower than its surroundings. Normally, a closed refrigerant circuit (system) is used to move the heat from inside the space or material and deposit it where it is not objectionable. As the refrigeration system continues to operate, more heat is removed from inside the space and both the space and its contents are further cooled. The more heat removed from a space the colder the objects inside that space become.

Refrigeration is the process used for cooling perishable foods, vegetables, and other products so that they remain usable for a longer period of time. Other processes such as air-conditioning and process cooling use refrigeration equipment to remove some of the moisture from the air and at the same time cool the air to a desirable temperature.

Refrigeration is made possible by the circulation of a fluid, called the refrigerant, through a series of pipes and devices, known as the system. During the circulation process, the refrigerant is compressed, cooled, and evaporated (changed to a vapor). The compressor causes the refrigerant to circulate through the system by causing an increase in pressure on one side of the system and a reduction of pressure on the other side. See Figure 1–1.

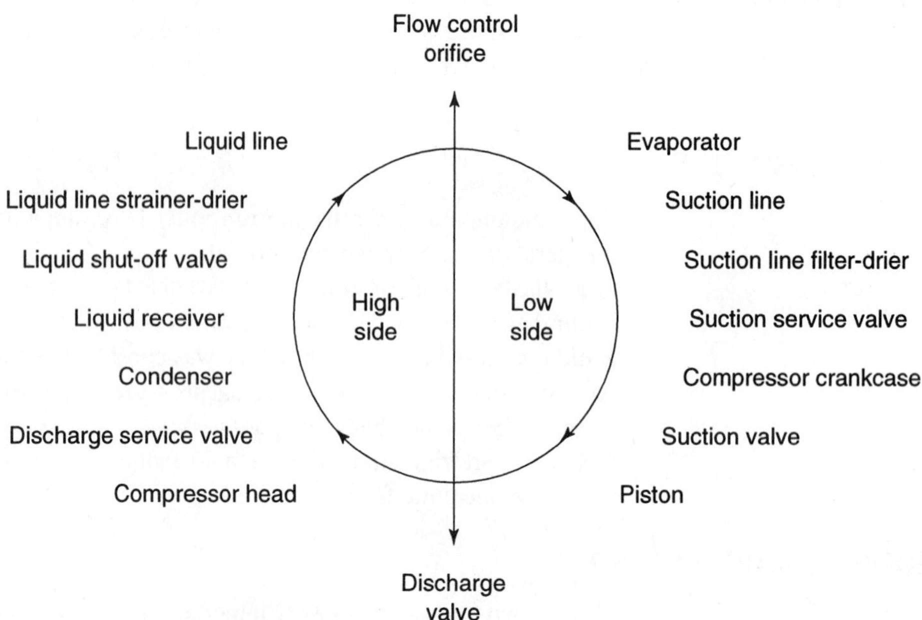

Figure 1–1 Refrigerant circuit diagram

During the operating cycle, the compressor discharges the compressed refrigerant into a device called the condenser. Here the refrigerant is cooled and liquefied. The liquid then flows through the piping to the flow control device, where the flow is metered and the pressure reduced. The refrigerant under lower pressure then flows into the evaporator where it evaporates, absorbing heat from the evaporator and its surroundings. In this process, the refrigerant absorbs heat and changes to a low pressure vapor. The compressor then pumps the refrigerant vapor from the evaporator through the suction line to the compressor, where the cycle is started over again.

This is a very brief explanation of the refrigeration system. However, for any degree of success in this industry, it should be apparent that people involved will need a basic understanding of physics and mechanics. The following sections in this book will provide more theories upon which to build your knowledge and help with advancement in your career.

SUMMARY 1–2

- Refrigeration is defined as the process of removing heat from a space or material and maintaining that space or material at a temperature lower than its surroundings.
- Normally a closed refrigerant circuit (system) is used to move this heat.
- Processes such as air conditioning and process cooling use refrigeration equipment to remove some of the moisture from the air and at the same time cool the air to some desirable temperature.
- As refrigerant flows through the circuit it is compressed, cooled, and evaporated.
- The compressor discharges the compressed refrigerant gas into the condenser.
- The gas is cooled and liquefied in the condenser.
- The flow control device meters the refrigerant into the evaporator.
- The compressor then pumps the refrigerant from the evaporator through the suction line to the compressor.
- People involved in this industry will need a basic understanding of physics and mechanics.

REVIEW QUESTIONS 1–2

1. What is the name of the closed circuit that is used to move heat from one place to another?
2. What is the name of the fluid that is circulated inside a closed system?
3. What must be done to a refrigerant when changing it from a gas to a liquid?
4. What happens to the refrigerant as it passes through the flow control device?
5. Briefly, what happens to the refrigerant in the evaporator?

1–3 COMMON ELEMENTS

There are currently more than 100 basic elements known to mankind. Of these, 92 of them are natural elements, with the remaining being synthetic, or manmade. Most substances are a combination of more than one element. The following is a discussion of the most common elements.

Aluminum, Cadmium, Chromium, Copper, Gold, Iron, Nickel, Silver, Tin, Tungsten, and Zinc. These elements are generally considered to be metals and are most often used alone in objects. There are, however, some found in mixtures, or compounds, commonly known as alloys. These elements are usually present in the solid form.

Calcium, Potassium, Silicon, Sodium, and Sulphur. These elements are found in many materials; however, they are almost always found in chemical combination with other elements. They are generally found in the solid form.

Carbon. This is the major element found in coal, cloth, gasoline, natural gas, oil, and paper. It may also be found in carbon dioxide, methyl chloride, and the fluorocarbon refrigerant that is presently used in refrigeration and air conditioning systems. It exists as a solid at atmospheric pressures and temperatures.

Nitrogen. Nitrogen constitutes about 78 percent of the air around us. It is very important in the growth and life of plants and it is found in nature as a gas.

Oxygen. The atmosphere is made up of about 21 percent oxygen. Oxygen is essential to all animal and human life. Oxygen is an active element that combines readily with most other chemicals to form oxides or more complex chemicals. It is found in nature as a gas.

Air. Air is made up of about 21% oxygen and 78% nitrogen. The other 1% is made up of other gases, namely argon, carbon dioxide, helium, hydrogen, krypton, ozone, and xenon.

Hydrogen. Hydrogen is commonly found in many compounds. It is especially important in the structure of acids, fuels, and oils. It is seldom found alone in nature. It is an extremely light gas form. When it is burned, water is formed. Burning is the process of properly combining hydrogen and oxygen. It is the process used in combustion furnaces and boilers.

SUMMARY 1–3

- Nitrogen constitutes about 78% of the air around us.
- The atmosphere is made up of about 21% oxygen. Oxygen is essential to all animal and human life.
- Air is made up of about 21% oxygen and 78% nitrogen. The other 1% is made up of other gases.
- Hydrogen is especially important in the structure of acids, fuels, and oils.

REVIEW QUESTIONS 1–3

1. What percentage of the air around us is nitrogen?
2. What component in the atmosphere is about 21%?
3. What element in the atmosphere combines with other chemicals to form more complex chemicals?
4. What atmospheric element is important in the structure of acids, fuels, and oils?
5. What is formed when hydrogen is burned?

1–4 ATOMS, MOLECULES, CHEMICAL COMPOUNDS, AND MOLECULAR MOTION ATOMS

Atoms are the particles that form to make up each element. They number into the millions in each element. An atom is the smallest particle that makes up an element. An atom is so small it cannot be seen even with a very strong microscope. For the purposes of this text we will consider an atom as invisible and unchangeable. It cannot be divided by ordinary means. The atoms that make up all elements in the universe are different; thus, iron is made up of iron atoms, and hydrogen is made up of hydrogen atoms.

Scientific study has revealed many things about atoms. However, how they are known is beyond the scope of this text. We must accept some things as being true if we are to understand the fundamentals of refrigeration.

Molecules. The molecule is just larger than an atom and is the next larger particle of a substance. A molecule is made up of one or more of only one kind of atom. They are generally referred to as a molecule of that element. It is possible for a molecule to contain more than one kind of atom. It must be remembered, however, that a molecule can contain several of the same type of atom. As an example, a molecule of iron contains only one iron atom, while a molecule of sulphur will contain eight sulphur atoms.

A very small piece of any element is made up of billions of molecules. Each of these molecules is made up of one or more atoms of that same element.

Chemical Compounds. The molecules that combine to form a chemical compound are made up of two or more atoms from different elements. This combining of the different elements causes the substance to become something entirely different. The new substance probably has none of the characteristics of either of the other elements that combine to make it. For example, a molecule of water is made up of two atoms of hydrogen and one atom of oxygen, both of which are a gas.

The refrigerants that are used in refrigeration systems are prime examples of some of the chemical compounds that are found in this industry.

Many of the common substances that we use in our everyday lives are chemical compounds, such as table salt, baking soda, and calcium.

Molecular Motion. From the previous information it should be understood that all matter is made up of very small particles known as molecules. These molecules may exist in one or a combination of three states. These three states are solids, liquids, and gases. Molecules can be broken down into their components, which are known as atoms. The chapter on basic electricity will discuss atoms in much more detail.

However, in this unit we will discuss the theory of molecular motion and its action as it is used in refrigeration and air conditioning. The amount of movement or vibration of the molecules is what determines the amount of heat present in any given body, or substance. This heat is produced through the friction of the molecules rubbing against each other. As the temperature is increased, the attraction of these molecules to each other is reduced. When heat is removed from a substance, the molecular motion decreases. If a substance is cooled to absolute zero, all molecular motion stops because at absolute zero the substance contains no heat.

Each molecule has a different weight, shape, and size. Like molecules will cling together to form a substance. Thus, the substance has the same characteristics as the molecules that form together to make it. All molecules are capable of moving around, and the substance formed will, to a degree, depend on the amount of space between them. The molecules in a solid will be closer together than those in either a liquid or a gas. A liquid will have more space between the molecules than a solid. But, it will have less space between the molecules than a gas. A gas has the greatest amount of space between the molecules. Any particular substance can be made to exist in any one of the three forms, a solid, liquid, or a gas, by the addition or removal of heat.

SUMMARY 1–4

- Atoms are the particles that form to make up each element. Each element contains millions of atoms.
- The atoms that make up all the elements in the universe are different.
- Molecules are larger than atoms, the next larger particles of a substance.
- A molecule can contain several of the same type of atom.
- The molecules that combine to form a compound are made up of two or more atoms from different elements.
- When a compound is formed, the new substance probably has none of the characteristics of any of the other elements that combine to make it. Refrigerants are prime examples of this phenomenon.
- All matter is made up of very small particles known as molecules.
- Molecules may exist in solids, liquids, or gases.
- Molecules can be broken down into each element or atom.
- The amount of movement or vibration is what determines the amount of heat present in any given body or substance. The heat is caused by the friction between the molecules.
- Each kind of molecule has a different weight, shape, and size.
- Like molecules will form together to form a substance.
- All molecules are capable of moving around, and the substance formed will, to a degree, depend on the amount of space between them.

REVIEW QUESTIONS 1–4

1. What is the smallest particle in an element?
2. How many kinds of atoms make up molecules?
3. How many kinds of atoms does it take to form a compound?
4. Is it possible that a chemical compound will not resemble any of its components?
5. In a refrigeration system, what component is a compound?
6. Name the three states of matter.
7. By what is the heat in any given body determined?
8. What determines the characteristics of a substance?
9. In what type of material are the molecules the closest together?
10. In what type of substance do the molecules have the greatest amount of friction?

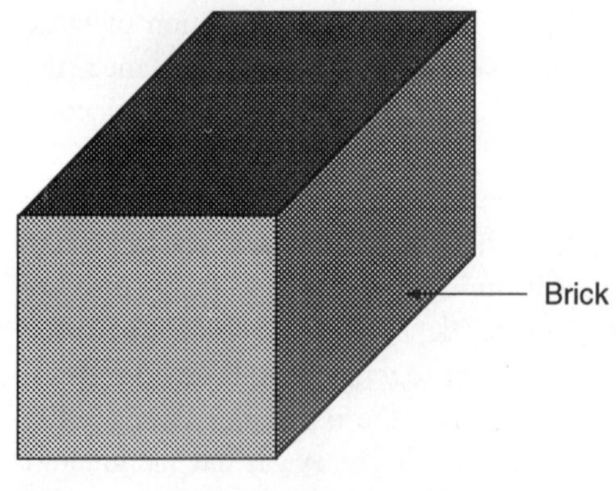

Figure 1–2 A solid

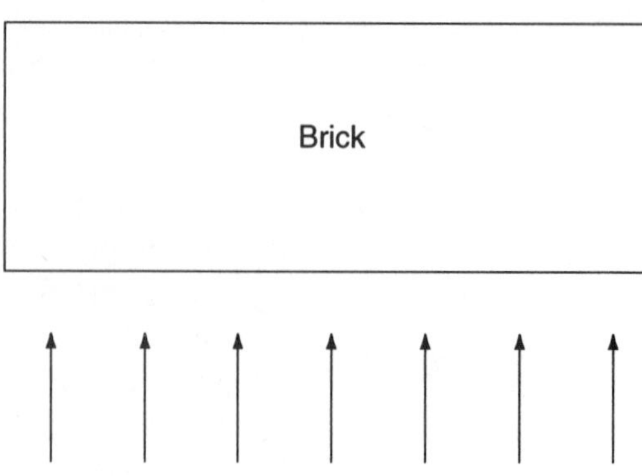

Figure 1–3 Force supporting a solid

1–5 SOLIDS, LIQUIDS, GASES, AND CHANGE OF STATE

As stated earlier, the three states of matter are solids, liquids, and gases. These three forms are important in understanding refrigeration and air conditioning fundamentals.

Solids. A solid is a substance that has the ability to keep its shape when it is supported. See Figure 1–2.

All of the molecules in a solid are identical in size and shape and will keep their relative position inside the substance. As in all substances, the molecules are constantly vibrating. The speed at which the molecules vibrate is dependent on the type of substance and its temperature. Remember, at higher temperatures the molecules vibrate at a faster speed than at lower temperatures.

The gravity of the earth has a stronger pull on solids than it does on either liquids or gases. Because of this, a solid must be supported or it will fall. The force that supports a solid is always in the upward direction. See Figure 1–3.

When a solid has stopped all movement and its weight is being fully supported by a force, its weight and the supporting force are said to be in total equilibrium.

Liquids. When a substance is a liquid, it has the ability to assume the shape of its container. See Figure 1–4.

The force exerted by a liquid on its container is in both an outward and downward direction. The weight of the liquid and the magnetic pull of the earth will cause a stronger force toward the bottom of the container.

As in any substance, all the molecules in a liquid are constantly vibrating. The vibration of the liquid molecules are greater than in a solid because the liquid molecules have more space between them and, therefore, less attraction to each other than those in a solid. The speed with which these molecules vibrate is in direct relation to the temperature of the substance; the higher the temperature the higher the rate of vibration. As the liquid becomes warmer the molecules take up less space, become lighter, and move toward the top of the liquid where the force is less. At this point, some of the molecules will break through the surface of the liquid and escape into the atmosphere. This escapement of molecules is known as *evaporation.*

Gases. A gas has the ability to escape to the atmosphere if it is not confined in a sealed container. See Figure 1–5.

Gaseous molecules have very little attraction for each other or for other substances. Thus, they are free to move around and will sometimes move quite violently.

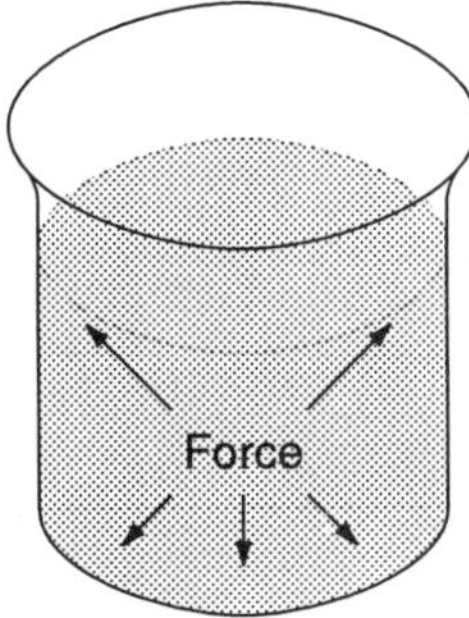

Figure 1–4 Liquid force inside a container

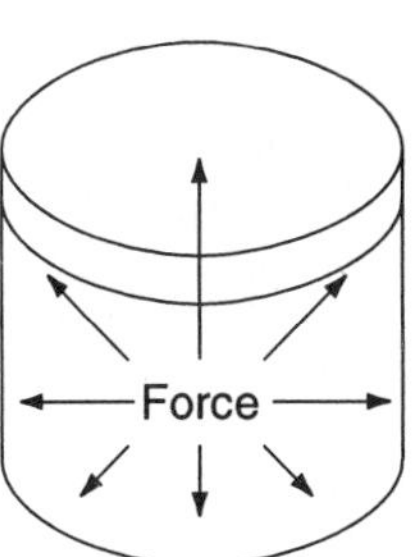

Figure 1–5 Gas inside a container

When free to move, these molecules travel in a straight line, bouncing off one another and the walls of the container during their movement.

Change of State. When a sufficient amount of heat is added to or removed from a substance, it may be changed from one state to another. As an example, the three states of water are ice, water, and steam. When heat is added to the ice, it will change to water. Then, with the addition of a sufficient amount of heat, the water will change into steam. This process is also reversible. When the same amount of heat is removed the steam will change to water. When an additional, specific amount of heat is removed from the water it will change into ice.

SUMMARY 1–5

- The three states of matter are solids, liquids, and gases.
- A solid is a substance that has the ability to keep its shape when supported.
- All the molecules in a solid are identical in size and shape and will keep their relative position in the substance.
- At high temperatures, molecules vibrate at a faster speed than at lower temperatures.
- The force that supports a solid is always in the upward direction.
- A liquid has the ability to assume the shape of its container.
- The force exerted by a liquid is in both the outward and downward direction.
- The vibration of the molecules in a liquid is greater than in a solid because they have more space between them.
- The speed with which the molecules vibrate is in direct relation to the temperature.
- A gas has the ability to escape to the atmosphere if it is not confined in a sealed container.
- Gaseous molecules travel in a straight line, bouncing off one another and the walls of the container.
- When sufficient heat is added to or removed from a substance it may be changed from one state to another.

REVIEW QUESTIONS 1–5

1. In what type of substance are the molecules identical in shape and size?
2. At what temperature do molecules have the greatest amount of activity?
3. In what direction is the force on a container of liquid?
4. In what substance are the molecules constantly vibrating?
5. In what substance do the molecules have the least attraction for each other?
6. In what substance, when it gets warmer, do the molecules become lighter?
7. What is the process known as when a liquid changes to a solid?

CHAPTER

2

HEAT

OBJECTIVES

Upon completion of this chapter, you should be able to:

- Understand heat movement
- Know the definition of Btu
- Know the ways that heat is transferred
- Understand the definitions of each of the different kinds of heat
- Understand the relationship of heat and temperature
- Be familiar with the heat calculation procedure
- Know why the mechanical equivalent of heat is applied to a refrigeration system
- Understand the principles of humidification
- Know why the dehydration of a refrigeration system is important

INTRODUCTION

There are two words that are very common in our everyday language. They are "hot" and "cold." A boiling teakettle is considered to be very hot and ice is considered to be cold. This is because of the lack of understanding of the term cold.

Actually, there is no such thing as cold. "Cold" is a relative term meaning the lack of heat. When something is said to be cold, it is really meant that there is an absence of heat in the substance. If we sit close to a window during cold weather we say that the cold is coming into the room. Actually, the lower temperature that we feel is caused by heat going out the window and creating a lower temperature in that area. It is this lower temperature that we are feeling. Thus, to make something cold, we must remove the heat from the substance. This is the purpose of refrigeration and air conditioning.

2–1 HEAT MOVEMENT

Heat is measured by its effects on certain substances. If we place a container of water on a stove and light the burner, heat will be transferred from the flame to the water. See Figure 2–1.

The water in the bottom of the container will quickly become warmer than the water at the top of the container. As more heat is applied, the water in the bottom of the container will remain close to the same temperature. However, the water in the top of the container will gradually approach the temperature of the water in the bottom. See Figure 2–2.

This is the same process for all materials. A solid will transfer heat in much the same manner. It must be remembered that all materials will have a different heat transfer coefficient that must be taken into account during the heating process.

The continued application of heat to a substance may, in addition to causing an increase in the temperature, also cause a change of state of that substance. A solid may be changed into a liquid or a liquid changed into a gas.

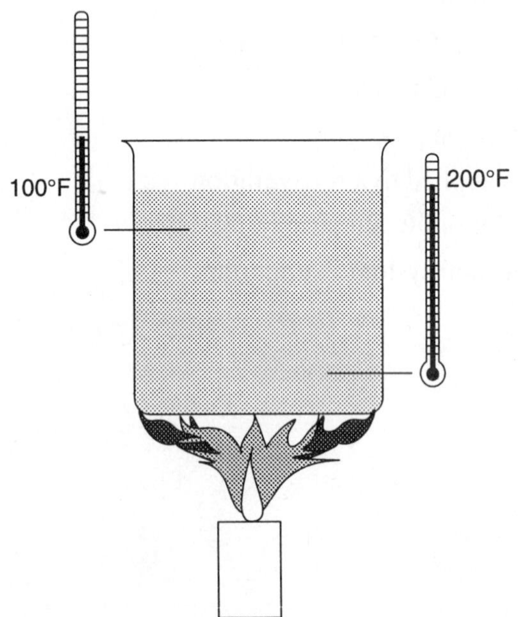

Figure 2–1 Heat movement in water

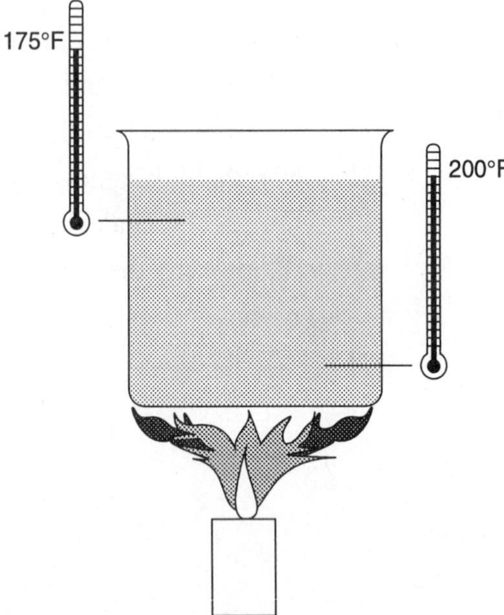

Figure 2–2 Water shortly after being heated

It must be remembered that there is a difference between the words “heat” and “temperature.” *Heat measurement*, by definition, is the measure of quantity, and *temperature measurement* is a measure of degree or intensity. Thus, the words “heat” and “temperature” should not be confused. Two bodies may have the same temperature, but because one is larger than the other and because of their heat-absorbing capacity, their heat content may actually be quite different. As an example, a gallon of water and a quart of water at the same temperature will contain two different amounts of heat. In this example, the gallon will contain four times more heat than the quart. See Figure 2–3.

British Thermal Unit (Btu)

The British thermal unit is the standard for measuring a quantity of heat. The Btu is defined as the amount of heat necessary to raise the temperature of one pound of pure water one degree Fahrenheit. It should be noted that this measurement has nothing to do with the degree or intensity of the heat. The Btu is a measure of the amount of heat in a given body, and degrees Fahrenheit is a measure of the intensity of the heat (temperature).

SUMMARY 2–1

- Heat is measured by its effects on certain substances.
- It must be remembered that all materials have a different heat transfer coefficient that must be taken into account during the heating process.
- The continued application of heat to a substance may, in addition to causing an increase in the temperature, also cause a change of state of that substance.
- Heat measurement, by definition, is the measure of the quantity, and temperature measurement is a measure of degree or intensity.
- The British thermal unit is the standard for measuring a quantity of heat.
- The Btu is defined as the amount of heat required to raise the temperature of one pound of pure water one degree Fahrenheit.

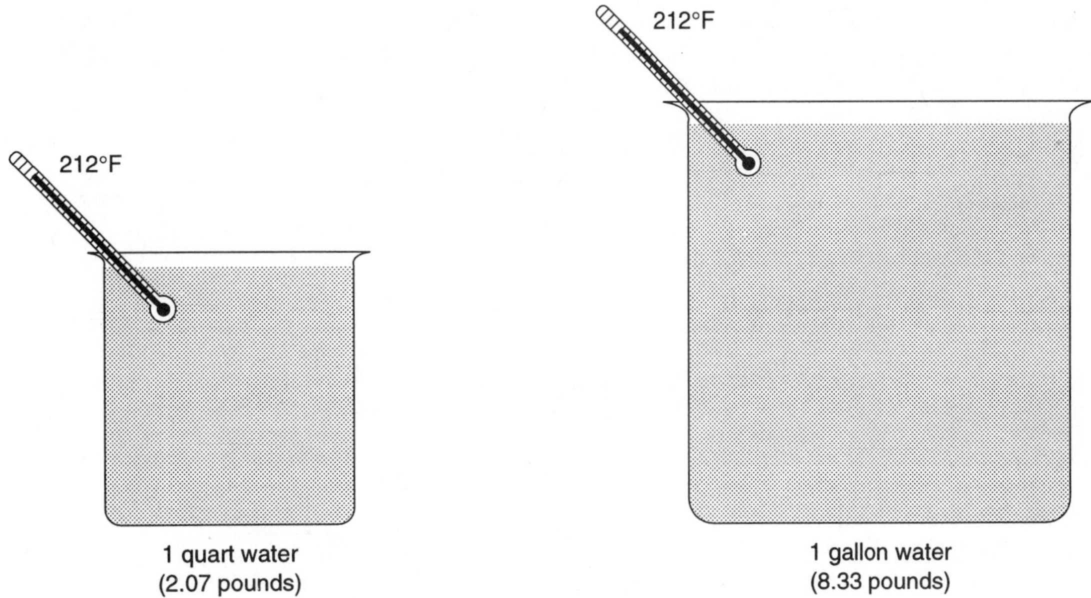

Figure 2–3 Heat content of two different quantities of water

REVIEW QUESTIONS 2–1

1. By definition, what is heat measurement?
2. By definition, what is temperature measurement?
3. If we have a pound of copper at 70°F and two pounds of copper at 70°F, which will contain the most heat?
4. What is the standard unit for measuring heat?
5. Do all materials have the same heat transfer coefficient?

2–2 HEAT FLOW

It must be understood that heat always flows from a warmer to a cooler object. This can be related to the way that water flows downhill. How fast the heat flows depends on the temperature difference between the two objects. The greater the temperature difference, the greater the rate of heat flow between the two.

Example: If two objects of exactly the same type of material were lying next to each other and insulated from their surroundings, one weighing 1 lb with a temperature of 400°F and the other object weighing 1,000 lb with a temperature of 300°F, the heat content of the larger object will be much greater than the heat content of the smaller. But, because of the temperature difference between them, the heat will flow from the smaller object to the larger object until the two have exactly the same temperature. At the beginning, the temperature flow would be the greatest. As the two objects approach the same temperature, the flow would be reduced to a mere trickle and would progressively be reduced until it completely stopped.

There are three ways that heat can be transferred from one object to another. They are: (1) conduction, (2) convection, and (3) radiation.

Conduction

Conduction is the flow of heat through an object or between two objects that are touching. See Figure 2–4.

During this method of heat transfer very little heat is lost, making it a very efficient method of heat transfer. Conduction can be illustrated by heating one end of a metal bar while holding the other end with the bare hand. The heat will travel from the heated end to the cooler end and can be felt with the hand. If the bar is heated enough and it is held long enough, the hand may eventually be burned. The heat is transferred from the heated end to the cooler end by conduction.

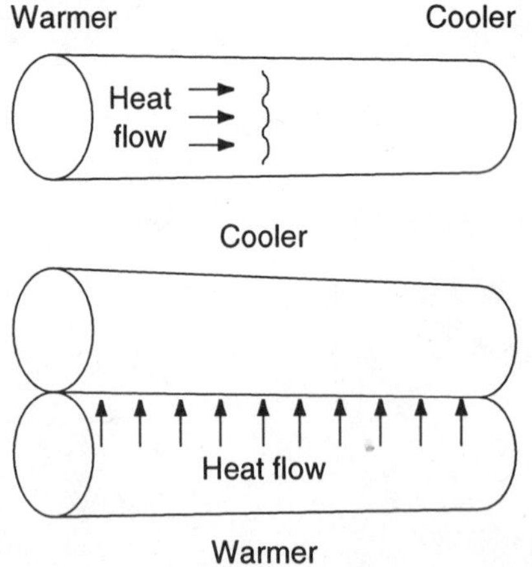

Figure 2–4 Heat transfer by conduction

Convection

Heat flows by convection through a fluid, either a gas or a liquid. The two most popular fluids used in this method are air and water. As the fluid is heated it becomes less dense and will rise. At the same time, the cooler fluid will contract, become heavier, and fall, creating a continuous movement within the fluid. Perhaps a more familiar example of convection is with a warm air furnace. As the air is heated and blown into the room, the objects in the room become warmer because the heat is being transferred by convection. See Figure 2–5.

Radiation

In this method, heat is transferred by wave motion. Heat can be radiated by light waves or radio frequency waves. We are all familiar with how the rays from the sun will warm us. During the radiation process, the air between the heat source and the objects being heated remains unheated. This can be noticed when a person steps from the shade into the direct rays of the sun. In either location the air temperature is very close to the same, but the rays cause the person to feel warmer. See Figure 2–6.

It should be noted that there is very little radiation occurring at low temperatures and when the temperature differences between the objects is very small. Therefore, heat transfer by radiation is of little consequence in refrigeration, unless the refrigerated space is located in the direct rays of the sun. This will then cause an additional load on the refrigeration unit, in some instances, requiring a larger refrigeration unit than if the space were located out of the sun's rays.

It should be noted that heat will travel in a combination of these three methods at the same time. Remember that heat transfer cannot occur without a temperature difference. Each material has a different capability to transfer heat. Metal is a good conductor of heat while fiberglass is a poor conductor. A poor conductor is considered to be an insulator. The capability of a refrigeration unit to transfer heat is known as the overall rate of heat transfer.

The factors having the greatest effect on heat transfer are (1) temperature difference, (2) amount of surface area, and (3) type of material.

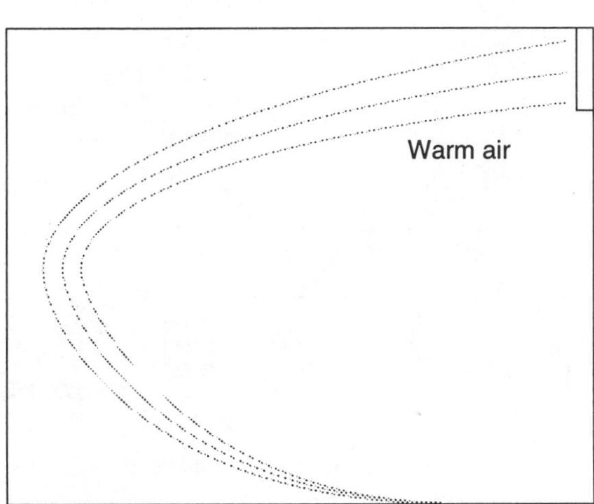

Figure 2–5 Heat transfer by convection

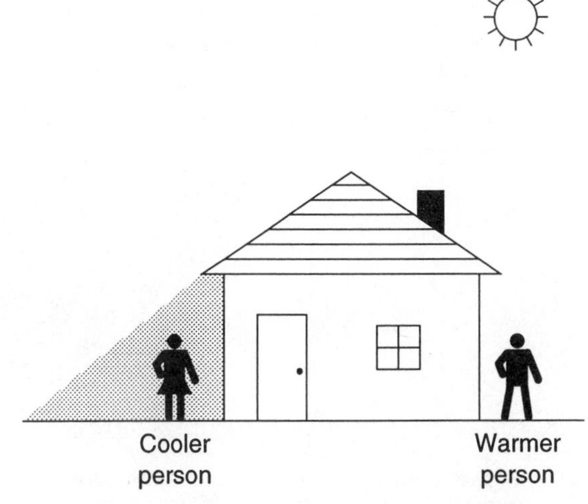

Figure 2–6 Heat transfer by radiation

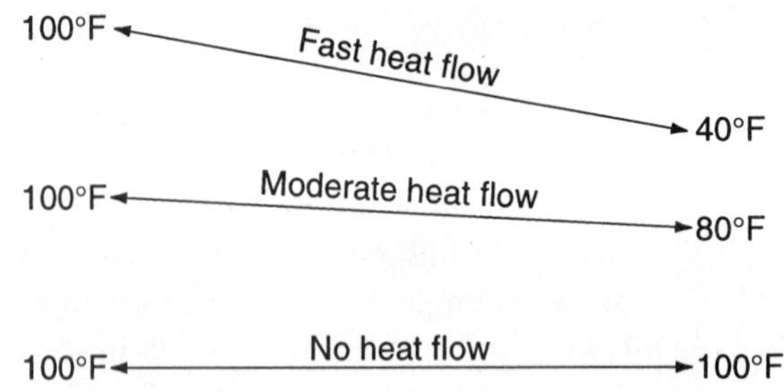

Figure 2–7 Heat flow due to temperature differences

Example 1: The greater the temperature difference, the greater the heat flow. See Figure 2–7.

The heat conducted through the rod heated to 100°F on one end and 50°F on the other end will conduct heat much faster than the rod heated to a temperature of 100°F on one end and 90°F on the other. Notice that when both ends are the same temperature there is no heat flow.

Example 2: A large surface area will transfer more heat than a smaller surface area. See Figure 2–8.

The heat flow from the ice to the liquid will be much greater in the glass with the crushed ice than it will be in the glass with the larger ice cubes because of the greater surface area of the larger ice cubes. The larger ice cubes will last longer, but the temperature of the liquid will also be higher.

Example 3: A material that has a high resistance to the flow of heat will allow less heat to flow through it than one having less resistance. A building that is insulated with fiberglass or rock-wool insulation between the joists and rafters will allow much less heat to flow through the walls than a building that has little or no insulation. See Figure 2–9.

SUMMARY 2–2

- Heat always flows from a warmer object to a cooler object.
- The greater the temperature difference the greater the rate of heat flow between the two objects.
- Three ways that heat can be transferred are: (1) conduction, (2) convection, and (3) radiation.

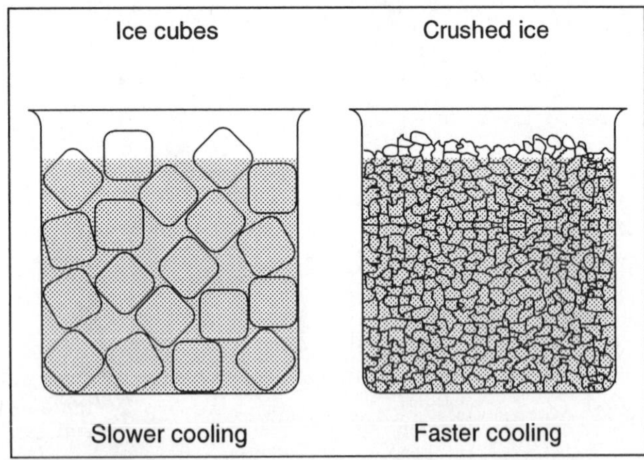

Figure 2–8 Result of cooling surface area

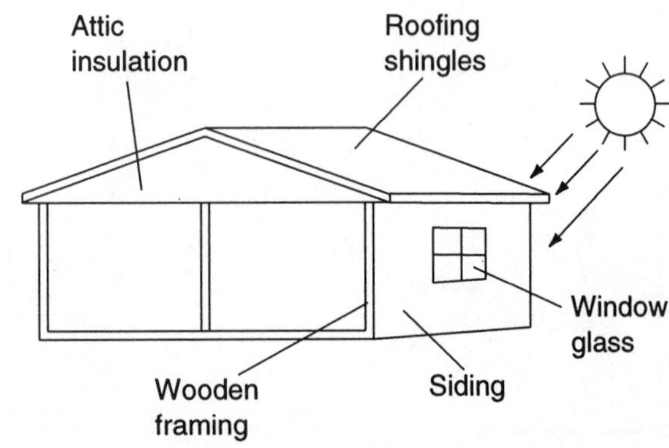

Figure 2–9 Heat flow through building insulation

- Conduction is the transfer of heat through an object or between touching objects.
- Heat flows by convection through a fluid, either a gas or a liquid.
- In radiation, heat is transferred by wave motion.
- Heat will travel in a combination of these three methods at the same time.
- The factors having the greatest effect on heat transfer are: (1) temperature difference, (2) amount of surface area, and (3) type of material.

REVIEW QUESTIONS 2–2

1. In what direction does heat flow?
2. Will heat always flow from a larger object to a smaller object?
3. What process involves conduction, convection, and radiation?
4. Which is the best conductor of heat: iron or glass?
5. Upon what do temperature difference, surface area, and the type of material have an effect?

2–3 SPECIFIC HEAT

Specific heat is defined as the amount of heat required to raise the temperature of one pound of any substance one degree Fahrenheit. Each material has its own specific heat. It is also considered to be the ratio between the quantity of heat required to change the temperature of a substance one degree Fahrenheit and the amount of heat required to change an equal amount of water one degree Fahrenheit.

When we consider that and the definition of the Btu, we can see that the specific heat of water must be 1 Btu per pound. The specific heat values of some of the more popular foods are given in Table 2–1.

A more complete listing of the specific heat of foods can be found in the ASHRAE Handbook. Notice that after foods are frozen, their specific heat values are considerably less than fresh foods. In most cases, the specific heat after being frozen is just about one-half of the value of fresh foods.

SUMMARY 2–3

- Specific heat is defined as the amount of heat required to raise the temperature of one pound of any substance one degree Fahrenheit.
- Each material has its own specific heat value.
- The specific heat of water is taken to be one Btu per pound.

FOOD	SPECIFIC HEAT (UNFROZEN) BTU	SPECIFIC HEAT (FROZEN) BTU
Veal	0.70	0.39
Beef	0.68	0.38
Pork	0.57	0.30
Fish	0.82	0.43
Poultry	0.80	0.42
Eggs	0.76	0.40
Butter	0.55	0.33
Cheese	0.64	0.37
Whole Milk	0.92	0.47

Table 2–1 Specific heat of foods

REVIEW QUESTIONS 2–3

1. Define specific heat.
2. What is the specific heat of frozen poultry?
3. Which is the greater, the specific heat of a frozen substance or an unfrozen substance of the same type?

2–4 SENSIBLE HEAT

Sensible heat is the heat added to or removed from a substance, resulting in a change in temperature but no change of state. It is termed "sensible heat" because it is heat that can be felt with the hand or measured with a thermometer.

Example: If we heat a container of water from 70°F to 212°F, a temperature change of 142°F will occur. This temperature is sensible heat. It can be felt with the hand and it can be measured with a thermometer. See Figure 2–10.

SUMMARY 2–4

- Sensible heat is heat added to or removed from a substance, resulting in a change in temperature but no change of state.
- It is termed "sensible heat" because it can be felt with the hand or measured with a thermometer.

REVIEW QUESTIONS 2–4

1. Why is sensible heat so named?
2. What type of heat is required to raise a pound of water from 50°F to 70°F?
3. Is sensible heat added to water to change it to steam?

2–5 LATENT HEAT

Latent heat is also known as hidden heat. It cannot be felt nor can it be measured with a thermometer. It is the heat that is added to or removed from a substance during a change of state but with no change in temperature. Basically, there are four types of latent heat; (1) latent heat of fusion, (2) latent heat of condensation, (3) latent heat of vaporization, and (4) latent heat of sublimation.

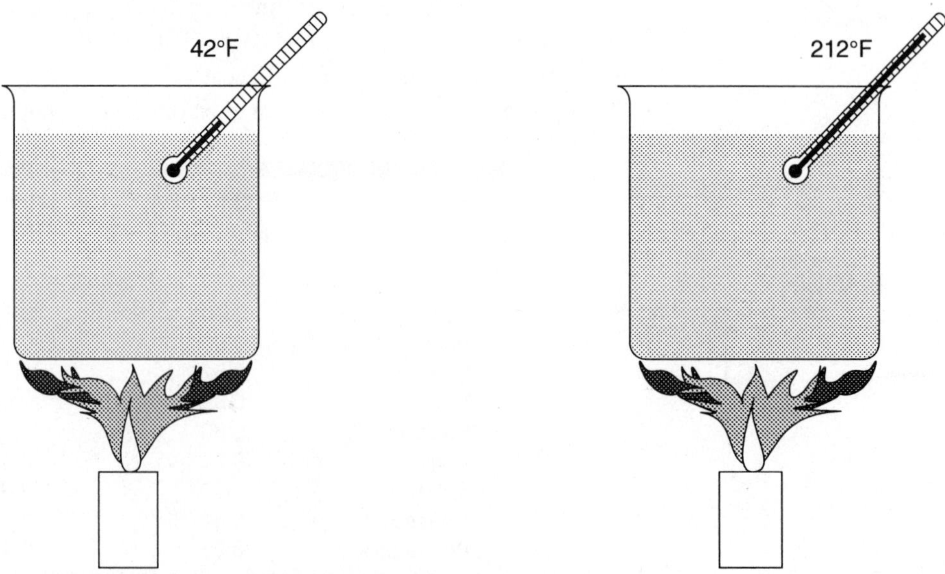

Figure 2–10 Sensible heat example

Latent Heat of Fusion

This is the amount of heat required to change a solid to a liquid or a liquid to a solid at a constant temperature. The value is the same regardless of which is being done. A familiar example of the latent heat of fusion is the changing of water to ice or ice to water.

Latent Heat of Condensation

This is the amount of heat that must be removed from a vapor to change to a liquid (condensation) at a constant temperature.

Example: The moisture collecting on a cold surface such as a cold drink container is caused by the latent heat of condensation. The latent heat is removed from the surrounding vapor (air) causing it to condense on the cooler surface of the container (glass).

Latent Heat of Vaporization

Latent heat of vaporization is the amount of heat that must be added to a liquid to cause it to change into a vapor (vaporization) at a constant temperature.

Example: If we heat a container of water until it reaches the boiling temperature and then continue to add heat causing some of the water to boil off, or change to steam, we are adding the latent heat of vaporization to the water, causing it to evaporate.

Latent Heat of Sublimation

This is the amount of heat that must be added to a solid to change it to a gas with no visible evidence of it going through the liquid state. Not all substances will go through this process.

Example: If we placed some dry ice in a container and watched it, we would not see it go through a liquid state, but it would all eventually disappear. Another example would be ice that is left in the freezer for long periods of time. Some of it will disappear. This is because of the sublimation process. The latent heat of sublimation is equal to the sum of the latent heat of fusion and the latent heat of vaporization of that substance.

SUMMARY 2–5

- Latent heat is also known as hidden heat. It cannot be felt nor can it be measured with an ordinary thermometer.
- Latent heat is the heat that is added to or removed from a substance during a change of state but with no change in temperature.
- The latent heat of fusion is the amount of heat required to change a solid to a liquid or a liquid to a solid at a constant temperature.
- The latent heat of condensation is the amount of heat that must be removed from a vapor to change it to a liquid (condensation) at a constant temperature.
- The latent heat of vaporization is the amount of heat that must be added to a liquid to cause it to change into a vapor (vaporization) at a constant temperature.
- The latent heat of sublimation is the amount of heat that must be added to a solid to change it to a gas with no visible evidence of it going through the liquid state.

REVIEW QUESTIONS 2–5

1. What is hidden heat known as?
2. When 144 Btu are added to a pound of water at 32°F, what occurs?
3. What does the latent heat of fusion cause?
4. In a refrigeration system, where does the latent heat of condensation occur?
5. In a refrigeration system, where does the latent heat of vaporization occur?
6. What process occurs when a solid changes to a vapor with no visible liquid state?

2–6 RELATIONSHIP OF HEAT AND TEMPERATURE

When heating or cooling a substance, there are certain characteristics that are definite for each substance at given points along the way. To help in understanding this, we will work with one pound of water to illustrate what occurs during the heating and cooling process. Remember that each substance will have its own set of characteristics, but they will always be the same for that substance.

If we would heat and cool a pound of water and plot on a chart the different temperatures and what happened at that temperature, we would have the chart shown in Figure 2–11.

If we would heat a pound of ice starting at 0°F and finish with 212°F steam, and then cool the steam back to ice at 0°F, we would prove that the process is exactly reversible.

We can follow the process, starting at the bottom left-hand corner of Figure 2–11 and following the arrows as the ice is heated and then cooled. Remember that all substances have a different specific heat. Each substance will have a different specific heat value in each state that they can be in. These values are located in reference tables that should be reviewed when making heat load calculations. In the following example, ice has a specific heat value of 0.5 Btu per pound; water has a specific heat value of 0.1 Btu per pound, and steam has a specific heat value of 0.5 Btu per pound.

Step 1. The process shows that when we add 16 Btu of heat to the ice, its temperature will change from 0°F to 32°F. This is sensible heat because it can be measured with a thermometer. But notice that it did not change the state of the ice.

Step 2. We add another 144 Btu of heat to the ice. The temperature will remain at 32°F, but the ice will now be changed to water at 32°F. This represents a change of state of the ice; that is, ice to water.

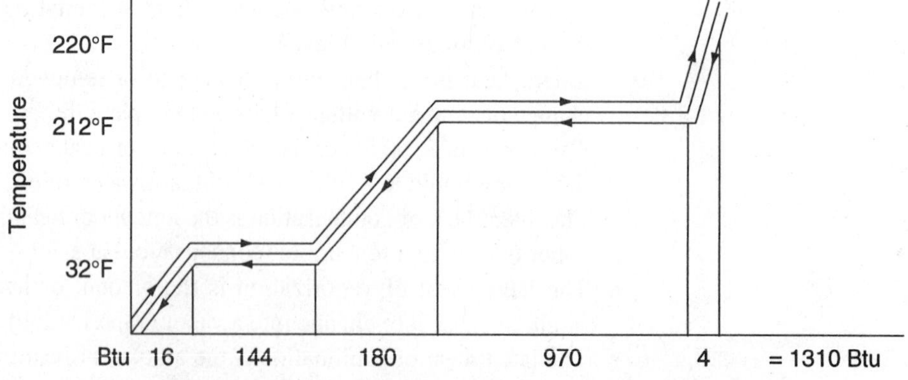

Figure 2–11 Relationship of heat and temperature

Step 3. When we add another 180 Btu of heat to the water, the temperature of the water will increase to 212°F. This is sensible heat because it can be measured with a thermometer. This is also the boiling temperature of water at sea level and atmospheric pressure.

Step 4. When we add another 970 Btu to the pound of water, it will change from water at 212°F to steam at 212°F. These 970 Btu are latent heat and are known as the latent heat of vaporization because there is no change in temperature, only a change in state—liquid to vapor.

Step 5. When another 4 Btu are added to the steam at 212°F, the temperature of the steam will increase from 212°F to 214°F. This is sensible heat because it can be measured with a thermometer. It resulted in a change in temperature but no change in state.

If we follow the arrows in the other direction we can see that the process is reversible. That is, it requires exactly the same number of Btu for each step when cooling as it did when heating the water. Instead of adding heat to the water we would need to remove it.

SUMMARY 2–6

- When heating or cooling a substance, there are certain characteristics that are definite for each substance at given points along the way.
- As an example, if a pound of water is heated from 0°F to 216°F, a definite amount of heat must be added to cause this change. If this same pound of water is cooled from 216°F to 0°F, the exact same number of Btu must be removed. This process is the same for every substance known to man.

REVIEW QUESTIONS 2–6

1. Will a given set of characteristics be the same for all substances?
2. Why is only 0.5 Btu per pound of ice required to change its temperature when 1 Btu per pound of water is required to change its temperature?
3. What are the 970 Btu required to change a pound of water at 212°F to steam at 212°F known as?

2–7 HEAT OF COMPRESSION

Heat of compression is the process done to a vapor when it is mechanically compressed. This is the process that occurs in a refrigeration compressor. The heat of compression adds heat to the refrigerant vapor, increasing its temperature high enough so that it can be condensed in the condenser. The only heat that is added to the vapor is that caused by the compression process. The vapor temperature is increased because the molecules of the vapor are being squeezed closer together, and their friction as they rub together is what causes the higher temperature.

In an operating system, the low-temperature, low-pressure refrigerant vapor is pumped from the evaporator to the compressor. At this point its volume is reduced and its temperature is increased to a temperature higher than the condenser cooling medium. When the high-pressure, high-temperature refrigerant vapor is cooled it changes back to a liquid. It condenses.

SUMMARY 2–7

- The heat of compression is the process done to a vapor when it is mechanically compressed.

- The heat of compression adds heat to the refrigerant vapor, increasing its temperature high enough so that it can be condensed in the condenser.

REVIEW QUESTIONS 2–7

1. Why is the heat of compression important?
2. What causes the heat of compression?

2–8 SUPERHEAT

By definition, superheat is the heat that is added to or removed from a vapor at a temperature above its boiling temperature at that pressure. In refrigeration work, this is taken to be a vapor that is no longer in contact with its liquid.

Example: When a liquid has been heated enough to cause all of it to change into a vapor, any additional heat will be in the form of sensible heat, known as superheat. If we boiled some water until it all turned to steam, then added enough heat to increase its temperature from 212°F to 214°F, as in our previous example, the additional heat would be superheat. See Figure 2–12.

SUMMARY 2–8

- By definition, superheat is the heat that is added to or removed from a vapor at a temperature above its boiling temperature at that pressure.
- In refrigeration work, this is generally taken to be a vapor that is no longer in contact with its liquid.

REVIEW QUESTIONS 2–8

1. What kind of heat is contained in a vapor that is at a temperature above its boiling point?
2. If we heat steam at 220°F and 0 psig pressure, what is the amount of superheat?

2–9 HEAT CALCULATION

When we are considering the proper amount of heat to be transferred when cooling or heating a substance or a space, it is necessary that an accurate calculation of the amount of heat to be transferred be completed. As we learned earlier, all substances have their own particular specific heat.

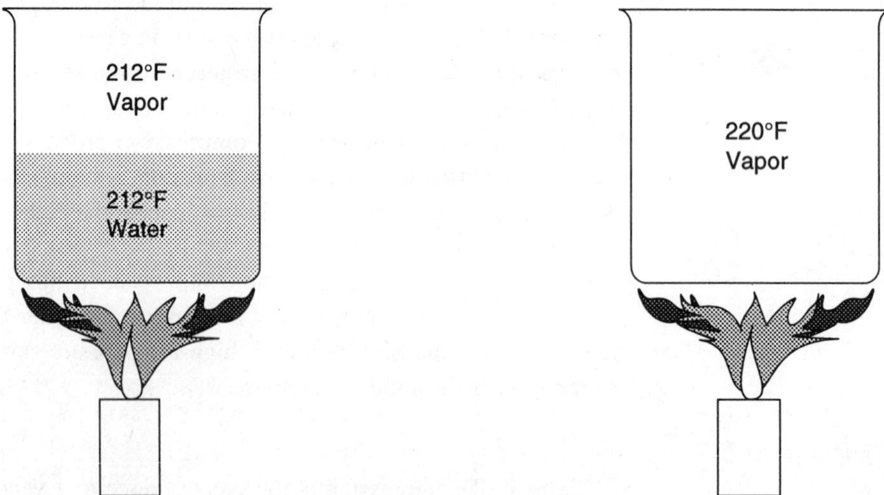

Figure 2–12 Example of superheat

Heat calculation is the determination, by formula, of the amount of heat in Btu that is either gained or lost by a substance during the changing of its temperature. Because each substance has a different specific heat, a list of substances must be on hand before accurate calculations can be accomplished. They are available through several sources, such as ASHRAE.

The formula used for this purpose is fairly simple to use, in most cases. The formula uses the difference in temperature of the substance between the time it is stored until it reaches the desired storage temperature, DT (delta T); multiplied by the weight of the substance, W; multiplied by the specific heat of the substance, SH. Use the formula:

$$\text{Btu} = \Delta T \times W \times SH$$

Where:

W = Weight of the substance in pounds
SH = specific heat of the substance
ΔT = change in temperature, °F

Example: If we were to cool 50 pounds of water from 120°F to 60°F, how many Btu are required to cool this water? The specific heat of water is considered to be 1.

$$\begin{aligned}\text{Btu} &= \Delta T \times W \times SH\\ &= (120 - 50) \times 60 \times 1\\ &= 70 \times 50 \times 1\\ &= 3{,}000 \times 1\\ &= 3{,}000\ \text{Btu}\end{aligned}$$

Example: If we were to heat 50 pounds of water from 60°F to 120°F, how many Btu would be required to heat this water?

$$\begin{aligned}\text{Btu} &= \Delta T \times W \times SH\\ &= (120 - 60) \times 50 \times 1\\ &= 70 \times 50 \times 1\\ &= 3{,}000 \times 1\\ &= 3{,}000\ \text{Btu}\end{aligned}$$

Notice that in these two examples the amount of heat involved was exactly the same. In the first example, the heat was removed; in the second example, the heat was applied.

When a heat load is to be calculated that involves different types of substances, each substance must be calculated separately for an accurate heat load calculation. Also, when the substance is to be frozen, the difference between the frozen and the nonfrozen state must be considered. In practice, this may not always be possible because of the different uses of refrigerated storage units. In such instances, an approximate calculation is made and the person doing the storing of the substances must use proper judgment to prevent overloading the refrigeration equipment. It is possible to take an average of the weights and the specific heat values of the substances to make this calculation, if they are relatively close together.

SUMMARY 2–9

- Heat calculation is the determination of how much heat in Btu that is either gained or lost by a substance during the changing of its temperature.
- Because each substance has a different specific heat, a list of substances must be on hand before accurate calculations can be accomplished.
- When a heat load is to be calculated that involves different types of substances, each substance must be calculated separately for an accurate heat load calculation.
- It is possible to take an average of the weights and specific heat values of the substances to make this calculation.

REVIEW QUESTIONS 2–9

1. Write the formula for calculating heat to be transferred to cool an object.
2. Calculate the amount of heat that must be removed to cool 5 pounds of water from 70°F to 50°F.
3. What must be taken into consideration when cooling substances below the freezing temperature?

2–10 ENTHALPY

Enthalpy, by definition, is the *total heat* of a substance. That is, it is the sum of both the sensible heat and the latent heat of the substance. Enthalpy is measured in Btu per pound of that substance and is used in both refrigeration and air conditioning applications.

Theoretically, enthalpy is measured from absolute zero of the temperature scale in use: –460°F and –273°C. In practice, because of the large numbers encountered when using absolute zero, other reference points have been set. As an example, the reference point for water is 32°F and for refrigerants is –40°F. When making calculations involving a reference point, any enthalpy above the reference point is considered to be a positive enthalpy and any enthalpy below that point is considered to be a negative enthalpy. When latent heat is involved in the heat transfer, use:

$$E = W \times SH \times \Delta T + LH$$

When no latent heat is involved in the heat transfer, use:

$$E = W \times SH \times \Delta T$$

Where:

E = enthalpy
W = weight of the substance
SH = specific heat of the substance
ΔT = temperature change of the substance
LH = latent heat of the substance, if a change of state is encountered

Example: Calculate the enthalpy of one pound of steam at 212°F. (Use 32°F as the reference point.) This formula requires the use of the first formula because latent heat is involved in the calculation. Note that the SH of steam is 1 and the latent heat of water is 970 Btu/lb. Therefore:

$$\begin{aligned} E &= W \times SH \times \Delta T + LH \\ &= 1 \times 1 \times (212 - 32) + LH \\ &= 1 \times 1 \times 180 + 970 \\ &= 180 + 970 \\ &= 1{,}150 \text{ Btu/lb} \end{aligned}$$

Example: Calculate the enthalpy of 1 lb of water at 70°F. (Use 32°F as the reference point.) Because no latent heat is involved, use the formula:

$$\begin{aligned} E &= W \times SH \times \Delta T \\ &= 1 \times 1 \times (70 - 32) \\ &= 1 \times 1 \times 38 \\ &= 1 \times 38 \\ &= 38 \text{ Btu/lb} \end{aligned}$$

SUMMARY 2–10

- Enthalpy, by definition, is the total heat of a substance. It is the sum of both the sensible heat and the latent heat.
- Theoretically, enthalpy is measured from absolute zero of the temperature scale in use.
- The reference point for water is 32°F and for refrigerants it is 40°F.

REVIEW QUESTIONS 2–10

1. Define enthalpy.
2. When is an enthalpy measurement considered to be positive?

2–11 MECHANICAL EQUIVALENT OF HEAT

Mechanical equivalent of heat has been determined by scientific experiments to be the amount of heat that is produced by the expenditure of a certain amount of mechanical energy. Theoretically, if the heat energy produced by the expenditure of 1 Btu could be changed into mechanical energy without any energy loss, it would be equal to 778 ft-lb of work. A relationship is represented in this conversion. Thus, one Btu of heat energy is theoretically equivalent to 778 ft-lb, and 1 ft-lb of mechanical energy is equal to 1/778 Btu or 0.00128 Btu.

To determine the equivalent amount of heat in Btu gained from ft-lb, divide the ft-lb by 778. Use the formula:

$$\text{Btu} = \frac{\text{ft-lb}}{778}$$

To determine the ft-lb from heat in Btu, multiply the Btu by 778. Use the formula:

$$\text{Ft-lb} = \text{Btu} \times 778.$$

SUMMARY 2–11

- The mechanical equivalent of heat has been determined by scientific experiments to be the amount of heat that is produced by the expenditure of a certain amount of mechanical energy.

REVIEW QUESTIONS 2–11

1. What is the expenditure of a certain amount of mechanical energy known as?
2. To how many Btu is 1 ft-lb of mechanical energy equivalent?

2–12 COOLING

Cooling, by definition, is merely the removal of heat from a substance. This heat removal can be accomplished by use of several different methods; however, we will discuss only the evaporation and expansion methods.

Evaporation

Evaporation is the process that causes a liquid to change to a vapor. The most common form of evaporation occurs when a container of water is left open to the atmosphere and some, or all, of it disappears. The rate of evaporation of water is dependent on two things: temperature and moisture (humidity) in the air. The higher the temperature and the lower the humidity, the faster the rate of evaporation of the water. Should the temperature drop, and/or the humidity rise, the evaporation process will be slowed. This is the reason that on a muggy or humid day we seem to sweat much more than on a dry, warm day. The moisture does not readily evaporate from our skin, making us feel warmer than normal. Also, if the humidity were to rise, the rate of evaporation would decrease. Evaporation occurs at temperatures down to absolute zero.

It takes heat to cause moisture to evaporate. When the humidity is low and we sweat, the moisture is quickly evaporated from our skin. This evaporation requires heat that is taken from our skin, making us feel cooler. The heat that causes this evaporation is known as the latent heat of vaporization. The required heat can be taken from the object being cooled or from the moisture itself.

Expansion

Expansion occurs when a vapor is compressed and the pressure is suddenly reduced. When vapor is compressed, heat is generated in a quantity equal to the amount of work done during the compression process. This is known as the heat of compression. The greater the compression required, the higher the temperature of the compressed gas. On a hot day, the vapor in a refrigeration system requires more work to compress it than is required on a cooler day. This will add more heat to the vapor and cause its temperature to rise still further. When this occurs, the system must work harder to deliver the same amount of refrigeration.

The cooling of the vapor experienced when the compressed vapor is allowed to expand is just the opposite of the compression process described above. Expansion is what occurs in a refrigeration system when the refrigerant is passed through the flow control device, or expansion device, as it is sometimes called.

The three steps that occur in a compression-expansion refrigeration cycle are:

1. The vapor is compressed to a high pressure by the compressor.
2. The heat of compression caused by the compression process along with the heat picked up in the evaporator is removed in the system condenser where the refrigerant is condensed into a liquid.
3. The refrigerant is expanded, causing its temperature to drop. This reduction in temperature is what causes the refrigeration effect.

SUMMARY 2–12

- Cooling, by definition, is merely the removal of heat from a substance.
- Evaporation is the process that causes a liquid to change to a gas.
- The rate of evaporation is dependent on two things: temperature and moisture (humidity) in the air.
- Expansion occurs when a vapor is compressed and then suddenly reduced.
- When vapor is compressed, heat is generated in a quantity equal to the amount of work done during the compression process. This is known as the heat of compression.
- Expansion is what occurs in a refrigeration system when the refrigerant is passed through the flow control device, or expansion device, as it is sometimes called.

REVIEW QUESTIONS 2–12

1. What is the removal of heat from a substance known as?
2. What is the process that causes a liquid to change into a vapor?

3. Name two factors that will slow the evaporation process.
4. What is the heat caused by the compression of a vapor known as?
5. Where does the expansion of the refrigerant in a refrigeration system occur?

2–13 SUBCOOLING

Subcooling is the process of lowering the temperature of a liquid refrigerant below its condensing temperature. A liquid refrigerant at a temperature below its saturation temperature is said to be subcooled.

In refrigeration systems, subcooling takes place in the bottom of the condenser after the refrigerant has been condensed. It may also occur to some extent in the receiver and the liquid line if they are exposed to temperatures below the refrigerant condensing temperature.

Subcooling is used to reduce the amount of flash gas that occurs at the flow control device. Reducing the amount of flash gas will tend to increase both the efficiency and the capacity of the unit. In high-efficiency systems, subcooling is very important to their satisfactory operation.

Sometimes a separate subcooler is installed in the liquid line to aid in the reduction of flash gas. Subcoolers of this type are installed so that the liquid refrigerant will pass through them before entering the evaporator. Subcoolers may be either water cooled or air cooled, depending on the nature of the system and the available water supply and disposal possibilities. In most systems, a subcooler will pay for itself through the increased efficiency and capacity of the unit.

SUMMARY 2–13

- Subcooling is the process of lowering the temperature of a liquid refrigerant below its condensing temperature.
- Subcooling is used to reduce the amount of flash gas at the flow control device.
- Sometimes a separate subcooler is installed in the liquid line to aid in the reduction of flash gas.

REVIEW QUESTIONS 2–13

1. Define subcooling.
2. In a refrigeration system, where does the greatest amount of subcooling occur?
3. In what type of system is subcooling most important?

2–14 HUMIDITY

By definition, humidity is the water vapor contained in air within a given space. Humidity is the general term used when describing the moisture contained in a given quantity of air. It is expressed in two ways: absolute humidity and relative humidity.

The actual amount of moisture, or water vapor, that air can hold is dependent on the vapor pressure and the temperature of the air. When the air temperature is lowered, it will absorb and hold less moisture. Likewise, as the temperature is raised, the air will absorb and hold more moisture. Humidity measured in this way is generally referred to as relative humidity and is expressed as a percentage of the moisture that the air could actually hold at that temperature and pressure.

Example: We have a given quantity of air at a given temperature and it has a relative humidity of 60%; i.e., the air is holding 60% of the moisture that it can hold at that temperature.

Absolute Humidity

By definition, absolute humidity is a measure of the amount of moisture actually present in a given quantity of air. The temperature and pressure has nothing to do with absolute humidity. It is measured in grains of moisture per pound of dry air, or as pounds of water vapor per pound of dry air, and is all the moisture present in that quantity of air at that time.

Example: When a sample of air contains 30 grains of moisture, the absolute humidity is 30 grains.

Relative Humidity

By definition, relative humidity is the percentage of humidity that a given quantity of air can hold at that given pressure and temperature. It is expressed as a percentage of the amount that the air could hold under those same conditions. When the air cannot hold any more moisture, it is said to be saturated and the relative humidity is 100%. If the air should contain some lesser amount of moisture it is said to have a relative humidity equal to that percentage.

Example: If a quantity of air contains 45% of the moisture it could hold at those same conditions, it is said to have a relative humidity of 45%.

When a given quantity of air is at its saturation point, it has 100% relative humidity. If the air temperature is lowered, some of the moisture will condense out and form water beads on some cooler surface, or it may be present as fog in the air when no surface is available. In a domestic refrigerator these water droplets appear as frost or ice on the evaporator. In an air conditioning system, these water droplets appear as condensation on the evaporator coil. This may be illustrated better if we consider a glass containing some liquid and ice and having a temperature lower than the surrounding dew point temperature. See Figure 2–13.

When the moisture-laden air comes into contact with the colder surface of the glass, the moisture will condense out of the air because of the lowered temperature, and form as droplets on the surface of the glass.

Effects of Humidification

The effects that humidity has on the operation of an air conditioning or refrigeration system must be emphasized as much as possible. If we consider the operation of a walk-in refrigerated box when the relative humidity is low, the moisture will be removed from the products inside. If the owner sells the products by the pound, then

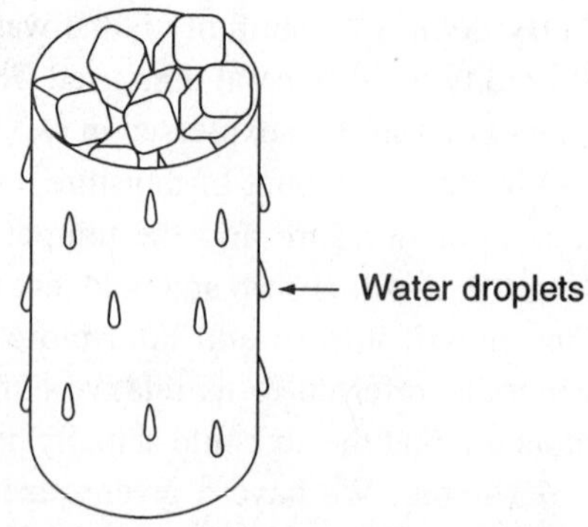

Figure 2–13 Effects of a cold surface on humidity

he will have less pounds to sell. Also, in some products a lack of humidity will cause the products to spoil. When the humidity is too high, the formation of mold will occur. This will also cause the product to become less desirable, and may prevent its being sold.

When a building is air conditioned, the relative humidity is also important. If the relative humidity is too low, the building structure will dry out and crack. The occupants will complain of feeling cold and complain about drafts. When the relative humidity is too high, the occupants will feel clammy. There will be condensation on the windows and colder surfaces. Condensation will also form inside the walls and cause rotting and mildew to form.

SUMMARY 2–14

- Humidity is the amount of water vapor contained in a given quantity of air within a given space.
- Absolute humidity is a measure of the amount of moisture actually present in a given quantity of air.
- Relative humidity is the percentage of humidity that a given quantity of air can hold at that given pressure and temperature.
- When a given quantity of air is at its saturation point, it has 100% relative humidity.
- If the relative humidity is low, the building structure will dry out and crack. The occupants will complain of feeling cold and complain about drafts.
- When the relative humidity is too high, the occupants will feel clammy.

REVIEW QUESTIONS 2–14

1. To what is the amount of moisture air can hold dependent?
2. Will air at a lower temperature hold more moisture than warmer air?
3. If we have a pound of air containing 40 grains of moisture, what is the amount of humidity?
4. A quantity of air at a given temperature is capable of holding 8 grains of moisture, and is holding 6 grains. What is the type and measure of humidity?
5. Why will a temperature that is too low in a fresh-food refrigerated box cost the owner more than the electric bill?

2–15 DEHYDRATION AND NONCONDENSABLES

Dehydration is the process of removing moisture, air, and noncondensables from a refrigeration system. It is also known as evacuation.

Manufacturers have done much research into the effects of moisture and noncondensables in refrigeration systems. However, there remains some effects of moisture and noncondensables that are not completely understood. It has been determined that the moisture in any foreign matter can cause a lot of damage if left inside a system. Some of the effects are copper plating, corrosion, oil breakdown, sludging, and carbon formations. Any one of these contaminants is capable of causing compressor failure and complete system breakdown.

Thus, it can be seen that the major reasons for evacuating a system as completely as possible are for the protection of the system and to keep it operating as economically as possible.

The two most common means of causing water to boil and evaporate are to reduce the pressure on it and to increase the temperature to which it is subjected. On larger systems, increasing the temperature on the complete system is almost impossible. Therefore, the best method of causing any water to change into a vapor for easier removal is to decrease the pressure on it; that is, to pump a vacuum on the interior of the system components causing the water to change to a vapor, then removing the vapor with the vacuum pump.

In refrigeration and air conditioning work, the equipment manufacturers are quite specific about the degree that a system should be exposed to a vacuum. Most manufacturers recommend that a vacuum of around 500 to 1,000 microns be pumped on the system for the best results.

To help understand the importance of proper evacuation techniques, remember the following: as the discharge temperature increases to 200°F, any contaminants left in the system have a dramatic effect. For each 18°F increase in temperature above 200°F, the chemical reaction of the contaminants to the system components doubles. This is a rather startling revelation that should cause every technician to be more careful during the evacuation, service, and charging of refrigeration systems. This is especially important when we remember that the discharge temperature of most refrigeration compressors is above 200°F during normal operation.

If air is left in the refrigeration system and the system is charged, it will always operate with a higher than normal discharge pressure. This higher discharge pressure causes the system to be less efficient and causes the discharge temperature to increase in direct relation to the discharge pressure. This causes the contaminants to react as mentioned in the previous paragraph.

Noncondensables

Noncondensables are any gases other than the refrigerant in a refrigeration system. The contaminants cannot be condensed at the pressures and temperatures normally encountered during normal operation of the system. During operation, the compressor will attempt to cause a pressure high enough to condense these contaminants, causing an increase in the discharge pressure. The noncondensables in the system will cause reduced efficiency in two ways: (1) by causing a higher discharge pressure, and (2) by taking up space in the system reducing the amount of refrigerant that can be charged into it. The smaller charge of refrigerant will reduce the refrigerating effect of the system.

SUMMARY 2–15

- Dehydration is the process of removing moisture, air, and noncondensables from a system.
- Some of the effects of contaminants in a system are copper plating, corrosion, oil breakdown, sludging, and carbon formations.
- The two most common methods of causing water to boil and evaporate are to reduce the pressure on it and to increase the temperature to which it is subjected.
- For each 18°F increase above 200°F, the chemical reaction of the contaminants to the system components doubles.
- If air is left in the refrigeration system and it is charged, the system will always operate with a higher than normal discharge pressure.
- Noncondensables are any gases, other than the refrigerant, in a refrigeration system.
- Two ways that noncondensables will cause reduced efficiency are: (1) by causing a high discharge pressure, and (2) by taking up space in the system, reducing the amount of refrigerant that can be charged into it.

REVIEW QUESTIONS 2–15

1. What does the copper coloring on an expansion valve seat indicate?
2. What will cause the compressor lubricating oil to sludge?
3. When the discharge gas temperature reaches 218°F, what happens if there are contaminants in the system?
4. What will be the results of noncondensables in a refrigeration system?
5. When a compressor operates with a higher than normal discharge pressure, what happens to the discharge gas temperature?

2–16 ENERGY, WORK, AND POWER

Energy

By definition, energy is the ability to do work. Energy may be in any form such as electrical, mechanical, and heat energy. Energy can be changed from one form to another, but it cannot be destroyed. A common example of this is the electric motor. During motor operation, electrical energy is changed into mechanical energy. Also, some of the energy is also changed into heat energy, from which there is no benefit as far as running the electric motor, but the energy is not destroyed.

The heat that is removed from the product in a refrigerated cabinet constitutes the load on the refrigeration system. This heat is removed through the use of several forms of energy. First, the electrical energy was converted to mechanical energy in the compressor. The mechanical energy was then used to remove the heat energy from the product inside the cabinet. Some of the efficiency was lost during all of these changes of forms, but the energy itself was not lost, just changed into a form from which we received no work.

Force

By definition, force is that which tends to, or actually does, produce motion. Force is generally measured by the pound. This force may be in any direction, as indicated by the simple definition of force. It can be seen that direction is not taken into account.

Motion

Motion is considered to be the movement or constantly changing position of a body. It may also be considered as the speed with which this body changes position per unit of time. The common measurement of motion is in feet per minute (ft/min) or feet per second (ft/sec).

Work

Work is defined as the force applied multiplied by the distance through which this force acts. Work is commonly measured by foot pounds (ft-lb). Thus, work is the amount of work done by a force of one pound through a distance of one foot. It is expressed by the formula:

$$\text{Work} = \text{Force} \times \text{Distance}$$

Example: If we wanted to move a 50-pound weight through a distance of 10 feet, how much work would we do?

$$
\begin{aligned}
W &= F \times D \\
W &= 50 \text{ lb} \times 10 \text{ ft} \\
&= 500 \text{ ft-lb}
\end{aligned}
$$

Power

Power is defined as the time rate of doing work. It is calculated by dividing the work done by the time required to do the work. This calculation adds the time dimension to the equation used for calculating work done.

Example: In our previous problem we moved a weight of 50 pounds through a distance of 10 feet. This gave us the amount of work done. However, it did not mention the time required to do the work. The power formula is:

$$\text{Power} = \frac{\text{Work}}{\text{Time}}$$

Example: Determine the power required to move the 50-pound weight through 10 feet in 5 minutes.

$$
\begin{aligned}
W &= F \times D \\
&= 50 \text{ lb} \times 10 \text{ ft} \\
&= 500 \text{ ft-lb} \\
P &= \frac{W}{T} \\
&= \frac{500}{5} \\
&= 100 \text{ ft-lb/min}
\end{aligned}
$$

Notice that the results are the same regardless of the direction of the applied force. Neither does the type of resistance enter into the equation.

Horsepower

In most energy calculations, the ft-lb is much too small a measure. Thus, the horsepower is used in most everyday calculations. One horsepower is defined as 33,000 ft-lb/min. Notice that in the basic horsepower formula time is included. The formula for calculating horsepower is:

$$\text{Horsepower} = \frac{(\text{Weight} \times \text{Distance})}{(\text{Time} \times 33{,}000)}$$

Example: If we wanted to lift a condensing unit weighing 1,000 pounds on to a roof 15 feet high and we needed to lift it in 1 minute, what would be the required horsepower?

$$
\begin{aligned}
H &= \frac{(W \times D)}{(T \times 33{,}000)} \\
&= \frac{(1{,}000 \times 15)}{(1 \times 33{,}000)} \\
&= \frac{15{,}000}{33{,}000} \\
&= 0.45 \text{ horsepower}
\end{aligned}
$$

Density

Density is defined as the weight per unit volume of a substance and is usually considered to be the weight per cubic foot.

Most substances will expand when exposed to heat and will contract when cooled. One exception to this rule is water. Water will contract when cooled down to a temperature of 39°F. When this temperature is reached the water starts to expand. This is why frozen water will burst pipes during cold weather. From this, we can see that water is at its greatest density at 39°F. Because of this phenomenon, ice has a lower density than water.

SUMMARY 2–16

- Energy is the ability to do work.
- Energy can be changed from one form to another but it cannot be destroyed.
- The heat that is removed from the produce in a refrigerated cabinet constitutes the load on the refrigeration system.
- Force is that which tends to, or actually does, produce motion.
- Force is generally measured by the pound.
- Motion is considered to be the movement, or constantly changing position, of a body. It may also be considered as the speed with which this body changes position per unit of time.
- Work is defined as the force applied multiplied by the distance through which this force acts.
- Power is defined as the time rate of doing work.
- One horsepower is defined as 33,000 ft-lb/min.
- Density is defined as the weight per unit volume of a substance and is usually considered to be the weight per cubic foot.

REVIEW QUESTIONS 2–16

1. Can energy be destroyed?
2. When changing electrical energy into mechanical energy, not all of the energy is changed to mechanical energy. What happens to it?
3. What constitutes the load on a refrigeration system?
4. Write the formula for calculating work.
5. What is power?
6. What happens to water when it is cooled below 39°F?

CHAPTER

3

TEMPERATURE

OBJECTIVES

Upon completion of this chapter, you should be able to:

- Have a better understanding of temperature
- Know how temperature is measured
- Know the different types of temperature measurement and how they apply to refrigeration and air conditioning

INTRODUCTION

In our everyday conversations we generally discuss temperature as it relates to our comfort, weather, food, etc. Most of the time the word temperature is used without a complete understanding of what it actually means.

In refrigeration and air conditioning it is generally referred to as being some level of heat measurement. Actually, temperature is the measure of motion of the molecules in a substance. The faster the molecules vibrate, or move, the greater will be their temperature. Water is generally taken as a standard for making most measurements or setting the standards by which we measure other things. Water is said to boil at exactly 212°F at sea level when the surrounding temperature is at 70°F. Water will freeze at exactly 32°F at sea level. We are all familiar with these two points on the ordinary thermometer. The other most used thermometer uses the Celsius scale, which can be related to the Fahrenheit scale used in the United States.

There are several different types of temperature that we must be familiar with. They will be discussed in the following paragraphs.

3–1 TEMPERATURE SCALES

The intensity of heat is measured with a thermometer and is indicated on a thermometer scale. The scale may be either Fahrenheit, Celsius, or one of the other scales used for temperature measurements. The Fahrenheit and Celsius (Centigrade) scales are the most popular in refrigeration and air conditioning work. These two scales are divided into degrees relating to each scale and are read as °F or °C, depending on the scale being used at the time. See Figure 3–1.

Notice that on the Fahrenheit scale, absolute zero is registered at –460°, while the Celsius scale shows absolute zero to be at –273°.

Notice that there is a considerable amount of difference in the temperature between each degree on the two scales. On the Fahrenheit scale there are 180° equally spaced between the freezing and boiling points. On the Celsius scale there

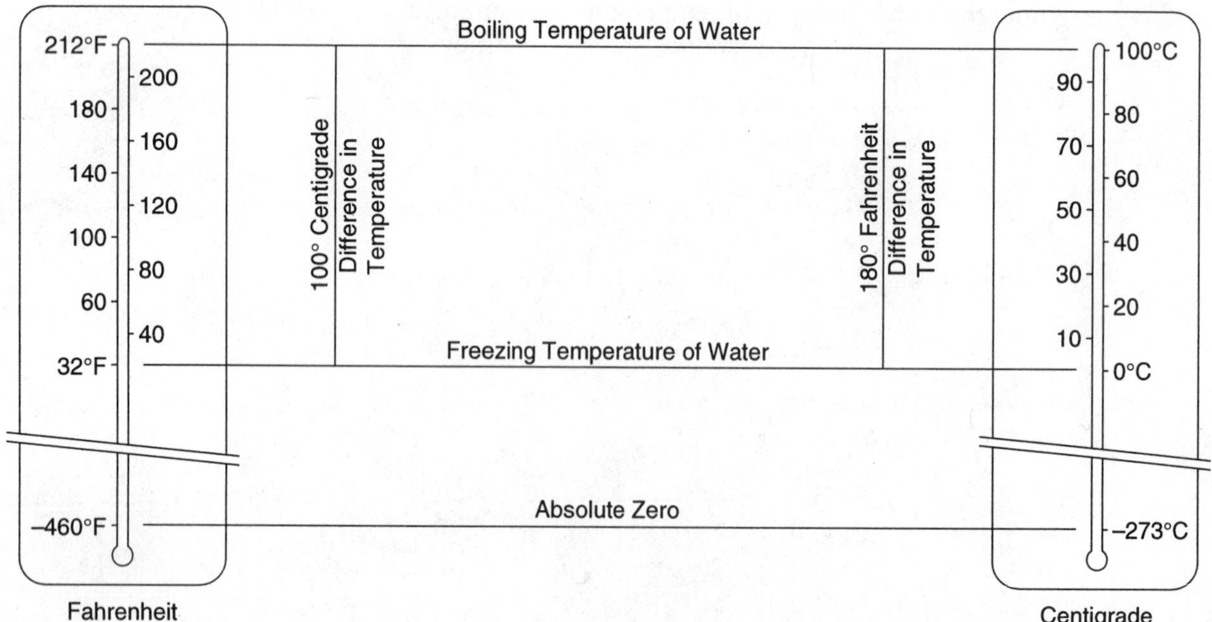

Figure 3–1 Fahrenheit and Celsius thermometer scales

are only 100° between the freezing and boiling points. On the Fahrenheit scale the freezing temperature is located at 32°F and the boiling temperature is located at 212°F. On the Celsius scale the freezing temperature is located at 0°C and the boiling temperature is located at 100°C. Sometimes it may be desirable to change from one scale to another. Notice that the distance between each degree on the Celsius scale is 1.8 times those on the Fahrenheit scale. When converting Celsius degrees to Fahrenheit degrees, multiply the number of Celsius degrees by 1.8, then divide the result by 32. (See the following formula.)

$$\text{Fahrenheit degrees} = (C^\circ \times 1.8) + 32$$

The 32 is added because the Fahrenheit zero degree is 32° below the freezing temperature of water. This value equals the same as zero degrees on the Celsius scale.

To convert Fahrenheit degrees to Celsius degrees, use the following formula.

$$\text{Celsius degrees} = \frac{^\circ F - 32}{1.8}$$

Example: Convert 95°C to Fahrenheit degrees. Use the formula:

$$\begin{aligned} \text{Fahrenheit} &= (^\circ C \times 1.8) + 32 \\ &= (95 \times 1.8) + 32 \\ &= 171 + 32 \\ &= 203^\circ F \end{aligned}$$

Example: Convert 45°F to Celsius degrees. Use the formula:

$$\begin{aligned} \text{Celsius} &= \frac{^\circ F - 32}{1.8} \\ &= \frac{45 - 32}{1.8} \\ &= \frac{13}{1.8} \\ &= 7.22^\circ C \end{aligned}$$

Recall that temperature is a measure of degree or intensity of heat. Temperature is measured with a thermometer having the appropriate scale for the system being used, Fahrenheit or Celsius (Centigrade). In refrigeration and air conditioning work we are primarily interested in five types of temperature; (1) critical temperature, (2) saturation temperature, (3) dry-bulb temperature, (4) wet-bulb temperature, and (5) dew-point temperature. We are also interested in the wet-bulb depression of air.

SUMMARY 3–1

- The two most popular temperature scales are Fahrenheit and Celsius.
- The Fahrenheit scale has 180° between the freezing and boiling temperatures of water.
- The celsius scale has 100° between the freezing and boiling temperatures of water.
- Absolute zero on the Fahrenheit scale is –460°F. On the Celsius scale absolute zero is –273°C.
- Temperature is a measure of degree or intensity of heat. It is measured with a thermometer.
- The temperatures with which we are most concerned are critical, saturation, dew-point, and wet-bulb.

- Wet-bulb depression also plays an important part of the calculations in refrigeration and air conditioning work.

REVIEW QUESTIONS 3–1

1. What causes the temperature in a substance?
2. On the psychrometric chart, where are the dew-point and wet-bulb scales located?
3. Convert –40°C to Fahrenheit.

3–2 CRITICAL TEMPERATURE

The critical temperature of any substance is considered to be the highest temperature at which a vapor may be changed into a liquid, regardless of the amount of pressure applied to it. When doing refrigeration and air conditioning work, the condensing temperature must be kept lower than the critical temperature. Should the vapor reach the critical temperature it cannot be compressed and changed into a liquid. This higher temperature causes an overload on the compressor and reduces or completely stops refrigeration.

SUMMARY 3–2

- The critical temperature is the highest temperature at which a vapor may be changed to a liquid.
- In refrigeration work the condensing temperature must be kept below the critical temperature.

REVIEW QUESTIONS 3–2

1. Define critical temperature.
2. What would happen in a refrigeration system if the critical temperature was reached?

3–3 SATURATION TEMPERATURE

In refrigeration and air conditioning work the saturation temperature is used every time we check the refrigerant pressures and the temperatures inside a refrigeration system. It is a condition of both pressure and temperature at which both liquid and vapor refrigerant can exist in the same container at the same time. When this condition exists, the vapor or liquid is at its boiling temperature and pressure. The saturation temperature will increase with an increase in pressure applied to it. It will also decrease with a decrease in the pressure.

SUMMARY 3–3

- The saturation temperature is considered every time we check the refrigerant pressures in a refrigeration system.
- The saturation temperature occurs when both liquid and vapor of a refrigerant exist in the same container.
- At the saturation temperature, the refrigerant is at its boiling temperature for that pressure.
- An increase in pressure will cause an increase in the saturation temperature.

REVIEW QUESTIONS 3–3

1. Define saturation temperature.
2. What will happen to the saturation temperature if the pressure is reduced on a cylinder of pure refrigerant?
3. Name two points in the refrigeration system where true saturation temperature normally occurs.

3–4 DRY-BULB TEMPERATURE

When we measure the temperature of the air around us we are using the dry-bulb thermometer. It is an ordinary thermometer that is used every day and its measurement is indicated by (DB). See Figure 3–2.

The temperature measured with this thermometer is sensible heat and can be felt with the hand or sensed in some other manner.

SUMMARY 3–4

- The dry-bulb temperature is measured with an ordinary thermometer and is indicated by (DB).
- A dry-bulb reading is an indication of sensible heat.

REVIEW QUESTIONS 3–4

1. What type of thermometer is used to measure the temperature shown on a bank sign?
2. With what type thermometer is a dry-bulb thermometer used to determine the wet-bulb depression?
3. When we measure the temperature drop through a refrigeration coil, what type thermometer is used?

3–5 WET-BULB TEMPERATURE

A wet-bulb thermometer is one that has its bulb covered with a sock for holding moisture while taking a measurement. The sock is soaked with distilled water. Be careful not to touch the sock with the bare skin because the oil in your skin will

Figure 3-2 Dry-bulb thermometer

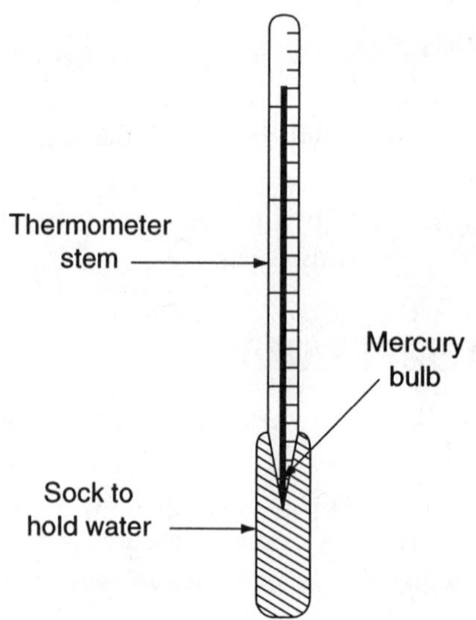

Figure 3-3 Wet-bulb thermometer

gradually affect the evaporation rate of the water from the sock, causing faulty temperature readings. See Figure 3–3.

The thermometer is exposed to a rapidly moving air stream. This may be accomplished by either twirling the thermometer around or holding it in an air stream until the reading has been taken. To get as accurate a reading as possible, the thermometer should be exposed to the rapidly moving air stream until the same temperature is indicated for at least two consecutive readings. The temperature is recorded after the temperature has stabilized and is indicated by (WB). The wet-bulb reading will always be at or below the dry-bulb temperature. The difference between the dry-bulb and wet-bulb readings is known as the wet-bulb depression. This lower reading is because the moisture in the wick (sock) evaporates and causes the bulb to feel cooler.

The temperature at which two temperatures have stabilized is known as the equilibrium temperature. It indicates the rate of sensible heat transfer from the air to the water in the wet-bulb sock and is equal to the amount of heat transferred to the air by the evaporation of the water. At every corresponding dry-bulb and wet-bulb temperature, and at any given relative humidity, the point of equilibrium will always be at the exact same point.

How fast the moisture evaporates from the wick is dependent on the amount of moisture in the air, the sensible heat, and the wet-bulb temperature. All three of these are an indication of the total heat (enthalpy) in the air.

SUMMARY 3–5

- A wet-bulb thermometer is one that has the bulb covered with a sock.
- Wet the sock with distilled water and avoid touching the sock with the bare skin.
- The thermometer is placed in a rapidly moving air stream to obtain the temperature reading.
- Two consecutive readings that are exactly the same are required for a satisfactory reading.
- At every corresponding dry-bulb temperature, wet-bulb temperature, and relative humidity, the point of equilibrium will always be exactly the same.
- The rate of evaporation of the water from the sock depends on the sensible heat, moisture in the air, and the wet-bulb temperature.

REVIEW QUESTIONS 3–5

1. With what is the sock on a wet-bulb thermometer wetted?
2. What factors determine the rate of evaporation of the water from the wetted sock?
3. What is the equilibrium temperature?

3–6 WET-BULB DEPRESSION

The difference between the dry-bulb and the wet-bulb temperatures is commonly referred to as the wet-bulb depression. It is an indication of how far below the dry-bulb reading the wet-bulb reading is. It should be remembered that when the air is at the saturation temperature, it is holding all of the moisture it can hold (100% humidity).

SUMMARY 3–6

- The difference between the wet-bulb and the dry-bulb temperatures is commonly referred to as the wet-bulb depression.
- The wet-bulb depression indicates how far below the dry-bulb temperature the wet-bulb reading is.
- When air is holding all of the moisture it can hold, it is said to have 100% relative humidity.

REVIEW QUESTIONS 3–6

1. Define wet-bulb depression.
2. At what temperature is air holding all of the moisture it can hold?

3–7 DEW-POINT TEMPERATURE

The temperature at which moisture will start to condense out of the air is termed the dew-point temperature and it is indicated by (DP). The amount of water contained in the air will always be exactly the same for every dew-point temperature. We can determine the amount of moisture in a given quantity of air simply by measuring the dew-point temperature and plotting it on a psychrometric chart. At the dew-point temperature the air is holding all of the moisture that it can hold at that temperature. The air is said to be at its saturation temperature and it is at 100% relative humidity. When the dew-point temperature remains constant, the amount of moisture in that sample of air will always be the same. Unless some of the latent heat is removed from the air, there will be no removal of moisture. The dew-point temperature will not change unless moisture is either added to or removed from the air. The dew-point temperature and the wet-bulb temperature will always be the same. They are both located on the instep of the psychrometric chart.

SUMMARY 3–7

- The dew-point temperature is the temperature at which moisture will start to condense out of the air.
- The amount of moisture in the air will always be the same for every dew-point temperature.
- Unless some of the latent heat is removed from the air there will be no moisture removal.

- The dew-point temperature will not change unless moisture is either added to or removed from the air.
- Both the wet-bulb temperature and the dew-point temperature are located on the instep of the psychrometric chart.

REVIEW QUESTIONS 3–7

1. At what temperature does moisture collect on a surface?
2. What is a simple way to determine the amount of moisture in a sample of air?
3. What is required to remove moisture from air?

CHAPTER

4

PRESSURE

OBJECTIVES

Upon completion of this chapter, you should be able to:

- Have a better understanding of pressure
- Know the different types of pressure as they apply to refrigeration and air conditioning
- Be more familiar with the laws that affect the pressures in a refrigeration system

INTRODUCTION

The pressure exerted on the refrigerant in a refrigeration system determines both the evaporating and condensing temperatures of the refrigerant. There is a definite temperature at which a given liquid will boil for each pound of pressure exerted on it. When the pressure over the liquid is changed the boiling temperature is also changed. If the pressure is increased, the boiling temperature is increased. If the pressure is decreased, the boiling temperature is also decreased. This is the phenomenon that the compression refrigeration system is based on. There are three pressures to which a refrigeration system is constantly exposed: (1) atmospheric, (2) gauge, and (3) absolute.

4–1 ATMOSPHERIC PRESSURE

Atmospheric pressure is exerted on the earth by the atmosphere above it. Atmospheric pressure is changed only by a change in the weather or the depth of the atmosphere at any given point. Normally, atmospheric pressure is taken to be 14.7 psi (pounds per square inch) at sea level. When the altitude is changed the atmospheric pressure will also change. Atmospheric pressure is determined by a column of mercury (Hg) measuring 29.92 inches in height. See Figure 4–1.

As the altitude is decreased, such as going up on a mountain, the height of the atmosphere is less and, therefore, the weight of it is less and the pressure on the mercury column would be less. For example, if we were to measure the atmospheric pressure on a mountain 5,000 feet high, the atmospheric pressure would be only 12.2 psi. This reduced pressure causes water to boil at a lower temperature.

SUMMARY 4–1

- Atmospheric pressure is exerted on the earth by the atmosphere above it.
- Atmospheric pressure is taken to be 14.7 psi (pounds per square inch) at sea level.
- When the altitude is changed the atmospheric pressure will also change.

REVIEW QUESTIONS 4–1

1. What causes a change in the atmospheric pressure?
2. When measuring atmospheric pressure with a mercury column, what is the normal reading at sea level?
3. Why will water boil at a lower temperature when heated on a mountain top?

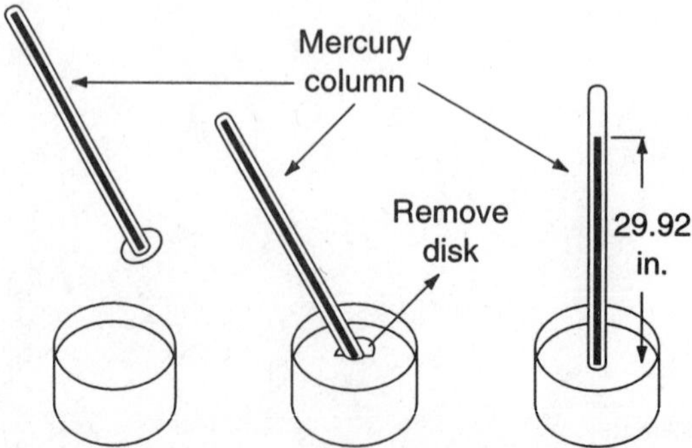

Figure 4–1 Measuring the atmospheric pressure

Figure 4–2 Refrigeration compound gauge (Henry Valve Company)

4–2 GAUGE PRESSURE

The pressure reading shown on a gauge not connected to a source of pressure is called gauge pressure. When it is not connected to a pressure source, it is commonly read as zero psig (pounds per square inch gauge). The pressures below zero psig on a gauge are read in inches of vacuum pressure, or inches of mercury pressure. The low side gauges used in refrigeration and air conditioning work are called compound gauges and will measure pressure both above and below zero psig. The readings above zero are considered to be pressures while the readings below zero are considered to be inches of mercury. See Figure 4–2.

If we change inches of mercury to psi we can see that 29.92" Hg is equal to 14.7 psi. Thus we can see that 1 psi is equal to 2" Hg. We can determine this as follows: 29.92 ÷ 14.7 = 2.03" Hg. We must note that gauge pressures are relative to absolute pressures. See Table 4–1.

SUMMARY 4–2

- The pressure reading shown on a gauge not connected to a source of pressure is called gauge pressure.
- The pressures below zero psig on a gauge are read in inches of vacuum or inches of mercury (Hg).
- One psi is equal to 2" Hg.
- The low side gauges used in refrigeration and air conditioning work are called compound gauges and will measure pressures both above and below zero psig.

ALTITUDE	PSIA	PRESSURE IN INCHES (HG)	BOILING POINT OF WATER (°F)	REFRIGERANT BOILING POINTS (°F)		
				R–12	R–22	R–502
0 FT	14.7	29.92	212	–21	–41	–50
1,000 FT	14.2	28.85	210	–23	–43	–51
2,000 FT	13.7	27.82	208	–25	–44	–53
3,000 FT	13.2	26.81	206	–26	–45	–54
4,000 FT	12.7	25.84	205	–28	–47	–56
5,000 FT	12.2	24.89	203	–29	–48	–57

Table 4–1 Comparison of atmospheric and absolute pressures at varying altitudes

REVIEW QUESTIONS 4–2

1. Two inches of mercury vacuum is equal to how much pressure?
2. If a gauge is just lying on a workbench, what should the pressure reading be?
3. What would be the boiling temperature of an open container of R–22 at 4,000 feet?

4–3 ABSOLUTE PRESSURE

Absolute pressure is measured from a perfect vacuum. Thus, it can be seen that atmospheric pressure and absolute pressure are the same measurement. It is expressed in terms of pounds per square inch absolute (psia). Atmospheric pressure at sea level is taken to be 14.7 psia. It is equal to gauge pressure plus atmospheric pressure. Absolute pressure can be determined from a gauge reading by simply adding 14.7 to the gauge reading. Use the formula:

$$A_p = G_p + AT_p$$

Where:

A_p = Absolute pressure
G_p = Gauge pressure
AT_p = Atmospheric pressure

Example: We have a gauge reading of 15 psig. What is the absolute pressure reading?

$$A_p = G_p + AT_p$$
$$= 15 + 14.7 = 29.7 \text{ psia}$$

SUMMARY 4–3

- Absolute pressure is measured from a perfect vacuum. It is expressed in terms of pounds per square inch absolute (psia).
- Atmospheric pressure at sea level is taken to be 14.7 psia. It is equal to gauge pressure plus atmospheric pressure.

REVIEW QUESTIONS 4–3

1. Change a gauge reading of 15 psig to absolute pressure.
2. From what pressure is absolute pressure measured?

4–4 CRITICAL PRESSURE

By definition, the critical pressure of a liquid is the pressure at or above which the liquid will remain a liquid regardless of the changes applied to it. Thus, when the critical pressure is applied to a liquid, the liquid cannot be changed to a vapor by the addition of heat.

SUMMARY 4–4

- The critical pressure of a liquid is the pressure at or above which the liquid will remain a liquid regardless of the changes applied to it.

REVIEW QUESTIONS 4–4

1. At what pressure can a liquid not be evaporated?

4–5 PRESSURE MEASUREMENT

Measuring the refrigerant pressure inside a refrigeration system is very important in understanding how the system is operating. There are several different pressures that must be determined and there are different means of determining these pressures. There is the high-pressure reading that is taken at the compressor discharge, there is the low-side pressure that is normally taken at the compressor suction, and then there is the pressure induced by an evacuation unit. All of these pressures determine to some extent the operating characteristics of the system.

High (Discharge) Pressure

This pressure is taken as close to the compressor discharge valve as possible. Different conditions will cause this pressure to vary according to the type of refrigerant being used, the temperature of the cooling medium for the condenser, the condition of the compressor, and the cleanliness of the condenser, just to mention a few. A pressure-temperature chart for the type of refrigerant being used and the cooling medium temperature must be known for this reading to have much meaning. With air-cooled condensers, this reading should be approximately 30°F to 35°F higher than the cooling medium. With a water-cooled condenser, the discharge pressure should correspond with a leaving water temperature of about 90°F + 10°F, or at about 100°F.

Low (Suction) Side Pressure

The low side pressure is an indication of the evaporating temperature of the refrigerant inside the evaporating coil. This pressure is generally taken as close to the compressor suction service valve as possible. Different conditions will cause this pressure to vary according to the type of refrigerant being used, the temperature of the fixture being cooled, the cleanliness of the evaporator, and the condition of the compressor, just to mention a few. A pressure-temperature chart for the type of refrigerant being used and knowledge of the temperature inside the cabinet being refrigerated must be known before this reading has much meaning. The temperature drop of the air flowing over the evaporator will vary according to the application and the temperature of the cabinet. The cabinet manufacturer should be able to provide this information.

Evacuation Pressures

This is the pressure that the system is subjected to before it is charged with refrigerant. Because these pressures are so small, special gauges must be used to measure them. Normally evacuation pressures are measured in microns. The micron is a metric measurement of length and is used for measuring vacuums induced in a refrigeration system by a vacuum pump. These are absolute pressures and are measured with a micron meter. See Figure 4–3.

One micron has the linear length of 1/1,000 of a millimeter. There are 25.4 millimeters in one inch. Thus, one micron equals 1/25,400 of an inch. Most manufacturers recommend that their refrigeration systems be evacuated to around 500 microns. This is equal to an absolute pressure of 0.02" Hg. From this it can be readily seen that a standard refrigeration compound gauge cannot be used for this purpose.

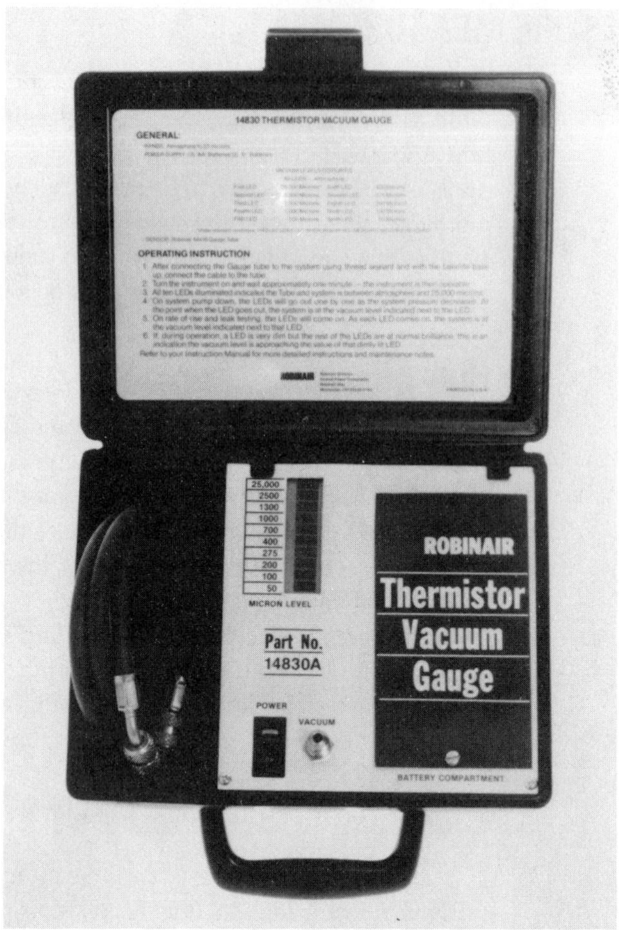

Figure 4–3 Thermistor vacuum (micron) gauge (Courtesy of Robinair Manufacturing Company)

SUMMARY 4–5

- Measuring the pressure inside a refrigeration system is very important in understanding how the system is operating.
- The discharge pressure is taken as close to the compressor discharge valve as possible.
- Different conditions will cause the discharge pressure to vary.
- With air-cooled condensers the discharge pressure should be about 30°F to 35°F higher than the cooling medium.
- With water-cooled condensers the discharge pressure should correspond to a temperature of about 10°F higher than the leaving water temperature.
- The low-side pressure is an indication of the evaporating temperature of the refrigerant inside the evaporating coil.
- The suction pressure is taken as close to the compressor suction valve as possible.
- Different conditions will cause the suction pressure to vary.
- The temperature drop of the air as it flows through the evaporator will vary according to the application and the temperature of the cabinet.
- The evacuation pressure is the pressure the system is subjected to before it is charged with refrigerant. One micron is the linear length of 1/1,000 of a millimeter. There are 25.4 millimeters in one inch.

REVIEW QUESTIONS 4–5

1. Where should the discharge pressure be taken?
2. Will bad compressor discharge valves affect the discharge pressure?
3. How can the evaporating temperature of a refrigerant in a system be determined?
4. Will bad compressor discharge valves affect the suction pressure?
5. Will a high fixture temperature cause a low suction pressure?
6. What should be used to measure the vacuum inside a refrigeration system?

4–6 LAWS AFFECTING PRESSURE

The effect that temperature has on the refrigerant pressure inside a refrigeration system must be understood before an adequate understanding of how these systems operate can be realized. The laws dealing with these effects are: (1) Boyle's Law, (2) Charles' Law, (3) Dalton's Law of Partial Pressures, (4) Pascal's Law, and (5) The General Gas Law.

Boyle's Law

Boyle's Law states that with the temperature constant, the volume of a gas is inversely proportional to the absolute pressure applied to it. In mathematical form this is:

$$P_1V_1 = P_2V_2$$

Where:

P_1 = old pressure
P_2 = new pressure
V_1 = old volume
V_2 = new volume

Example: If there is 5 cubic feet of gas in an enclosed vessel, the pressure is increased from 30 psig to 60 psig, and the temperature remains constant, what will be the new volume of the gas? Note that the gauge pressure must be converted to absolute pressure.

Use the formula:

$$V_2 = \frac{P_1V_1}{P_2}$$

$P_1 = 30 + 14.7 = 44.7$ psia
$P_2 = 60 + 14.7 = 74.7$ psia
$V_1 = 5$ cu. ft.

Then:

$$V_2 = \frac{(44.7 \times 5)}{74.7}$$

$$= \frac{223.5}{74.7}$$

$$= 2.99 \text{ cu ft}$$

This can be proven by using a properly fitted piston inside a cylinder. See Figure 4–4.

As the gas is slowly compressed so that there is no temperature change of the gas, and simultaneous pressure and temperature readings are taken, each side of the equation will always be equal. This is possible because there will always be a decrease in volume when the pressure is increased.

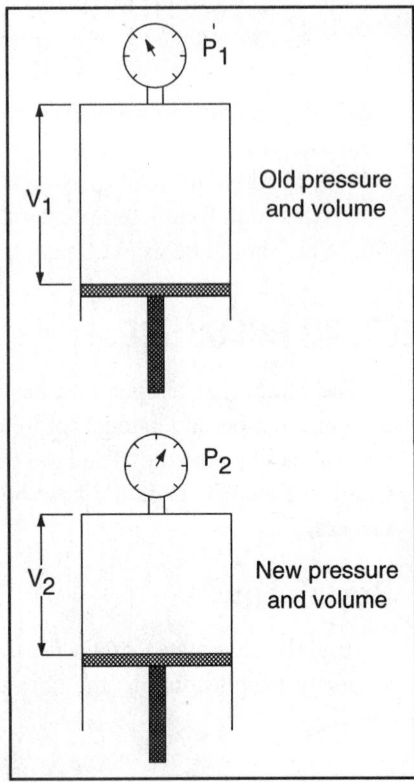

Figure 4–4 Boyle's Law example

Charles' Law

Charles' Law states that with the pressure constant, the volume of gas is directly proportional to the absolute temperature. Mathematically this may be stated:

$$\frac{V_1}{V_2} = \frac{T_1}{T_2}$$

Where:

V_1 = old volume
V_2 = new volume
T_1 = old temperature
T_2 = new temperature

Example: If we increase the temperature of 5 cu ft of gas from 60°F to 100°F with no change in pressure, what would be the new volume of the gas? Note that absolute temperatures must be used. Use the formula: (above formula converted.)

$$V_2 = \frac{(V_1 \times T_2)}{T_1}$$

V_1 = 5 cu ft
T_1 = 60 + 460 = 520
T_2 = 100 + 460 = 560

$$V_2 = \frac{(5 \times 560)}{520}$$

$$= \frac{2{,}800}{520}$$

$$= 5.38 \text{ cu ft}$$

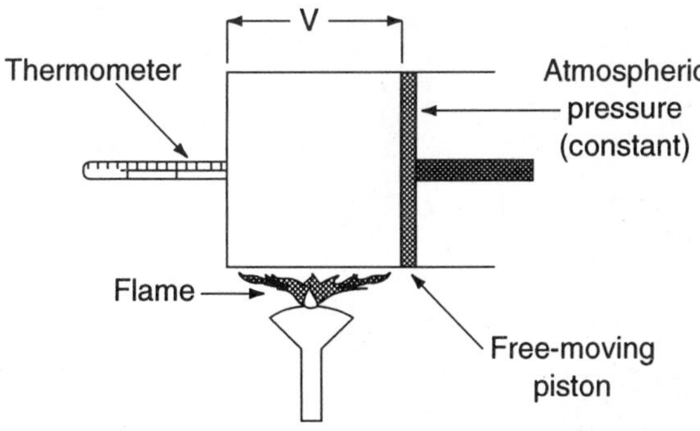

Figure 4–5 Charles' Law example I

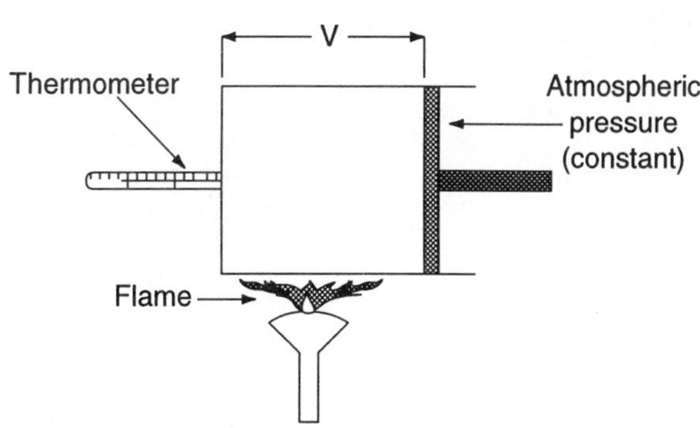

Figure 4–6 Charles' Law example II

This law can be proven by using two pistons properly fitted within two cylinders. See Figure 4–5.

Here the cylinder is properly fitted with a sliding piston. The cylinder is completely full of a gas at atmospheric pressure. When heat is applied to the cylinder it causes the temperature to increase. Because the piston is properly fitted, it will move easily as the volume of gas increases because of the increased temperature, but the pressure remains the same. See Figure 4–6. If we cool the gas, its volume will decrease as the temperature drops. See Figure 4–7.

If we could lower the temperature to absolute zero (–460°F), the volume would be reduced to zero because there would be no movement of the molecules to cause heat in the gas.

Dalton's Law of Partial Pressures

Dalton's Law of Partial Pressures states that each gas occupying the same space will fill the volume and act as though the other gases were not there. This law, in combination with Boyle's Law and Charles' Law, forms the basics of the psychrometric properties of air.

Dalton's Law explains that the total pressure inside a cylinder of compressed air, which is a mixture of oxygen, nitrogen, water vapor, and carbon dioxide, is determined by adding the total pressures that are applied by each of the individual gases present.

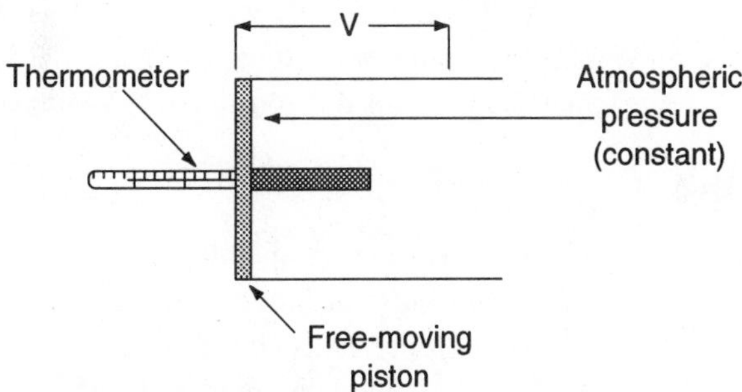

Figure 4–7 Charles' Law example III

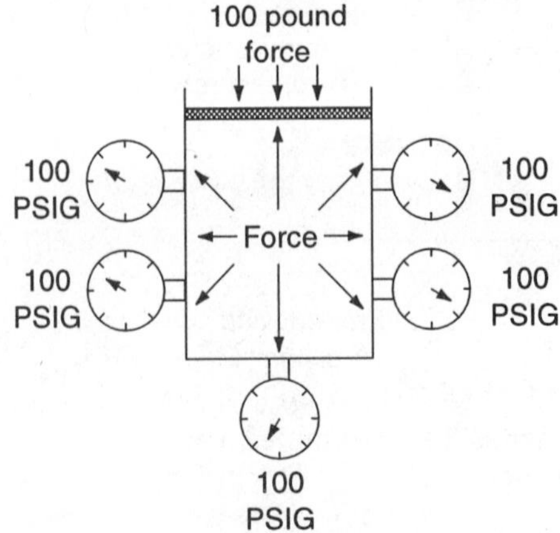

Figure 4–8 Pascal's Law example

Pascal's Law

Pascal's Law states that the pressure exerted on a confined fluid is transmitted equally in all directions. An example of Pascal's Law can be shown with a cylinder filled with a liquid having a fitted piston. See Figure 4–8.

The pressure applied on the piston is 100 psig and the piston has a cross-sectional area of 2 sq. in. If we have pressure gauges installed on the cylinder, we can see that the 100 psig is equally exerted in all directions.

The General Gas Law

The General Gas Law combines both Boyle's Law and Charles' Law. Mathematically it is expressed as:

$$\frac{(P_1V_1)}{T_1} = \frac{(P_2V_2)}{T_2}$$

A more simple form is:

$$PV = WRT$$

Where: P = absolute pressure in pounds per square foot
V = volume of the given quantity of gas in cubic feet
W = weight of the given amount of gas in pounds
R = universal gas constant of 1,545.3 divided by the weight of the gas
T = absolute temperature of the gas

The General Gas Law is generally used when studying the changes caused by the changing conditions of a given gas. Absolute pressures and absolute temperatures must be used for this formula to provide correct assessments.

SUMMARY 4–6

- The effect that temperature has on the refrigerant pressure inside a refrigerating system must be understood before an adequate understanding of how these systems operate can be realized.
- Boyle's Law states that with the temperature constant, the volume of a gas is inversely proportional to the absolute pressure applied to it.

- Charles' Law states that with the pressure constant, the volume of gas is directly proportional to the absolute temperature.
- Dalton's Law of Partial Pressures states that each gas occupying the same space will fill the volume and act as though the other gases were not there.
- Pascal's Law states that the pressure exerted on a confined fluid is transmitted equally in all directions.
- The General Gas Law combines both Boyle's Law and Charles' Law.
- The General Gas Law is generally used when studying the changes caused by the changing conditions of a given gas.

REVIEW QUESTIONS 4–6

1. By what gas law do we know that air in a refrigeration system will cause a higher than normal discharge pressure?
2. If a compressor cylinder full of refrigerant vapor is moved to increase the pressure, in what direction is the resulting pressure directed?
3. When working with gas laws, what must be done to the pressure and temperature readings?

4–7 PRESSURE-TEMPERATURE RELATIONSHIPS

Pressure and temperature relationships are very vital to the overall operation of a refrigeration or air conditioning unit. As learned earlier, the temperature at which a liquid will boil is dependent on the amount of pressure applied to it. Likewise, the pressure under which a liquid will boil is also dependent on the temperature applied to it. Thus, it should be known that, for each pressure applied to a liquid, there is also an accompanying temperature at which it will boil. The liquid must be in an uncontaminated container.

Because an uncontaminated liquid will act exactly the same each time it is placed under the same set of conditions, refrigerant pressure is a very reliable method of regulating the temperature inside of the evaporator. When an evaporator is placed inside an enclosed space and is not subjected to the surrounding atmospheric conditions, a pressure corresponding to the desired evaporating temperature can be applied to the liquid to control the temperature at which the liquid will boil. The liquid will always boil at that same temperature under the same set of conditions, and as long as heat is being absorbed, the refrigeration process will take place.

The reverse process is also used in the refrigeration cycle. If the pressure on the gas is high enough to cause the high-pressure gas to be at a temperature higher than the surrounding cooling medium, heat will be taken from the gas and given to the cooling medium. When the gas has given up a sufficient amount of heat it will change to a liquid. This is the process that is used in the condenser of a refrigeration system.

SUMMARY 4–7

- Pressure and temperature relationships are vital to the overall operation of a refrigeration or air conditioning unit.
- The temperature at which a liquid will boil is dependent on the pressure applied to it. Likewise, the pressure under which a liquid will boil is dependent on the temperature applied to it.
- It should be known that for each pressure applied to a liquid there is also an accompanying temperature at which it will boil.

REVIEW QUESTIONS 4–7

1. Why is the pressure temperature relationship of refrigerants important in refrigeration work?
2. What causes the refrigerant to condense in the condenser?
3. When an evaporator is fully refrigerated and the system pressure is lowered, what will happen?

CHAPTER

5

VAPOR COMPRESSION REFRIGERATION SYSTEMS

OBJECTIVES

Upon completion of this chapter, you should be able to:

- More thoroughly understand how a refrigeration system operates
- Have a better understanding of the purpose of the various components in a refrigeration system
- Know the difference between flooded and dry systems
- Know more about the purpose of flow control devices

INTRODUCTION

The vapor compression refrigeration system is the most popular type of system in use today. It uses the principle of compression, cooling, evaporating, and then compression again to accomplish the refrigeration process.

As we learned earlier, refrigeration is the process of removing heat from one place or material and depositing it in another and maintaining that space or material at a temperature lower than its surroundings.

There are two ways to cause a liquid to boil: (1) apply a sufficient amount of heat to the liquid, and (2) lower the pressure applied to the liquid. If the liquid used is a refrigerant, it will usually exist as a gas at atmospheric pressures and temperatures, and will absorb heat and boil when exposed to pressures that are equal to atmospheric pressures. During this boiling process it will absorb heat from its surroundings (because of the latent heat process studied earlier). In this process, heat will be removed from the space and the material to be refrigerated.

5–1 REFRIGERATION BY EVAPORATION

We have been accustomed to the fact that anything that is boiling is hot. This is not always true. We are most familiar with water boiling at 212°F. However, most refrigerants used have boiling points below 0°F when subjected to the normal operating pressures of a refrigeration system. Probably the easiest way to obtain a refrigeration effect is to place an open container of refrigerant inside a space to be cooled. The refrigerant will absorb heat from the surroundings and boil. The liquid changes to a gas that escapes out the top of the container. See Figure 5–1.

If we use HCFC–22 as the refrigerant, the liquid will absorb heat and boil as long as it is exposed to temperatures above –41.6°F and is subjected to atmospheric pressure. Above this temperature, HCFC–22 liquid will boil very rapidly, changing into a gas. At this point it is absorbing the latent heat of vaporization from the space and material, changing from a liquid to a vapor at a constant temperature of –41.6°F and atmospheric pressure.

It should be noted that the liquid refrigerant will remain at this temperature (–41.6°F) until all of the liquid has been changed into a gas. This changing from a liquid to a gas will continue as long as there is any liquid HCFC–22 in the container

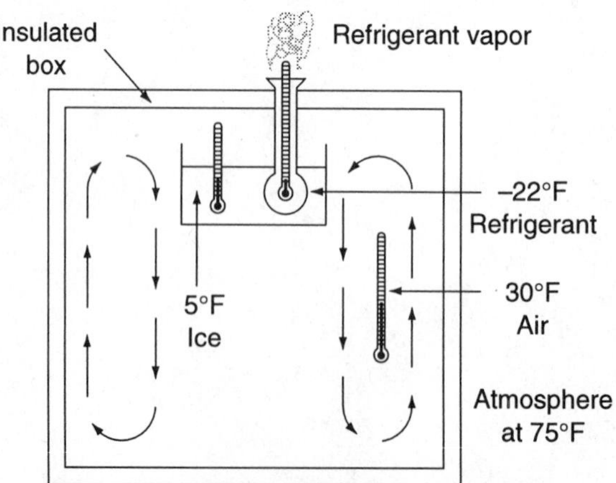

Figure 5–1 Refrigeration by evaporation

and the pressure remains at atmospheric. The liquid continues to absorb heat from inside the space, lowering its temperature. Circulation of the air, which is caused by the cooling and heating of it, are also shown in Figure 5–1. Notice that the air that is cooled by the boiling refrigerant becomes more dense and falls to the bottom of the space, where it begins to absorb heat from the space and material and will begin to rise because it is less dense at higher temperatures. Thus, the air circulation is caused and maintained by the alternate cooling and heating process. If we place a stopper in the outlet of the refrigerant container we can partially control the boiling by controlling the pressure exerted on the container and the liquid. This is the same principle that is used with the flow control device used on a refrigeration system.

This is not a practical refrigeration system because to keep the refrigerant replaced would be cost prohibitive and it is not automatic in operation. Also, the environmental problems would be astronomical.

SUMMARY 5–1

- We are most familiar with water boiling at a temperature of 212°F; however, most refrigerants used have boiling points below 0°F when subjected to the normal operating pressures of a refrigeration system.
- Liquid refrigerant, when boiling, will absorb heat from inside the space, lowering its temperature.

REVIEW QUESTIONS 5–1

1. When liquid refrigerant boils inside an evaporator, where does the heat come from?
2. Why does cool air fall?
3. Name two ways to cause a liquid to boil.
4. At atmospheric pressure, at what temperature does HCFC–22 boil?

5–2 COMPRESSION REFRIGERATION SYSTEM PRINCIPLES

When the refrigeration system components are connected with leak-proof tubing, the problem of always replenishing the refrigerant would be eliminated and the system would cost much less to operate. Also, because the system is completely enclosed, the pressures can be more easily controlled. The compressor will cause the higher pressures required to condense the refrigerant vapor in the condenser. The higher pressure vapors will also have a higher temperature because the heat is concentrated in the compressed vapor. The vapors have been made much hotter without appreciably increasing the amount of heat they contain because of the compression process. See Figure 5–2.

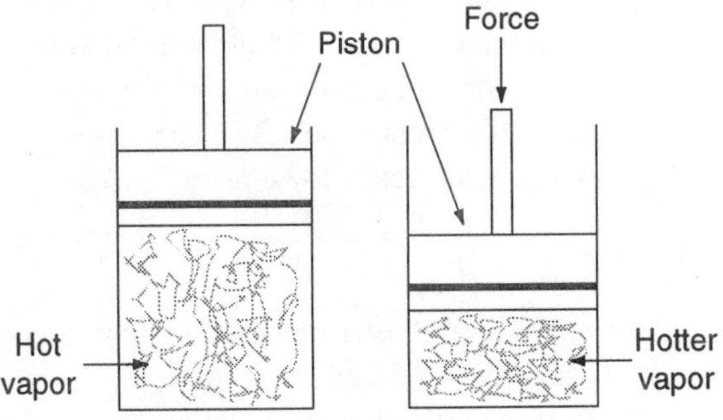

Figure 5–2 Compression principle

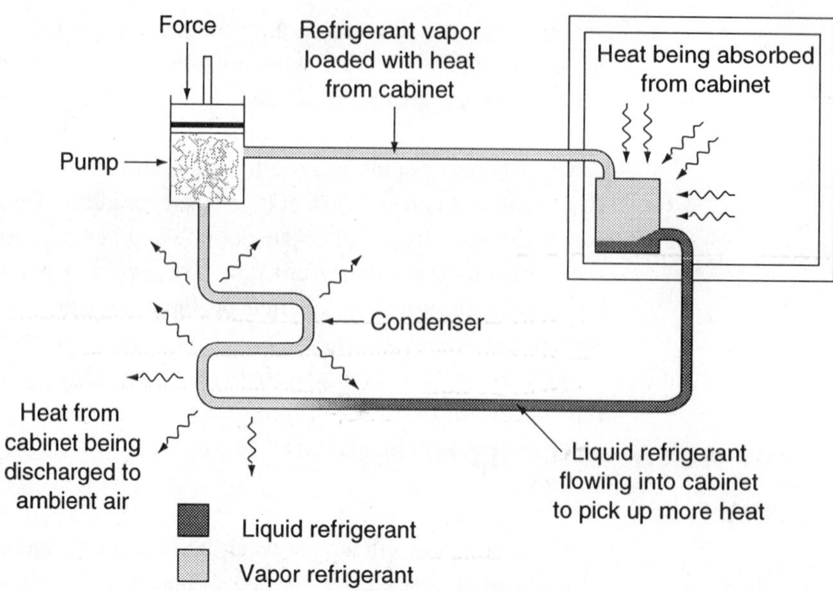

Figure 5–3 Elementary compression refrigeration system

These are the basic scientific rules governing the operation of a compression refrigeration system. It may be well to review these principles as they were discussed in the previous material. However, if the following major points are kept in mind the basic refrigeration cycle can be understood:

1. Liquids, when changing into a vapor, absorb tremendous amounts of heat without getting any hotter.
2. The refrigerant vapors can be made to change back into a liquid by the application of pressure.

Refrigerants used in this manner can be used over and over again. They never wear out and never lose their refrigerating effect as long as they are not contaminated or they do not leak out of the system.

Elementary Compression Refrigeration System

We know from our previous discussion that if a container of HCFC–22 refrigerant was placed inside an insulated cabinet, it will absorb heat from the box and the contents and boil because of its low boiling temperature at atmospheric pressures.

Rather than exhaust these vapors to the atmosphere, which would require that they be replaced continuously, they could be piped out of the insulated cabinet and back to the compressor. The pipes could carry the heat-laden refrigerant vapor back to the compressor so that it could be used again. See Figure 5–3.

As the compressor applies the pressure, the heat is literally squeezed out of the cold refrigerant vapor. By removing this heat in the condenser, the vapor returns to a liquid and is ready to be piped back inside the insulated cabinet and used again.

SUMMARY 5–2

- Because the refrigeration system is completely closed, the pressures can be more easily controlled.
- If the following major points are kept in mind the basic refrigeration cycle can be understood:

1. Liquids, when changing into a vapor, absorb tremendous amounts of heat without gaining any temperature.
2. The refrigerant vapors can be made to change back into a liquid by the application of pressure.

- As the compressor applies pressure, the heat is literally squeezed out of the cold refrigerant vapor.

REVIEW QUESTIONS 5–2

1. Why are the refrigerant pressures in a system easily controlled?
2. Why are the compressor discharge vapors at such a high temperature?
3. Why do liquid refrigerants absorb a lot of heat without gaining in temperature?

5–3 THE BASIC REFRIGERATION CYCLE

Refrigeration is accomplished by constantly compressing, condensing, evaporating, and circulating a fixed supply of refrigerant through a closed system. The refrigerating effect is made possible by the proper use of pressure, temperature, latent heat of vaporization, latent heat of condensation, condensation, and evaporation. Evaporation of the refrigerant is accomplished at a low pressure and temperature. Condensation is accomplished because of the high temperature and high pressure applied to the refrigerant. Through the condensation and evaporation process, it is feasible to transfer heat from a cool space and deposit it at a space with a higher temperature.

When considering the refrigeration cycle the process generally is started at the compressor discharge. However, it may be started at any place in the system. We can number the starting point number 1 to make it easier to follow the diagram in Figure 5–4.

The refrigerant vapor is compressed by the compressor, 1, to a high-temperature, high-pressure vapor and discharged into the condenser through the discharge line.

In the condenser, 2, heat is removed from the vapor by the cooling medium. It is cooled and condensed to a high-pressure, high-temperature liquid. In the bottom of the condenser, some subcooling is done to the liquid, cooling it about 5° to 10° cooler than the condensing temperature.

The liquid refrigerant then flows through the liquid line to the liquid line filter-drier, 3. The filter-drier removes contaminants from the refrigerant, helping to prevent plugging of the flow control device with foreign particles.

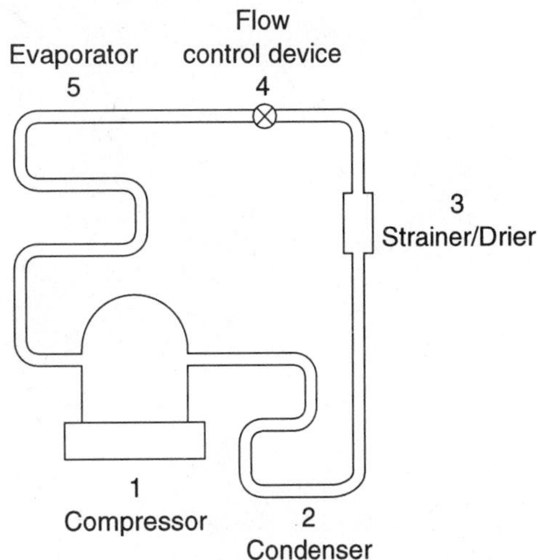

Figure 5–4 Compression refrigeration cycle

The liquid flows on through the liquid line to the flow control device at the inlet of the evaporator, 4. At this point the refrigerant encounters a restriction that causes the pressure to drop as it enters the evaporator. The purpose of the flow control device is to control the flow of refrigerant into the evaporator.

When the high-pressure, high-temperature liquid passes through the flow control device and into the evaporator, the pressure is reduced from the liquid line pressure to the pressure desired in the evaporator. This low pressure is created by the sucking action of the compressor and the restriction caused by the flow control device. This reduction in pressure is also accompanied by a reduction in temperature. At this point about 20% of the liquid refrigerant flashes into a gas that is used to cool the remaining liquid down to the evaporator temperature. This amount produces no refrigeration effect to the cooled space. The refrigerant starts picking up heat and evaporating as it flows through the evaporator.

When the low-pressure, low-temperature refrigerant vapor leaves the evaporator, it enters the suction line to the compressor, 5. At this point the liquid refrigerant has picked up enough heat to be completely evaporated. Thus, there is no liquid entering the suction line. There is a certain amount of superheat in the vapor at this point. The amount of superheat will depend on the type of system.

The low-pressure, low-temperature refrigerant vapor then travels down the suction line to the compressor. At the compressor, the compression process is repeated and the cycle is started again.

After the system has operated for a sufficient amount of time, the refrigerated space reaches the desired temperature and the thermostat (temperature control) will stop the compressor, ending operation of the unit. When the temperature of the refrigerated space rises to the cut-in point of the thermostat the contacts will close, starting the compressor, and the unit again produces refrigeration.

Inside the compression refrigeration system there are two distinctly different pressures. The dividing points are the discharge valve in the compressor and the needle seat in the flow control device. See Figure 5–5.

The two pressure levels are termed the high side and the low side.

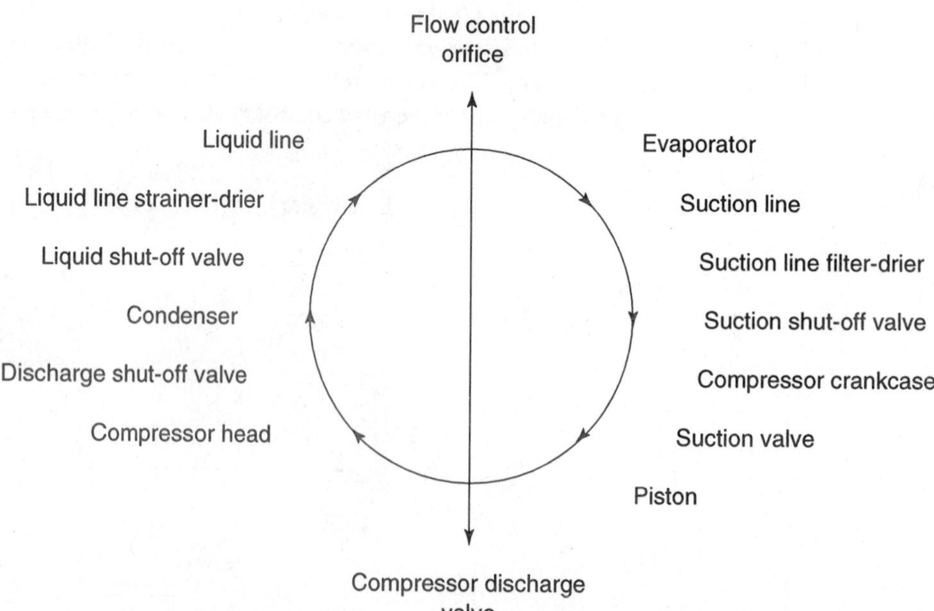

Figure 5–5 Pressure differential parts

The high pressure side of the system includes all the components that are subjected to the high-pressure refrigerant. Referring to Figure 5–5, the high side of the system includes the discharge side of the compressor, the compressor discharge service valve, the condenser, the liquid receiver (when used), liquid line shutoff valve (when used), the liquid line filter-drier, and all connecting lines. The complete compressor is generally considered to be in the high side of the system.

The low-pressure side of the system includes all the components subjected to the low-pressure refrigerant. Referring to Figure 5–5, the low-pressure side includes the outlet of the flow control device, the evaporator, the suction line, the compressor suction service valve, the compressor crankcase, and the suction valves inside the compressor. The flow-control device is generally considered to be in the low side of the system.

There are several different types of refrigeration systems. They may be based on either the type of compressor or the type of flow control device. However, regardless of the type of system, the refrigerant cycle in all of them is the same. In summary, the compression refrigeration cycle is as follows:

1. The low-pressure refrigerant vapor is compressed and discharged into the condenser.
2. The hot high-pressure refrigerant vapor is cooled and caused to condense into a liquid in the condenser.
3. The liquid refrigerant then passes through the flow-control device and into the evaporating coil.
4. When the pressure is reduced in the evaporator, the refrigerant vaporizes. About 20% of the refrigerant is changed to flash gas to cool the remainder of the refrigerant to evaporator temperature. Heat is then absorbed by the low-pressure, low-temperature refrigerant liquid during the vaporization process.
5. The heat-laden, low-pressure, low-temperature refrigerant vapor is then drawn into the compressor and the cycle is again repeated.

The service technician must be able to analyze every component in the system and understand its intended use before attempting to determine if there is any type of malfunction of the unit.

SUMMARY 5–3

- Refrigeration is accomplished by constantly compressing, condensing, and circulating a fixed supply of refrigerant through a closed system.
- The refrigerating effect is made possible by the proper use of pressure, temperature, latent heat of vaporization, latent heat of condensation, condensation, and evaporation.
- The pressure in the low side of the system is caused by the sucking action of the compressor and the restriction caused by the flow control device.
- The dividing points between the high and low pressures in the system are the compressor discharge valve and the needle seat in the flow control device.
- The type of refrigeration system may be based on the type of flow control device or the type of compressor.

REVIEW QUESTIONS 5–3

1. Name the six (6) factors that control the refrigerating effect of a system.
2. Is evaporation or condensation of a refrigerant accomplished at a low pressure and high temperature?

3. What generally happens to the refrigerant in the bottom of the condenser?
4. What device removes contaminants from the refrigerant?
5. Where is the flow control device located?
6. Define flash gas.

5–4 TYPES OF FLOODED SYSTEMS

The two types of systems based on the condition of the refrigerant in the evaporator are flooded and dry. The flooded systems have a puddle of liquid refrigerant in the evaporator. The dry systems have mostly vapor containing droplets of liquid refrigerant in the evaporator.

Flooded Systems

Flooded refrigeration systems normally operate with a definite amount of liquid refrigerant in the evaporator. This liquid level is maintained by the flow control device. See Figure 5–6.

The flooded type of system has several advantages over the dry type. Flooded systems operate with higher efficiency, a higher average suction pressure, shorter operating time, lower operating costs, less cycling, higher rate of heat transfer, and closer temperature control.

The liquid refrigerant in the evaporator allows a greater amount of the evaporator surface to be in contact with the refrigerant. This flooded condition allows for greater heat transfer into the evaporator.

The three flow control devices commonly used on flooded type systems are low-side float, high-side float, and capillary tube or restrictor.

Capillary Tube or Restrictor. Because of their simplicity in operation and their low cost, these types of systems are very popular in most domestic refrigerator and freezer applications and in most small residential air conditioning units.

The method used to control the flow of refrigerant into the evaporator is the major difference between these types of systems and other flooded systems. The capillary tube or restrictor controls the flow of refrigerant through it by friction and pressure difference.

The capillary tube is simply a small diameter copper tube with a very small inside diameter. It is basically the inside diameter, length of the capillary tube, and the system refrigerant charge that determine the amount of refrigerant that can flow through it.

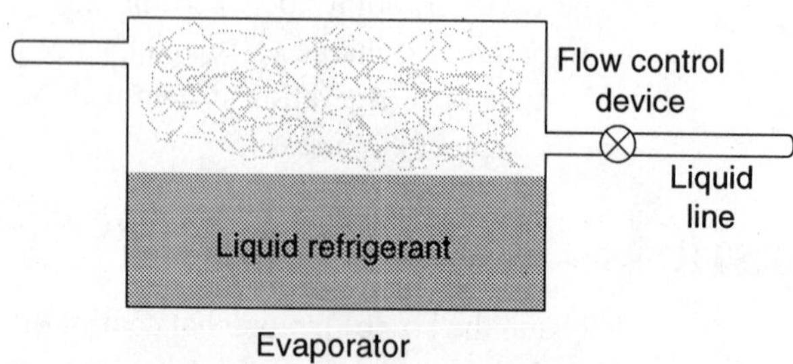

Figure 5–6 Flooded evaporator

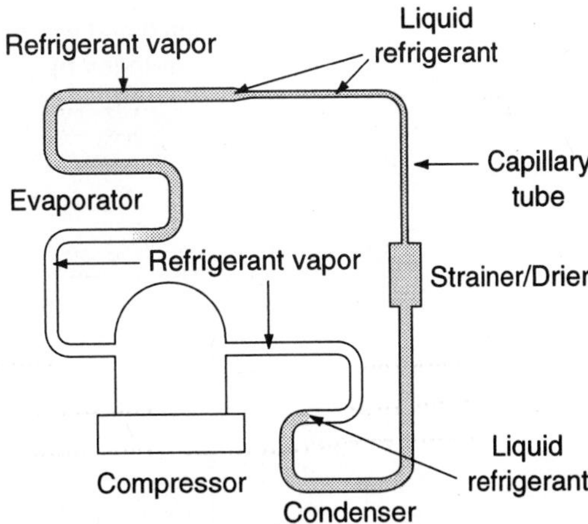

Figure 5–7 Capillary tube or restriction system

The capillary tube is located between a strainer-drier and the inlet of the evaporator. The strainer-drier must be located before the capillary tube to prevent foreign particles from plugging the very small orifice. See Figure 5–7.

As the compressor compresses the refrigerant and causes a simultaneous increase in the high-side pressure and a drop in the low-side pressure, refrigerant is allowed to flow through the capillary tube because of the difference in pressure between the two sides of the system. When the compressor first starts, the flow is very limited because the pressure difference is very small and the refrigerant is mostly in the vapor state. After the compressor has caused the discharge pressure to rise and the suction pressure to fall, and some of the refrigerant has changed to a liquid, the refrigerant flow is increased. The difference in the two pressures will eventually reach the desired operating conditions and will remain basically at this point until the temperature control stops the compressor.

These systems require a critical refrigerant charge. There are several reasons for this, such as: (1) any overcharge of refrigerant will be located in the low-side of the system and possibly result in floodback to the compressor, and (2) when the compressor stops, all of the refrigerant remaining in the high-side of the system will eventually pass into the low-side of the system until the refrigerant pressures are equalized. Liquid refrigerant in the low-side will also cause a probable floodback condition when the compressor starts for the next cycle.

Dry Systems

These types of systems normally operate with the refrigerant in a saturated vapor condition in the evaporator. These systems use either an automatic expansion valve (AXV) or a thermostatic expansion valve (TXV) as the refrigerant flow control device. In operation, as the refrigerant flows through the flow control device, about 20% of the liquid is immediately evaporated to cool the remaining liquid refrigerant down to the evaporator temperature. This 20% is called flash gas and it occurs in almost every system. It is not really wasted, but very little refrigerating effect is gained from it. If there were some means of eliminating this flash gas the system would operate much more efficiently. This is the reason that so much emphasis is placed on subcooling of refrigerant in the condenser. All of the liquid refrigerant is evaporated as it flows through the evaporator. There is always some superheat at the outlet of the evaporator to prevent floodback to the compressor. See Figure 5–8.

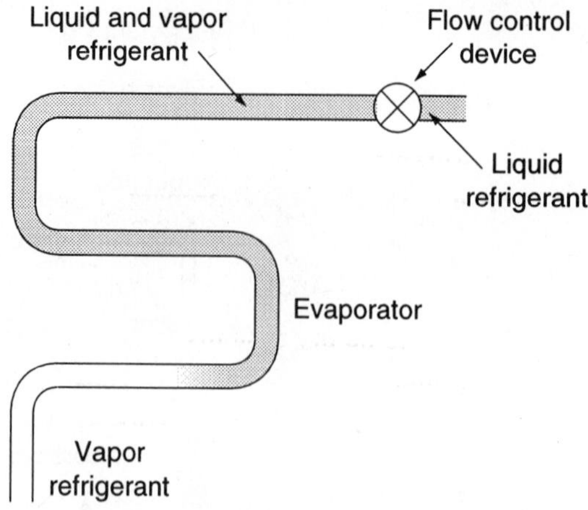

Figure 5–8 Dry system evaporator

Dry-type evaporators are generally made from tubes and do not have a chamber in which the liquid refrigerant can collect. Smaller evaporators may have only one tube through which the refrigerant flows. However, most evaporators are made of multiple circuits through which the refrigerant flows. These circuits increase the evaporator efficiency.

Automatic Expansion Valve (AXV). Automatic expansion valves are pressure-reducing devices that control the flow of liquid refrigerant into the evaporator. Automatic expansion valves are designed to provide a constant refrigerant pressure in the evaporator and low-side of the system.

The compression refrigeration cycle is the same as those systems used in flooded type systems. See Figure 5–9.

As the refrigerant flows through the expansion valve, it is atomized. The expansion valve regulates the flow of refrigerant into the evaporator in response to the pressure in the evaporator and the low-side of the system. The refrigerant is changed into a low-pressure, low-temperature saturated vapor (small droplets of liquid in the vapor) as it passes through the expansion valve. The saturated refrigerant then begins to absorb heat from the cabinet and its contents. It should be noted that there is no puddle of liquid refrigerant in the evaporator on these systems. The atomized

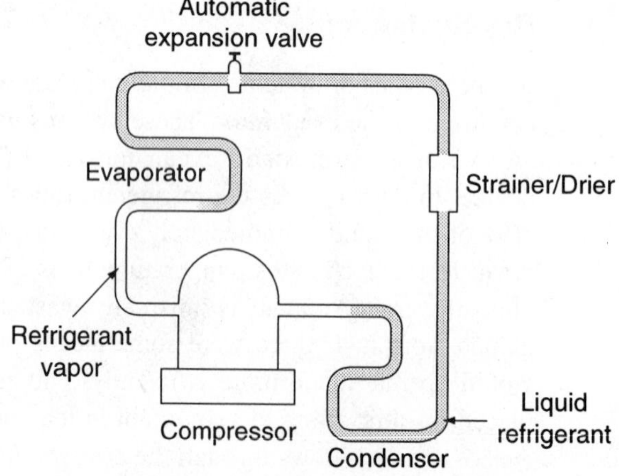

Figure 5–9 Automatic expansion valve system

refrigerant then flows through the evaporator tubing to the outlet. As the refrigerant gets closer to the end of the evaporator, there are fewer liquid droplets in the vapor. The compressor draws the refrigerant vapor through the suction line back to the compressor, where the cycle is started again.

Automatic expansion valve systems are not critically charged. However, it is never a good idea to overcharge any refrigeration system because the overcharge will reduce the efficiency and lower the capacity of the system.

Thermostatic Expansion Valve (TXV). These types of systems are also considered to be dry systems. In addition to pressure regulation like the AXV, the TXV also makes use of a thermostatic control that is used in combination with the evaporator pressure. This type of control provides a more accurate control of the refrigerant flowing into the evaporator. See Figure 5–10.

The TXV system is considered to be a dry system. However, there is more liquid allowed into the evaporator than in the AXV system because of the added temperature control by the TXV. These extra droplets of liquid refrigerant in the evaporator provide an efficiency that approaches that of the flooded systems.

The operation of the thermostatic expansion valve is almost identical to the automatic expansion valve except that the evaporator outlet temperature is controlled, along with the evaporator pressure. The purpose of the temperature element is so that the evaporator can be fully refrigerated during the complete operating cycle, rather than the evaporator being slightly starved at the beginning of the cycle, as with the AXV system.

Thermostatic expansion valve systems are not considered to be critically charged. However, it is never a good idea to overcharge any refrigeration system because an overcharged system will not operate to full efficiency or capacity.

SUMMARY 5–4

- The two types of refrigeration systems, based on the condition of the refrigerant in the evaporator, are flooded and dry.
- Flooded systems normally operate with a puddle of liquid refrigerant in the evaporator.
- The capillary tube, or restrictor, controls the flow of refrigerant through them by friction and pressure difference.

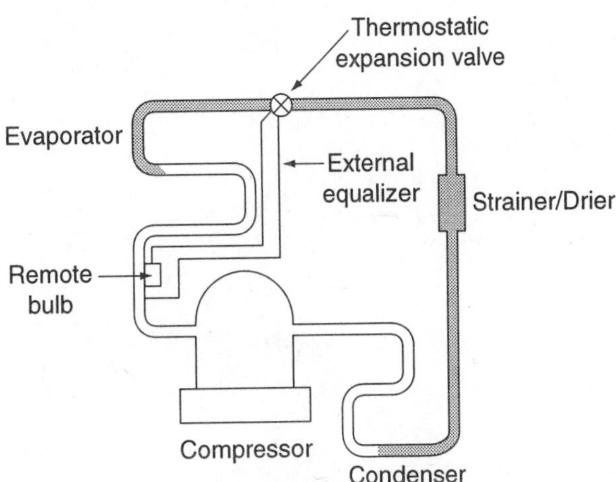

Figure 5–10 Thermostatic expansion valve system

- The capillary tube is simply a small diameter tube with a very small inside diameter.
- The capillary tube is located between the strainer-drier and the inlet of the evaporator.
- Capillary tube systems require a critical refrigerant charge.
- Dry type systems normally operate with the refrigerant in a saturated vapor condition in the evaporator.
- Dry systems usually use an automatic expansion valve (AXV) or a thermostatic expansion valve (TXV) as the flow control device.
- About 20% of the system charge changes to flash gas.
- There is always some superheat at the outlet of the evaporator to prevent floodback to the compressor.
- Automatic expansion valves are pressure-regulating valves that control the flow of refrigerant into the evaporator.
- AXVs are designed to provide a constant refrigerant pressure in the low-side of the system.
- AXVs are not critically charged systems.
- In addition to pressure regulation, TXVs also use a thermostatic control, which is used in combination with the evaporator pressure.
- This type of control provides more accurate control of the refrigerant flowing into the evaporator.
- The TXV is considered to be a dry type system.
- TXVs are not critically charged systems.

REVIEW QUESTIONS 5–4

1. Name the two types of systems based on the condition of the refrigerant in the evaporator.
2. Which type of system is the most efficient?
3. How does a capillary tube control the flow of refrigerant through it?
4. On a capillary tube system, where should the drier-strainer be located?
5. In a capillary tube system, what could happen if there is too much refrigerant charge in the system?
6. In dry systems, in what condition is the refrigerant in the evaporator?
7. Which is the most efficient, an AXV or a TXV?
8. Do expansion valve systems have a critical refrigerant charge?

CHAPTER

6

REFRIGERATION HAND TOOLS AND TEST INSTRUMENTS

OBJECTIVES

Upon completion of this chapter, you should be able to:

- More thoroughly understand the purpose of hand tools and test instruments
- Know more about the use of the various hand tools used in the refrigeration and air conditioning trade
- Know more about the use of the various test instruments used in the refrigeration and air conditioning trade
- Be more familiar with the purpose of the welding and brazing procedures used in the refrigeration and air conditioning trade

INTRODUCTION

During the installation and service procedures involved when working on refrigeration and air conditioning systems, certain tools and test instruments that are common to the trade must be used by the technician. They must be properly selected, cared for, and used to obtain the most accuracy when working on refrigeration and air conditioning systems. The use of the proper tools and test instruments will save much time and expense, and help to reduce personal injury.

6–1 HAND TOOLS

The specialized hand tools used by the refrigeration and air conditioning technician are in addition to the hand tools normally found in a mechanic's tool box.

Tubing Cutters. These tools are used to cut tubing and leave it with a square end. The square end will make the joining of pieces of tubing together much easier, and the joint will be stronger. It will be almost impossible to make flare type joints without a square end on the tubing. To use the tubing cutter, back the cutter wheel away from the rollers and place the cutters around the tube. To prevent damage to the cutter wheel, do not tighten it against the rollers. See Figure 6–1.

Then tighten the cutter wheel to touch the tubing. Do not tighten the cutter wheel too much or the tube may be squeezed out of round, making a good fitting almost impossible. See Figure 6–2.

After the cutter wheel touches the tubing, do not tighten the adjustment knob more than 1/2 turn. See Figure 6–3.

Turn the handle of the cutters around the tube at least one revolution, then tighten the cutter wheel another 1/2 turn. Repeat these steps until the tubing is completely separated.

When using the tubing cutters to remove an old flare from the end of a piece of tubing, place the flare in the groove on the cutter rollers. Center the tubing between the rollers with the flare in the groove and follow the above cutting procedure. See Figure 6–4.

If the old flare is not placed in the groove, the tubing will probably be threaded rather than cut off.

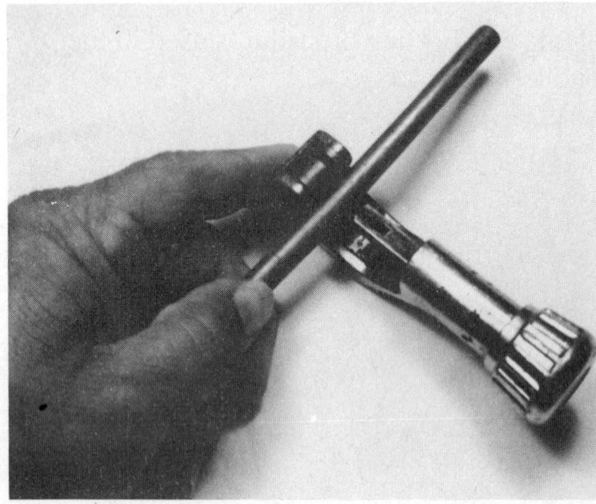

Figure 6–1 Use of tubing cutter (Photo by Billy C. Langley)

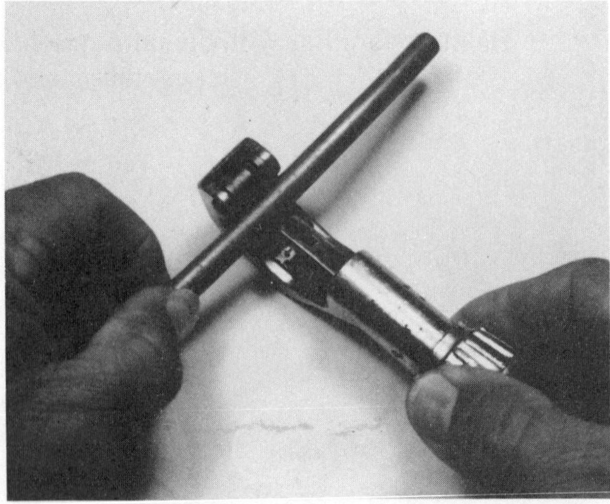

Figure 6–2 Tightening tubing cutter onto tube (Photo by Billy C. Langley)

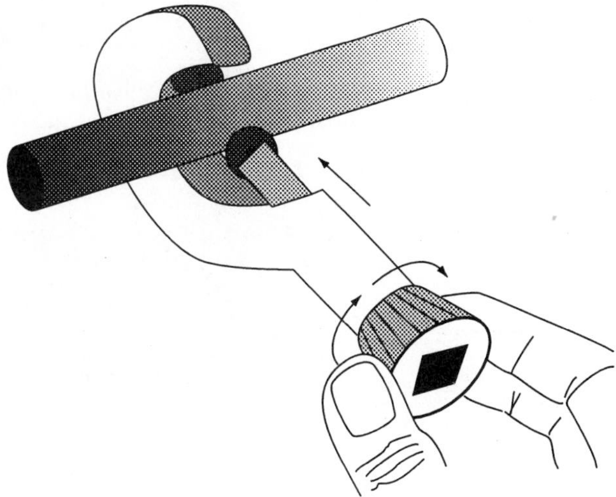

Figure 6–3 Tubing cutter adjustment

When tubing has been cut with a tubing cutter, there will be a burr on the inside of the tubing that must be removed. If it is not removed it will restrict the flow of refrigerant through the fitting. Some tubing cutters are also equipped with a reaming blade. See Figure 6–5.

To use the blade, simply loosen the screw and rotate the blade until it is in the position shown in Figure 6–5. This blade is used to remove the burr from the tube by inserting it into the tube until it touches and then turning the tubing cutters to remove the burr. Be sure to prevent the reamed-off material from falling into the tubing. Copper shavings left in the tubing will collect in the strainers and driers and cause a restriction to the refrigerant flow. The plugged components will then need to be replaced before the unit will operate satisfactorily.

Flaring Tools. Flaring tools are used to enlarge the end of copper tubing so that flare fittings can be used when joining two pieces of tubing together. See Figure 6–6.

These are generally made up of two pieces, the flaring block and the yoke. Before attempting to make a flare on the tubing, the tubing must be cut with a square end and reamed to remove all the burrs. The tubing is then inserted into the flaring block with about 1⁄4 inch of the tubing extended above the block face. See Figure 6–7. Some flaring tools have a tube height gauge that can be used for gauging the amount of tubing extending out of the block.

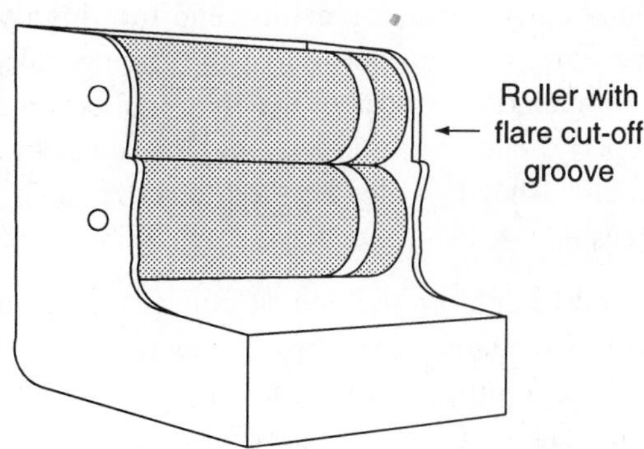

Figure 6–4 Tubing cutter rollers

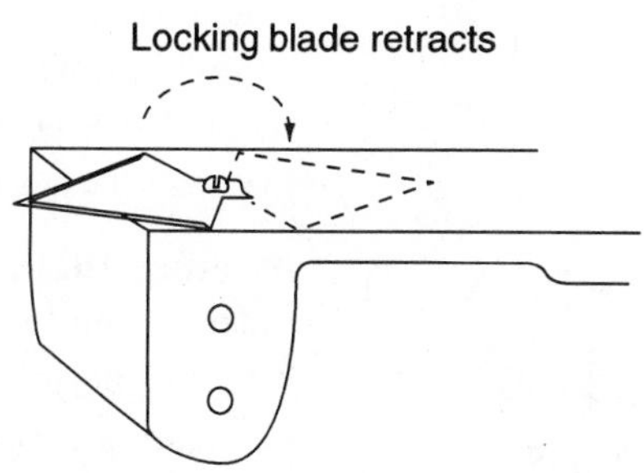

Figure 6–5 Tubing cutter reaming blade

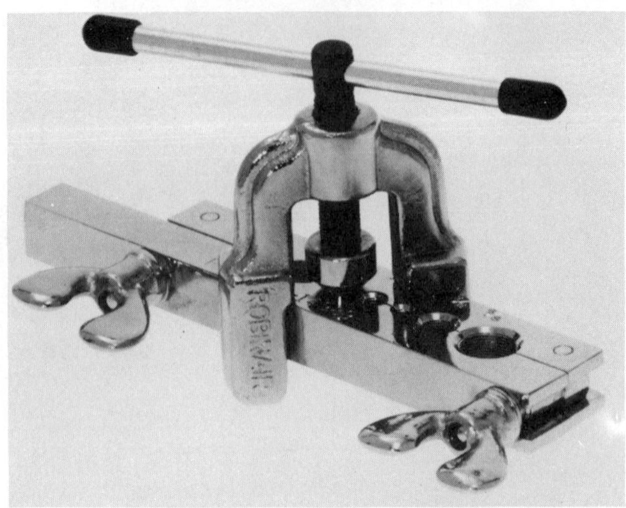

Figure 6–6 Flaring tool (Courtesy of Robinair Manufacturing Company)

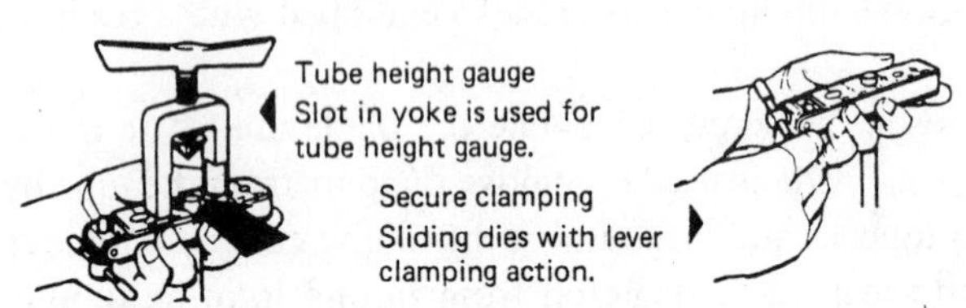

Tube height gauge
Slot in yoke is used for tube height gauge.
Secure clamping
Sliding dies with lever clamping action.

Figure 6–7 Placing tube in flaring block (Courtesy of Imperial Eastman)

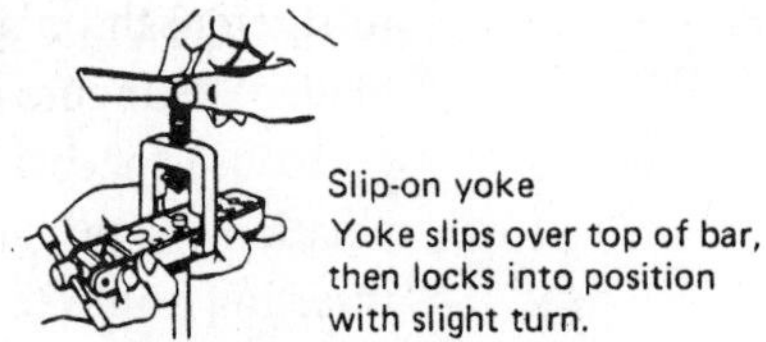

Slip-on yoke
Yoke slips over top of bar, then locks into position with slight turn.

Figure 6–8 Place the yoke on the flaring block (Courtesy of Imperial Eastman)

The clamping part of the block is moved into place and tightened against the tube enough to prevent it from slipping. The yoke is then slipped over the block and tubing and turned slightly to lock it into place. See Figure 6–8.

The yoke handle is then turned to move the cone-shaped tip toward the tubing. It is tightened very slowly to shape the flare and to prevent splitting of the material. See Figure 6–9.

Sometimes a drop of refrigeration oil between the flaring cone and the tube is helpful in making flares and burnishing the tubing end. Reversing the turning direction of the flare cone will automatically burnish the tube end to a highly polished finish. To prevent ruining the flare, do not overtighten the flaring yoke handle. When the yoke is tightened too much, the copper in the flare will become "dead" and will not seal the fitting properly. The flaring block can also be used to reround tubing that is slightly out of round. They can also be used to resize tubing that may have been enlarged for some reason.

Swaging Tools. Swaging tools are used to enlarge the end of copper tubing in order to join two pieces together. With this method, two pieces of tubing the same size can be joined without the use of a coupling, thus reducing one joint and eliminating a potential leak at that point. There is a swaging tool for each size of tubing. To make a proper swage, the proper size must be used. See Figure 6–10.

Figure 6–9 Making the flare (Courtesy of Imperial Eastman)

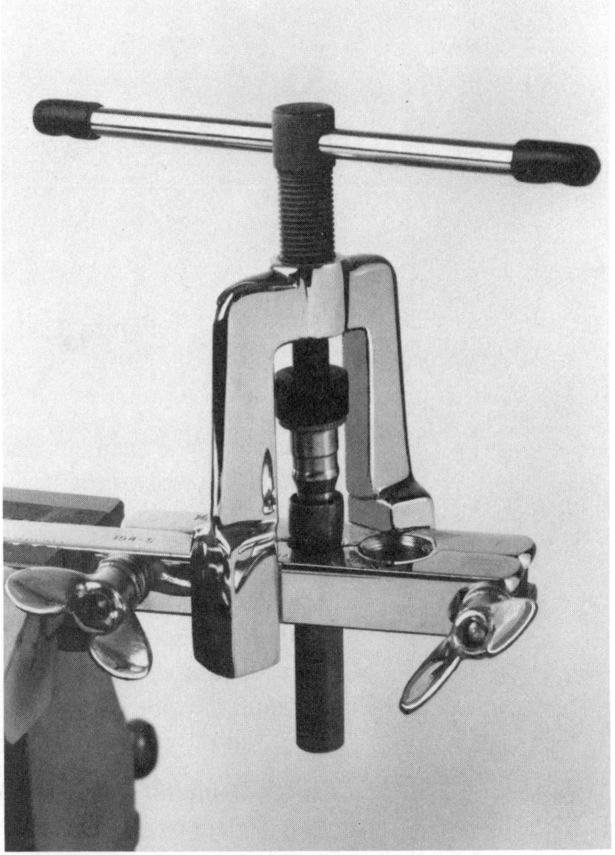

Figure 6–10 Swaging tool (Courtesy of Imperial Eastman)

To make a swage, insert the tube into the correct-sized die in the flaring block, with enough extending through to make a proper swage. If there is too much tubing extended past the flare block, the tube may bend during the swaging process and ruin the swage. On the other hand, if too little tubing is extended through the flare block, the swage may be cut off when the swaging tool is forced into the block face. When the tubing has been clamped in the flare block, place the swaging tool into the end of the tubing. Turn the swaging tool handle to force the swaging tool into the tubing. Force the swaging tool until it hits the bottom of the swage. Do not overtighten to prevent cutting off the swage.

Another type of swaging tool is one that must be driven into the tube with a hammer. With this type the tube is again placed in the flaring block. The swaging tool is then placed inside the tubing and struck with a hammer. During the swaging process, be sure to turn the swaging tool slightly between each blow of the hammer. This will prevent the swaging tool from sticking inside the tube, thus causing difficulty in removing it. When the swage is completed, tap the swaging tool with a sideways blow to slightly enlarge the swage and aid in its removal.

Tubing Reamers. Reamers are used to remove any burrs made when the tubing was cut. See Figure 6–11.

When the cut has been completed, place the reamer over the tube end and rotate it to remove any burrs. The reamer can be reversed to remove any burrs on the other side of the tubing. Be sure to remove the burrs from both the inside and the outside of

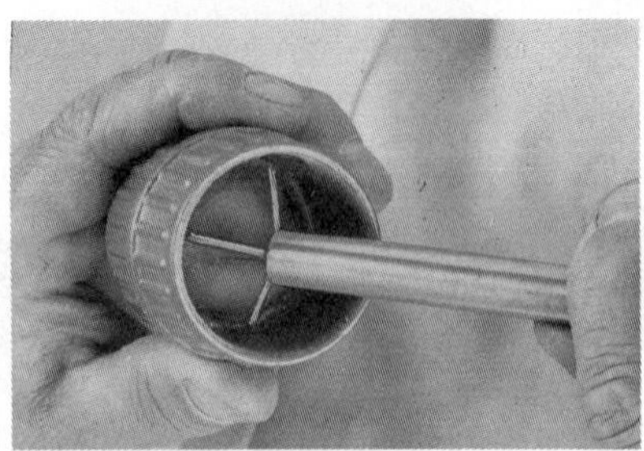

Figure 6–11 Tubing reamer (Courtesy of Imperial Eastman)

the tubing. These are multisized tools that will fit several sizes. There are several sizes available. Be sure to turn the tubing so that the shavings will not fall into the tube.

Pinch-Off Tools. These tools are used to place an intentional crimp in the tubing to prevent the escape of refrigerant. The tool is placed around the tubing and then tightened against it. See Figure 6–12.

Pinch-off tools are also provided with resizing dies so that the tubing may be resized if desired. When the service operation is completed, the pinch-off tool is used to seal off the tube. The service tools are then removed from the tube. The tubing is sealed with silver solder, then the pinch-off tool is removed.

Tubing Benders. Tubing benders are used to make bends in tubing to take the place of flare or sweat fittings where a leak may occur. They can be used to make bends to almost any degree desired. To make a bend, the tubing is first measured and marked at the place where the bend is desired. The tubing is then placed in the bender of the proper size. (Be sure to use the proper sized bender to prevent flattening the tube.) Place the mark at the 0° mark on the bender. Make the bend to the desired angle as indicated on the bender, plus about 3° to 5° extra. This extra bend is to allow for any restraightening of the tubing when the bender is removed. When a piece of tubing has been bent in a tubing bender, it can never be completely restraightened again.

Reversible Ratchets. Reversible ratchets are used to turn the service valve stems when servicing a refrigeration unit. Service valves permit the installation of different service tools for testing and servicing the unit. They are available in several different

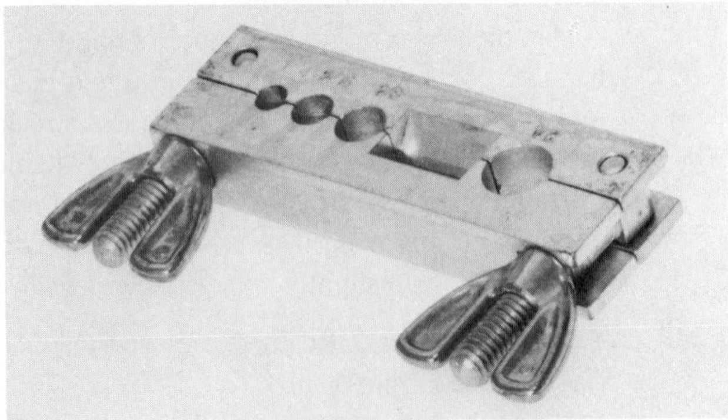

Figure 6–12 Pinch-off tool (Courtesy of Robinair Manufacturing Company)

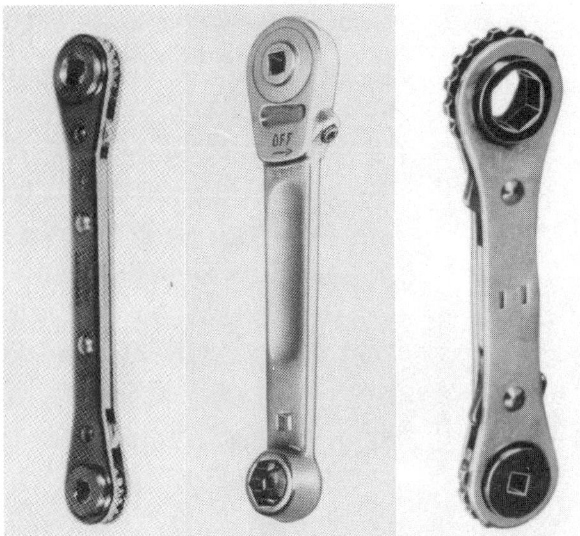

Figure 6–13 Reversible ratchets (Courtesy of Robinair Manufacturing Company)

sizes and configurations. Some have square valve stem fittings on both ends and some have the square fitting on one end and a hex socket on the other end. See Figure 6–13.

The intended use determines the type to buy. Also, there are sockets available that permit the use of one ratchet on several different-sized valve stems. See Figure 6–14.

There are also a variety of different-sized packing gland nut sockets available that are placed in the reversible ratchet. These sockets are used to tighten the packing glands on service valves to prevent the refrigerant from leaking.

Gauge Manifolds. Gauge manifolds are used to determine the operating pressures inside a refrigeration system and to perform other service operations involving the refrigerant and oil. They consist of a valve manifold, compound retard gauge, pressure gauge, and charging hoses. See Figure 6–15.

Valve Manifold. The valve manifold has openings that may be opened or closed so that various service operations can be accomplished. See Figure 6–16.

The proper use of the hand valves will permit almost any refrigerant or oil service operation required by the service technician. When the valves are cracked off the back seat, and the charging hoses are connected to the refrigeration system, the gauges will indicate the refrigerant pressures inside the system. The center hose is

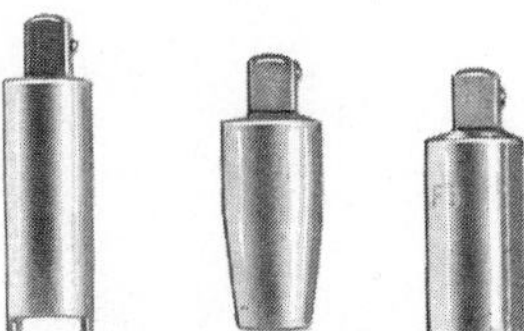

Figure 6–14 Valve stem sockets (Courtesy of Robinair Manufacturing Company)

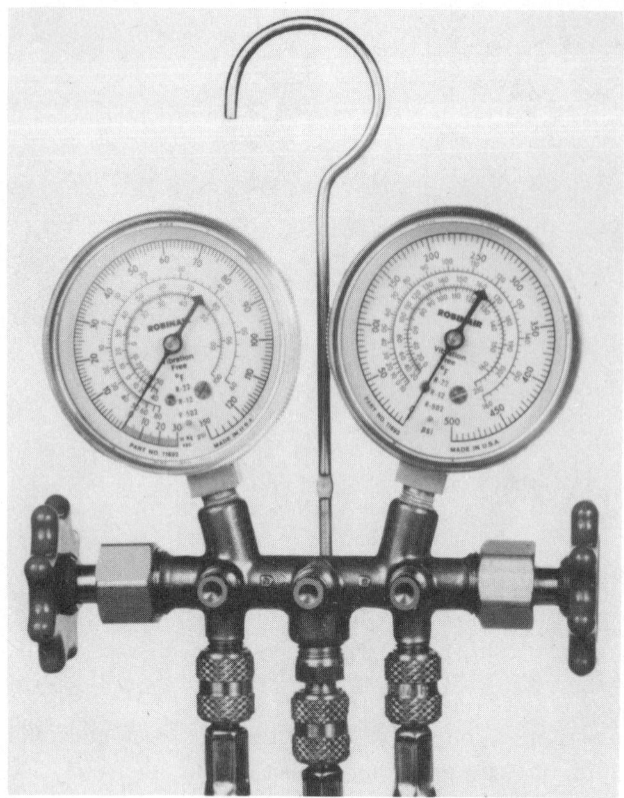

Figure 6–15 Refrigeration gauge manifold (Courtesy of Robinair Manufacturing Company)

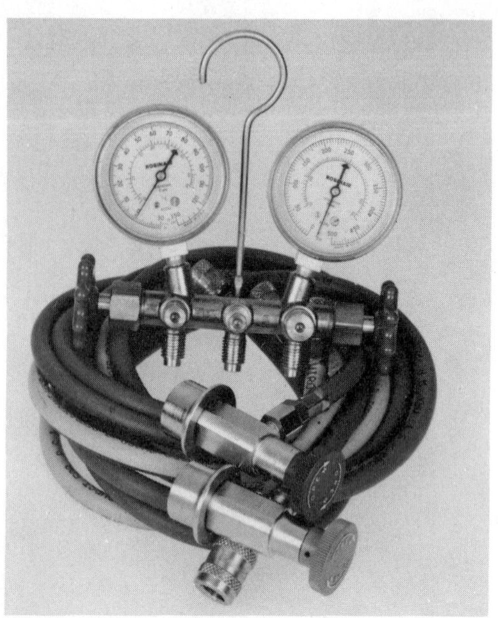

Figure 6–16 Valve manifold (Courtesy of Robinair Manufacturing Company)

usually connected to a refrigerant cylinder, vacuum pump, or to some other external source of material or function.

Compound Gauge. Compound gauges will indicate pressures both above and below atmospheric pressure. See Figure 6–17.

When properly connected to the system, these gauges will indicate the pressure in the low side of the refrigeration system. The outside scale indicates the pressure in psig. The inside scales are calibrated to indicate the saturation temperature of some popular refrigerants at that pressure.

Compound Retard Gauge. Compound retard gauges are equipped with a retarder that allows accurate pressure readings within a given range. Normally this pressure range is between 0 psig and 100 psig for refrigeration and air conditioning work. These gauges have graduations that change above the retard range. In Figure 6–17 the retard range starts at 120 psig.

Figure 6–17 Compound gauge (Courtesy of Henry Valve Company)

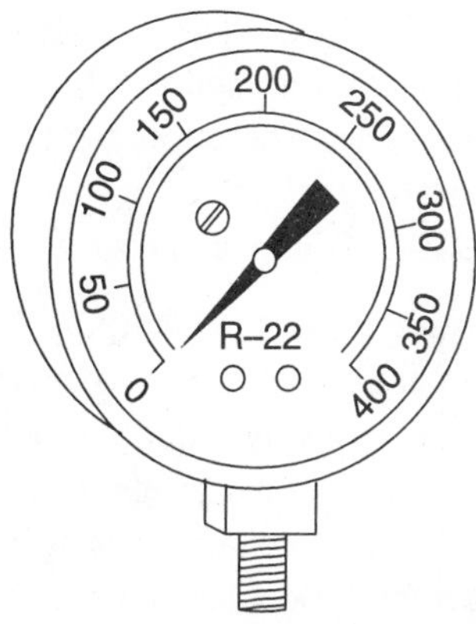

Figure 6–18 Pressure gauge

Pressure Gauges. Pressure gauges indicate the refrigerant pressure in the high side of the system. See Figure 6–18.

This gauge also has the pressure scale on the outside of its face. The inside scale indicates the saturation temperatures at the different pressures for the various refrigerants indicated. It should be remembered that all pressure gauges are not protected at pressures below atmospheric. Therefore, when drawing a vacuum on the system, it would be closed so it will not be subjected to pressures below atmospheric that could damage the gauge mechanism.

Refrigeration gauges indicate refrigerant pressures by the action of a bourdon tube. See Figure 6–19.

A bourdon tube is made from a flattened, curved metal tube that is sealed on one end. The other end is connected to the gauge pressure fitting so that pressure is transmitted to it from the pressure connection. As the pressure inside the Bourdon tube increases, the element has a tendency to try to straighten. When the pressure is decreased, the element tends to go back to its original shape. This movement is detected by a connection to the gauge needle. The needle then moves in a calibrated motion to indicate the pressure on the gauge face by the movement of the Bourdon tube.

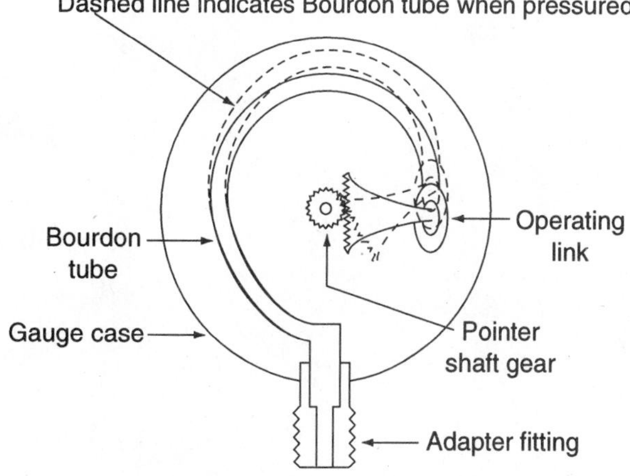

Figure 6–19 Bourdon tube principle

Digital Gauges. There are also digital gauges available that indicate the pressure readings using numbers rather than moving a needle through the scale on the gauge face (analog). These gauges use pressure-sensing transducers to signal the correct pressure to the gauge mechanism. The pressures are indicated in numbers on the gauge face. Digital gauges are also very sensitive to pressure.

Charging Hoses. Charging hoses provide the flexible connection between the system, the gauge manifold, and other tools and equipment. See Figure 6–20.

These hoses are equipped with 1⁄4 inch flare connections on each end for connecting them to the gauge manifold and to the system. Usually one end is equipped with a valve core depressor to push the valve core, if used, from its seat and allow refrigerant to flow through the fitting. The 1⁄4 inch connectors are also equipped with a pliable neoprene gasket to prevent refrigerant leaking at this point during service operations. These hoses may be purchased in a variety of colors and lengths. If a special length is needed it may also be specially ordered. The colors are a help when making connections to the system. Usually, a blue one is used for the low pressure side, a red one for the high pressure side, and a white or yellow one for the center connection to other tools and equipment.

The new refrigerants require that special hoses be used to prevent the escape of refrigerant through the pores of the hose material. These hoses are required by EPA and must be used on specific refrigerants. That is, the hoses used for CFC–12 and HFC–134a cannot be interchanged, etc. These hoses have a special liner to reduce the permeability of the material and prevent leakage of refrigerant.

Also, there is a special type of hose available that is better suited for working with below-atmospheric pressures. Some of the standard hoses will collapse under a vacuum and cause a false vacuum reading. The vacuum will be only on the part of the hose between the collapsed area and the vacuum pump. The system will not have as deep a vacuum as is indicated on the gauge or the micron meter. These special

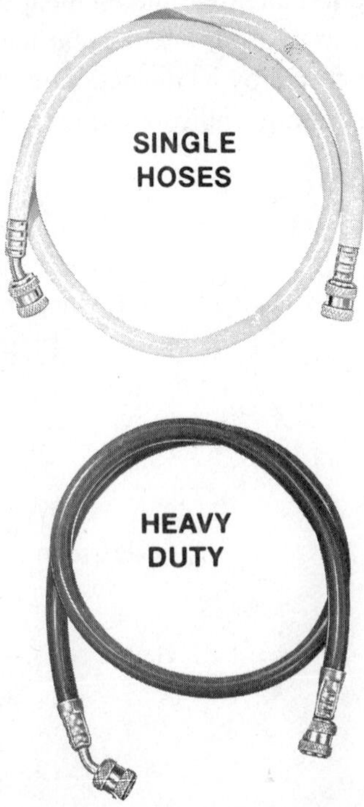

Figure 6–20 Charging hose (Courtesy of Robinair Manufacturing Company)

hoses are usually colored black. However, the black color is not always a good indication that the hose is good for use on a vacuum. Make certain when making the purchase what type it is.

Low-Loss Fittings. EPA requires that all refrigeration hoses and recovery equipment be equipped with low-loss fittings within six (6) inches of the equipment end of the hose. The purpose of the fitting is to reduce refrigerant loss during service operations.

Pocket Thermometers. Through the use of pocket thermometers, the service technician will save much time. The system operating temperatures can be checked to determine if the system is operating properly. If not, the thermometer will usually give a good indication of what the problem might be. The two popular types in use are mercury and bimetal. The bimetal type is the most durable and is therefore the most popular. See Figure 6–21.

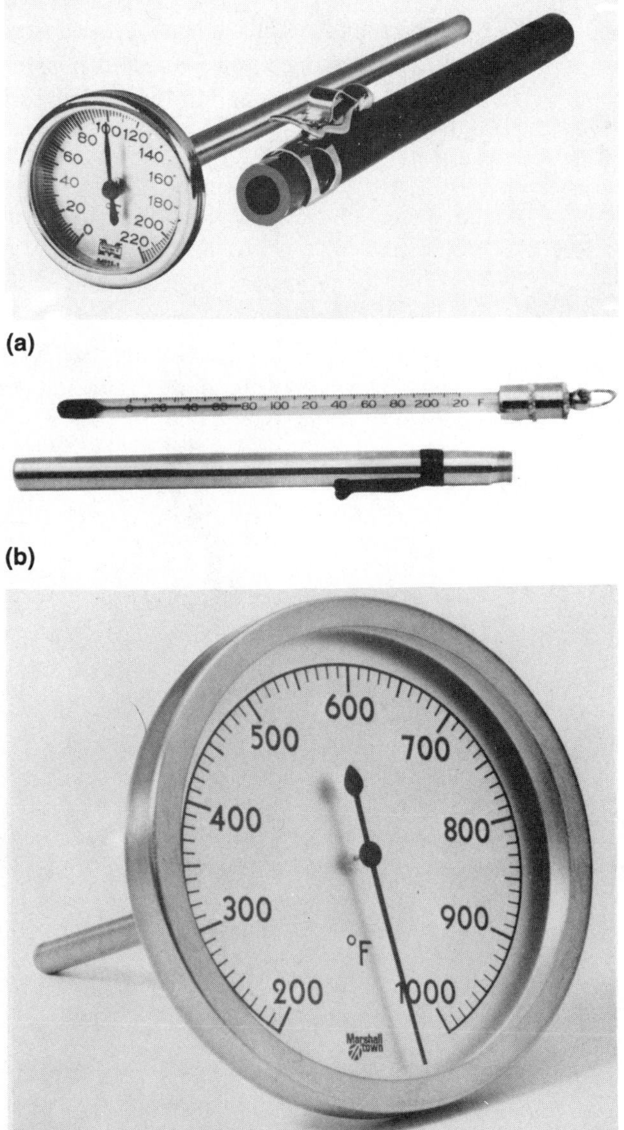

(c)

Figure 6–21 Various types of thermometers (a) mercury; (b) bimetal; (c) enlarged bimetal thermometer face (Courtesy of Marshalltown Instruments)

These tools are available in a protective case that is fastened to the shirt pocket. Most technicians carry thermometers in their shirt pocket continuously so that they will be handy when needed.

Charging Cylinders. Charging cylinders are used to charge the exact amount of refrigerant into the system being serviced. See Figure 6–22.

Usually, a vacuum is drawn on the charging cylinder and the system at the same time. The cylinder is then loaded with the manufacturer's recommended refrigerant charge for the system, plus a couple of ounces for line and other losses. Then, while the system is still under a vacuum, the complete refrigerant charge is dumped into the system at one time. Some charging cylinders are equipped with a small electric heater to heat the cylinder and charge so that the charge will flow from the cylinder into the system much faster and more completely. Do not turn on the heater while attempting to charge the cylinder. The resulting pressure will prevent the proper amount of refrigerant from entering the cylinder.

Never apply a direct flame to the charging cylinder. A direct flame may melt the cylinder causing a blow-out and possible personal injury or property damage. Charging cylinders should never be stored in the direct sunlight or stored with refrigerant in them.

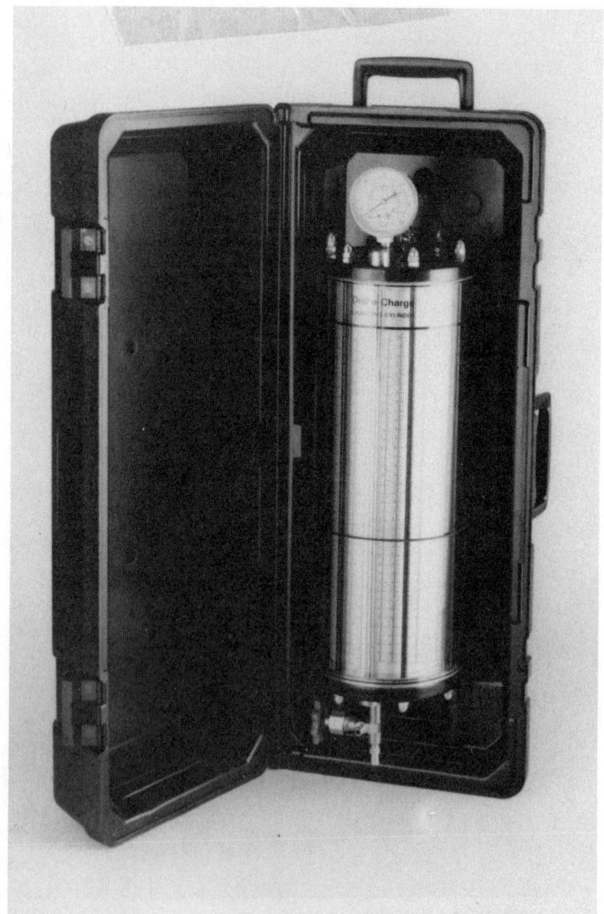

Figure 6–22 Charging cylinder (Courtesy of Robinair Manufacturing Company)

Fitting Brushes. Fitting brushes are used to clean the inside of a copper fitting that is to be soldered. The fitting should be cleaned to a shiny bright color. There is a brush for every size fitting. If one that is too large is used it will quite possibly be ruined, especially if it is forced into the fitting. If one is used that is too small, the fitting probably will not be cleaned properly. The brush is pushed into the fitting and then rotated until the fitting is properly cleaned. If the direction of rotation is reversed, the brush may be ruined.

SUMMARY 6–1

- Tubing cutters are used to cut tubing with a square end for making joints.
- Flaring tools are used to enlarge the end of copper tubing so the flare fittings can be used to join two pieces of tubing together.
- Swaging tools are used to enlarge the end of copper tubing in order to join two pieces together.
- Tubing reamers are used to remove any burrs made when the tubing was cut.
- Pinch-off tools are used to place an intentional crimp in the tubing to prevent the escape of refrigerant.
- Tubing benders are used to make bends to almost any degree desired.
- Reversible ratchets are used to turn the service valve stems when servicing a refrigeration unit.
- Gauge manifolds are used to determine the operating pressures inside a refrigeration system.
- Compound gauges will indicate pressures both above and below atmospheric pressure.
- Compound retard gauges are equipped with a retarder that allows accurate pressure readings within a given range.
- Pressure gauges indicate the refrigerant pressure inside the high side of the system.
- Charging hoses provide the flexible connection between the system, the gauge manifold, and other tools and equipment.
- Low-loss fittings are used in the end of charging hoses to reduce the loss of refrigerant when servicing the refrigeration system.
- Pocket thermometers are used to determine the system operating temperatures to see if it is working properly.

REVIEW QUESTIONS 6–1

1. Why should the tubing cutter wheel not be tightened against the rollers?
2. When tightening the cutter wheel to cut a piece of tubing, how tight should it be?
3. What should be done to the tubing before making a flare on it?
4. Why is a swaging tool turned while making a swage?
5. When reaming a piece of tubing, what should be done to the tubing?
6. When using a tubing bender, why should the tube be bent a little extra?
7. For what purpose are the hand valves on a gauge valve manifold?
8. What type of pressure gauge allows for accurate pressure readings within a given range?
9. What tool provides a flexible connection between the system and the service tools?
10. What is required to reduce the amount of refrigerant lost when checking refrigerant pressures?

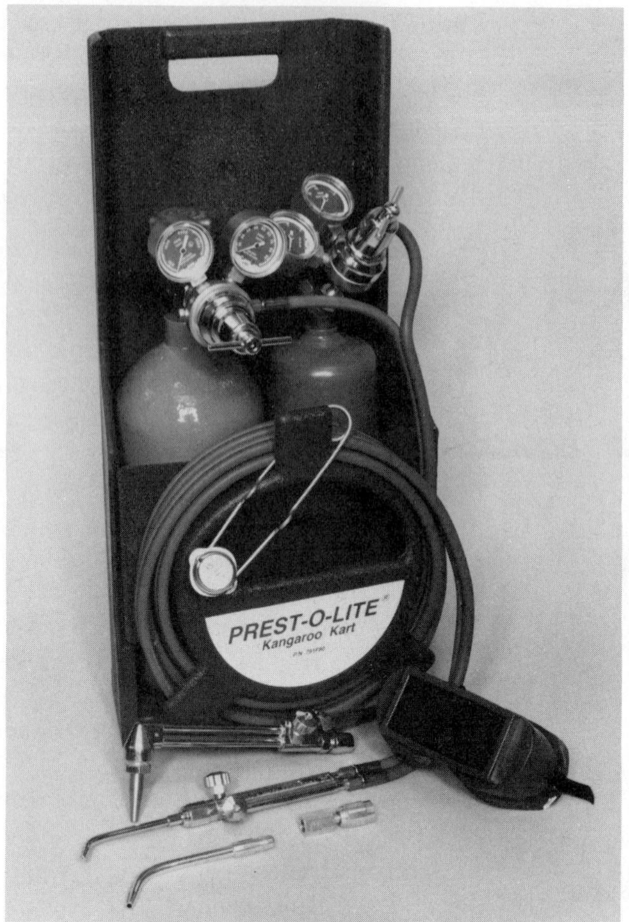

Figure 6–23 Portable welding unit (Courtesy of The Esab Group, Inc.)

11. Why should the heater on a charging cylinder not be used when charging the cylinder with refrigerant?
12. What is used to clean fittings before soldering them?

6–2 WELDING UNIT

Welding units are a necessity for installing and servicing refrigeration systems. A portable unit is the most popular because it can be easily carried around the job site. See Figure 6–23.

These units have the heating capabilities of the larger units and, because of their being smaller in size and weight, are very portable. They can produce a flame temperature of around 6,000°F.

The R oxygen tank and the MC acetylene tank are the most popular tanks for use in these units. There is also a larger B acetylene tank available. However, it is not as popular as the smaller one because of its size and weight. See Figure 6–24.

A disposable propane tank (DP) is also available but it provides a somewhat lower temperature than the oxyacetylene unit. Also, it will not last as long as the oxyacetylene. Welding unit tank specifications are shown in Table 6–1.

Figure 6–24 Acetylene and oxygen tanks (Courtesy of Praxair, Inc.)

STYLE	CAPACITY	HEIGHT	DIAMETER	WEIGHT, FULL
Disposable Propane (DP)	1 pt, 10.7 fl oz	10 1/2 in	2 3/4 in	2 lb, 2 oz
Type MC Acetylene	10 cu ft	14 in	4 in	8 lb
Type B Acetylene	40 cu ft	23 in	6 1/4 in	26 lb
Type R Oxygen	20 cu ft	14 in	5 3/16 in	13 1/2 lb

Table 6–1 Tank specifications (Courtesy of The Esab Group, Inc.)

Safety Precautions

Caution: Always study the manufacturer's instruction sheets and safety data sheets provided with the equipment before operating it. The safe and effective use of a welding unit is dependent on the user having a complete understanding of the unit and carefully following standard safety and operating instructions to prevent personal injuries and possible property damage.

Safety requires that the user be completely aware at all times that the temperature of the flame is almost 6,000°F, and that the piece being welded can reach almost 3,000°F. These temperatures can produce flying sparks, slag, fumes, and intense light rays. All can be harmful to persons working close by if proper precautions are not followed.

Protective clothing must also be worn by those welding and those working nearby. The clothing should be nonflammable, and all flammable materials must be removed to prevent fires.

Adequate ventilation must also be provided in confined areas. Ventilation is required to remove harmful fumes and provide proper air supply to the workers.

Important: The burning rate of almost any lighted material, especially oil and grease, is rapidly increased in the presence of oxygen. Because of this, oxygen should never be allowed to concentrate in confined work areas. An explosive mixture is formed when either oxygen-fuel or air-fuel mixtures in concentration are allowed in closed areas. A torch of any type should never be used on a container or pipe until a thorough purging, cleaning, and ventilation procedure has been completed.

Important: To prevent a dangerous pressure imbalance and possible cylinder contamination from accidental reverse flow of the welding gases, never allow cylinders, especially oxygen cylinders, to become empty during use. Always check for an adequate supply of welding gases before starting the job. It is recommended that reverse-flow check valves be installed in the hoses according to the manufacturer's recommendations.

Checking Out the Equipment. It should be made a standard safety practice to check out and leak-test all connections starting at the cylinder valves, and including the welding tips to make certain that it is a safe, leaktight system before using the equipment. A safety check is especially necessary when the equipment has been exposed to dirt, dust, oil, grease, and any possible damage. The following guidelines are recommended before the equipment is placed in use. They are designed to aid in checking out the equipment before use in normal working conditions and should be completed before each use.

Regulators: When working with regulators use the following procedures:

1. Use the regulators only on the type of gas for which they were designed.
2. Clean all dirt, dust, and debris from the regulator connections. If any of the fittings have been exposed to oil or grease, have them properly cleaned at a regulator service facility because oxygen and either of these products form an explosive mixture. Never use cylinders with dirty valves. Return cylinders with dirty values to the supplier.
3. Do not open welding cylinder valves near a source of ignition. The area must be well ventilated. Always stand clear of the cylinder valve outlet when opening to prevent possible personal injury if the regulator should rupture, and to prevent any sudden release of pressure from striking the person.
4. Crack the cylinder valves only slightly and then reclose to blow out any debris that may have accidentally entered the valve. Crack each cylinder valve separately in this manner.
5. Install the regulator on the cylinder. Tighten the nut with the proper wrench. Do not overtighten to prevent damage to the valve threads or to the nut.
6. When shutting down the welding unit, turn the regulator adjustment screw in a counterclockwise direction until the spring pressure is released. The regulator should always be shut off when not in use, thus preventing gas losses should there be a leaking connection.
7. When opening the cylinder valve, always stand to one side, never in front or to the rear of the regulator. Crack the cylinder valve and allow the pressure to build up gradually until the gauge indicates cylinder pressure, then open the oxygen valve completely. Open the acetylene valve only 1/2 to 1 turn. This will aid in closing down the fuel cylinder in case of fire.

Cylinders:

1. Secure all cylinders to prevent their falling over and possibly becoming damaged.
2. Never lay fuel gas cylinders on their side. They should only be used in the upright position.
3. Store cylinders only in well-ventilated areas—never near an open flame or other source of ignition.
4. Always completely close the cylinder valves when the cylinder is not in use or when the cylinder is empty.
5. Never allow cylinders to become completely empty when in use. This will prevent cross-contamination of the cylinders due to reverse-flow of the gases. Use reverse-flow check valves in the welding hoses.
6. Replace the cylinder valve protectors when the cylinders are not in use, or when the regulators are removed for any reason.
7. Never heat a cylinder with a torch flame or an electric arc. Most external sources of heat will weaken the cylinder wall and cause a hazard.
8. Never use a welding cylinder as a roller or support for moving heavy objects. The cylinder walls may be damaged and weakened, causing a hazardous condition.
9. Always use a pressure regulator on a cylinder.
10. Never allow oil or grease to come into contact with the regulator, cylinder, or any of the hose connections. Oxygen and oil or grease can form an explosive mixture. Never handle welding cylinders with oily gloves or oily hands.
11. Have all leaking valves repaired. When a leaking valve is found, take it to a well-ventilated area or outside and notify the supplier.
12. Always protect the cylinder valves from damage.

Pressure Settings. The following are the recommended steps for setting regulator pressures for normal welding procedures:

1. Open the fuel gas hand valve. Set the fuel gas (acetylene) regulator to provide not less than 5 psig. Close the fuel gas hand valve.
2. Open the oxygen hand valve. Set the oxygen regulator to provide 30 to 35 psig. Close the oxygen hand valve.

WELDING AND BRAZING PROCEDURES

The following procedures are recommended for welding and brazing.

Welding Steel. Oxyacetylene equipment must be used when steel is to be welded. Clean the parts and assemble them together. Light the torch and place the tip of the flame just above the surface to be welded until a puddle forms and penetrates into the base material. At this point, dip the welding rod into the puddle while moving the flame along the joint at a steady pace. The application of the rod is to fill the joint. When the base material is 1⁄8 inch thick or less, the rod should be melted in front of the flame. The flame may be tilted either directly toward the metal or away from it. This will automatically adjust the amount of heat applied to the base metal. It may be necessary to move the flame away from the metal to avoid overheating. When welding metal thicker than 1⁄8 inch, the joint edges should be beveled to about 30° for good penetration.

Brazing. The brazing process makes use of a metal that will flow at a temperature below the melting point of the base metal. This process uses temperatures in the range of 1,150°F to 1,600°F. The proper alloy rod and flux must be used during this process. The joints must also be cleaned and properly assembled with a good fit. Proper ventilation must be provided during the brazing process because of the harmful fumes given off. It is recommended that the tip of the brazing rod be heated, then dipped into the flux while it is still hot enough to melt the flux so that it will adhere to the rod. The metal is then preheated to a dull red color. The fluxed rod is then applied to the heated base metal. When a part of it is melted off onto the joint, the brazing temperature is right. Continue to heat and melt the rod into the joint while moving the flame along the joint.

Silver Brazing. The silver brazing alloys melt at about 1,200°F. The joints must be cleaned before this procedure can be successfully completed. The joints must have a relatively close fit, approximately 0.002 to 0.003 inches for maximum strength. Coat the joint with flux and heat it evenly. When the flux bubbles and turns clear, the temperature is right to proceed. Apply the rod to the joint. It will flow into and around the joint when everything is just right. Caution must be exercised to prevent overheating and burning the flux. If the flux is burned, the alloy will not adhere to it properly. Remove the flame from the joint when the alloy has filled the space.

Soft Soldering. Generally, soft solders melt at around 400°F to 500°F. Soft soldered joints must be especially clean and have a proper fit. They must also be adequately covered with flux to clean the joint further. Apply heat until the flux boils, then apply the alloy to the joint. If the heat is right, the solder will flow into the joint. Do not overheat the joint and burn the flux. If the joint is overheated, it must be taken apart and cleaned again before the alloy will adhere to it. When the alloy has completely filled the joint, remove the flame and allow the joint to cool before moving it.

SUMMARY 6–2

- Welding units are a necessity for installing and servicing refrigeration and air conditioning units.
- There are several sizes of cylinders available for use on the portable type units.
- Safety requires that the user be completely aware at all times that the temperature of the flame is almost 6,000°F, and that the piece being welded can reach almost 3,000°F.
- Protective clothing must also be worn by those welding and working nearby.
- The burning rate of almost any lighted material, especially oil and grease, is rapidly increased in the presence of oxygen.
- It should be made a standard safety practice to check out and leak test all connections starting at the cylinder valves and including the welding tips to make certain that it is a safe, leaktight system before using the equipment.
- Secure all cylinders to prevent their falling over and sustaining possible damage.
- Set the fuel gas regulator to provide not less than 5 psig.
- Set the oxygen regulator to provide 30 to 35 psig.
- When welding, the application of the rod is to fill the joint.
- The brazing process makes use of a metal that will flow at a temperature below the melting point of the base metal.
- The silver brazing alloys melt at about 1,200°F. The joints must be cleaned before this procedure can be successfully completed.
- Generally, soft solders melt at around 400°F to 500°F. Soft soldered joints must be especially clean and have a proper fit.

REVIEW QUESTIONS 6–2

1. On a welding unit, in what position should the fuel gas cylinder be used?
2. At what minimum pressure should the fuel gas regulator on a welding unit be set?
3. Welding unit oxygen regulators should be set at what pressure?
4. At what temperature do soft solders usually melt?
5. What two indicators are used to determine when a fitting is hot enough to silver braze?

6–3 TEST INSTRUMENTS

Test instruments are a valuable asset to the technician who uses them properly. The technician should be just as familiar with test instruments as with any other tool in the tool chest. Those who do not learn to properly use test instruments can cost their employers and the customers many dollars each year because of wasted parts and effort.

Selection

Test instruments are expensive only if they are not properly purchased, used, and maintained. The purchase of test instruments must be approached properly so that unnecessary purchases are not made. It is generally recommended that a check list be made listing the necessary features, those features that are not essential, and those that will most likely never be used.

The matter of cost is one of great importance. There are those instruments that have many functions in a single case that will be less expensive than instruments in individual cases. The instrument must not be any larger than is absolutely necessary. The use of and the complexity in using the instrument must be kept in mind. An easy-to-use instrument will be used more than one that is difficult.

Some of the advantages of each instrument having its own case over the multiple instrument cases are: they are much easier to use, the cost is less per instrument unit, and they are much more flexible because if one needs repair the other instruments are still available for use.

Repair

Test instruments should be returned to the manufacturer for repairs rather than attempting to repair them in the field because the internal circuitry is much too complicated. Also, the necessary components are not generally available for field repair operations.

Almost all instruments have a means of adjusting their sensitivity or have some type of calibration designed for fine adjustments. Instrument calibration should be periodically checked to make certain that the instrument is indicating accurately. Always follow the manufacturer's instructions when making adjustments or calibrations of any type.

Care

The proper care of test instruments is very important. If the instrument does not indicate the proper reading, the test will not be good and a wrong diagnosis is possible. Meter movements are usually position sensitive. That is, they must be used in the position indicated by the manufacturer. Meter indicators have a position that is called the null point. The needle should rest at the null point when it is not in use.

The correct test leads must be used with the meter, otherwise incorrect readings may result. The leads must be properly cared for to prevent damage and unnecessary deterioration. When the insulation on a lead becomes worn or cracked the lead should be replaced. Be sure to choose the correct replacement lead. Make certain that the lead makes a good solid contact with the meter connections. Improperly fitting leads can cause inaccurate readings.

Almost all test instruments are provided with a fuse or a circuit breaker to protect the instrument. It is a good idea to have some extra fuses on hand, just in case.

Test instruments should not be stored in the tool box because they would be constantly receiving a beating. Also, they should not be stored in extreme temperatures. Periodically clean any oil and dust that may accumulate on the instrument. Never subject them to abnormally high moisture conditions without proper protection and maintenance. Always store test instruments in their protective cases.

Use good quality batteries in test instruments. Do not store the instrument with batteries in them because they may leak and damage the inside of the instrument. Be sure to remove any corrosion caused by a leaking battery because it will damage other parts of the instrument.

Do not leave the selector switch in a position that will demand battery output during storage. When the instrument does not have a selector switch, remove the leads from the instrument.

Application

Each instrument has a specific application. If it is used for some other application it will probably be ruined. At best it will produce faulty readings. Always use the instrument for which it is intended.

Multimeter. Multimeters are test instruments that are designed to provide several functions. The numbers and types of functions are dependent on the instrument design. See Figure 6–25.

Most of these instruments are position sensitive and should be placed in the proper position when in use. Otherwise an incorrect reading may result. These instruments have many uses because of their flexibility. They have several scales that can be used to make the correct measurements.

When using an instrument of this design, the selector switch must be placed on the proper scale and position. When the correct voltage or amperage is not known, place the selector switch on a higher scale than may be needed. Then move the selector switch to the correct scale when the voltage or amperage is known. The most accurate reading is at the midscale position.

Voltmeter. Voltmeters are used for measuring voltages to a refrigeration or air conditioning system. They are probably the most frequently used instruments. Voltmeters are single function meters, which makes them easy to use and read. See Figure 6–26.

When doing service and installation work, the technician will encounter voltages that range from the millivolt range to more than 5,000 volts AC. To prevent personal injury, it is necessary that caution be used when working with these higher voltages. The selector switch must be placed in the proper position to prevent damage to the instrument and to obtain the correct reading. When the voltage or amperage is not known, it is usually best to start measuring on a scale higher than the anticipated reading and then move to the proper scale. The midscale reading is usually the most accurate.

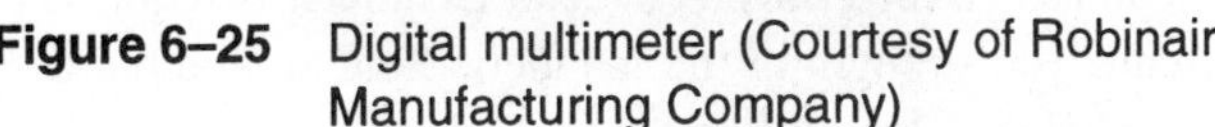

Figure 6–25 Digital multimeter (Courtesy of Robinair Manufacturing Company)

Figure 6–26 Digital multimeter (Courtesy of Robinair Manufacturing Company)

Ohmmeter. Ohmmeters are used for measuring resistances and checking the continuity of electrical components, including wiring, coils, motor windings, etc. See Figure 6–27.

When checking the continuity of an electrical circuit, make certain that the electrical power is turned off. If an ohmmeter is subjected to line voltage electrical power, the meter will most likely be ruined beyond repair. To prevent damage, the battery must be used to power these instruments. When measuring an unknown

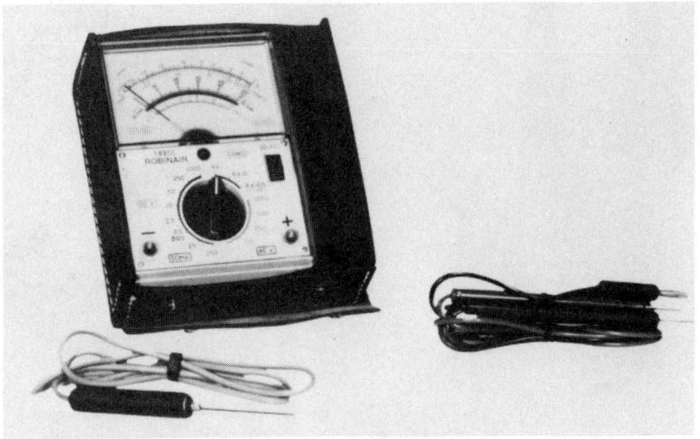

Figure 6–27 Ohmmeter (Courtesy of Robinair Manufacturing Company)

resistance, start with the lowest scale. Then change scales until the reading is at about the midscale position. Be sure to zero the meter when the correct scale is determined.

When the resistance is known, select the correct scale, then install the meter leads into the proper jacks for the reading. Hold the probes together and adjust the meter to indicate a zero resistance. If the scale is changed, the meter must be zeroed again. When the meter cannot be zeroed, replace the batteries and then try again to zero the meter.

When checking a switch or a wire that has very little, if any, resistance, the meter should read zero resistance. Zero resistance indicates a closed circuit. When a reading is obtained, the part has continuity. If the meter needle does not move from the at-rest position, the circuit is open. What this reading indicates is dependent on the nature of the circuit being tested.

When an open circuit is being tested, the meter will indicate an infinity reading. The part being tested does not have a complete circuit through it.

Ammeter. The ammeter is used to check the current flow through a wire at a given point. The current draw will indicate if the unit is operating within the amperage limits as set by the manufacturer. The clamp-on type ammeter is the most popular for measuring amperage in AC circuits, because the amperage can be checked without disconnecting any wiring. See Figure 6–28.

The amperage is measured by simply clamping the ammeter tongs around the wire and observing the reading. See Figure 6–29.

A more accurate reading is obtained when the wire is in the center of the tongs and the tong ends are clean and properly matched. These types of meters are designed to read the amperage in AC circuits only. A special DC meter is required to read current flow in a DC circuit. The strength of the magnetic field around the wire serves as the indicator of current flow. This will be studied in more detail in a later chapter. The ammeter tongs actually act as the primary side of a transformer. The circuit feeding the meter movement acts as the secondary side of a transformer.

In circuits where the amperage flow is small, wrap the wire several times around one tong of the ammeter. This will increase the amperage reading to a more readable

Figure 6–28 Digital clamp-on meter (Courtesy of Robinair Manufacturing Company)

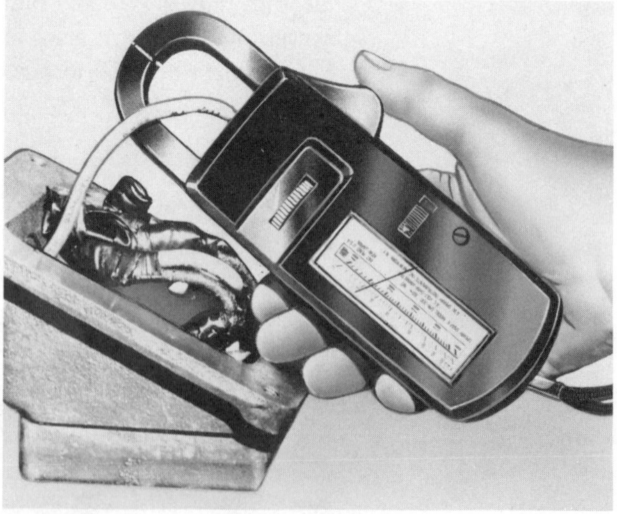

Figure 6–29 Use of clamp-on ammeter (Courtesy of A.W. Sperry Co.)

number. Then divide the meter reading by the number of turns taken with the wire. Some manufacturers have a device available called a multiplier for this purpose. Most people make ten turns of the wire on the meter tong to make the mathematics easier.

Wattmeter. Wattmeters are used to determine the amount of current being used by the unit. Some of these instruments are equipped with a voltmeter so that the voltage to the unit and the wattage reading can be more accurately determined at the same time. See Figure 6–30.

If the approximate wattage is not known when connecting the instrument to the unit, set the meter on the highest scale possible. Then select the proper scale after the unit is running. Select the scale that allows the needle to be at midscale for the most accurate reading.

Capacitor Analyzer. Capacitor analyzers are used to determine the condition of a starting or running capacitor. The use of this instrument will allow the technician to determine the condition of the capacitor. This is preferable to simply replacing the capacitor to see if that is the problem. See Figure 6–31.

Most of these instruments will indicate whether a capacitor is open, shorted, the microfarad capacity, and the power factor. These meters have two scales. One is for the voltage supplied to the capacitor, and the other indicates the microfarad reading of the capacitor.

Electronic Temperature Tester. Electronic temperature testers are used to determine the temperature of a substance or a process that is being tested. These are very

Figure 6–30 Wattmeter (Courtesy of Robinair Manufacturing Company)

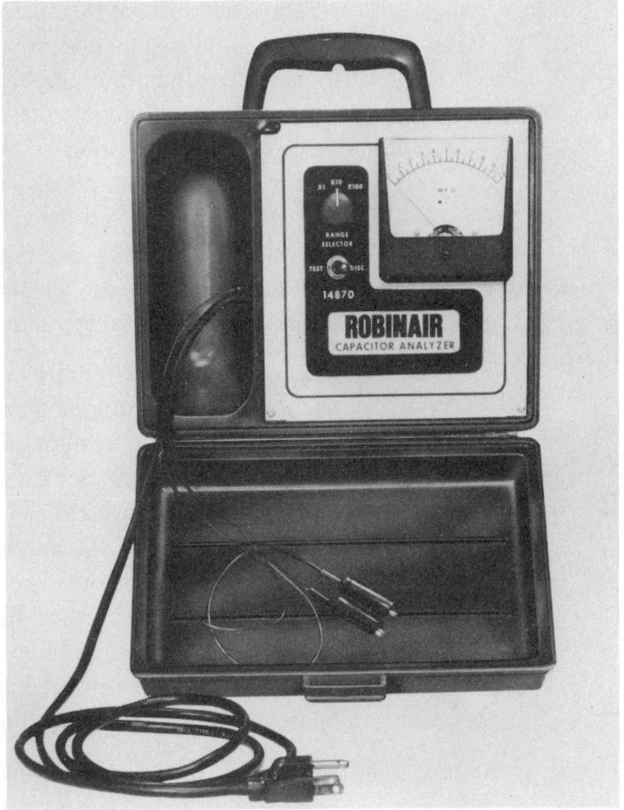

Figure 6–31 Capacitor analyzer (Courtesy of Robinair Manufacturing Company)

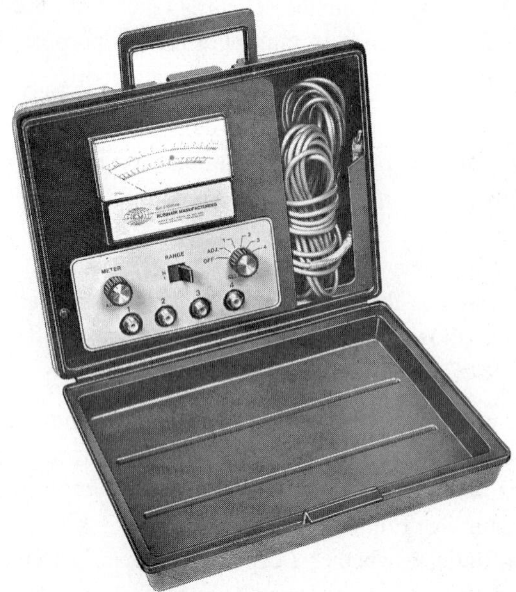

Figure 6–32 Electronic temperature tester (Courtesy of Robinair Manufacturing Company)

Figure 6–33 Recording temperature-time meter (Courtesy of Airserco Manufacturing Company)

fast operating instruments and are generally very accurate. Most temperatures measured with this type of thermometer are under 200°F. See Figure 6–32.

The leads include thermistors and they must be properly cared for if their accuracy is to be maintained. Never attempt to alter these leads. If they malfunction, replace them with the proper type. These instruments are available in both single lead models and models that will measure several different locations at the same time without interruption.

These instruments are also available with a recording function. When recording, they will indicate both the time and the temperature on a piece of paper for later viewing and analyzing. See Figure 6–33.

Recording electronic temperature testers are extremely valuable when attempting to find intermittent temperature problems.

Micron Meter. Micron meters are thermistor meters that accurately measure a vacuum. The thermistor measures the thermal conductivity of the gases being removed from the system and, therefore, can indicate the effectiveness of the evacuation procedure. See Figure 6–34.

These instruments are electrically powered and measure the last 25,400 of an inch. An inch has 25,400 microns.

The sensing element is not designed to measure pressures above atmospheric. Because of this, some means must be used to protect the sensor when greater pressures are expected. Also, the element is most sensitive when it is installed not more than 15° off top vertical center. The element must also be protected from lubricating oil because the oil will destroy the sensitivity of the element.

Millivolt Meter. Millivolt meters are used to measure the low DC voltage output of a thermocouple, such as those used in pilot safety control circuits. One millivolt is equal to 1/1,000 of a volt DC. See Figure 6–35.

Sometimes it is necessary to measure the closed circuit voltage of a thermocouple. When this is required, there is a special adaptor that can be used.

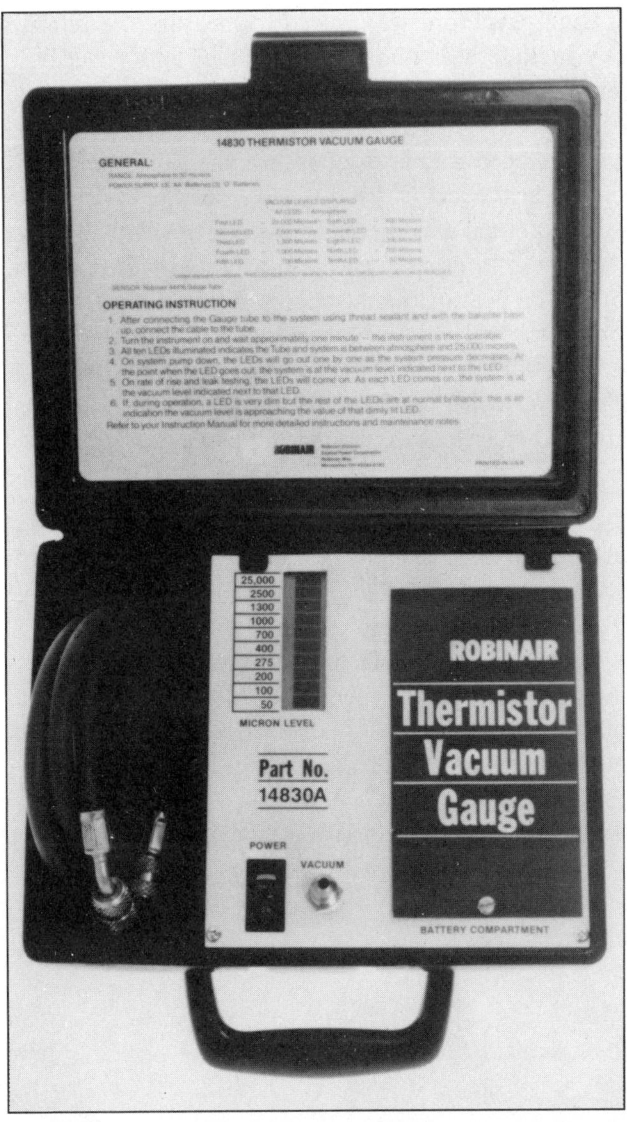

Figure 6–34 Micron meter (Courtesy of Robinair Manufacturing Company)

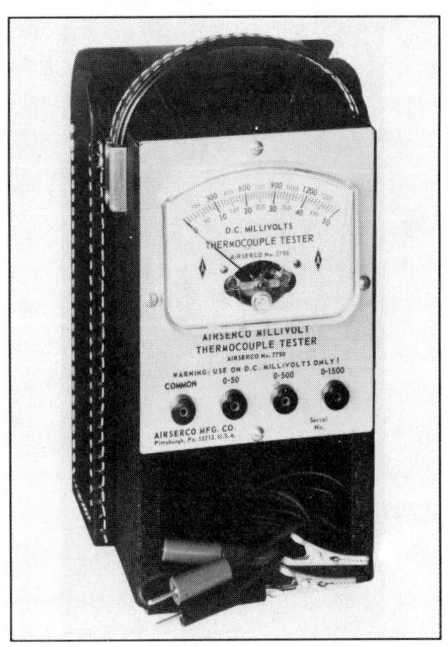

Figure 6–35 Millivolt meter (Courtesy of Airserco Manufacturing Company)

Hermetic Analyzer. Hermetic analyzers are used to determine the condition of a hermetic compressor. Some of the uses of these instruments are (1) rocking stuck or frozen compressors, (2) to reverse the rotation of a compressor motor, (3) as an auxiliary capacitor, (4) as a bank of starting capacitors from which the proper capacitance can be chosen, and (5) to check the continuity of circuits such as motor windings. Some analyzers are equipped with an ammeter, voltmeter, ohmmeter, and a relay analyzer.

Potential Relay Analyzer. These instruments are used to check the condition of potential starting relays. A potential relay analyzer can, when properly used, save much time and money by indicating if the relay actually needs to be replaced. It can also be used to check the relay to see if it is open, closed, or short circuited. It will also indicate the pickup and the drop-out voltage of the relay coil. The analyzer can be used to manually operate the relay. It can also be used as a continuity tester when an ohmmeter is not readily available.

Sling Psychrometer. The sling psychrometer is used to measure the relative humidity in a building. They include a dry-bulb thermometer and a wet-bulb thermometer, both mounted on a common base so that they can be twirled around to speed up the measurement process. See Figure 6–36.

The wet-bulb thermometer is equipped with a sock covering the bulb. The sock is wetted with distilled water and the instrument is then twirled around in the air to be measured. Do not touch the sock with the hand because oils on the skin will affect the capillary action of the sock and ruin its use. The instrument is twirled around until the same reading is indicated on both thermometers for at least two successive readings. The wet-bulb thermometer will almost always indicate a lower reading than the dry-bulb thermometer. The difference between the two is called the wet-bulb depression. The wet-bulb depression is what indicates the relative humidity. Plot the two temperatures on a psychrometric chart to determine the air properties.

SUMMARY 6–3

- Test instruments are a valuable asset to the technician who uses them properly.
- Test instruments are expensive only if they are not properly purchased, used, and maintained.
- Test instruments should be returned to the manufacturer for repairs.
- Instrument calibration should be periodically checked to make certain that the instrument is indicating properly.
- The correct test leads must be used with the instrument to ensure accurate readings.
- Do not leave the selector switch in a position that would demand battery output during storage.
- Each instrument has a specific application.
- Most instruments are position sensitive.

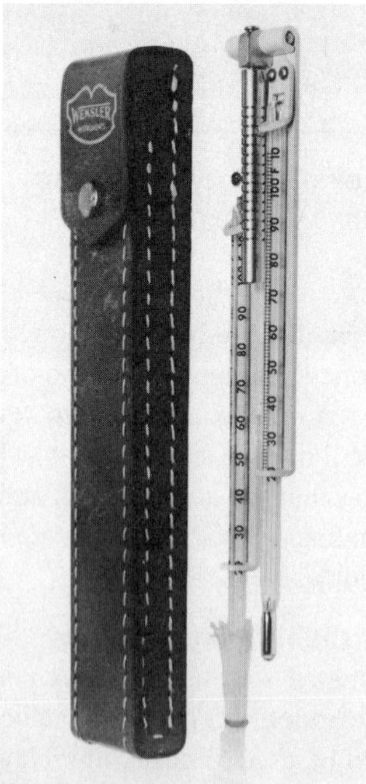

Figure 6–36 Sling psychrometer (Courtesy of Robinair Manufacturing Company)

REVIEW QUESTIONS 6–3

1. Under what conditions are test instruments expensive?
2. What should be done to a test instrument before it is stored?
3. What must be done to a position-sensitive instrument?
4. When testing for an unknown voltage, what should be done to the meter?
5. Can a single meter scale be used to measure both AC and DC electricity?
6. What is preferable to simply replacing a capacitor?
7. When are recording temperature testers desirable?

6–4 LEAK DETECTORS

Electronic leak detectors are used to locate leaks or to warn the user that a leak is present. Some are equipped with a pick-up tube that is placed near the suspected area to draw refrigerant through the detector to indicate a leak. Other types detect the noise made by a gas escaping through a hole. There are also types that use a fluorescent dye placed inside the refrigeration system to indicate a leak. The type of leak detector used will depend on personal preference and the type of refrigerant in the system.

Electronic Leak Detectors. Electronic leak detectors are used to locate leaks in refrigeration systems. They are very sensitive instruments that operate by moving the refrigerant vapor across an ionizing element in the sensing probe. See Figure 6–37.

They alert the user to a leak by a built-in warning device. Their warning device may be one, or a combination of, lights, buzzers, or high-pitched screeches. These instruments are not effective when ammonia, sulphur dioxide, or some of the newer refrigerants are used.

To use these instruments, refer to the following list:

1. The refrigeration system must be charged with sufficient refrigerant to produce at least 50 psig static pressure (not in operation).
2. Visually check the system for any obvious leaks. Wipe any suspected surfaces with a shop towel to remove any dirt or oil accumulation. Do not use cleaners or solvents because they may falsely indicate a leak.

Figure 6–37 Electronic leak detector (Courtesy of Robinair Manufacturing Company)

3. Start at one point and completely leak test the entire system to prevent overlooking an area. Do not stop when a leak has been located. Test the complete system to make certain that there are no more leaks.
4. Move the sensing probe at the speed of about 1 to 2 inches per second. Hold the sensing probe not more than 1/4 inch from the surface.
5. Blow compressed air into any area that a leak is indicated. This will remove any refrigerant that may have accumulated. Then leak test again to make certain that a leak is present. Do not blow air into a concentration of HFC–134a because it is possible to cause an explosive mixture.

Ultraviolet Leak Detectors. These leak detectors are used to locate a leak with almost any type of refrigerant. Dye is placed into the system prior to leak detecting. A black light is used to locate any dye that leaks out of the system. The dye that is installed into the system will stay there and be effective until the oil is drained. The system is operated for a sufficient period of time, usually about 15 minutes, to allow the dye to seep out of the system with the refrigerant and oil. The black light is then directed toward any suspected leak. The dye will turn yellow under the blacklight beam.

It is desirable to leak test the system in as little light as possible—the darker the better. Move the light beam along and across each component in the system. It is not necessary to stop the unit when using this leak detector.

Ultrasonic Leak Detectors. Ultrasonic leak detectors may be used when testing any type of refrigerant for leaks. These detectors locate leaks by sound rather than by chemical action or the use of dyes. The theory is that if there is a movement of gas or air through a hole the movement will make some type of noise. The ultrasonic leak detector will sense this sound and alert the technician. The system can be tested for leaks as long as there is a pressure difference between the inside and the atmosphere. There is nothing placed inside the system. The sensing probe is moved along the surface of the piping and the system components. When a leak is detected, the detector will alert the technician through a set of headphones that are used to aid in hearing the leak.

SUMMARY 6–4

- Electronic leak detectors sense a leak by moving the refrigerant vapor across an ionizing element in the sensing probe.
- The system should be charged to at least 50 psig when leak testing with an electronic leak detector.
- To leak test any system, first complete a visual check of the system for any obvious leaks.
- The entire system should be tested before repairing any leaks.
- An ultraviolet dye and a black light is another method used to sense leaks.
- Ultrasonic leak detectors detect leaks by sensing sound.

REVIEW QUESTIONS 6–4

1. How do electronic leak detectors detect a leak?
2. What should be the first step in determining the location of a leak?
3. Why should air not be blown into concentrations of HFC–134a?
4. How do ultrasonic leak detectors find leaks?
5. What type of leak detector uses a dye in the system?

CHAPTER

7

REFRIGERATION MATERIALS

OBJECTIVES

Upon completion of this chapter, you should be able to:

- Know the types and uses of piping and tubing needed in the refrigeration and air conditioning trade
- Know how to properly cut copper tubing
- Be more familiar with the types of fittings used in the refrigeration and air conditioning trade
- Know how to properly make tubing connections for use in refrigeration systems
- Be more familiar with the procedures used in refrigeration system piping design
- Be more familiar with the types and uses of the different solders used on refrigeration lines

INTRODUCTION

A refrigeration technician is required to know many things in order to become proficient in this field. A knowledge of the theories that are a very important part of this industry is required, along with the knowledge of how to accomplish certain mechanical skills. This section will present the basic knowledge of the materials used in refrigeration and air conditioning work.

7–1 PIPING AND TUBING

A series of piping and tubing is used to connect the various components of a refrigeration system so that the refrigerant can move from one component, phase, and function to another component, phase, or function.

The major difference between piping and tubing is that piping has a thicker wall than tubing. The difference in the wall thickness is the reason for some of the different working procedures. Piping can be threaded for joining pieces together but tubing cannot be threaded because of its relatively thin wall. Another difference is that piping is normally sized according to the inside (ID) diameter and tubing is sized according to the outside diameter (OD).

For the most part, the tubing used in refrigeration systems is made of copper. However, some domestic refrigerator and freezer manufacturers are using steel tubing in the condenser and aluminum to make the evaporator. However, copper is still the most preferred in field built-up systems. Also, copper is generally much easier to work with than either aluminum or steel.

Seamless Copper Tubing. Seamless copper tubing is used for the refrigerant lines that connect the various components of the system. However, it must be remembered that copper cannot be used when the refrigerant is ammonia. Ammonia will corrode copper and eventually cause it to disintegrate, even if the vapor is in contact with it for only short periods of time. Copper has the greatest heat conductivity of any of the common materials. Because of its extremely good heat conductivity factor, copper is used in most larger coils and line installations. Because of the relative softness of copper tubing, solder fittings are used rather than threaded fittings. Soldered joints are very strong and are leak resistant to common refrigerants. Copper is easy to heat and bend, eliminating the need for elbows and their accompanying joints, which are liable to leak.

Refrigeration tubing is available in either rolls or as straight pieces of tubing. The rolls are soft-drawn tubing and the straight pieces are hard-drawn tubing. The soft-drawn tubing may be purchased in rolls of 25-, 50-, and 100-ft lengths, ranging in size from 1/8 inch to 1 3/8 inch in diameter (OD). The larger size soft-drawn tubing is available in straight lengths of 20 ft. The 50-ft roll is the most popular because it is easier to handle and work with. It will usually be long enough so that the entire line can be run without any unnecessary connections or fittings. Care must be taken when working with soft-drawn tubing to prevent kinking it, causing a restriction. The most common methods of joining copper tubing are by silver soldering, soft soldering, flaring, and epoxy material.

The hard-drawn tubing is available in 20-ft lengths, ranging in size from 1/4 inch to 6 1/8 inch OD. Hard-drawn tubing is not easily formed by hand. It will usually kink, but it can be bent if it is annealed. Annealing is done by heating the tubing to a cherry red and allowing it to cool in the ambient air. It can then be bent if care is taken. This type of tubing is very useful when long lengths that cannot be supported are required and when the appearance of the job is very important. Hard-drawn tubing may be joined by silver soldering, soft soldering, or using epoxy procedures.

TUBE OD	WALL THICKNESS	WT PER FOOT	150 (°F) PSI	250 (°F) PSI	350 (°F) PSI	400 (°F) PSI
1/4	0.030	0.0804	1230	1130	970	720
3/8	0.032	0.134	860	700	670	500
1/2	0.032	0.182	630	580	490	370
5/8	0.035	0.251	540	500	430	320
3/4	0.035	0.305	440	400	350	260
7/8	0.045	0.455	500	460	390	300
1 1/8	0.050	0.655	430	400	340	250
1 3/8	0.055	0.884	390	360	300	230
1 5/8	0.060	1.140	370	340	280	220

Table 7–1 Safe working internal pressures of soft-drawn ACR tubing

Safe Working Pressures. The size of the tubing and the temperature of the material it is carrying will mostly determine the safe working pressures of copper tubing. The pressures of soft-drawn tubing are shown in Table 7–1.

As the temperature of the fluid inside the tubing increases, the safe working pressure decreases.

It should be noticed that the safe working pressures for hard-drawn tubing are different from soft-drawn tubing. See Table 7–2.

The safe working pressures of tubing should never be exceeded. To do so may result in property damage and personal injury.

Classification of Seamless Copper Tubing. The S.A.E. and type L copper tubing are the types mostly used in refrigeration and air conditioning systems. This tubing is generally indicated by the designation of "ACR" tubing. Type M tubing is thin walled and is usually used for condensate drains lines and for other uses where the working pressure does not exceed 150 psig.

In accordance with the ASTM 3280 and ANSI-B9.1-1971 refrigeration standards, tubing with the ACR designation must be properly cleaned, dried, and capped to prevent the entrance of moisture and foreign particles. If a complete length of tubing is not used, what is left over must be capped to retain its cleanliness for future use.

TUBE OD	WALL THICKNESS	WT PER FOOT	150 (°F) PSI	250 (°F) PSI	350 (°F) PSI	400 (°F) PSI
3/8	0.030	0.126	900	870	570	380
1/2	0.035	0.198	800	770	500	330
5/8	0.040	0.285	740	720	470	310
3/4	0.042	0.362	650	630	410	270
7/8	0.045	0.455	590	570	370	250
1 1/8	0.050	0.655	510	490	320	210
1 3/8	0.055	0.884	460	440	290	190
1 5/8	0.060	1.14	430	420	270	180
2 1/8	0.070	1.75	370	360	230	150
3 1/8	0.090	3.33	330	320	210	140
3 5/8	0.100	4.29	320	310	200	130
4 1/8	0.110	5.38	300	290	190	120
5 1/8	0.123	7.61	280	270	180	120

Table 7–2 Safe working internal pressures of hard-drawn ACR tubing

Hard-drawn tubing is generally pressurized with dry nitrogen and the ends sealed with reusable plugs.

Seamless Steel Tubing. Steel tubing is sized according to the outside diameter (OD). The walls are of a standard gauge number thickness. Seamless tubing will have a higher working pressure than welded type tubing. It may be purchased in straight lengths of 20 ft. Steel tubing is generally joined by soft soldering or silver soldering. Sometimes a reversed flare fitting is used.

Steel tubing is almost exclusively used in the condensers of domestic refrigerators and freezers. Steel tubing does not have the heat transfer capabilities that copper tubing has. But in these units, the steel tubing is tightly fastened to the refrigerator or freezer shell, which, in effect, increases the heat transfer capabilities of the condenser. The lower cost of the steel, when compared to the cost of copper, will more than offset the expense of fastening the tubing to the appliance shell.

Aluminum Tubing. Aluminum is used in the manufacture of many of the coils used in refrigeration and air conditioning equipment. It is very popular in domestic refrigerator evaporators. The heat transfer of aluminum is not as good as that of copper tubing. However, when the cost of aluminum is compared to the cost of copper, any extra cost in the manufacturing process will be more than offset. Aluminum tubing is soft and may be formed by hand. This eliminates the cost of many fittings. Special alloys are used to join pieces of aluminum tubing. Epoxy is also used for repairing holes and in the joining processes.

Cutting Copper Tubing

The cutting of copper tubing is a relatively routine procedure used by service and installation technicians. The tubing must be cut properly so that proper joints can be made to prevent refrigerant leakage and restrictions.

There are two methods used to cut copper tubing: (1) with a hacksaw and sawing fixture, and (2) with hand-held tubing cutters. There are hand-held cutters available that will cut tubing ranging in size from 1/8 in OD to about 4 1/8 in OD. See Figure 7–1.

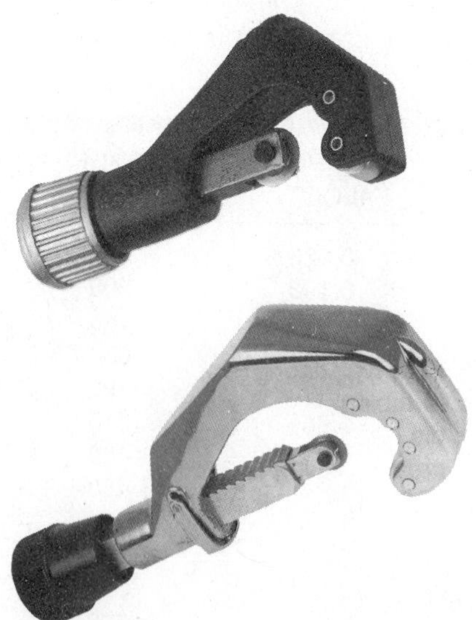

Figure 7–1 Typical tubing cutters (Courtesy of Robinair Manufacturing Company)

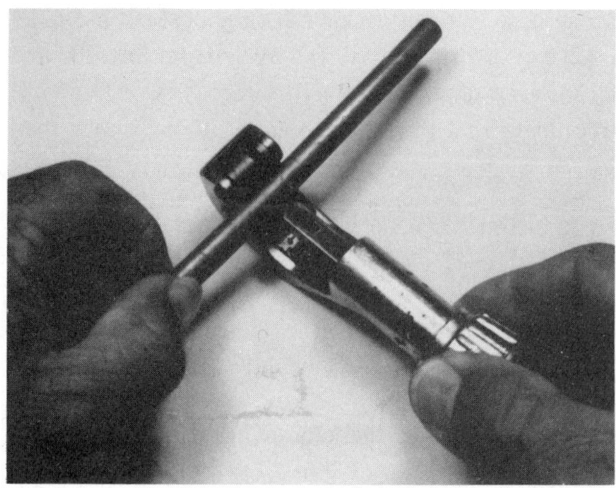

Figure 7–2 Appling tubing cutter to tube (Photo by Billy C. Langley)

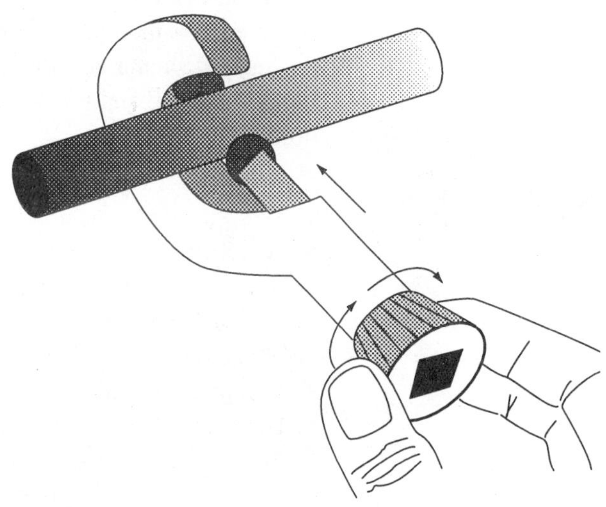

Figure 7–3 Tubing cutter adjustment

In use, the hand-held cutter is positioned on the tubing at the place where the tubing is to be cut. The cutter blade is adjusted so that it touches the tubing. See Figure 7–2.

The cutter is then tightened onto the tubing by turning the adjustment knob about 1/4 turn only. See Figure 7–3.

Then the cutter is rotated around the tubing. The cutter adjustment knob is tightened another 1/4 turn after each revolution of the cutter around the tube. More than 1/4 turn may cause the tube to go out of round as well as leave a heavy burr inside the tube. Continue this procedure until the tube is separated. Even when every caution is taken, there will still be a burr left inside the tube that must be removed. This burr is removed with the reamer that is part of most tubing cutters. See Figure 7–4.

It is generally easier to cut the larger sized tubing with a hacksaw and fixture. The fixture is required to make a square cut. See Figure 7–5.

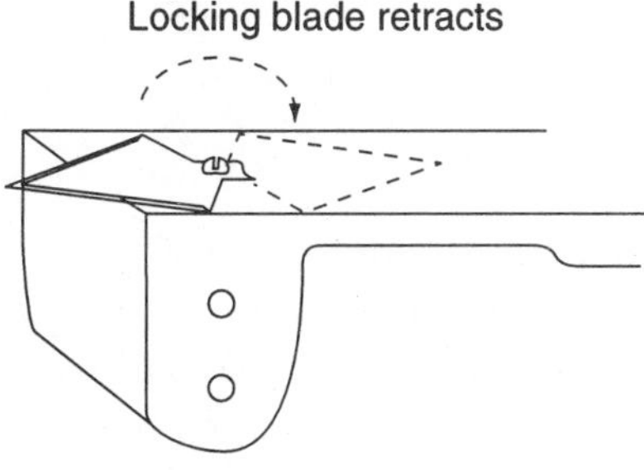

Figure 7–4 Location of tubing cutter honing blade

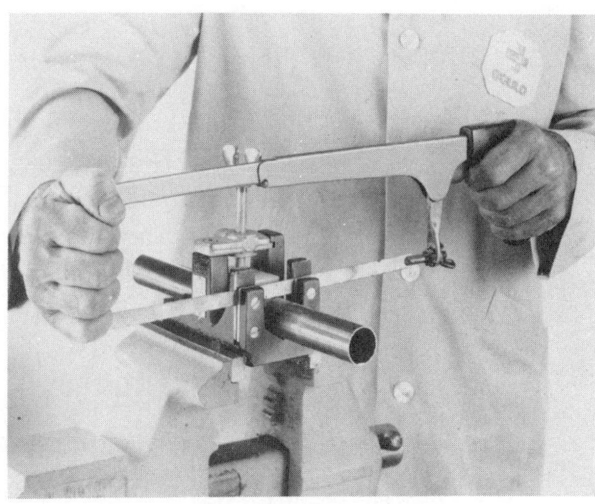

Figure 7–5 Cutting tube with hacksaw and sawing fixture (Courtesy of Imperial Eastman)

The hacksaw blade should have at least 32 teeth per inch for the best cut possible. Be sure to position the tubing so that the saw filings will fall away from the tube and not into it. Should any filings fall into the tube, they must be removed or they will block the strainers and filters, requiring that they be replaced much sooner than normal.

SUMMARY 7–1

- A refrigeration technician is required to know something about a lot of things in order to become proficient in this field of employment.
- The major difference between piping and tubing is that piping has a thicker wall than tubing.
- Piping is normally sized according to the inside diameter (ID) and tubing is sized according to the outside diameter (OD).
- For the most part, the tubing used in refrigeration systems is made of copper.
- Refrigeration tubing is available in either rolls or as straight pieces of tubing.
- Copper has the greatest heat-conducting capabilities of any of the common materials.
- Tubing to be used in refrigeration systems is designated with "ACR."
- Steel tubing is used almost exclusively in domestic refrigerator and freezer condensers.
- Aluminum is very popular in domestic refrigerator evaporators.
- There are two methods used to cut copper tubing: (1) with a hacksaw and a sawing fixture, and (2) with hand-held tubing cutters.
- It is generally easier to cut the larger size tubing with a hacksaw and fixture.

REVIEW QUESTIONS 7–1

1. What is the purpose of the piping and tubing used in a refrigeration system?
2. How is air conditioning tubing sized?
3. Should copper tubing be used in an ammonia system?
4. Where is steel tubing most commonly used?
5. Name the two methods used to properly cut copper tubing.
6. When cutting tubing with a cutter, how much is the adjustment knob tightened after each revolution?

7–2 SWEAT FITTINGS

There are several different types of fittings available for use in the refrigeration industry. However, the sweat or solder type is the most popular for joining copper tubing together. Flare fittings are a close second in use for joining copper tubing together. There are other types available but they are more commonly used in the automotive air conditioning segment of the industry. Several types of sweat fittings are shown in Figure 7–6.

Sweat Fittings

This type of fitting may be used with either hard-drawn or soft-drawn tubing and may be either silver soldered or soft soldered. These types of fittings will allow almost any kind of turn or connection to be made that may be required for the particular installation. The long radius types offer a very low pressure drop to the refrigerant flow. To use these fittings the tubing must be cut to the desired length and the proper fitting then used. The tubing must be cut square so that the fitting can be

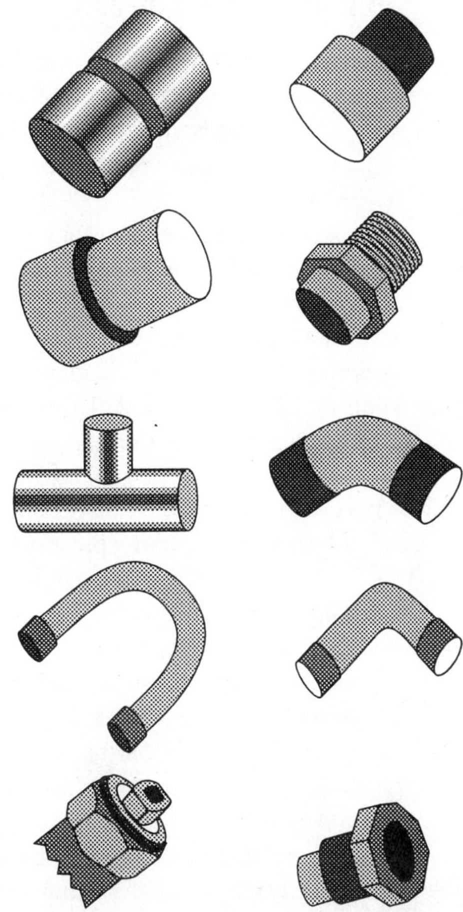

Figure 7–6 Wrot (Sweat) copper fittings

installed completely to the stop inside the fitting. When it fits properly, a stronger, leak-free joint can be made. See Figure 7–7.

When the tubing is not cut squarely, it is possible that the solder will flow through the crack and into the system, causing a restriction to the refrigerant flow at this point. Should the solder not stick to the inside of the tubing, it could break loose and enter other system components, causing possible damage and obstruction to the refrigerant flow.

The service and installation technician must be familiar with all the sweat fittings available. There is one for almost any need and their proper use can make a job much simpler. However, when they are used in the wrong manner many problems can occur.

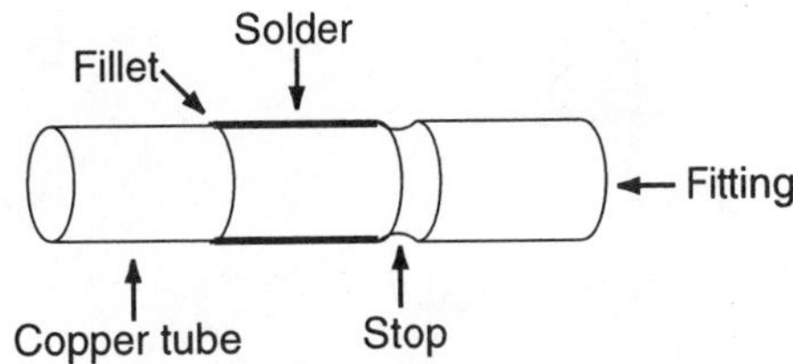

Figure 7–7 Proper joint between tube and fitting

Joining Sweat Fittings

The joining of tubing together is a very common procedure in the service and installation of refrigeration and air conditioning systems. Many people tend to minimize the importance of tubing installation because it is so routinely done. But in order to have a leak-free, properly operating system, the steps outlined next must be followed.

Basically, there are six steps used for both silver-soldering and soft-soldering copper tube fittings. These steps must be followed so that strong leak-free joints are made: (1) properly cut the tubing, (2) make certain that the fit between the tubing and the fitting is correct, (3) properly clean both the tube and the fitting before soldering, (4) proper application of the flux to the joint, (5) proper heating of the joint so that the solder will flow into it properly, and (6) removing the flux and excess solder from the finished joint. The following is a discussion of these steps:

1. Properly cutting the tubing is of vital importance because if the joint does not fit properly, the soldering process cannot make a strong joint. See Figure 7–8.

   When a hacksaw is used to cut the tubing, make certain that the open end of the tube is pointed downward so that the saw filings will fall from the tube rather than into it.

   Be sure to remove any burrs from the inside of the tube with a tubing reamer, or some other suitable tool.
2. Placing the fitting on the tube to check the fit is an important step that is often overlooked. There should be a close fit between the fitting and the tubing wall and between the tubing and the fitting stop. If the tubing is out of round, use the proper sizing tool to bring it back to size and round.
3. Properly clean both the fitting and the tubing to be soldered. All oil, grease, rust, or oxidation must be removed from all the surfaces to be soldered. See Figure 7–9.

   Both the tube and the fitting must look shiny and new. Use sand cloth to polish the tube and fitting. Some types of cleaning solvents may be used to remove paint, rust, or other such residue. Be careful to prevent any of the solvent from entering the tube. The solvent must be completely removed from the surfaces to be soldered so that the soldering process can be properly completed.

   The inside of the fitting can be cleaned with a wire brush, fitting brush, or a piece of rolled-up sand cloth. See Figure 7–10.

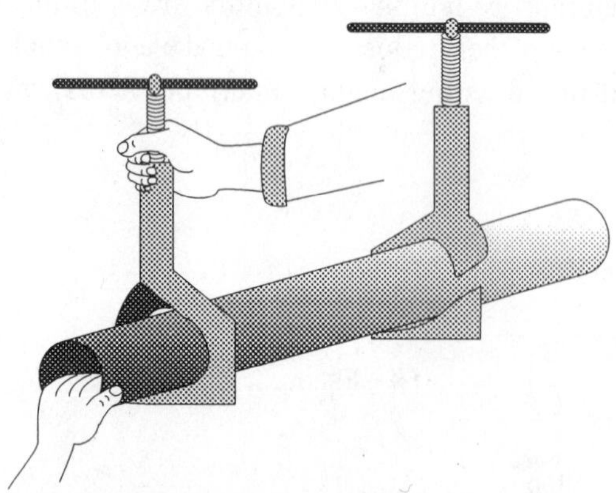

Figure 7–8 Cutting the tube

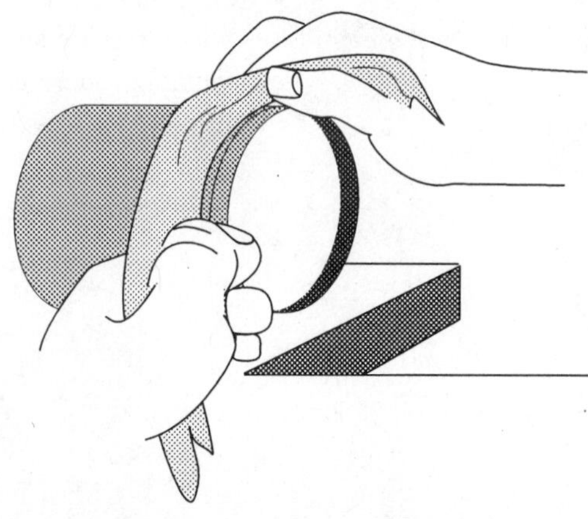

Figure 7–9 Cleaning the tube

Figure 7–10 Cleaning the fitting

Figure 7–11 Stirring flux before using

Never use steel wool or emery cloth to clean these joints. Steel wool will break apart and possibly enter the tube and cause problems in the future. Emery cloth may have oil or other contaminants on it, as well as other abrasives. The abrasives are more likely to break off and enter the tube than when sand cloth is used. Emery cloth abrasives are small and can enter the compressor lubricating oil and cause severe damage to the compressor bearings.

Always clean the joints just before they are to be soldered. Cleaning at this time prevents oxidation from forming on the newly cleaned surfaces before being soldered. Never touch a cleaned surface that is to be soldered with the bare hand or with dirty shop towels. The oils in the skin are enough to upset the soldering process.

4. Flux the joint properly to aid in the cleaning process. The application of flux prevents air from coming in contact with the cleaned surfaces and the heated metal. The air will cause the surfaces to oxidize if not protected. Any oxidation will prevent the solder from sticking to the surface, resulting in a weak, and possibly a leaking, joint. Solder flux is available in both paste and liquid forms. The one to use is determined by the user. Both types produce excellent results. Use the following steps when applying flux to a joint:
   - A. The proper flux must be used for the material and the type of solder that is being used. Silver-brazing flux will not work on soft-solder projects. Neither will soft-solder flux work on silver-brazing projects. If the wrong flux is used the job will surely fail.
   - B. It is always a good idea to stir the flux before using it. See Figure 7–11. If the flux is allowed to stand, especially during hot weather, the chemicals have a tendency to settle to the bottom of the container.
   - C. Insert the tube end part of the way into the fitting.
   - D. Using a brush, apply the flux to the complete joint. If fingers are used to apply the flux, oil from the skin may prevent the solder from sticking to the joint. Use only enough flux to coat the tube because any excess could possibly enter the tubing and cause damage to the inside of the system, especially to the motor windings of hermetic and semihermetic compressors.
   - E. Insert the tubing all the way into the fitting. If at all possible, rotate the tube or the fitting to distribute the flux evenly over the complete joint. See Figure 7–12.
5. Properly heat the joint so that the solder will melt and flow all the way around the fitting. When making a soft-solder joint, apply the torch flame to the fitting and direct the flame so that the tube will also be heated. See Figure 7–13.

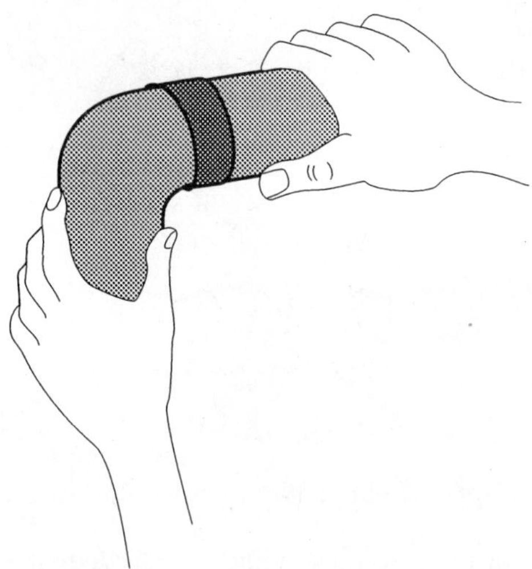

Figure 7–12 Tube inserted in fitting

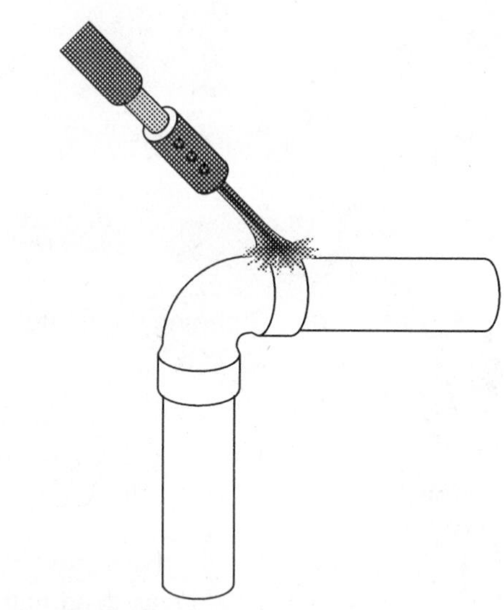

Figure 7–13 Properly applying heat to a soft solder joint

To prevent overheating the flux, be sure to keep the flame from striking the fluxed area. When this happens the joint must be separated, recleaned, and refluxed before the solder will stick to the joint. The heaviest part of the fitting should be heated first, then move the flame toward the joint. If possible, move the flame around the fitting so that the complete joint will be heated evenly, causing the solder to flow more evenly. During the initial heating process, periodically take the flame from the fitting and touch the solder to the joint at the point where the tube and fitting meet. See Figure 7–14.

Never melt the solder with the flame. When the tube has been properly heated, it will melt the solder. If the solder is melted by the flame a cold, weak joint will result. When the tube melts the solder, capillary action will draw the solder into the joint. When the joint is filled, a fillet of solder will appear where the tube and the fitting are joined. This is sufficient solder for the joint and the joint is finished.

On larger fittings of 2" or more, a special Y-shaped torch tip is required to properly heat the joint. See Figure 7–15.

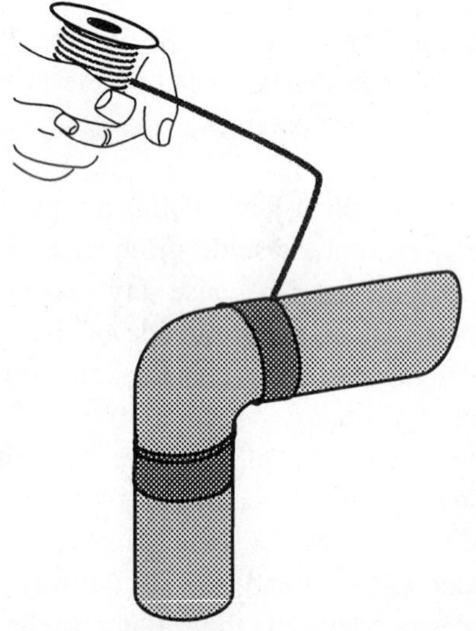

Figure 7–14 Application of solder to the joint

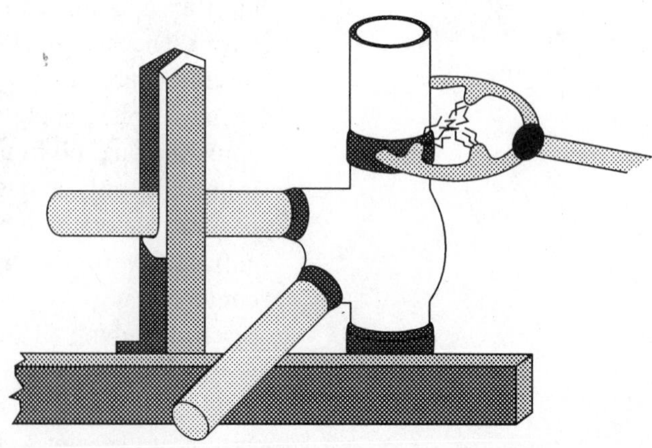

Figure 7–15 Four-shaped torch tip for heating large fittings

When a Y-shaped torch tip is not available, two regular torches may be used, one on each side of the fitting. When soldering larger fittings, it is sometimes wise to tap the fitting with a small mallet or hammer at two or three places around the fitting during the soldering process. The jarring helps to remove any trapped gases inside the joint that will interfere with the proper flow of the solder.

A soft oxyacetylene flame is usually best when silver-brazing a joint. The flame should be adjusted to provide a slightly reducing type of flame. A reducing flame uses a smaller amount of oxygen than a neutral flame.

The initial heat should be directed at the tube about 1/2 to 1 inch away from the fitting. The tube should be heated evenly all the way around. When the flux turns to a clear liquid, the heat should be moved to the fitting. See Figure 7–16.

Be careful not to overheat the joint. If the joint is overheated it must be separated, recleaned, and refluxed. Many times this is impossible when silver-solder is used.

During the silver-soldering process, keep the heat moving back and forth between the tube and the fitting, with the flame continuously directed toward the tubing to keep both tube and fitting heated. If the flame remains in one place too long, the spot can become overheated and possibly ruin the fitting.

Watch the flux. When it turns to a clear liquid, slightly move the flame away from the joint and apply the silver-solder to the joint where the tube and the fitting meet. If the joint is sufficiently heated, the silver-solder will flow into the joint.

When excessively large fittings are to be silver-brazed, it is usually desirable to heat a small section of the joint at a time, then move to the next section. Overlap a small part of each successive section each time.

Use only enough silver-solder to fill the joint. When excessive solder is used it will either be wasted or it will possibly flow inside the system where it can cause problems. It is recommended that a piece of solder the length of the diameter of the tubing is enough to make a strong, leak-tight joint.

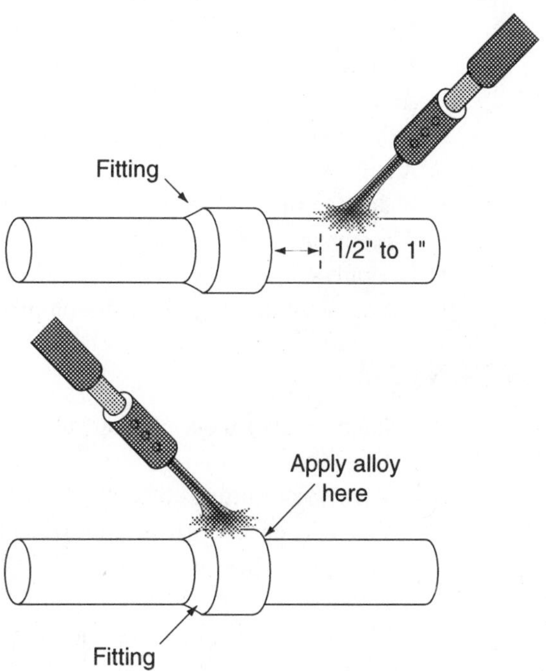

Figure 7–16 Properly heating silver brazed joint

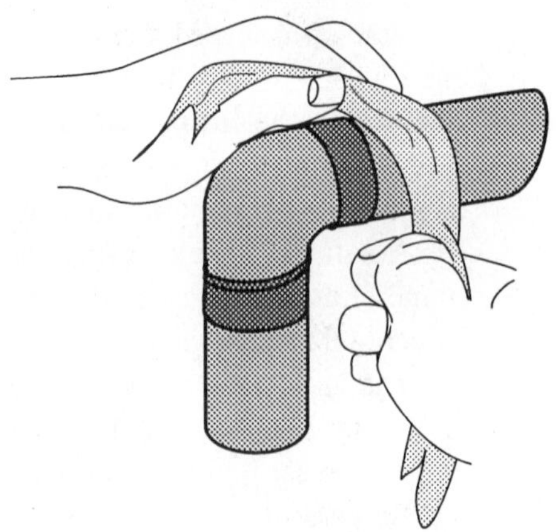

Figure 7–17 Cleaning a solder joint

6. When the silver-soldering process is completed be sure to clean the joint. This will make a more professional-looking job. This is done by wiping the joint clean with a clean cloth before the silver-solder has hardened. See Figure 7–17.

SUMMARY 7–2

- The sweat or solder type fittings are the most popular in refrigeration and air conditioning work.
- Flare fittings are a close second in popularity.
- Sweat fittings are the most popular for joining copper tubing together.
- When the tubing is not cut squarely, it is possible that the solder will flow through the crack and into the system, causing a restriction at this point.
- Basically, there are six steps used for both silver-soldering and soft-soldering tube fittings.
- Never use steel wool or emery cloth to clean solder joints.
- Always clean the joints just before they are to be soldered.
- The proper flux must be used for the material and the type of solder to be used.
- To prevent overheating the flux, be sure to keep the flame from striking the fluxed area.
- If the solder is melted by the flame a cold, weak joint will result.
- A soft oxyacetylene flame is usually best when silver-brazing a joint.
- Watch the flux. When it turns to a clear liquid, slightly move the flame away from the joint and apply the silver-solder to the joint.
- Use only enough silver-solder to fill the joint.
- When the soldering process is completed be sure to clean the joint.

REVIEW QUESTIONS 7–2

1. Name the two most popular tubing fittings used in refrigeration and air conditioning work.
2. What type of fittings offer the least restriction to a refrigeration system?
3. What happens in a sweat joint when the tubing is not cut squarely?
4. What two types of materials should never be used to clean refrigeration line joints?
5. What will happen to a sweat joint if it becomes oxidized?
6. What will usually be required when the torch touches the fluxed area of a joint?
7. How much solder should be used to make a sweat joint?

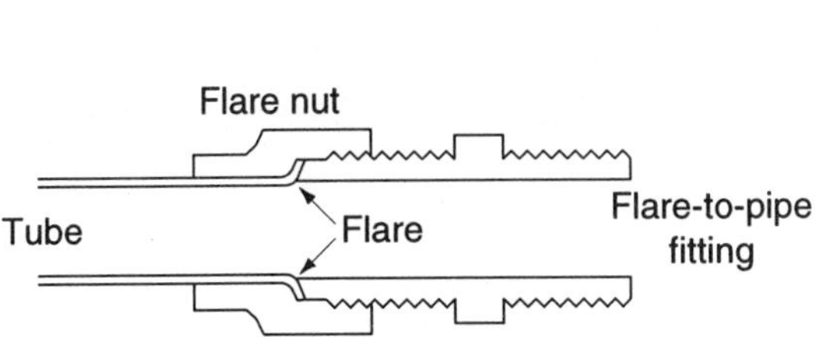

Figure 7–18 Flare connection

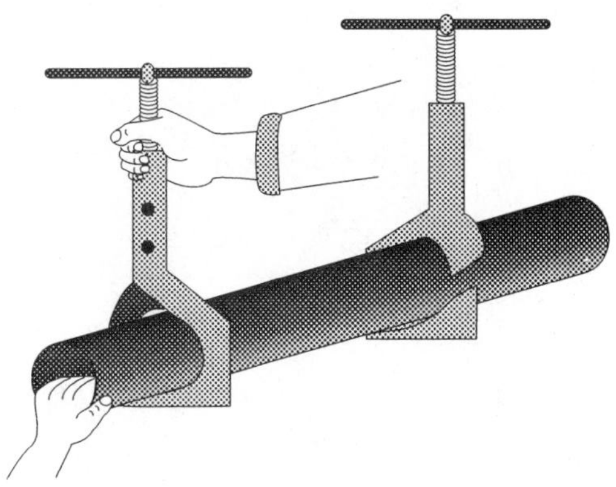

Figure 7–19 Properly cutting the tube

7–3 FLARE FITTINGS

In refrigeration and air conditioning installations, the 45° flare is standard. These types of fittings are usually easier to install than the sweat type because no flame is needed. To make a flare fitting, place the flare nut over the tube. Flare the tube and make the connection to the fitting by tightening the flare nut onto the flare fitting. See Figure 7–18.

The flare should be very close to the same size as the chamfer on the fitting. Sometimes a drop of refrigeration oil will allow the mating parts to fit better and prevent twisting the tube when the connection is tightened. The tubing that is squeezed between the mating parts is what acts as the gasket. If the fitting is tightened excessively, the gasket may be either cut off or squeezed too much. If the flare is squeezed too much, an effective seal will be almost impossible. It is recommended that the fitting be tightened hand-tight plus one turn. When flare fittings are used in places where the temperature varies several degrees with each cycle of the unit, a leak will sometimes occur. The expansion and contraction of the metal components of the joint will cause the flare nut to back off of the fitting. The joint must be retightened in order to stop the leak before adding more refrigerant to the system.

Joining Flare Fittings

The steps involved in making leak-free, strong flare joints are: (1) cut the tube properly, (2) install the flare nut onto the tubing, (3) install the flare block on the tube, (4) install the flare yoke on the flaring block, (5) make the flare, (6) remove the yoke from the flare block, (7) remove the flare block from the tube, and (8) check the size of the flare.

The following is a discussion of these steps:

1. Cut the tube to the proper length using the proper cutting tool. See Figure 7–19.

   Make certain that the end of the tube is cut square so that a strong, leak-free joint is possible. If a hacksaw is used, be sure to tilt the end of the tube downward so that the filings will fall away from the tube rather than inside it.

   After the cutting has been completed, remove the burr from inside the tube. Always tilt the tube with the end downward so that the filings will fall away from the tube rather than inside it. See Figure 7–20.
2. Place the flare nut over the end of the tube with the threads toward the end of the tube. Make certain that there is enough room for the flare nut and the flaring

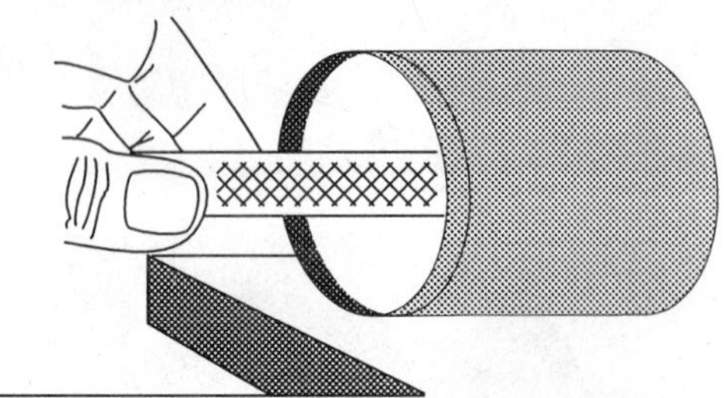
Figure 7–20 Deburring the tube

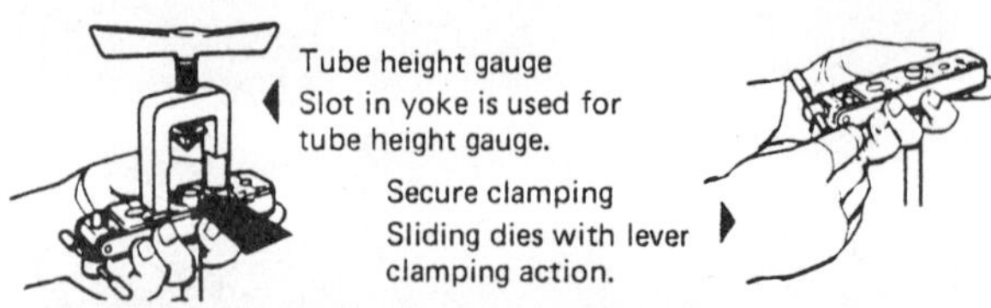

Tube height gauge
Slot in yoke is used for tube height gauge.
Secure clamping
Sliding dies with lever clamping action.

Figure 7–21 Installing the flaring block on tube end (Courtesy of Imperial Eastman)

tools between the end of the tube and the next bend or fitting. It may be necessary to make the flare before making any other joints required if enough room is not available.

3. Place the flaring block on the end of the tube with the proper amount of tube extended from the block so that the flare can be properly made. See Figure 7–21.

   Be sure that the tubing is inserted into the proper die in the block. If too much tube extends from the block, too large a flare will be made and the tube could possibly split. If too little is extended through the block, too small a flare will be made and a weak, leaking joint will be the result. Tighten the flare block sufficiently on the tube to prevent it from slipping when the yoke is tightened on the block.
4. Install the flare yoke on the flare block with the cone over the tube to be flared. See Figure 7–22.

   It is sometimes advisable to place a drop or two of refrigeration oil on the flare cone so that it can turn inside the tube easier.
5. When the yoke is properly in place, start turning the feed screw. See Figure 7–23.

   When it has touched the tube turn it another five or six turns very slowly to avoid ruining the flare. Do not tighten the yoke too much. To do so will cause the copper tube to become dead (lose its resiliency), resulting in a weak joint.
6. When the flare has been completed, remove the yoke from the flare block by turning the feed screw in the opposite direction until the cone is released from the copper flare. When the feed screw is turned in this direction, the cone automatically burnishes the flare.
7. After the yoke has been removed, remove the flare block from the tube by loosening the clamp screws enough to allow removal of the block from the tube.

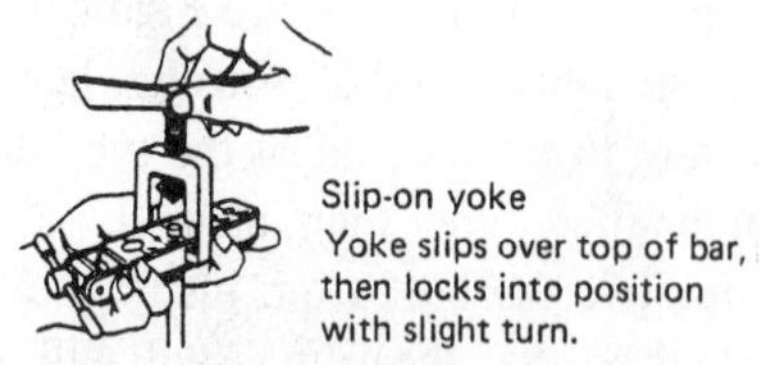

Slip-on yoke
Yoke slips over top of bar, then locks into position with slight turn.

Figure 7–22 Installing the yoke on the flaring block (Courtesy of Imperial Eastman)

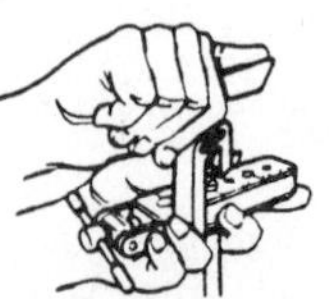
Figure 7–23 Making the flaring block (Courtesy of Imperial Eastman)

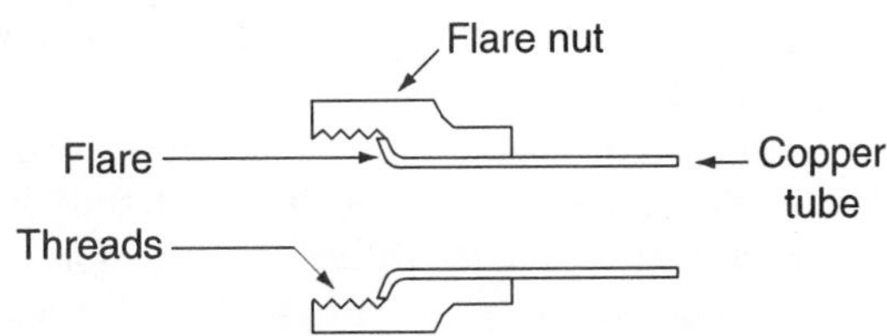

Figure 7–24 Checking the flare size

8. When the block has been removed, check the size of the flare. This is done by moving the flare nut over the flare. See Figure 7–24.

   The edge of the flare should just barely pass through the threads of the flare nut and completely fill the chamfer in the bottom.

SUMMARY 7–3

- In refrigeration and air conditioning work, the 45° flare is standard.
- The flare should be very close to the same size as the chamfer on the fitting.
- If the flare is squeezed too much, an effective seal will be almost impossible.
- It is recommended that flare fittings be hand-tightened plus one turn.
- Make certain that the end of the tubing is cut square so that a strong, leak-free joint is possible.
- It is sometimes advisable to place a drop or two of refrigeration oil on the flare cone so that it can turn inside the tube easier.
- The edge of a completed flare should just barely pass through the threads of the flare nut and completely fill the chamfer in the bottom.

REVIEW QUESTIONS 7–3

1. What is the degree of the standard flare fitting used in refrigeration and air conditioning systems?
2. When making a flare joint, what should be done to the tubing after cutting it?
3. When making a flare fitting, why is it desirable to place a drop of refrigeration oil on the flare cone?

7–4 COMPRESSION, HOSE, AND "O"-RING FITTINGS

These types of fittings are popular for making quick connections between lines and the various components of the system. The type of application will usually determine what type of fitting is best used.

Compression Fittings

Compression fittings are used to make quick-connect connections between the various components of a refrigeration system and the refrigerant lines. See Figure 7–25.

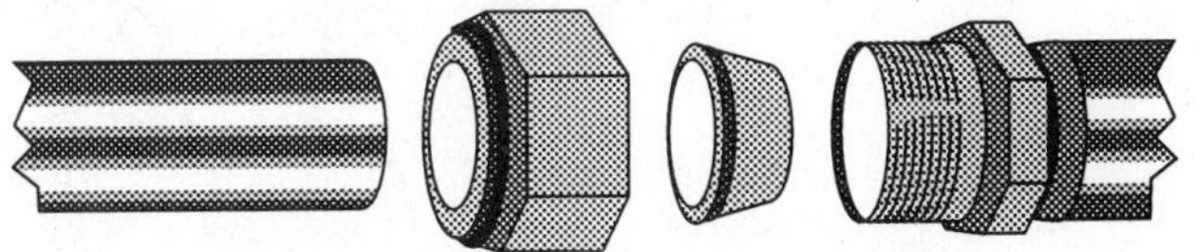

Figure 7–25 Compression fitting

These fittings are more commonly used with hard-drawn tubing because it is usually more rigid than soft tubing, it is completely round, and it is stiff enough to allow the fitting to be properly tightened onto it. Soft-tubing may also be used if it is perfectly round and straight. The nut and the ferrule are placed over the end of the tube. The tube is then inserted into the fitting until it bottoms out. The nut is then tightened hand-tight plus 1½ turns to effect a leak-tight joint. Do not overtighten these fittings. To do so would probably damage the tubing and cause a leak. If this type of joint leaks it is usually best to cut the old ferrule off the tube and install a completely new fitting.

Hose Fittings

Hose fittings are most commonly used in automotive air conditioning and transport refrigeration systems. They are used to make connections between the neoprene hose and the screwed fittings used for the rest of the system components. When installing these fittings, the screwed fitting is connected to the component, then the hose end is placed over the barbed part of the fitting with a clamp in place over the hose. The clamp is then tightened on the hose and fitting. These fittings will usually provide a leak-proof joint. However, once the clamp has been tightened, removing the hose requires that it be cut off the fitting.

"O"-Ring Fittings

"O"-ring fittings are sometimes used on special service valves. The "O"-ring is placed into matching, machined grooves in the mating parts. See Figure 7–26.

When assembling these fittings, caution must be used to see that the "O"-ring is inside the grooves in the fitting parts. Hand-tightening plus ½ turn will effect a leak-tight seal. If soldered joints are to be made close to these fittings, the fitting must be disassembled before soldering, and reassembled after the soldering is completed. When the heating is to be done some distance from the fitting it may be possible to keep from ruining the "O"-ring by cooling the fitting with a wet cloth. The cloth must be kept wet during the soldering process and until the fitting has cooled.

SUMMARY 7–4

- Compression fittings are used to make quick connections between the system tubing and the various components.
- These fittings are commonly used with hard drawn tubing.
- Do not overtighten these fittings.
- Hose fittings are most commonly used in automotive air conditioning and transport refrigeration systems.
- "O"-ring fittings are sometimes used on special service valves.

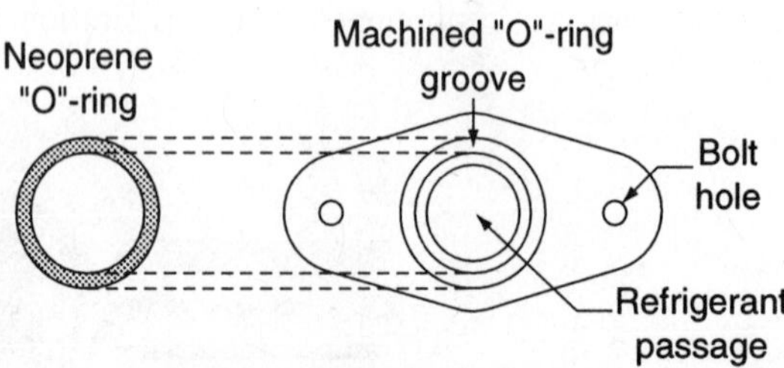

Figure 7–26 "O"-ring fitting

REVIEW QUESTIONS 7–4

1. What is the name of quick connect fittings?
2. Can soft tubing be used with quick connect fittings?
3. When assembling an "O"-ring fitting, what caution must be used?
4. When heating tubing close to a joint with some type of gasket, what can be done to prevent ruining the gasket?

7–5 BENDING AND CHANGING TUBE SIZE

Tubing benders are used to make bends and turns in tubing when it is desired or needed to eliminate as many connections as possible. This makes the system virtually leak-free. When it is desirable to make a bend in tubing, it is recommended that tubing benders be used, even when soft-drawn tubing is used. Benders will make a much nicer looking bend that will offer the least resistance to the refrigerant flow. There is a tubing bender for each size of tubing. If the wrong size bender is used, the tube will probably be damaged and will usually need to be replaced.

The tubing is measured to the required length and marked where the bend is desired. The bender is then placed on the tubing with the mark lined up with the desired angle of bend. See Figure 7–27.

It is important that the tubing be placed in the bender in the exact position because, once the tubing is bent, it cannot be effectively straightened again. The bender handles are then pulled together to make the required bend. The bend should be about 3° to 5° greater than that desired because the tubing will straighten out some when the benders are removed.

Another method of bending tubing is hand-bending. Hand-bending works very well on the smaller size soft-drawn tubing. The larger sizes of soft-drawn and hard-drawn tubing are very difficult to bend by hand. When using the hand-bending method, start bending the tubing in a large radius and work with the tubing, bending it just a small amount at a time until the desired angle is reached. See Figure 7–28.

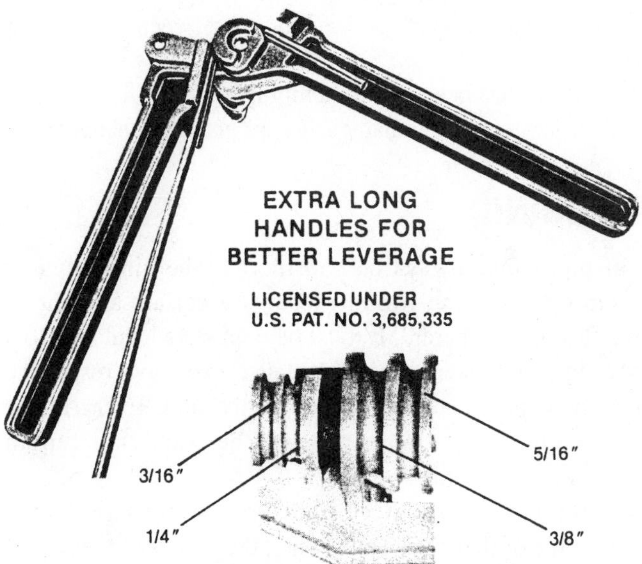

Figure 7–27 Bending tubing (Courtesy of Robinair Manufacturing Company)

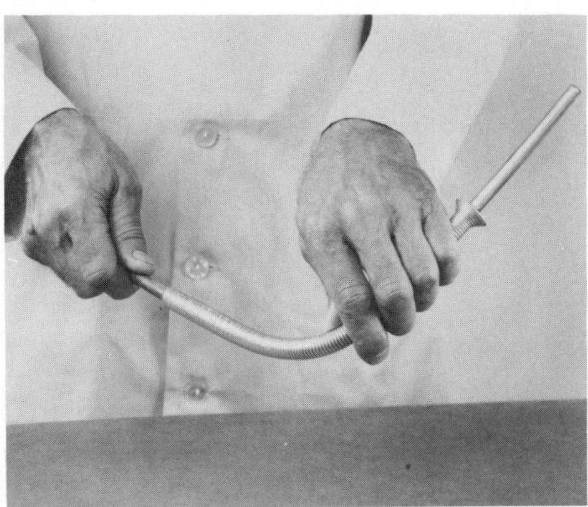

Figure 7–28 Hand bending tubing (Courtesy of Imperial Eastman)

Many times when an attempt is made to bend the tubing to the desired radius the first time, it will kink. The recommended bending radius when hand-bending tubing is about 5 times the diameter of the tube for the smaller sizes. When the larger sizes are to bend, this radius is about ten times the tube diameter.

Changing Tube Sizes

It is not an uncommon practice to change the size of tubing when installing or repairing refrigeration units. The recommended method of changing tubing size is to use a reducing fitting. Many times, however, when on a job site some distance from the supply house or shop where the supplies are kept, the needed fitting is not available. This is not a very big problem when the tubing size is to be changed only one size because the tubing used in refrigeration lines is sized in 1⁄8 inch increments. This type of sizing allows a smaller tube to be placed inside the next larger size. This does not make a joint that meets the desired tightness. However, it will serve the purpose very well. This type of fitting must be silver-soldered because soft-soldering will not provide the necessary strength. This type of fitting is not uncommon when doing service and installation work.

SUMMARY 7–5

- Tubing benders are used to make bends in tubing to reduce the number of fittings used.
- Benders will make a much nicer looking bend that will offer less resistance to the flow of refrigerant.
- It is important the tubing be placed in the bender in the exact position because, once the tubing is bent, it cannot be effectively restraightened.
- Two methods of bending tubing are with a bender and hand-bending.
- The recommended method of changing tubing size is to use a reducing fitting.
- The tubing used in refrigeration systems is sized in 1⁄8 inch increments. Therefore, a smaller size can be placed inside the next larger size and silver-soldered.

REVIEW QUESTIONS 7–5

1. Why are tubing benders used?
2. Is it easy to straighten tubing that has been bent in a tubing bender?
3. In a joint being made without a fitting, how many sizes are generally changed?

7–6 EQUIVALENT LENGTHS OF PIPE (TUBING)

The equivalent lengths of pipe must always be considered when installing or servicing refrigeration system tubing. Each fitting presents a certain amount of friction (pressure drop) to the flow of refrigerant. It must be considered and kept to a minimum when designing refrigerant system piping, because excessive resistance (pressure drop) reduces both the capacity and efficiency of the unit. Every fitting installed represents a given amount of resistance that will be equal to a certain number of feet of tubing of that size. See Table 7–3.

When consulting pressure-drop tables for a given line size, the pressure drop is given for each 100 ft of straight pipe of that size. Thus, when the equivalent length of pipe is used, the data can be used directly.

When excessively long lines or when some other reason requires that a very accurate estimate of the pressure drop of a system be made, the equivalent length of

OD LINE SIZE (IN)	GLOBE VALVE	ANGLE VALVE	90° ELBOW	45° ELBOW	TEE LINE	TEE BRANCH
1/2	9	5	0.9	0.4	0.6	2.0
5/8	12	6	1.0	0.5	0.8	2.5
7/8	15	8	1.5	0.7	1.0	3.5
1 1/8	22	12	1.8	0.9	1.5	4.5
1 3/8	28	15	2.4	1.2	1.8	6.0
1 5/8	35	17	2.8	1.4	2.0	7.0
2 1/8	45	22	3.9	1.8	3.0	10.0
2 5/8	51	26	4.6	2.2	3.5	12.0
3 1/8	65	34	5.5	2.7	4.5	15.0
3 5/8	80	40	6.5	3.0	5.0	17.0

Table 7–3 Equivalent length in feet of straight pipe for valves and fittings

each fitting should be carefully calculated. Most experienced design engineers are capable of making an educated estimate of the piping resistance of a system. Therefore, when a system requires 100 ft of pipe or more, the engineer will usually estimate an overall percentage allowance of about 20 to 30%. When the length of the lines is relatively short, an allowance of about 50 to 75% is not uncommon. It takes experience and knowledge to accurately design refrigerant lines. Many times the engineer will make periodic checks using actual calculations to determine if his estimates remain accurate.

SUMMARY 7–6

- The equivalent length of tubing must be considered when installing or servicing refrigeration system tubing.
- Each fitting presents a certain amount of friction to the flow of refrigerant.
- When consulting pressure drop tables for a given line size, the pressure drop is given for each 100 ft of straight pipe of that size.
- It takes experience and knowledge to accurately design refrigerant lines.

REVIEW QUESTIONS 7–6

1. How is the pressure drop presented in tables?
2. What is the equivalent length of tubing represented by a 90° × 1 1/8" elbow?
3. What does it take to accurately design refrigerant lines?

7–7 MATERIALS

There are many materials that are unique to the refrigeration and air conditioning industry. There are still several new materials being presented each year to aid the technicians in their work. Because there are so many we will discuss only the most commonly used in this text.

Soft Solder

This is a low temperature adhesion method of joining metals in refrigeration and air conditioning systems. Adhesion means that the metals being joined are not melted. Instead, the joining material is melted and adheres (sticks) to the other surfaces being joined. Soft solder is made of a mixture of tin and lead. This includes tin and antimony. The percentage of each determines the strength of the solder. In most refrigeration systems, the soft solder used is 95/5, meaning that the solder is 95% tin and 5% antimony. The 95/5 mixture has a melting temperature of 460°F; a 50/50 mixture has a melting temperature of 421°F. The melting temperature of copper tubing is 1,984°F and, therefore, either of these soft solders can be used in installation and repair work. However, with the higher operating temperatures of the compressor discharge gas, it is desirable to use a higher melting temperature material in these areas. Soft solder also has a low tensile strength that reduces its ability to withstand the vibration encountered in refrigeration systems.

Soft solder is commonly used in condensate drain lines and other low pressure applications. It is much less expensive than silver-solder.

Silver Solder

Silver solder is most commonly used on refrigeration systems today. Silver soldering is often referred to as silver brazing. The two terms are used interchangeably in the refrigeration and air conditioning industry. Silver solder is available in several different sizes, lengths, and strengths.

When silver soldering different metals together, it is recommended that a 45% silver solder be used. This percent silver solder has a melting temperature of 1,145°F. It is very strong and ductile; that is, it can withstand heavy vibration without breaking down. See Figure 7–29.

There are cadmium-free silver solders available that are recommended because cadmium has been determined to be hazardous to human beings. Some of these types of solder contain the proper amount and type of flux to allow a leak-free joint to be made without using additional flux. Often, new copper will not need any flux because it is already sufficiently clean for silver soldering.

Caution. Some of the older types of silver solders contain cadmium. Cadmium is a product that produces harmful fumes when heated; therefore, as a precaution, make certain that sufficient ventilation is present when using any type of silver solder.

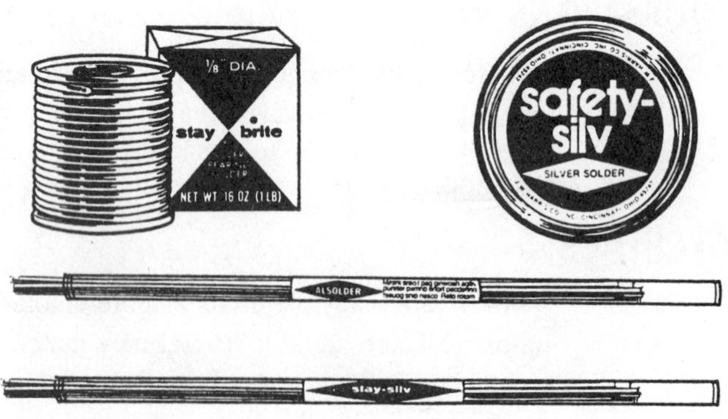

Figure 7–29 Silver solder (Courtesy of J.W. Harris Co., Inc.)

Figure 7–30 Soldering flux (Courtesy of J.W. Harris Co., Inc.)

Flux

Flux is a cleaning agent that is used when making sweat joints. It may be purchased in either the liquid or paste form. See Figure 7–30.

There is a flux designed for every type of solder. One type flux cannot be used with another type of solder. Always use flux sparingly. When a joint is overfluxed any excess could enter the system and cause problems. The problems occur because the flux is an acid and will attack the internal components of the system, especially the hermetic and semihermetic compressor motor windings.

Caution. Soldering flux has a high acid content and must be kept away from clothing, eyes, or open cuts. Flux can be removed with soap and water. If it gets into an eye, see a physician immediately for treatment.

Sand Cloth

Sand cloth is used for cleaning surfaces to be soldered. It can be purchased at the supply house in rolls. In use, strips are torn from the roll and then wrapped around the tubing. Each end of the sand cloth is then alternately pulled to move the cutting face across the surface to remove any oxidation. The surface is polished to a shiny new appearance. When cleaning the tube, be sure to point the open end downward to prevent the sanding particles from entering the tube or fitting. The inside of a fitting can be cleaned with sand cloth by rolling the short strip very tightly and inserting it into the fitting to be cleaned. The sand cloth is then rotated around until the fitting is properly cleaned.

SUMMARY 7–7

- Soft solder is a low-temperature adhesion method of joining metals in refrigeration and air conditioning systems.

- With the higher operating temperatures of the compressor discharge, it is desirable to use a high melting temperature material in these areas.
- Soft solder has a low tensile strength that reduces its ability to withstand the vibration encountered in refrigeration systems.
- When silver soldering different metals together, it is recommended that 45% silver solder be used.
- Flux is a cleaning agent that is used when making sweat joints.
- There is a flux designed for every type of solder.
- Sand cloth is used for cleaning surfaces to be soldered.
- When cleaning the tube, be sure to point the end downward to prevent the sand particles from entering the tube or fitting.

REVIEW QUESTIONS 7–7

1. What is a low-temperature adhesion method of joining metals?
2. In what areas of refrigeration and air conditioning is soft solder most commonly used?
3. In what application is it recommended that 45% silver solder be used?
4. When silver soldering a joint, what type of flux must be used?
5. What is removed from copper tubing when cleaning it to be soldered?

CHAPTER

8

COMPRESSORS AND LUBRICATION

OBJECTIVES

Upon completion of this chapter, you should be able to:

- Be more familiar with the types of compressors used in refrigeration systems
- Better understand the lubrication requirements of refrigeration compressors and systems
- Better understand how a refrigeration compressor operates
- Understand the purpose and function of the different compressor components
- Know more about the factors that control compressor output

INTRODUCTION

The compressor is generally considered to be the heart of the system. All compression-type refrigeration systems make use of some type of compressor. The purpose of the compressor is to compress and circulate the refrigerant from component to component. When working, the compressor draws the refrigerant vapor from the evaporator, lowering the pressure in the evaporator so that the refrigerant can boil and absorb heat at the desired operating pressure and temperature. The compressor then causes an increase in the refrigerant pressure and causes the refrigerant to flow to the condenser. It is the compressing action on the vapor that causes the refrigerant pressure to have a saturation temperature higher than the temperature of the medium used to cool the condenser and condense the refrigerant.

8–1 COMPRESSOR TYPES

There are several different types of compressors used. The most popular types are the reciprocating, rotary, centrifugal, and screw. Some of the newer, high-efficiency are the discus and the compliant scroll types. The reciprocating is the most popular type in use today.

Reciprocating Compressors

This type of compressor is used in air conditioning and refrigeration equipment up to about 100 tons of capacity. They are also used in commercial and industrial refrigeration equipment of almost all sizes.

The reciprocating compressor is similar in design to the automobile engine. They have a crankshaft that is driven by a motor. The crankshaft in turn drives a piston, or pistons, depending on the compressor size. The piston, through a series of valves, makes alternating suction and discharge strokes inside of a cylinder. See Figure 8–1.

In operation, when the crankshaft pulls the piston down in the cylinder, refrigerant vapor is drawn into the cylinder by the sucking action of the piston. A reed valve is

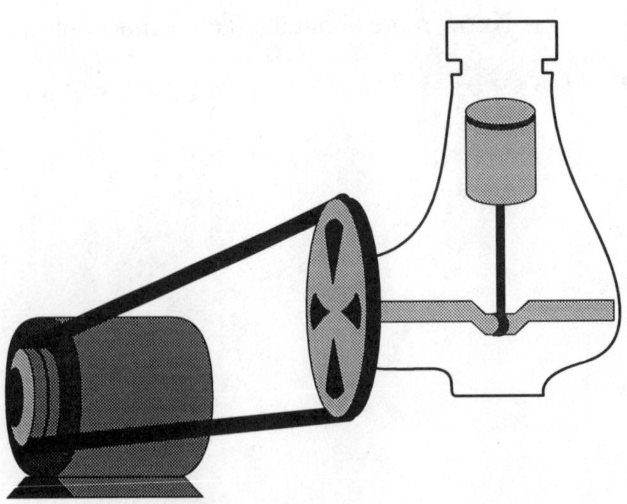

Figure 8–1 Reciprocating compressor

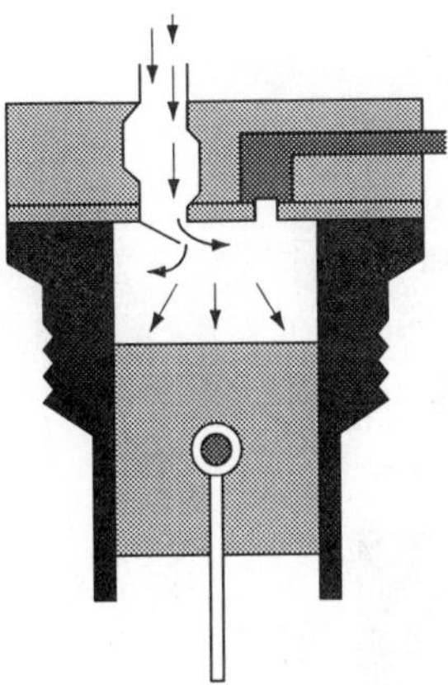

Figure 8–2 Suction stroke of the piston

located in the compressor cylinder head and opens to allow the refrigerant vapor to enter the cylinder. It then closes to prevent the vapor from leaving the cylinder through that port. See Figure 8–2.

When the refrigerant vapor inside the cylinder, plus the spring action of the valve material, equals the refrigerant pressure in the suction line, the valve will close to prevent the escape of refrigerant through that port. It is actually the spring action of the valve material that causes the valve to close.

When the piston reaches the bottom of its stroke, no more refrigerant vapor can enter the cylinder. The piston then starts an upward motion, causing the space between it and the cylinder head to become smaller. This reduction in space is what causes the refrigerant to be compressed. Because the suction valve has that port blocked so that no refrigerant can escape through it, the refrigerant is compressed more as the piston moves to the top of the cylinder.

The discharge reed valve, located in the valve plate, opens in an opposite direction than the suction valve. However, it is equipped with springs that cause it to remain closed under a somewhat higher pressure than the suction valve. When the pressure of the refrigerant vapor in the cylinder reaches a point greater than the combination of discharge pressure and the closing action of the valve springs, the valve will open and allow the refrigerant vapor to escape the cylinder through the port and enter the compressor discharge line that leads to the condenser. See Figure 8–3.

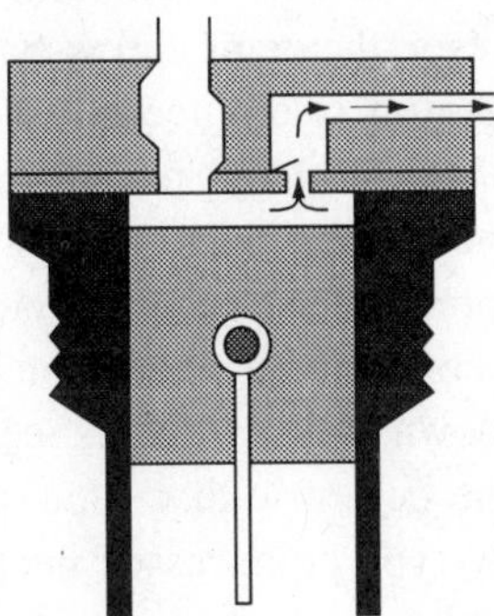

Figure 8–3 Discharge stroke of the piston

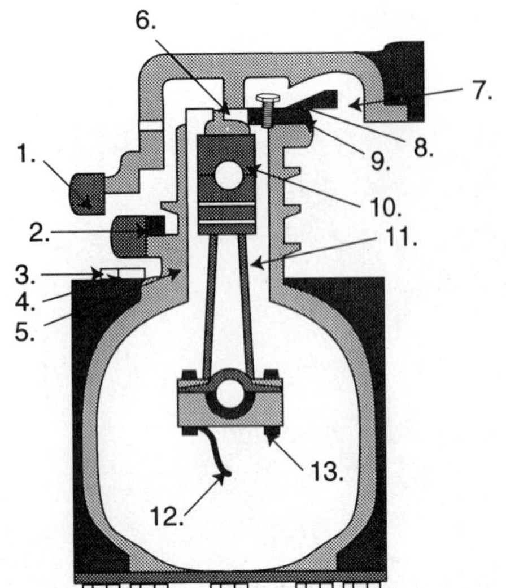

1. Suction port
2. Suction-chamber baffle
3. Oil plug
4. Oil plug gasket
5. Crank case port
6. Suction valve
7. Discharge port
8. Cylinder head gasket
9. Valve plate gasket
10. Piston
11. Connecting rod
12. Oil dipper
13. Removable cap
14. Valve plate
15. Piston pin button
16. Cocking fins
17. Crankshaft outer pushing
18. Crank case cover gasket
19. Crank case cover
20. Crank shaft
21. Seal equalizing
22. Bull ring
23. Seal spring
24. Seal bellows
25. Seal guide
26. Bearing oil plug
27. Base gasket
28. Bearing oil plug
29. Base plate
30. Discharge valve
31. Cylinder block
32. Piston pin
33. Crankshaft inner bushing
34. Thrust plate gasket
35. Crank shaft thrust bolt seat
36. Crank shaft thrust bolt
37. Crank shaft thrust plate

Figure 8–4 Crank type reciprocating compressor

As the compressor reaches the top of its stroke, and the refrigerant pressures equal that in the lines plus the spring pressure, the valves close and stop the flow of refrigerant. The piston is now ready to start the downward (suction) stroke and repeat the compression cycle. This procedure occurs for every piston in the compressor.

The reciprocating compressor is a positive displacement pump. It is very efficient at high condensing pressures and temperatures and where high-compression ratios are normally encountered. They may be used for several different types of refrigerants, they are very simple in design, and they are extremely durable.

The reciprocating motion of the piston can be achieved in several different ways. There have been several different methods used in the past to accomplish this action. However, the crank-throw-type crankshaft is the most popular to date. In this model, the piston is connected to the crankshaft through a connecting rod, which is connected to the crankshaft. See Figure 8–4.

The eccentric disc type is also popular in smaller sized compressors. This type of compressor uses a straight crankshaft that is fitted with eccentric discs. The discs are then connected to the piston by an eccentric strap that causes the piston to move up and down in the cylinder. See Figure 8–5.

The eccentric-disc-type compressor is cheaper to manufacture than the crank-throw-type compressor because of the machining requirements. The refrigerant vapor is compressed and circulated in exactly the same manner as with the crank-throw-type compressor.

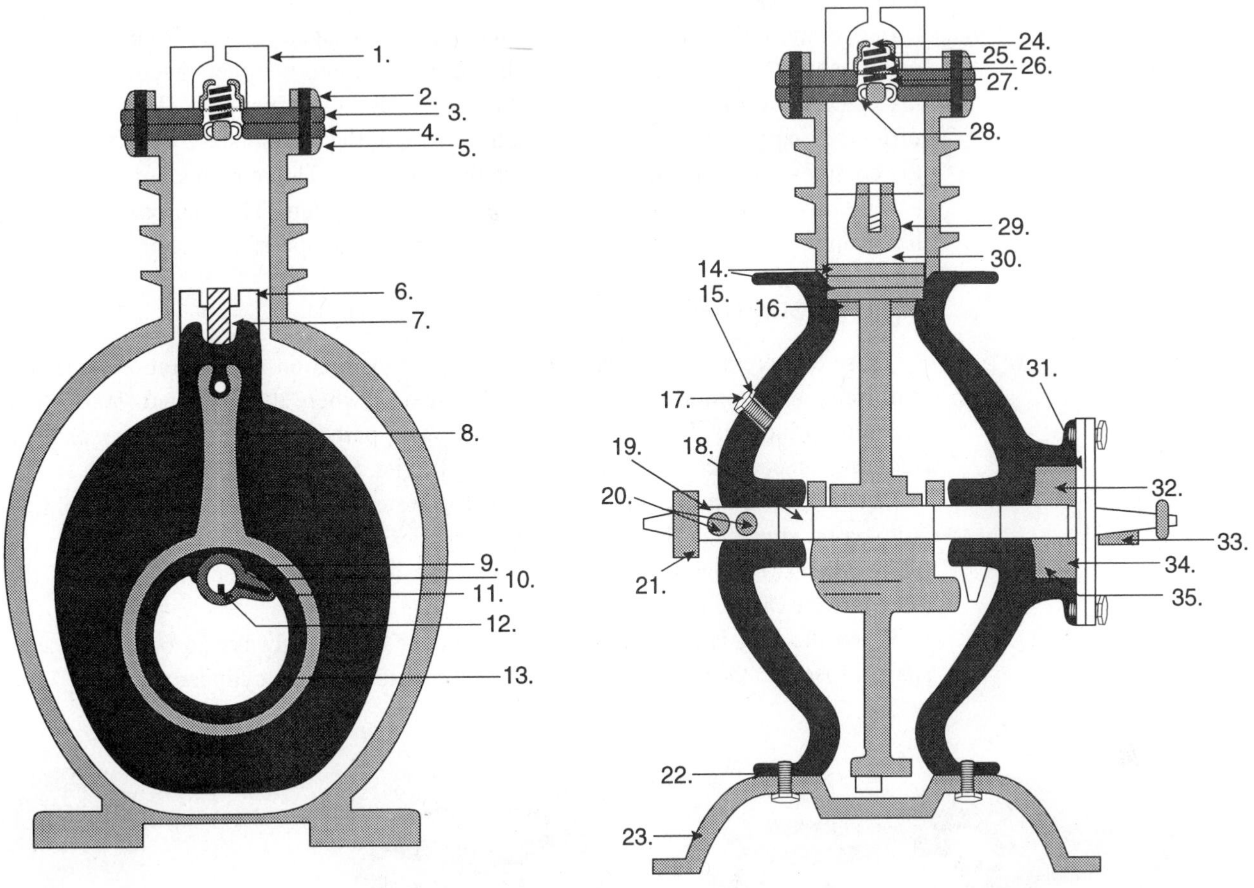

1. Cylinder head
2. Discharge valve plate gasket, upper
3. Discharge valve plate
4. Discharge valve plate gasket, lower
5. Cylinder block
6. Suction valve guide
7. Suction valve
8. Eccentric strap and rod
9. Lock washer
10. Lock nut
11. Set screw
12. Woodruff key
13. Eccentric disc
14. Piston ring
15. Oil plug gasket
16. Piston
17. Oil plug
18. Crankshaft
19. Crankshaft thrust plug
20. Crankshaft thrust ball
21. Crankshaft thrust plug gasket
22. Base gasket
23. Base
24. Discharge valve safety spring retainer
25. Discharge valve safety spring
26. Discharge valve spring
27. Discharge valve guide
28. Discharge valve
29. Piston pin
30. Piston pin button
31. Seal bellows
32. Seal guide
33. Woodruff key
34. Seal spring
35. Seal

Figure 8–5 Eccentric disc type reciprocating compressor

Two-Stage Reciprocating Compressors

This type of compressor is generally used in low-temperature refrigeration applications. The refrigerant compression process may be accomplished in one of two methods; (1) using two compressors with one discharging directly into the suction of the second, and (2) one compressor having one cylinder discharging into another. It is generally preferred not to use the two separate compressors because of the problem of proper oil return to their separate crankcases.

Two-stage compressors are generally divided into separate stages inside the compressor. See Figure 8–6.

In operation, the refrigerant vapor enters the first-stage cylinders from the suction line. It is then compressed, discharged, and metered into the interstage manifold. At this point the refrigerant is used to help cool the compressor motor. The compressed refrigerant vapor then passes to the suction ports of the second-stage cylinders where it is compressed again and discharged to the condenser. The remainder of the cycle is just like any other type of compression refrigeration system. Their purpose is to keep the compression ratio as low as possible on ultra-low temperature systems.

Rotary Compressors

These compressors are relatively simple in construction and operation. It is this simplicity that sometimes causes the confusion when attempting to learn their operating principles. In essence, the only moving parts of a rotary compressor are a steel ring, an eccentric or cam, and a sliding barrier. See Figure 8–7.

The ring and cam are both located inside the steel cylinder. The rotating steel ring is just a bit smaller in diameter than the cylinder. The rotating ring is manufactured off-center just enough to allow it to always be in touch with the inside wall of the cylinder. See Figure 8–8.

When the ring is rotating inside the cylinder there is always an open, crescent-shaped space on the side opposite where the ring touches the cylinder wall.

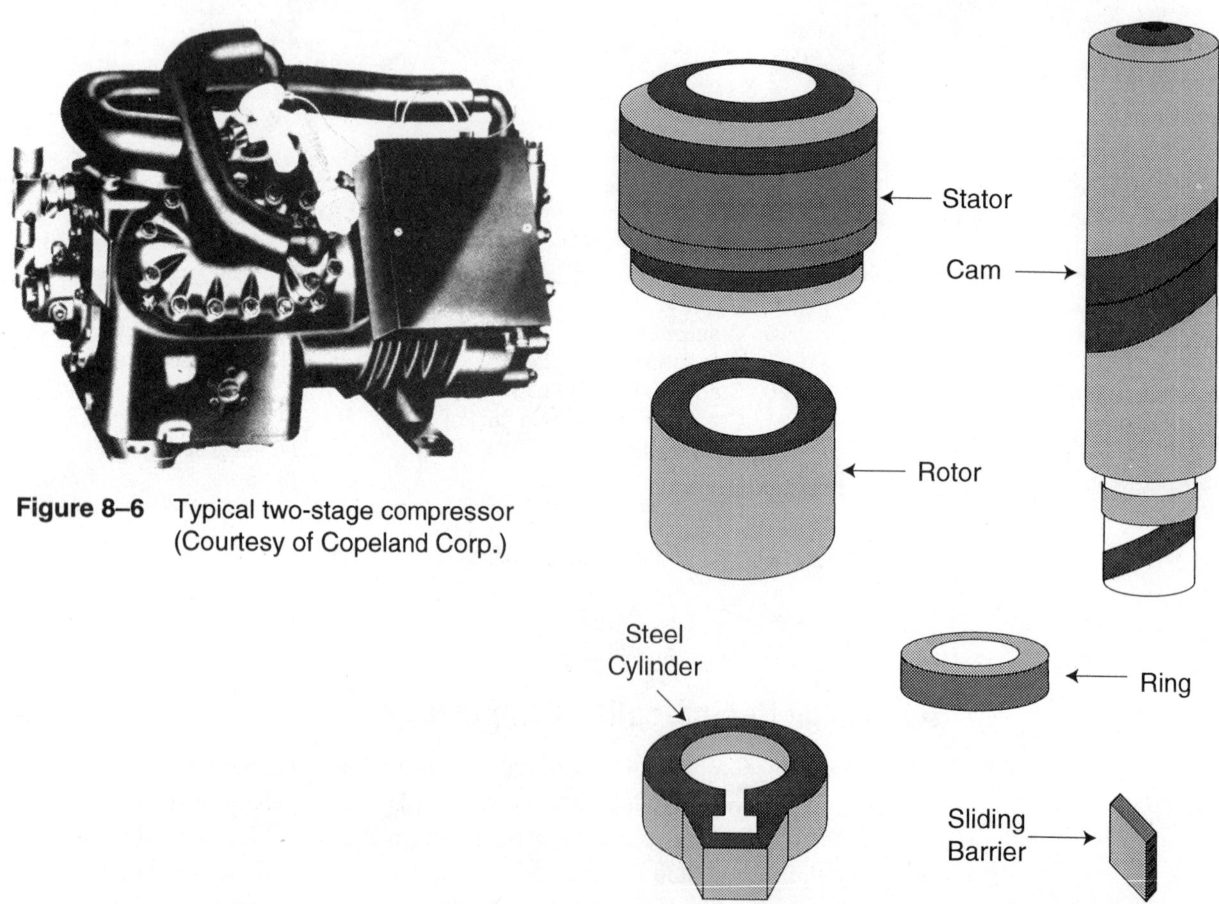

Figure 8–6 Typical two-stage compressor (Courtesy of Copeland Corp.)

Figure 8–7 Parts of a rotary compressor

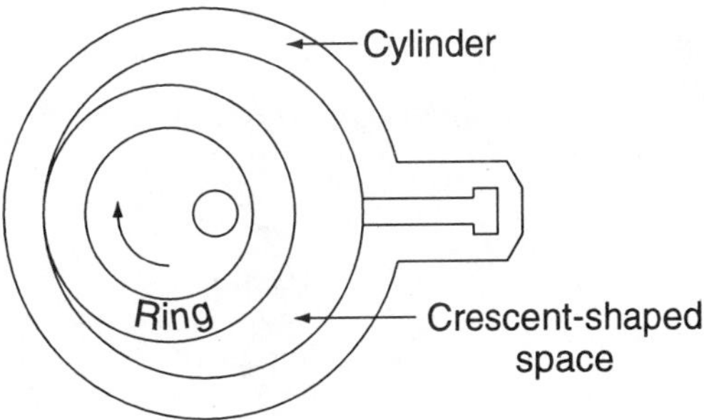

Figure 8–8 Rotary compressor component relationship

The cam is rotated by an electric motor connected to one end of the motor-compressor shaft. The rotating cam moves the steel ring along with it. This motion causes the steel ring to roll around the inside of the cylinder wall. See Figure 8–9.

If two ports are installed in the cylinder wall, the refrigerant vapor can be taken into the compressor through one and discharged out through the other. In this way the compressor will compress and move refrigerant throughout the system.

In operation, when the steel ring is moved just a small amount, both of the openings are open to the crescent shape inside the cylinder. Therefore, no compression action will be experienced. There must be something to cause a separation of the two ports. This device must form a barrier to prevent the discharge gas from entering the suction side of the system.

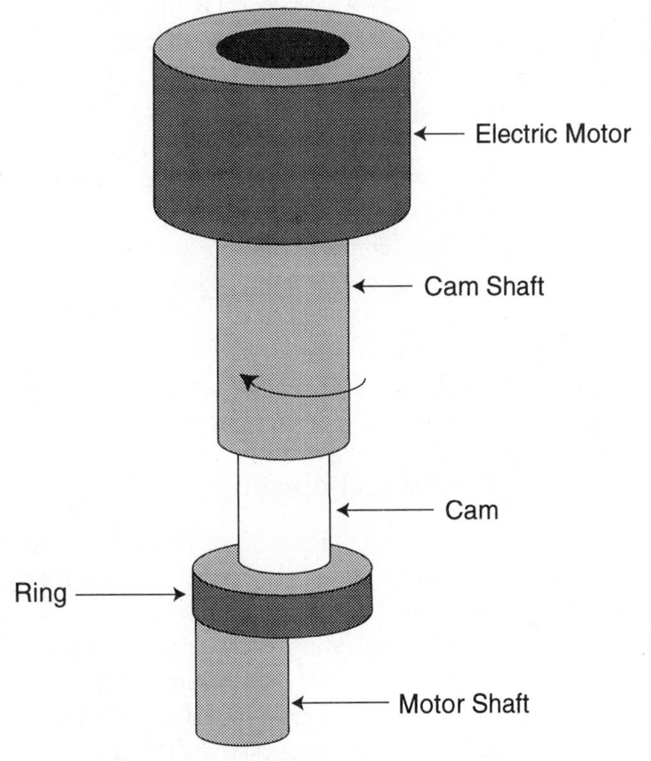

Figure 8–9 Rotary compressor operation

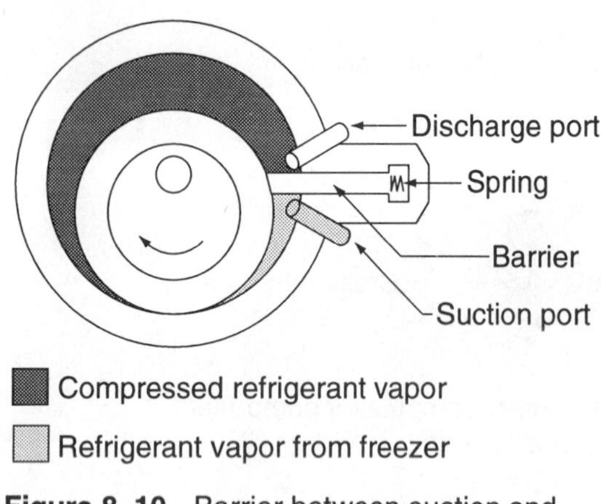

Figure 8–10 Barrier between suction and discharge ports

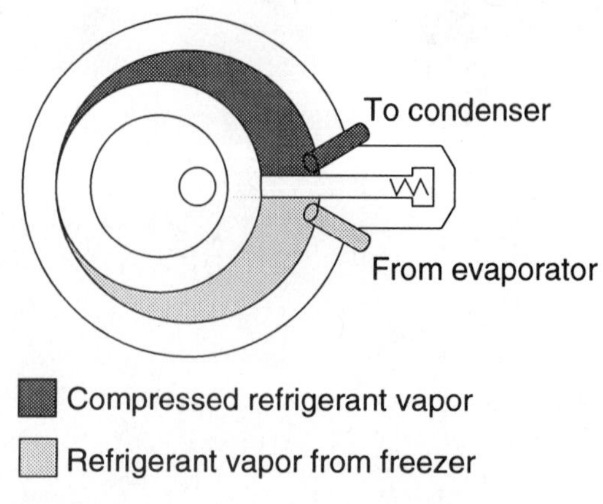

Figure 8–11 Cycle of a rotary compressor

Thus, a simple sliding barrier that is always in contact with both the rotating steel ring and the compressor cylinder wall will provide the desired function. See Figure 8–10.

A slot is machined into the cylinder wall with enough space to allow the barrier to slide into the groove until it is flush with the cylinder wall. A spring is placed behind the barrier to keep a slight pressure on the barrier, making certain that it will always be in contact with the rotating steel ring. See Figure 8–10. With this configuration, the sliding barrier will be in constant contact with the steel ring in its rotation around the cylinder wall.

When the movable barrier is operating properly, the refrigerant vapor trapped in the crescent-shaped space has only one direction to move as the steel ring rotates around the cylinder, pushing the vapor in front of it. See Figure 8–11.

With the barrier in place, the only means for the refrigerant vapor to escape the cylinder is through the discharge port. A check valve is usually placed in either the discharge line or the suction line to the compressor to prevent the discharge gas from reentering the suction line during the OFF cycle. Any liquid floodback must be prevented during the OFF cycle because both the suction and discharge lines enter directly into the cylinder.

These compressors are generally used in applications where large volumes of refrigerant are to be moved. They are also desired where a low compression ratio is used. Rotary compressors are positive displacement type pumps.

Rotary compressors are popular in domestic refrigerators and freezers and in the smaller-sized air conditioning systems. The precision machining procedures required make it necessary for these compressors to be manufactured in volume.

Screw Compressors

Screw compressors are relatively simple in construction. They consist of two matching screws between which the refrigerant vapor is compressed and discharged to the condenser. See Figure 8–12.

When operating, the refrigerant vapor fills the space between the lobes on the compressor screws. See Figure 8–13.

As the screws are turned, the space between the lobes is gradually reduced, causing the refrigerant to be squeezed into a smaller space. When the discharge port is opened by the rotating screw, the compressed refrigerant is discharged into the condenser.

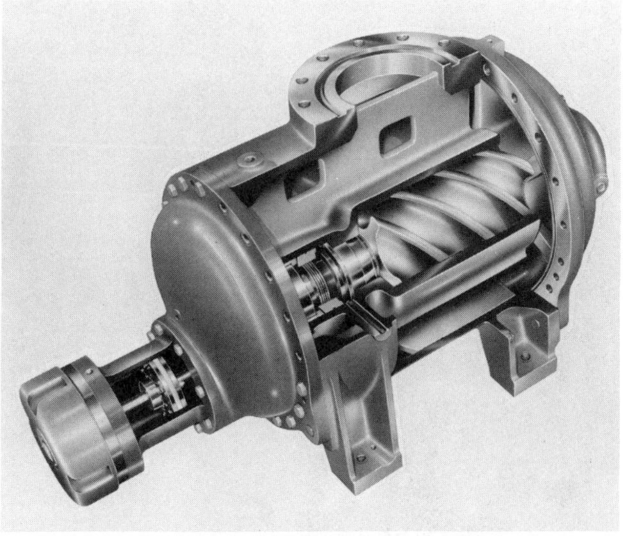

Figure 8–12 Screw compressor (Courtesy of York Division Borg-Warner Corporation)

These are positive displacement type compressors and are available in sizes ranging from about 100 to 700 tons capacity. They are generally used on chilled-water systems. Screw compressors will operate satisfactorily over a wide range of condensing temperatures. They will operate smoothly when the capacity is reduced to as low as 10%. In most instances, capacity control is accomplished by recirculating the refrigerant vapor inside the compressor.

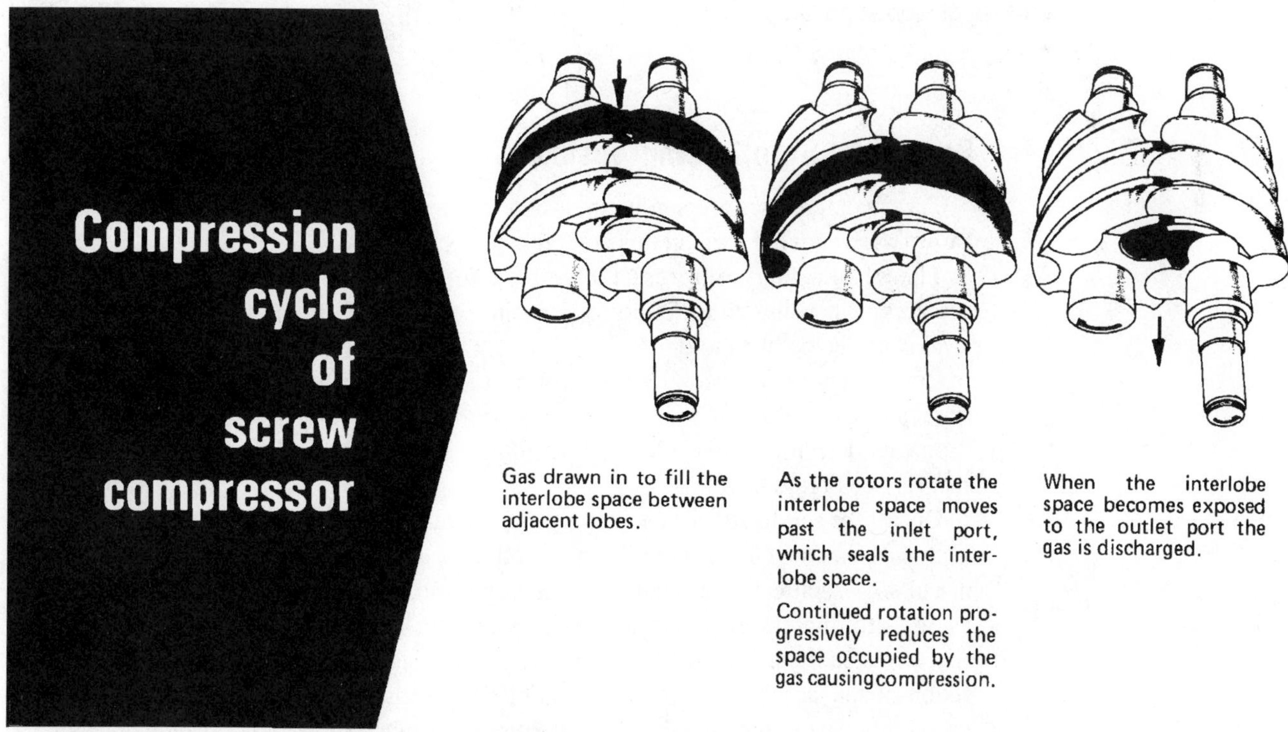

Compression cycle of screw compressor
Gas drawn in to fill the interlobe space between adjacent lobes.
As the rotors rotate the interlobe space moves past the inlet port, which seals the interlobe space.
Continued rotation progressively reduces the space occupied by the gas causing compression.
When the interlobe space becomes exposed to the outlet port the gas is discharged.

Figure 8–13 Screw compressor compression cycle (Courtesy of York Division Borg-Warner Corporation)

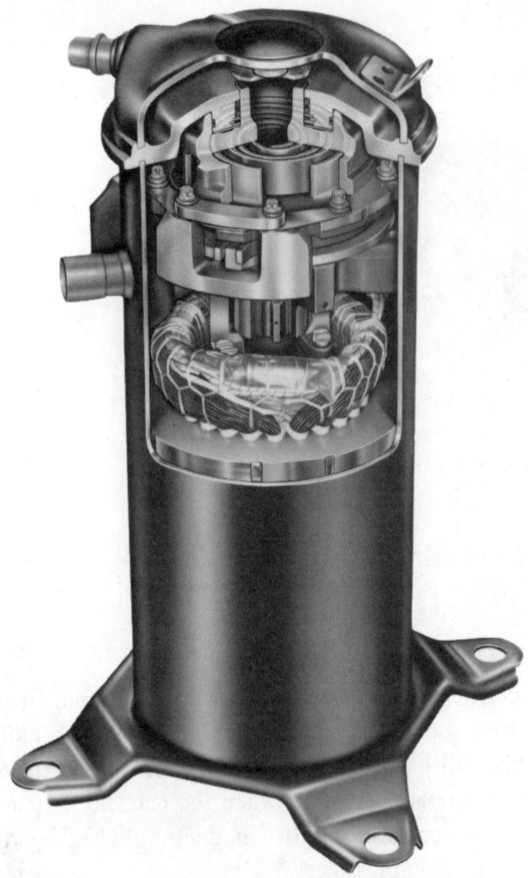

Figure 8–14 Cutaway of a scroll compressor (Courtesy of Copeland Corp.)

Figure 8–15 Scroll shape (Courtesy of Copeland Corp.)

Compliant Scroll Compressors

These compressors are simple in design and are very efficient in operation. There are very few moving parts in this compressor. See Figure 8–14.

The scrolls that compress the refrigerant are positioned in the top of the compressor housing and the motor is positioned in the bottom. The lubricating oil level is just below the motor.

Refrigerant compression by the scroll is a simple concept that centers around the spiral shape of the scroll and its inherent properties. See Figure 8–15.

In this design, the two identical scrolls are matched, forming concentric spiral shapes. See Figure 8–16.

One of the scrolls remains stationary and the other one "orbits." See Figure 8–17.

It should be noted that the orbiting scroll does not rotate or turn but simply orbits in a clockwise direction around the stationary scroll.

The orbiting scroll pulls the refrigerant vapor into the outer, crescent-shaped space created by the movement of the two scrolls. See Figure 8–17a. The centrifugal action of the scroll seals off the flanks of the scrolls, preventing the escape of the refrigerant. See Figure 8–17b. As the orbiting motion continues, the refrigerant vapor is forced toward the center of the scrolls and into a smaller space between them. See Figure 8–17c. As the scroll pushes the refrigerant vapor into the center of the two

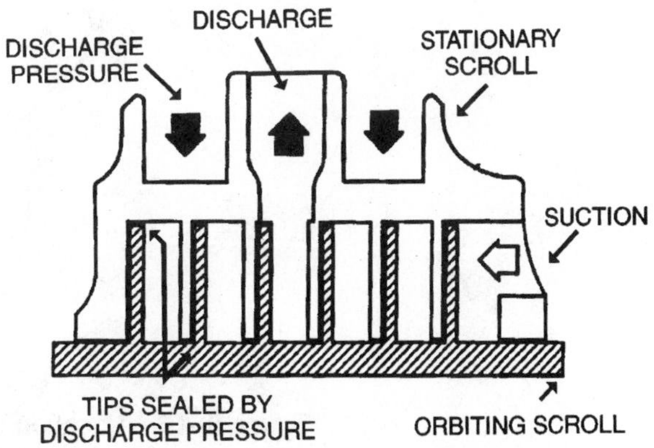

Figure 8–16 Cross section view of the scrolls (Courtesy of Copeland Corp.)

scrolls (see Figure 8–17d), it is discharged vertically upward into a chamber containing the discharge port in the top of the compressor. See Figure 8–14.

After the refrigerant has passed through the discharge port, the higher pressure refrigerant causes a downward pressure on the top scroll, helping to seal off both the upper and lower tips of the scroll. See Figure 8–16. While the compressor is

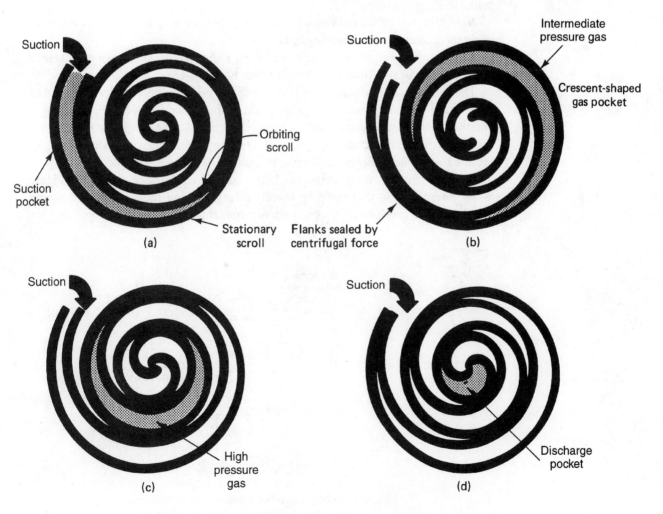

Figure 8–17 Scroll operation (Courtesy of Copeland Corp.)

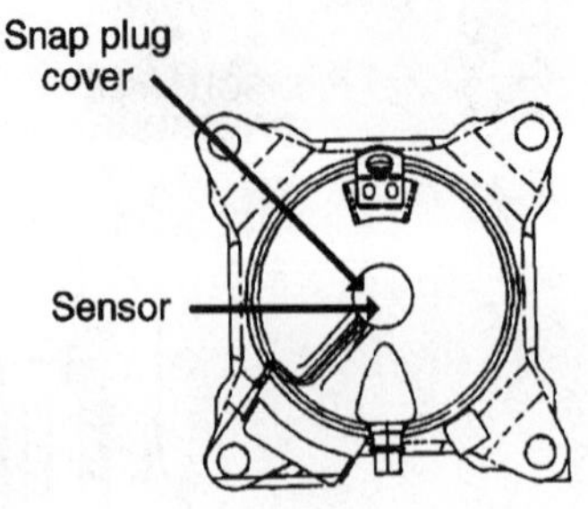

Figure 8–18 Location of scroll compressor discharge temperature sensor (Courtesy of Copeland Corp.)

completing one revolution, several pockets of refrigerant vapor are simultaneously compressed. This action produces a smooth, continuous compression cycle that reduces the amount of compressor vibration.

Liquid slugging does not seem to do any apparent harm to the scroll compressor. When liquid enters the scrolls, they are allowed to separate. The liquid is then worked toward the center of the scroll where it is discharged to the condenser without doing any harm.

Note: The top of the housing of a scroll compressor may be hot because it is in constant contact with the discharge gas.

Every scroll compressor is equipped with a discharge gas temperature sensor located on top of the compressor housing. See Figure 8–18.

The sensor is a single-pole single-throw (SPST) thermostatic switch designed to open when the compressor discharge temperature reaches 280 ± 8°F. The switch interrupts the control circuit to the compressor contactor to stop the compressor when an overheated condition occurs. The thermostatic switch automatically recloses when the compressor housing has cooled to 130 ± 14°F. The sensor can be found by prying up on the snap plug located on the compressor dome. See Figure 8–18. Make certain that the plug reseals after servicing the switch. The sensor terminals are located inside the compressor terminal junction box. See Figure 8–19.

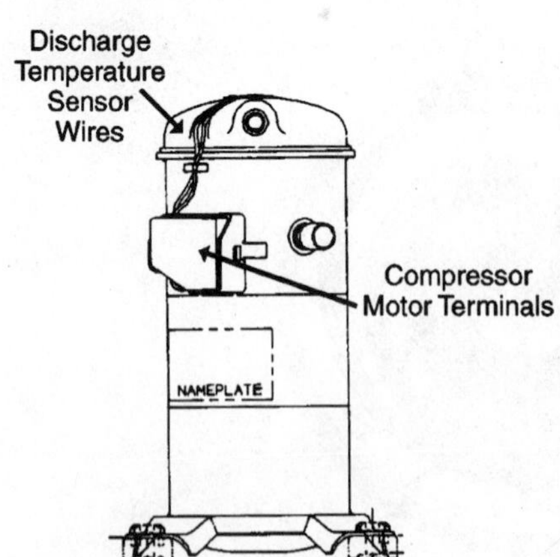

Figure 8–19 Scroll compressor terminal box (Courtesy of Copeland Corp.)

Centrifugal Compressors

Centrifugal compressors are most commonly used in the larger tonnage-type systems. They lend themselves very well to chilled water systems. They use the centrifugal force created by rapidly rotating vanes inside the compressor. The actual refrigerant compression is created when a mass of vapor is whirled at high speed, causing it to be thrown outward into a channel where it is caught.

This centrifugal action can be demonstrated by tying a ball on the end of a string and twirling the ball around at high speed. The ball tends to pull on the string and attempts to break loose and fly away. When a centrifugal compressor is turning, the molecules of refrigerant gas can be likened to the ball. The force on the molecules causes the same action as the twirling did to the ball. When the centrifugal compressor turns at a high rate of speed, the refrigerant molecules are thrown off the outer edge of the rotor. At this point they are caught in a channel where they are compressed by the following molecules. The molecules are actually squeezed together, causing the compressing action.

The centrifugal compressor is made up of a series of rotors or impellers. Each of these rotors is located in a separate compartment, or stage, of the compressor. See Figure 8–20.

This picture shows that there are five stages in this compressor. Thus, it is a five-stage centrifugal compressor. Each of the rotors consists of a series of vanes that are an integral part of each rotor. When the compressor is operating, the refrigerant vapor is drawn into the compressor from the suction line. It then flows through the suction passages close to the shaft and into the number one rotor. The rotation of the rotor produces a centrifugal force that throws the refrigerant vapor from the edge of the rotor at a much higher pressure than that in the suction line. The refrigerant vapor compressed in the first stage is discharged into the space labeled C in Figure 8–20, which is located between the first and second stage rotors. The vapor then flows back to the center of the rotor and enters the second stage rotor through the openings around the compressor shaft. Again, the rotating action of the rotor forces the vapor outward into space D. The refrigerant vapor follows this same procedure through all of the stages of the compressor until it is discharged out to the condenser through port G.

In each of the stages, the vapor is compressed further until it is compressed to the desired pressure for the application. Each of the stages is physically smaller than the previous stage, causing the vapor to be squeezed into a smaller space.

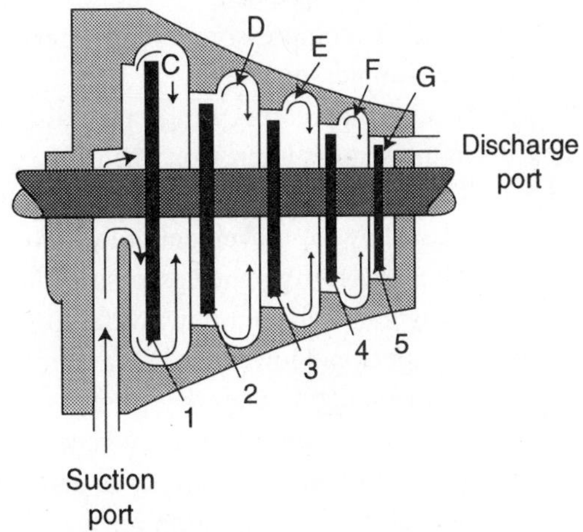

Figure 8–20 Longitudinal section of a centrifugal compressor

The refrigerant passages labeled C, D, E, F, and G are formed by metal labyrinths to prevent the refrigerant from leaking back into the last stage. These labyrinths are designed to close tolerances with the rotor, which allows the seal but prevents their touching.

The shaft end bearings are the only components that require lubrication. Because of this lubrication requirement, the refrigerant vapor in a centrifugal system is relatively oil-free. This prevents the accumulation of oil on the inside of the heat transfer surfaces such as the condenser and evaporator, allowing a greatly improved heat transfer coefficient of the overall system.

Centrifugal compressors are best suited for systems ranging from 250 to 3,000 tons capacity. They are designed to operate at very high speeds and are very efficient when operating with evaporating temperatures of 50°F to 120°F. They are not positive displacement compressors and are therefore very flexible under varying load conditions. They have good efficiency even when operating at 40% of their rated capacity.

SUMMARY 8–1

- Reciprocating compressors are used in systems up to about 100 tons of capacity.
- The reciprocating compressor is similar in design to the automobile engine.
- When the vapor inside the cylinder, plus the spring action of the valve material, equals the refrigerant pressure in the suction line, the valve will close to prevent the escape of refrigerant through that port.
- The discharge valve is located in the valve plate and operates just the opposite to the suction valve.
- The reciprocating compressor is a positive displacement pump.
- Two-stage reciprocating compressors are generally used in low-temperature refrigeration applications.
- The purpose of the two-stage compressor is to keep the compression ratio as low as possible.
- The only moving parts of a rotary compressor are a steel ring, an eccentric or cam, and a sliding barrier.
- In rotary compressors, any liquid floodback must be prevented during the OFF cycle because both the suction and discharge lines enter directly into the cylinder.
- Rotary compressors are positive displacement pumps.
- Screw compressors consist of two matching screws between which the refrigerant vapor is compressed.
- In screw compressors, the refrigerant vapor fills the space between the lobes on the compressor screws.
- Screw compressors are positive displacement pumps.
- In the scroll compressor, one of the scrolls "orbits" and the other is stationary.
- The orbiting scroll pulls the refrigerant vapor into the outer crescent-shaped space created by the movement of the two scrolls.
- After the refrigerant has passed through the discharge port, the higher pressure refrigerant causes a downward pressure on the top scroll, helping to seal off both the upper and lower tips of the scroll.
- Liquid slugging does not seem to do any apparent harm to the scroll compressor.
- Centrifugal compressors are generally used in larger tonnage applications.
- The centrifugal compressor is made of a series of rotors or impellers.
- In each of the stages, the refrigerant is compressed further until it is compressed to the desired pressure for the application.
- Centrifugal compressors are not positive displacement pumps.

REVIEW QUESTIONS 8–1

1. Name two of the more higher efficiency compressors.
2. To what is a reciprocating compressor similar?
3. What happens inside a compressor cylinder when the refrigerant vapor and spring action of the valve become equal?
4. What type of pump is a reciprocating compressor?
5. What type of compressor is used in low-temperature applications?
6. What type of compressor uses a steel ring, an eccentric or cam, and a sliding barrier?
7. In what type of compressor does the refrigerant fill the space between the lobes?
8. What type of motion does the moving scroll use in a scroll compressor?
9. Why does liquid slugging not harm the scroll compressor?
10. In what size range are centrifugal compressors most popular?
11. Are centrifugal compressors positive displacement pumps?

8–2 COMPRESSOR DESIGNS

Compressors are also typed by their body styles. There are three styles of compressors: open, semi-hermetic, and hermetic. The open type compressor may be further divided into belt-driven and direct-driven.

Open-Type Compressors

Open-type compressors were used almost exclusively at the start of the refrigeration industry. In this application, a separate motor was required to turn the compressor. There are two ways of connecting the compressor to the motor. One is to make the connection by use of a flexible belt. See Figure 8–21. The second is to connect the two shafts together through an in-line coupling, or direct drive.

Figure 8–21 Open type belt-driven compressor (Blissfield Mfg.)

All of the components inside the compressor body were connected to a shaft that extended through the compressor body. This type of compressor required that a shaft seal be used to prevent the leakage of refrigerant and oil from the system between the shaft and the compressor body.

Open-type compressors are very flexible in that the speed can be changed, within limits, to change the capacity and to allow the use of different types of refrigerants. Any required repairs were easily made because of the design of the compressor body.

Even with these advantages, they have several disadvantages. Their body is generally made from cast iron, making them very heavy. The shaft seal is an almost constant source of refrigerant and oil leakage. Aligning the shaft of the direct-drive models is usually very difficult and time consuming. They are costly to manufacture. Because of the belts and other external components, they are more noisy than the other types of bodies. The belts are another source of problems because they wear out and need replacing.

Open-type compressors are being replaced by hermetic and semi-hermetic body style compressors. They are used in only a few specialized applications such as transport refrigeration, automotive air conditioning, large commercial installations, and ammonia refrigeration systems.

Semi-Hermetic Compressors

These styles of compressors are driven by an electric motor that is mounted on the compressor crankshaft. The motor is also located inside the compressor body. See Figure 8–22.

When the motor and the compressor are sealed within a common housing, the external components such as the belts and shaft seal are not required, thus eliminating these sources of problems. The motors are generally sized for the specific load that the compressor is designed to handle.

The heads, stator covers, bottom plates, and the housing covers are removable, which allow for simple and easy field repairs.

Hermetic Compressors

Hermetic compressors are sometimes called welded hermetic motor-compressors. The purpose of this design was to further reduce the size and cost of manufacturing. This type of compressor is popular in all sizes up to about 25 tons capacity. The motor and compressor are mounted on a common shaft. The major difference is that the body is generally made from formed sheet metal and is hermetically sealed by welding the two halves together. See Figure 8–23.

Figure 8–22 Semi-hermetic compressor (Courtesy of Copeland Corp.)

Figure 8–23 Hermetic compressor (Courtesy of Copeland Corp.)

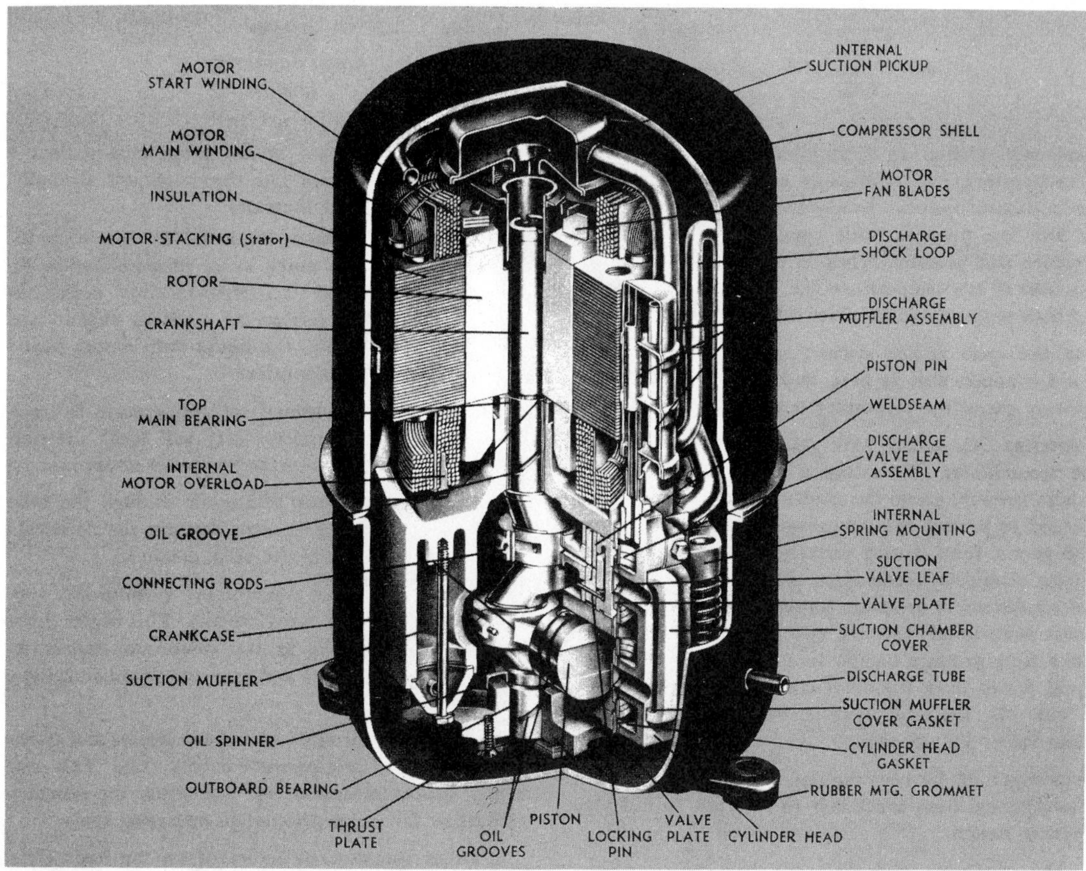

Figure 8–24 Internal view of a hermetic compressor (Courtesy of Tecumseh Products Company)

No field repairs can be made to the hermetic compressor because the shell must be physically cut open to gain access to the motor and compressor.

The internal components of a hermetic compressor are listed in the sequence of the refrigerant path as it flows through the compressor. See Figure 8–24.

The refrigerant vapor is drawn from the evaporator into the compressor shell. It then flows through the electric motor windings to remove some of the heat generated by the electricity. The crankshaft is designed to carry the lubricating oil from the oil sump in the bottom of the housing to all of the bearing surfaces. The refrigerant vapor is then drawn around the compressor crankcase and the motor windings and into the suction muffler, the suction valves, and then into the cylinder. The piston compresses the refrigerant vapor and discharges it through the discharge valves, the discharge muffler, and the compressor discharge tube.

SUMMARY 8–2

- In open-type compressors, a separate motor was required to drive the compressor.
- Open-type compressors required a shaft seal to prevent refrigerant and oil from leaking out of the system at this point.
- Open-type compressors are very flexible in that the speed can be changed, within limits, to change the capacity and to allow for the use of different types of refrigerants.
- Semi-hermetic compressors are driven by a motor that is mounted on the compressor crankshaft. The motor is also located inside the compressor body.

- The heads, stator covers, bottom plates, and the housing covers are removable, which allow for simple and easy field repairs.
- Hermetic compressors are popular in sizes up to about 25 tons capacity.
- The compressor and motor of a hermetic compressor are mounted on a common shaft inside the housing.
- No field repairs can be made to a hermetic compressor.

REVIEW QUESTIONS 8–2

1. Name two ways of connecting an open drive compressor to a motor.
2. What type compressors are the most flexible?
3. What types of compressors do not use a shaft seal?
4. What is used to cool the electric motor in semi-hermetic and hermetic compressors?

8–3 COMPRESSOR VALVES

Valves are used in modern refrigeration compressors to control the flow of refrigerant through the system. When the valves, either the discharge or the suction, are defective the compressor will not pump the refrigerant as it was designed. There is a discharge and a suction valve for each cylinder. The suction valve controls the flow of refrigerant vapor into the compressor cylinder and the discharge controls the flow of refrigerant from the cylinder.

The flapper valves, as they are usually called, are made from a very thin, flat strip of special type steel. The flapper valve operates efficiently at high rates of speed and their light weight prevents hammering of the valve seat. The steel strip offers very little resistance to the flow of refrigerant, resulting in a higher volumetric efficiency. Heavier valves, when used at high speeds, are very inefficient because their weight causes them to move slower than the flapper type valves.

Suction Valves

The purpose of the suction valve is to open and admit refrigerant vapor to the compressor cylinder when the piston is on the down (suction) stroke. When the piston is on the up (compression) stroke, the suction valve closes to prevent the refrigerant from returning to the suction line. The suction valve is operated automatically by the difference in pressure on each side of the steel strip and the weight of the strip itself. A spring is not generally used on suction valves. The flapper valve may take one of several shapes, such as a ring, a disc, or a reed. See Figure 8–25.

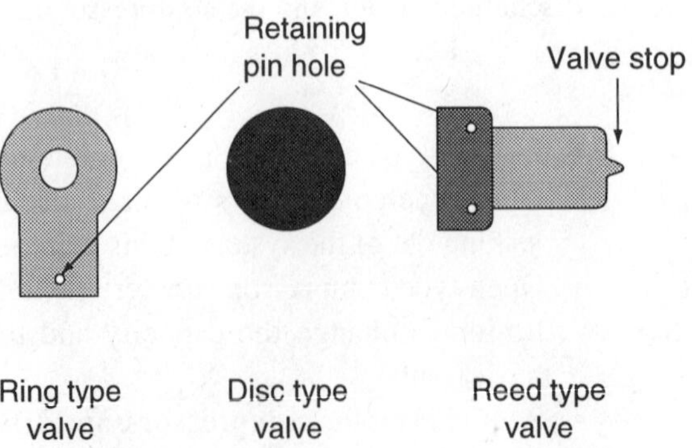

Figure 8–25 Types of suction valves

Valves are designed to cover a port or ports when in the closed position. When the difference in pressure causes the flapper valve to move off its seat, the ports are open.

Suction valves may be designed to fit into the compressor head or be located on a valve plate. The valve located in the piston is commonly referred to as a piston valve. See Figure 8–26.

All valves require some type of motion-limiting device. Should the valve reed move too much it could be broken, or at least damaged, requiring replacement. Also, if the valve moves too much it may not return to the seat quick enough to seal when the compressor is operating. There are several methods used to limit the motion of valves. For example, the ring-type piston valve uses a piston head plate, or cage. The head plate is screwed tightly to the piston head and keeps the various valve components in their proper position. See Figure 8–27.

The piston valve seat consists of a circular raised edge on top of the piston. The raised edge is machined smooth and lapped. Another lapped circular edge or seat is located concentrically inside the valve seat. There is an opening between these two lapped valve seats that allows the refrigerant vapor to pass through when the piston is on the suction stroke. An anchoring tongue extension with a hole in it that fits over the valve-retainer pin is a part of the valve. The movement of the valve ring is limited because it can only move between the edge of the valve seat and the bottom of the head plate. There is a small spiral spring fastened to the piston. Its purpose is to prevent abrupt opening of the valve, which would cause noise during compressor operation. The spring also helps to keep the valve closed during the compressor OFF cycle. The closed valve helps to prevent any refrigerant that has leaked through the discharge valve from entering the compressor crankcase.

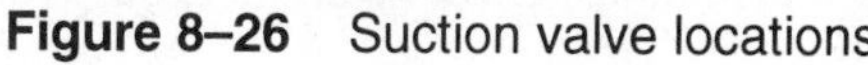

Figure 8–26 Suction valve locations

Figure 8–27 Ring-type suction valve stops

Figure 8–28 Valve plate location (Photo by Billy C. Langley)

There are many compressor designs that have the suction valve located in the compressor cylinder head. In such applications, the valves are placed in a valve plate located between the head and the compressor body. Usually, each valve plate will contain both the suction and discharge valves for that cylinder or cylinders. See Figure 8–28.

In this design, the cooler refrigerant vapor from the evaporator will help to cool the cylinder head and its related components. Also, the possibility of oil pumping is eliminated because the refrigerant does not pass through the compressor crankcase where it mixes with the oil. The valve designs are the same whether they are mounted in the valve plate or the piston head.

Suction valves of the reed type are made from very thin spring steel. This type of valve acts as a spring because of the type of material it is made from. The pressure of the refrigerant coming from the suction line causes the valve reed to bend downward and allow the refrigerant to enter the compressor cylinder. The reed valve has a tendency to hold itself in the closed position because of the spring action of the metal, except when the returning refrigerant (suction) pressure is sufficient to overcome the spring action of the valve.

Discharge Valves

The purposes of the discharge valves are to permit the compressed refrigerant vapor to be discharged into the condenser, and to provide a gas-tight seal between the valve and the valve seat, preventing leakage of the high-pressure refrigerant into the low side of the system. Discharge valves act as check valves by preventing any backward flow of the refrigerant.

The most popular location of the discharge valve is in the valve plate between the compressor body and the cylinder head. See Figure 8–28. The several designs of discharge valves are: the ring type, the disc type, and the reed type. Discharge valve designs usually incorporate a safety feature to prevent breakage of the valve should some liquid refrigerant or oil enter the discharge valve area. Liquids are not

Figure 8–29 Bridged flapper type discharge valve (Photo by Billy C. Langley)

compressible and some means must be provided to prevent accidental valve breakage if any should enter the compressor cylinder.

The bridged type flapper discharge valve is probably the most popular. The valve is named because the flapper valve actually bridges the discharge valve seat. See Figure 8–29.

The ends of the reed valve are fastened to the valve guides located on each end of the valve. The valve guides are specially shaped machine screws with threads that are screwed into the valve plate. The discharge valve itself is made of flexible spring steel. On top of the reed valve is a valve spring stop made with a slightly upward curved center to allow the reed to bend upward and allow the refrigerant to escape. During operation, should some liquid enter the cylinder, the reed will be forced up against the valve spring stop. The valve spring stop will also be forced upward if the pressure is high enough to cause this action. When the valve stop is forced upward, the valve springs are compressed, allowing the liquid to escape without damaging the compressor or the valve reed. The valve springs are a safety feature and are commonly referred to as safety springs. A safety wire is used on the screwed-in valve guides to prevent their loosening because of vibration.

Valve Sealing. The actual sealing between most surfaces in refrigeration systems is done by a thin film of oil between them. This is also true of valves and their seats. When the crankcase contains the proper charge of lubricating oil, there is a sufficient amount of oil flowing through the system to supply the oil for sealing purposes between any components requiring it. When the valves leak there is a proportional reduction in compressor efficiency. Thus, it is essential to maintain the correct oil level in the compressor crankcase.

Checking Valve Condition

It is essential to check both the suction and discharge valves to determine if either one or both are leaking. The following is a description of the procedures used for this operation.

Suction Valves. Install the gauge manifold on the unit. Front-seat (turn it all the way in) the suction service valve on the compressor. Allow the compressor to run for several minutes or until the low-side pressure has stopped dropping. The compressor should pump a vacuum of about 20 inches of mercury. If the compressor does not pump this amount of vacuum, stop it and allow it to stand for a few minutes. This is to allow any refrigerant in the compressor to completely vaporize. Then restart the compressor and determine if it will pump the desired vacuum. If the desired vacuum is not reached, the suction valves are leaking and must be replaced.

Discharge Valves. When determining the condition of the discharge valves, first connect the gauge manifold to the system. Front-seat (turn it all the way in) the suction service valve on the compressor. Run the compressor until as deep a vacuum as possible has been reached. Stop the compressor and observe the compound gauge on the gauge manifold set. The low-side pressure should not rise more than 3 or 4 inches of mercury. Should the pressure rise more than this, pump another vacuum, stop the compressor and again observe the compound gauge. An increase in pressure above 3 or 4 inches of mercury indicates that the discharge valves are leaking and must be replaced. If there is any question about the condition of these valves, there is one more step that should prove beneficial in condemning the valve. While the compressor is still idle and the vacuum remains, front-seat the discharge service valve on the compressor. If there is little or no further increase in the low-side pressure, the discharge valve is definitely leaking.

SUMMARY 8–3

- Valves are used in modern refrigeration compressors to control the flow of refrigerant through the system.
- The flapper valves, as they are sometimes called, are made from a very thin, flat strip of special steel.
- The purpose of the suction valve is to open and admit refrigerant vapor to the compressor when the cylinder is on the down stroke.
- The suction valve is operated automatically by the difference in pressure on each side of the steel strip and the weight of the strip itself.
- Valves are designed to cover a port or ports when in the closed position.
- All valves require some type of motion limiting device.
- Each valve plate contains both the suction and discharge valves for that cylinder or cylinders.
- The purposes of discharge valves are to permit the compressed refrigerant vapor to be discharged into the condenser, and to provide a gas-tight seal between the valve and the valve seat, preventing leakage of the high-pressure refrigerant into the low side of the system.
- The most popular location of the discharge valve is in the valve plate.
- The actual sealing between most surfaces in refrigeration systems is done by a thin film of oil between them.
- When the valves leak, there is a proportional reduction in compressor capacity.
- It is essential to check both the suction and discharge valves to determine if either one or both are leaking.

REVIEW QUESTIONS 8–3

1. What actually does the sealing in compressor valves?
2. What valve controls the flow of refrigerant into a compressor cylinder?
3. What causes the compressor suction valves to operate?
4. What do all valves require to keep from breaking?
5. What is the most popular location of compressor valves?
6. What happens to a compressor when the valves leak.
7. Approximately what amount of vacuum should a compressor with good suction valves pump?
8. What is indicated when the suction pressure rises in a compressor that has had a 20-inch vacuum pumped on it?

8–4 COMPRESSOR LUBRICATION

For the compressor and the other system components to be properly lubricated there must be a correct supply of oil in the crankcase. The compressor manufacturer designates the amount of oil that should be maintained in each specific model. In hermetic type compressors it is almost impossible to visually determine the oil level. However, when an oil sight glass is part of the compressor, the oil level is generally maintained just above the center line of the sight glass. See Figure 8–30.

Only the designated type of refrigeration oil should be used in compressors. If any other type of oil is used there can be many problems caused that will be quite expensive to remedy.

The compressor manufacturer will generally recommend the best type of oil to be used in the system. Usually, it will be recommended according to the type of refrigerant used and the temperature at which the system is designed to operate.

Lubrication Methods

Basically there are two methods of lubricating compressors: splash and forced. A combination of the two are used in almost all compressors.

Splash Lubrication. Splash lubrication is probably the simplest method. In this method, the lubricant is simply splashed onto the moving parts by the movement of the crankshaft. Some compressors use a dipper on the end of the connecting rod to aid in distributing the oil.

This method was extremely successful for use in the lower-speed compressors. However, when the higher-speed compressors were introduced, greater lubrication practices were needed. The lubricating oil is not only used for friction reduction between the components, but also for carrying away some of the heat generated by both the friction and compression inside the compressor.

Forced Lubrication. Forced lubrication is used almost extensively in modern compressor types. Two methods of forced lubrication are rifled passageways and a positive displacement oil pump.

Rifled passageways are used on smaller-sized compressors up to about 3 horsepower. The oil is forced through the passages, by splashing, to the desired points. When the compressor is operating, the oil is forced to the desired areas much like a bullet is caused to spin when traveling through a gun barrel.

When forced lubrication is used, the oil is forced to the desired areas by a positive-displacement type pump. The pump is generally mounted on the end of the crankshaft so that it will turn any time the compressor is turning. See Figure 8–31.

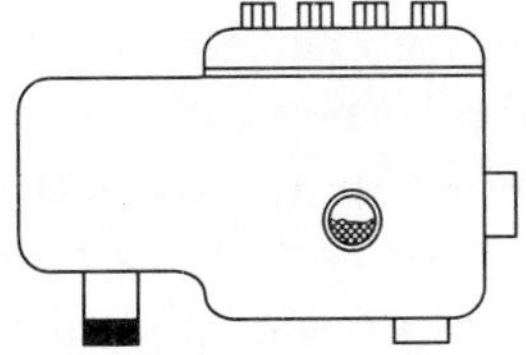

Figure 8–30 Oil level in sight glass

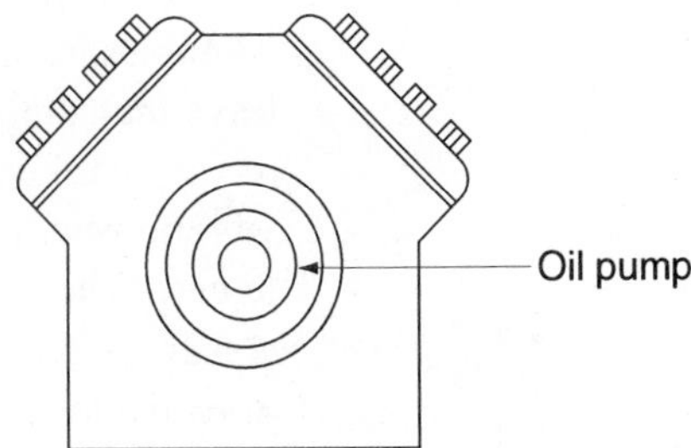

Figure 8–31 Location of oil pump

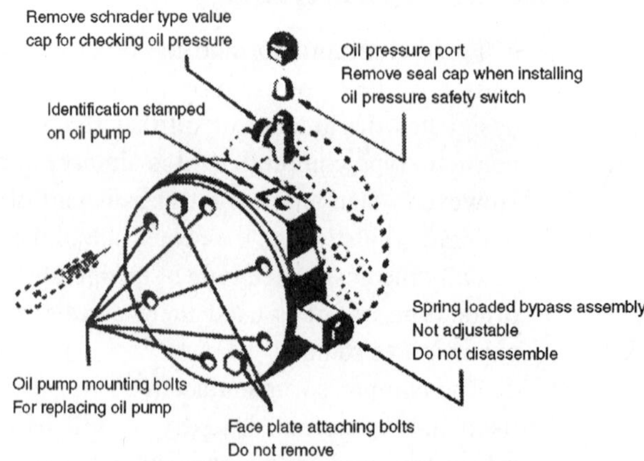

Figure 8–32 Location of spring-loaded bypass assembly (Courtesy of Copeland Corp.)

When the pump is turning, oil under pressure is forced through a hole in the crankshaft to the points needing lubrication. A spring-loaded pressure-relief valve allows the oil to bypass directly to the compressor crankcase in case the oil pressure becomes too high. See Figure 8–32.

The oil pump intake is located directly in the compressor crankcase. Therefore, the oil-pump inlet pressure will always be equal to the crankcase pressure. The oil pump outlet pressure will be the sum of the crankcase pressure plus the oil pump pressure. The net oil pressure will be equal to the pump outlet pressure minus the crankcase pressure. When the system is operating with a vacuum in the low side, the crankcase pressure is negative and must be added to the outlet pressure of the pump to calculate the net oil pressure. Below 0 psig readings, compound gauges are calibrated in inches of mercury. Remember that 2 inches of mercury is equal to approximately 1 psig.

Example: If we measured the crankcase pressure of a compressor and found it to be 50 psig and the pump outlet pressure was found to be 90 psig, the net oil pressure would then be: 90 – 50 = 40 psig.

The normal net oil pressure of an operating compressor is around 30 to 40 psig. However, adequate lubrication can be maintained with a net oil pressure of 10 psig. The spring-loaded bypass valve is usually set to open when the oil pressure exceeds 60 psig. The size of the compressor, the temperature and viscosity of the oil, and the amount of wear on the compressor bearings will affect the oil pressure of a normally operating compressor.

The oil pump is greatly affected by the presence of liquid refrigerant in the compressor crankcase. Violent oil foaming in the crankcase will cause the oil to leave the compressor, resulting in a loss of oil pressure until some of the oil returns to the crankcase. If liquid refrigerant is present in the crankcase, it can cause vibrations in the oil pump because of the flash gas caused by rapid evaporation of the liquid. When liquid refrigerant enters the crankcase and boils off it can cause the crankcase pressure to momentarily increase for short periods of time. This increase in crankcase pressure can possibly reduce the lubrication capacity of the oil pump. Thus, an oil failure device must always be used on systems when this is a possibility.

SUMMARY 8–4

- For the compressor and other system components to be properly lubricated, there must be a correct supply of oil in the crankcase.
- Only the designated type of refrigeration oil should be used in compressors.
- When splash lubrication is used, the lubricant is simply splashed onto the other moving parts by the movement of the crankshaft.
- Forced lubrication is used almost extensively in modern compressors.
- When forced lubrication is used, the oil is forced to the desired areas by a positive-displacement type pump.
- The oil pump outlet pressure will be the sum of the crankcase pressure plus the oil pump pressure.
- The net oil pressure will be equal to the pump outlet pressure minus the crankcase pressure.
- The normal net oil pressure of an operating compressor is around 30 to 40 psig.

REVIEW QUESTIONS 8–4

1. What type of lubricating oil should be used in refrigeration compressors?
2. Name the two methods used to lubricate refrigeration compressors.
3. Which type of lubrication method is needed for highspeed compressors?
4. On a forced lubricated compressor, what will be the pressure at the oil pump outlet?
5. Determine the net oil pressure when the crankcase pressure is 40 psig and the pump outlet pressure is 100 psig?

8–5 COMPRESSION RATIO

Compression ratio is very important to the service technician as well as the design engineer. It is important that the service technician know how to calculate compression ratio and what the calculation means to the longevity of the system.

Compression ratio can be defined as the absolute discharge pressure divided by the absolute suction pressure. To calculate the absolute pressure, add 15 psig to the gauge reading. When the suction pressure is below atmospheric, the absolute pressure is found by subtracting the gauge reading from 30 inches and dividing the result by 2. (Two inches of vacuum is equal to 1 psig.)

The chemical reaction that occurs with reactive materials such as oxygen, moisture, refrigeration oils, acid, heat, and pressure actually doubles with each 18°F temperature rise in the discharge temperature of a compressor. Because of this, a system operating with higher than normal discharge pressures will develop more problems than one operating within the desired limits. Also, the relationship between the discharge pressure and the suction pressure of a system should be within the industry standards of 10:1 for a normal compression ratio.

To calculate compression ratio use the following formulas:

$$CR = \frac{\text{Absolute discharge pressure}}{\text{Absolute suction pressure}}$$

For 0 psig suction pressure or above use:

$$\text{Absolute discharge pressure} = \text{Gauge reading} + 15 \text{ psig}$$
$$\text{Absolute suction pressure} = \text{Gauge reading} + 15 \text{ psig}$$

For a suction pressure reading below zero use:

$$\text{Absolute discharge pressure} = \text{Gauge reading} + 15 \text{ psig}$$

$$\text{Absolute suction pressure} = \frac{30 - \text{Compound gauge reading}}{2}$$

The following are examples of how to calculate compression ratio.

Example: The discharge pressure reading is found to be 250 psig and the suction pressure reading is found to be 70 psig. What is the compression ratio?

$$CR = \frac{\text{Absolute discharge pressure}}{\text{Absolute suction pressure}}$$

$$CR = \frac{250 + 15}{70 + 15}$$

$$CR = \frac{265}{85}$$

$$CR = 3.11{:}1$$

Example: The discharge pressure reading is found to be 225 psig and the suction pressure reading is found to be 10" vacuum. What is the compression ratio?

$$CR = \frac{\text{Absolute discharge pressure}}{\text{Absolute suction pressure}}$$

$$\text{Absolute discharge pressure} = 225 + 15 = 240 \text{ psia}$$

$$\text{Absolute suction pressure} = \frac{30 - 10}{2} = 10 \text{ psia}$$

$$CR = \frac{\text{Absolute discharge pressure}}{\text{Absolute suction pressure}}$$

$$CR = \frac{240}{10} = 24{:}1$$

This is a very high compression ratio that will probably cause problems in the near future. It is an indication of what can happen when a system is operating with a low suction pressure. A change in the discharge pressure does not cause such adverse effects. Therefore, the service technician must always be aware of what the compression ratio is. Every effort should be made to keep the system operating within the pressure ranges for which it is designed. Otherwise, problems will not be long in coming.

SUMMARY 8–5

- It is important that the service technician know how to calculate the compression ratio and what the calculations mean to the longevity of the system.
- The chemical reaction that occurs with the reactive materials such as oxygen, moisture, refrigeration oils, acid, heat, and pressure actually doubles for each 18°F temperature rise in the discharge gas temperature of a compressor.
- Every effort should be made to keep the system operating within the pressure ranges for which it is designed. Otherwise, problems will not be long in coming.

REVIEW QUESTIONS 8–5

1. What is the absolute discharge pressure divided by the absolute suction pressure?

2. If the gauge pressure is 15 psig what is the corresponding absolute pressure?
3. To how much vacuum is 1 psig equal?
4. What is the compression ratio when the discharge pressure is 225 psig and the suction pressure is 65 psig?
5. What will a high compression ratio cause?

8–6 CLEARANCE VOLUME

The clearance volume (pocket) is that space that is left between the piston and the valve plate when the piston has reached the top of its stroke. A clearance volume is necessary to prevent the piston from hitting the valve plate after the compressor has warmed to operating temperatures. There have been many efforts to reduce the clearance volume in order to increase compressor efficiency. The volumetric efficiency of a compressor is different for each compressor design. The two things that affect the volumetric efficiency of a compressor the most are properly sealing valves and the clearance volume. There is little that can be done about the clearance volume except to keep the discharge pressure as low as possible.

In operation, at the end of the compression stroke of the piston, there is a small space above the piston that cannot be cleared by the piston. See Figure 8–33.

Also, there is more space in the discharge valve ports that the piston cannot sweep. This space is also considered to be a part of the clearance volume. It is always filled with the discharge gas at the completion of the compression stroke.

On the downward stroke of the piston, any refrigerant that remains in this space is re-expanded back into the cylinder, reducing the amount of space available for new refrigerant from the evaporator. The pressure of this re-expanded refrigerant must be reduced to below that in the suction line before the suction valve can open and allow more refrigerant to enter the cylinder. In effect, the first part of the suction stroke of the piston is lost because of the re-expansion of the gas in the clearance pocket.

When the system is operating with a high suction pressure, the compression ratio is relatively low and the clearance volume is not so critical. However, when the system is operating with low suction pressure, such as in low-temperature applications, the clearance volume should be kept as small as possible. This may be done with the installation of low-temperature application valve plates. These plates have smaller discharge ports in an attempt to reduce the clearance volume.

A large clearance volume does, at times, have some good effects. The reduced velocity of the gas as it flows through the discharge ports in the valve plate helps to reduce wear, compressor noise, and the operating-power requirements for the compressor.

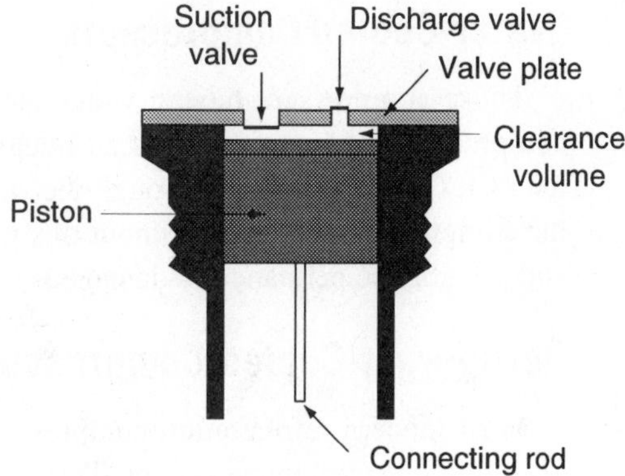

Figure 8–33 Clearance volume

SUMMARY 8–6

- The clearance volume (pocket) is that space left between the piston and the valve plate when the piston has reached the top of its stroke.
- A clearance pocket is necessary to keep the piston from hitting the valve plate after the compressor has warmed to operating temperatures.
- On the downward stroke of the piston, any refrigerant that remains in this space is re-expanded back into the cylinder, reducing the amount of space available for new refrigerant from the evaporator.
- When the system is operating with a relatively high suction pressure, the compression ratio is relatively low and the clearance volume is not so critical.
- A large clearance volume does, at times, have good effects. The reduced velocity of the gas as it flows through the discharge ports in the valve plate helps to reduce wear, compressor noise, and the operating-power requirements for the compressor.

REVIEW QUESTIONS 8–6

1. What is the space between the top of the piston and the valve plate known as?
2. Name two things that affect the volumetric efficiency of a compressor.
3. Why should the clearance volume and the discharge pressure be kept as low as possible?
4. What happens to the refrigerant that is left in the clearance pocket?
5. What can be done in low-temperature applications to reduce the clearance volume?

8–7 COMPRESSOR COOLING

Compressors require cooling to prolong their life and to help them operate as efficiently as possible. Some of the major methods used to cool refrigeration compressors are: air, water, refrigerant, and oil cooling.

Air-Cooled Compressors

Compressors that are cooled by air require that an adequate amount of air be circulated over the body to remove as much heat as possible. The air being discharged by the condenser fan should be directed at the motor-compressor. In most instances a draw-through fan will not provide sufficient cooling for a motor-compressor. These compressors will normally operate with a higher body temperature than when other types of cooling are used.

Water-Cooled Compressors

These compressors have a water jacket that is either mounted directly on the compressor body or is made as an integral part of it. During compressor operation, the water must be flowing through the water coil to provide the necessary cooling. If the compressor is operated without this water flow, it will rapidly become overheated and possibly be permanently damaged.

Refrigerant-Cooled Compressors

Most modern refrigeration compressors use some form of refrigerant cooling to keep them from running too hot. They are designed so that the suction gas from the

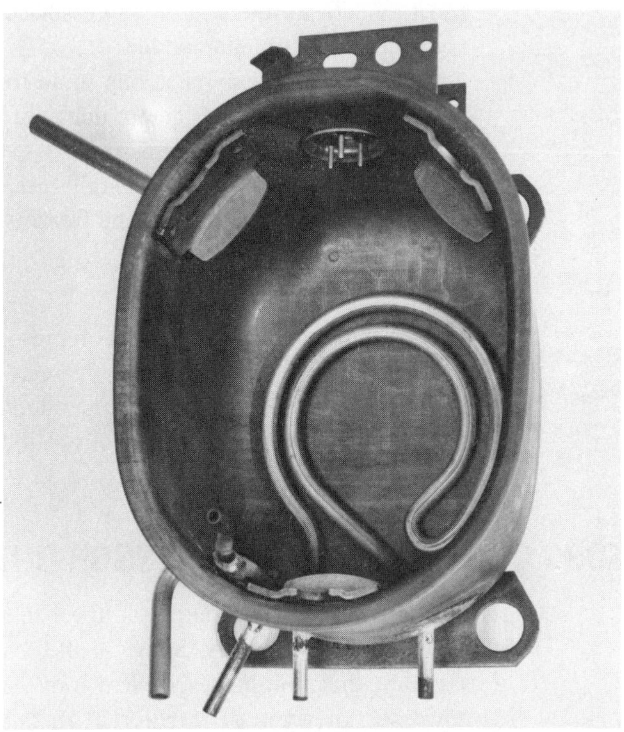

Figure 8–34 Oil cooler location (Courtesy of Tecumseh Products Company)

evaporator is directed over and through the compressor-motor windings and other components that tend to operate at higher than desired temperatures. On low-temperature applications, below 0°F, some form of additional compressor cooling is required because the refrigerant does not have sufficient heat-absorbing capacity to adequately cool the compressor. Usually this additional cooling is provided by forced-air fans provided specifically for this purpose, or condenser fans that are designed to include both condenser cooling and compressor cooling functions.

Oil-Cooled Compressors

Oil-cooled compressors are generally used in small, low-temperature applications such as domestic refrigerators and freezers. There are external connections to a coil located in the oil sump of the compressor. See Figure 8–34.

The coil is connected so that a part of the discharge gas passes through it after being cooled enough to cool the oil. The refrigerant vapor is then returned to the condenser where it is further cooled and liquefied.

SUMMARY 8–7

- Compressors require cooling to prolong their life and to help them operate as efficiently as possible.
- The air being discharged by the condenser fan should be directed at the motor-compressor.
- Water-cooled compressors have a water jacket that is either mounted directly on the compressor body or is made an integral part of it.

- Most modern refrigeration compressors use some form of refrigerant cooling to keep them from running too hot.
- On low-temperature applications some form of additional compressor cooling is required because the refrigerant does not have sufficient heat-absorbing capacity to adequately cool the compressor.
- Oil-cooled compressors are generally used in small, low-temperature applications such as domestic refrigerators and freezers.

REVIEW QUESTIONS 8–7

1. Name four methods of cooling compressors.
2. What is the best type fan to use on an air-cooled compressor?
3. What is required to cool low-temperature application compressors?
4. What type of compressor cooling do small, low-temperature compressors usually use?

8–8 FACTORS CONTROLLING COMPRESSOR OUTPUT

When the compressor output is lower than the designed output, the total system will not function properly. Some of the many factors that control the amount of refrigerant that a compressor can move are: the design, the condition of the compressor, the type of refrigerant in the system, and the conditions that the compressor is subjected to. The operating pressures have a great deal to do with compressor output. The following is a partial list of the results that pressures have on compressor output.

1. Compression ratio. Compression ratio is the ratio between the absolute discharge pressure and the absolute suction pressure.
2. The type of refrigerant used.
3. The volumetric efficiency. Volumetric efficiency is the ratio of the volume of gas actually drawn into the cylinder compared to the volume of the piston displacement. The piston displacement is the volume of space in a cylinder that is swept by the piston. It is calculated with the following formulas:

$$PD = \frac{\pi \times D^2 \times L}{4}$$

$$VE = \frac{\text{Actual volume}}{\text{Calculated volume}}$$

4. Cylinder cooling system. The volumetric efficiency will improve when more heat of compression is removed from the cylinder by a cylinder cooling system.
5. Cooling system efficiency. The more efficient the cooling system, the greater the volumetric efficiency, because the refrigerant temperature is lowered, reducing the re-expansion of the refrigerant. Thus, a larger amount of vapor can be drawn into the cylinder on each suction stroke.
6. Refrigerant pressure in the suction line. When the suction pressure is low, the volumetric efficiency of the compressor is also lowered because there is less refrigerant to be drawn into the cylinder on each suction stroke.
7. Compressor speed. When the compressor is operating at a high rate of speed, the compressor valves operate less efficiently than at lower speeds. Because of this there is less refrigerant drawn into the cylinder on the suction stroke, reducing the volumetric efficiency of the compressor.

8. Type and size of valves. The size of the valve ports, and the speed with which the valves operate, affect the amount of gas compressed by the compressor.
9. Refrigerant vapor friction. The amount of friction that the lines present to the refrigerant flowing through them reduces the amount of gas entering the cylinder. Also, the friction introduced by the openings into the compressor body has an affect on the flow of refrigerant into the compressor.
10. Compressor mechanical condition. When the rings, valves, and bearings are worn, the output of the compressor will be reduced accordingly.
11. Lubricants. The proper amount of lubricant flowing through the system aids in sealing the valves and the other compressor components, causing them to operate more efficiently while increasing the compressor output.

The compressor is considered by many to be the heart of the refrigeration system. The greatest difficulty lies in determining what causes it to malfunction. Generally the compressor fails due to the malfunction of another component. Because of this, the compressor is generally the first to get the blame for any malfunction without actually determining the reason for the failure. Many times the component that caused the malfunction is not found and corrected, resulting only in another compressor failure.

Any time a compressor or any other component fails in a system, the cause must be determined and corrected or there will usually be a repeated failure. To prevent such reoccurrences, the service technician must know the possible causes of failures and know how to correct them to prevent future failures.

SUMMARY 8–8

- When the compressor output is lower than the designed output, the total system will not function properly.
- Some of the many factors that control the amount of refrigerant that a compressor will move are: design, the condition of the compressor, the type of refrigerant in the system, and the conditions that the compressor is subjected to.
- The compressor is considered by many to be the heart of the system. Generally, the compressor fails due to the malfunction of another component.
- Any time a compressor or any other component fails in a system, the cause must be determined and corrected or there will usually be a repeated failure.
- To prevent such occurrences, the technician must know the possible causes of failures and how to correct them.

REVIEW QUESTIONS 8–8

1. What will bad valves in a compressor cause?
2. Why should the re-expansion of refrigerant in the clearance pocket be kept to a minimum?
3. Will a low lubricant level in the crankcase affect compressor capacity?
4. In most cases, why does a compressor fail?

CHAPTER

9

CONDENSERS AND RECEIVERS

OBJECTIVES

Upon completion of this chapter, you should be able to:

- Know the purpose of refrigeration system condensers
- Know how refrigeration system condensers operate
- Be familiar with the different types of condensers
- Know how to properly maintain refrigeration system condensers
- Understand condensing temperature
- Know why contaminants should not be allowed in a refrigeration system
- Know the purpose of the liquid receivers used in refrigeration systems

INTRODUCTION

Condensers are the components that change the state of the hot refrigerant vapor from the compressor to a warm liquid refrigerant ready for use in the evaporator. The receiver is simply a storage device that receives the liquid refrigerant from the condenser and stores it until it is needed by the evaporator.

9–1 PURPOSE OF THE CONDENSER

When the refrigerant is discharged from the compressor it is a hot, superheated vapor. It is ready to give up the heat that it collected in the evaporator, and that it gained during the compression process, to the medium used to cool the condenser.

The purpose of a condenser is to remove the heat from the hot refrigerant vapor and change it to a liquid. During this cooling process, the sensible heat must first be removed from the vapor by the condenser cooling medium, lowering its temperature until it has reached the condensing temperature. The latent heat of condensation is then removed from the vapor, changing it to a liquid (condenses). In later model condensing units, an extra row or rows are added to the condenser coil so that the liquid refrigerant can be cooled below the condensing temperature (subcooled). Subcooling is an essential part of any modern refrigeration condenser because it helps to increase the overall efficiency of the unit by reducing flash gas at the flow control device.

Condenser theory dictates that the amount of heat given up by the refrigerant in the condenser must always equal the amount of heat gained by the cooling medium. We can use a temperature-Btu chart to illustrate this. See Figure 9–1.

The horizontal line of the chart indicates the heat content of the refrigerant in Btu. The vertical line lists the temperature in degrees Fahrenheit. All the refrigerant to the right of line 1 is in the vapor state. All the refrigerant to the left of line 1 is in the liquid state. The area between lines 1 and 2 indicates that the refrigerant is in both the liquid and vapor state (a mixture). In this example, HCFC–22 at 100°F and 195.9 psig is used.

The refrigerant, when discharged from the compressor, enters the condenser at point A with a temperature of 120°F. At this point the refrigerant is superheated

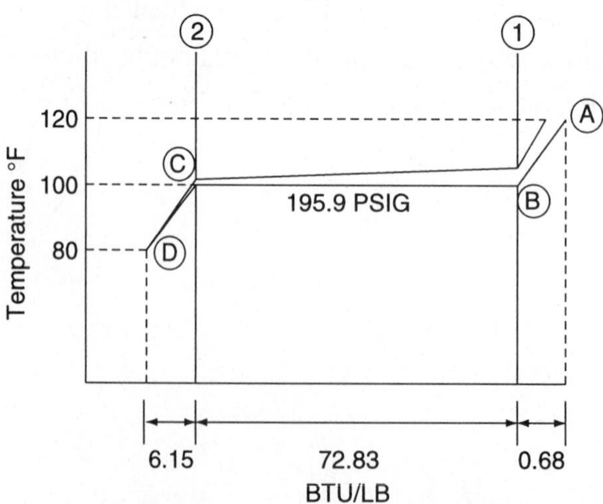

Figure 9–1 Heat content in Btu/lb

20°F. When the refrigerant vapor contacts the condenser tubes, heat is given up to the cooling medium. This first step is the removal of temperature only because the refrigerant is superheated. The temperature has been lowered from 120°F to 100°F and is represented by the line extending from point A to point B. The refrigerant vapor has now reached the saturation temperature corresponding to the given pressure. As more heat is removed, the vapor gradually changes to a liquid. At point C the refrigerant has essentially all been changed to a liquid. It should be noticed that between points B and C the temperature of the refrigerant has not changed. This is because only latent heat has been removed from the refrigerant. Since all of the refrigerant has now been changed to a liquid, any further cooling will result in subcooling the liquid. In an operating condenser, there will normally be some subcooling taking place. In this example, the liquid has been subcooled 20°F to point D. The refrigerant leaves the condenser at 80°F.

Notice that the amount of sensible heat removed from the refrigerant is very small when compared to the amount of latent heat removed. The total sensible heat removed from point A to point B is 0.68 Btu/lb and from point C to point D is 6.15 Btu/lb, totaling 6.83 Btu/lb sensible heat removed. The latent heat removed from point B to point C is 72.83 Btu/lb. Quite a difference. It can easily be seen that the majority of the cooling was done in changing the state of the refrigerant. This is an example of what occurs in the condenser and why we should be more interested in the amount of latent heat transferred than the sensible heat transferred.

SUMMARY 9–1

- The purpose of the condenser is to remove the heat from the hot discharge gas and change it to a liquid.
- Subcooling is an essential part of any modern refrigeration condenser because it helps to increase the overall efficiency of the unit by reducing the amount of flash gas at the flow control device.
- Condenser theory dictates that the amount of heat given up by the refrigerant in the condenser must always equal the amount of heat gained by the cooling medium.
- The amount of sensible heat removed in the condenser is very small when compared to the amount of latent heat removed.

REVIEW QUESTIONS 9–1

1. What is the purpose of the condenser in a refrigeration system?
2. What is subcooling?
3. What is the purpose of subcooling?
4. What is the amount of heat given up by the refrigerant in the condenser that must always equal the amount of heat gained by the cooling medium known as?
5. Is more sensible heat or more latent heat removed from the refrigerant in the condenser?

9–2 AIR-COOLED CONDENSERS

Air-cooled condensers are probably the most popular type of condenser used on small commercial, industrial, and residential refrigeration and air conditioning systems. The condensers are manufactured by attaching aluminum fins to the refrigerant tubing. The purpose of the fins is to increase the heat transfer charac-

Figure 9–2 Air-cooled condensing unit (Courtesy of Copeland Corp.)

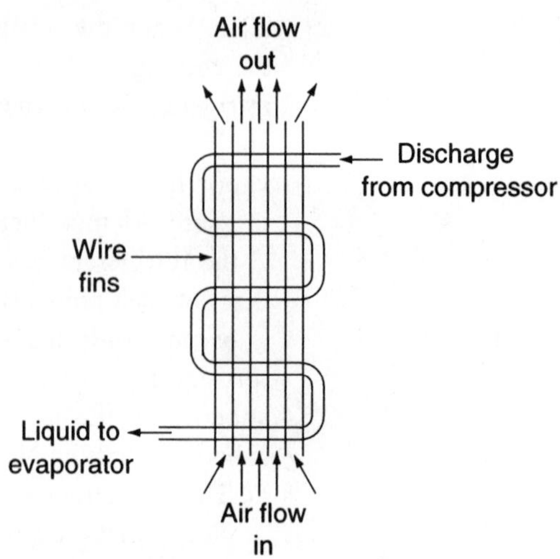

Figure 9–3 Natural draft condenser

teristics of the coil. The heat transfer is also increased by the use of a fan used to blow large quantities of air through the finned condenser coil. See Figure 9–2.

The major exception to this is in small residential refrigerator and freezer applications where natural draft condensers are very popular.

Natural-draft, or static-type, condensers are cooled by convection air movement over the surfaces. The air movement occurs when the air contacts the hot condenser coils and its temperature is increased. As the temperature is increased the air rises, drawing more cool air in at the bottom of the condenser to replace the warmer air that has risen. See Figure 9–3.

Basically, there are two types of natural-draft air-cooled condensers. They are the tube and fin type and the plate type. See Figures 9–4 and 9–5.

The plate type is manufactured by pressing two plates of metal together, leaving a path through them for the refrigerant to flow. They are then seam-welded together.

These condensers are restricted to smaller residential refrigeration applications, such as refrigerators and freezers, because of their limit capacity per unit size. However, they are inexpensive to manufacture, thus making them desirable for this very competitive market.

Air-cooled condenser capacity can greatly be increased by forcing large quantities of air through the coil. This is usually done by using fans to move the air. Both the

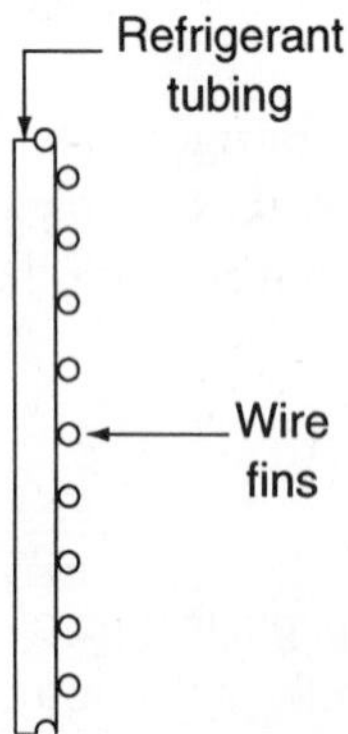

Figure 9–4 Tube and fin natural draft condenser

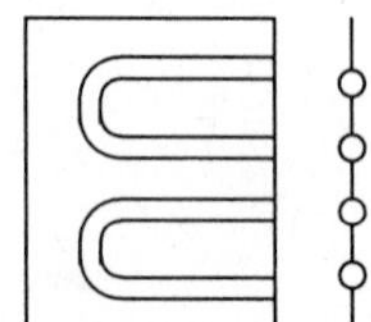

Figure 9–5 Plate type evaporator

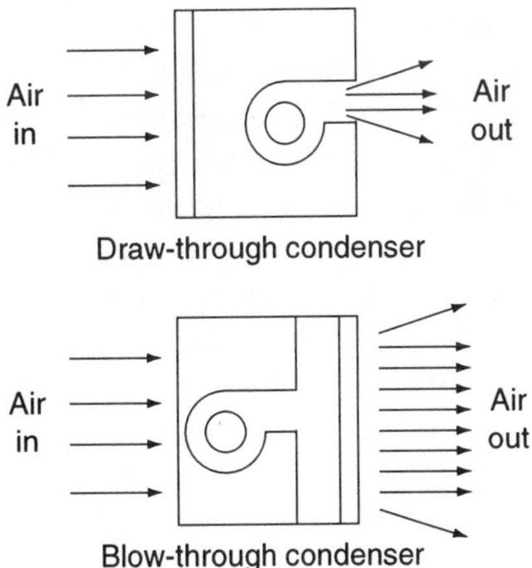

Figure 9–6 Draw-through and blow-through air-cooled condensers

propeller and the centrifugal type fans are used for this purpose. The major contributing factors when using forced draft air-cooled condensers are the available space and the economics of manufacture and operation.

These types of condensers are easy to install and maintain. They do, however, require a large quantity of fresh air. In larger installations, the fan may create a noise problem that must be dealt with. When the condenser is properly designed it will operate satisfactorily in almost any region, even the very hot ones.

Fans

There are two methods of moving the air through air-cooled condensers. They are the draw-through and the blow-through types. See Figure 9–6.

Draw-Through. In this type of coil, the fan is located so that it will pull the air through the coil and discharge it out of the unit. The draw-through is probably the most efficient method, because the air is naturally distributed over the entire surface of the coil. This occurs because the negative pressure created by the fan covers the complete coil face, causing the air to seek the path of least resistance and flow evenly over the face.

Blow-Through. The blow-through type coil uses a fan that blows the air directly on the coil face. The greatest quantity of air tends to pass through the coil at the point where the air strikes the surface. This type is not as preferred as the draw-through type because the efficiency is a little less. In most installations this is not noticeable. However, when energy use is monitored it is quite noticeable. Most manufacturers use various means of reducing this problem with varying degrees of success. It is not advisable to attempt to make any changes in the unit design to compensate for this loss. The manufacturer has probably already made the design as efficient as possible.

Maintenance

The major maintenance problem with air-cooled condensers is to keep the coil surfaces clean. This can be accomplished fairly easily by using a garden hose with a pressure spray nozzle attached. Force the water through the coil in the opposite direction to the air flow, if possible. Make certain to protect the fan motor and other electrical components from the water spray. If the coil is exceptionally dirty, there are cleaning solvents available that will not harm the aluminum fins or the atmosphere.

It is relatively simple to determine when the coil needs cleaning. Just simply look through the fins. It may be desirable to hold a light on the opposite side of the coil while looking through to see if it is plugged. If this is not possible check the discharge pressure. If the coil is dirty, it will be higher than normal. The discharge pressure should correspond to a refrigerant saturation temperature of approximately 25°F to 35°F higher than the ambient air temperature entering the condenser coil.

Should the discharge pressure be high, be sure to check the fan rotation, the fan belt, and the fan bearings to make certain that there is no problem with the air flow.

SUMMARY 9–2

- Air-cooled condensers are probably the most popular type of condenser used on small commercial, industrial, and residential refrigeration and air conditioning systems.
- Natural-draft, or static-type, condensers are cooled by convection air movement over the surfaces.
- Air-cooled condenser capacity can be greatly increased by forcing large quantities of air over the coil.
- The draw-through type of condenser is probably the most efficient method because the air is naturally distributed over the entire surface of the coil.
- The efficiency of the blow-through condenser is not as high as the draw-through type.
- The major maintenance problem with air-cooled condensers is to keep the surfaces clean.
- Should the discharge pressure be high on an air-cooled condenser application, be sure to check the fan rotation, the fan belt, and the fan bearings to make certain that there are no problems with the air flow.

REVIEW QUESTIONS 9–2

1. How can air-cooled condenser capacity be greatly increased?
2. Name two methods of air flow through an air-cooled condenser.
3. What is the major maintenance problem with air-cooled condensers?
4. What may be required to determine if an air-cooled condenser is dirty?

9–3 WATER-COOLED CONDENSERS

When there is an adequate supply of water available, water-cooled condensers are sometimes preferred because they operate with a lower condensing pressure and temperature. When the water is supplied from a well, it usually has a lower temperature than the ambient air temperatures used for air-cooled condensers. A cooling tower can be used when a sufficient quantity of water is not available, or it is too expensive to use for wastewater condenser cooling. The water can be cooled in a tower almost down to the ambient wet-bulb temperature. When a cooling tower is used, the water can be circulated continuously while keeping the water consumption to a minimum.

In order for any condenser to function as it was designed, it must have the required amount of cooling medium passing through it.

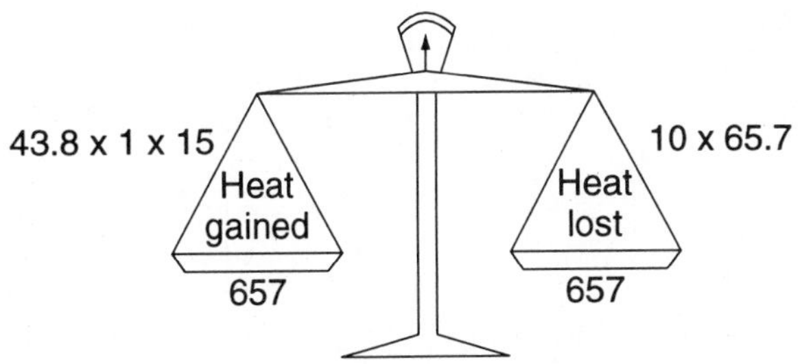

Figure 9–7 Balance of heat gained and lost in a water-cooled condenser

Example: We have 10 pounds of refrigerant circulating through a system using a water-cooled condenser. The condenser requires 43.8 pounds of water flowing through it, with a 15°F temperature rise, to remove the amount of heat gained during the evaporating process. See Figure 9–7.

From the figure we can determine that each pound of refrigerant in the system gives up 65.7 Btu to the water flowing through the condenser. In our example, there were 10 pounds of refrigerant flowing through the system. Therefore, the total amount of heat removed by the condenser is $10 \times 65.7 = 657$ Btu. We can also calculate how much heat was gained by the condenser during this process. We were given that 43.8 pounds of water passed through the condenser with a 15°F temperature rise. Thus, the amount of heat gained by the water is: 43.8 pounds of water × Specific heat of water × 15°F temperature rise. Then $43.8 \times 1 \times 15 = 657$ Btu, exactly the same amount that the refrigerant lost during the condensing process. We can also surmise that exactly the same amount of heat is given up by the water in the cooling tower by evaporation and radiation. This concept is valid for all types of condensers, both air-cooled and water-cooled.

Water-cooled condensers usually require less room than air-cooled condensers because of their excellent heat transfer characteristics. There are several types of water-cooled condensers on the market. They are generally divided into classes as follows: (1) shell and tube (condenser-receivers), (2) tube in tube, and (3) evaporative condenser. See Figure 9–8.

Shell-and-Tube Condensers

The shell-and-tube type condensers look very much like a liquid receiver tank. They may be made in either a horizontal or a vertical configuration. The horizontal

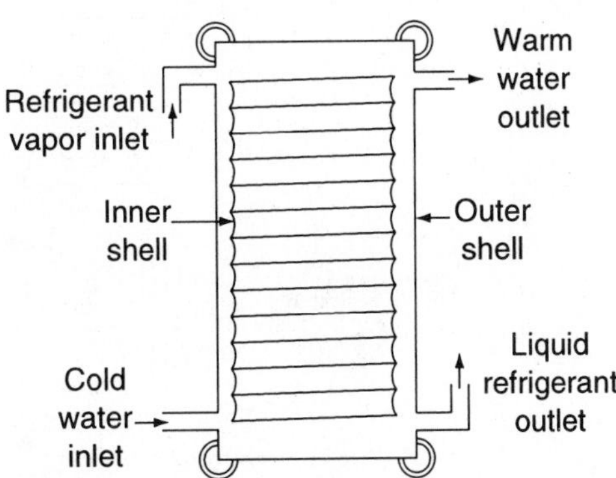

Figure 9–8 Water-cooled tube-in-shell condenser

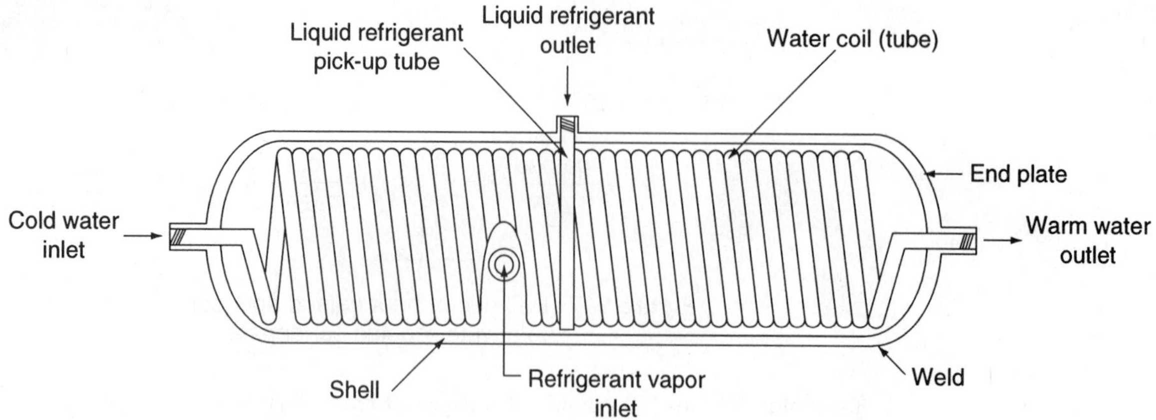

Figure 9–9 Shell and tube condenser

type is the most popular because it may be installed either under the compressor base or it may be installed in some remote location. Vertical type condensers take up space on the compressor base and therefore require more floor space than the horizontal type. In operation, a part of their function is to accumulate and store the liquid refrigerant as it comes from the condenser. See Figure 9–9.

They consist of a cylindrical tank equipped with a refrigerant inlet connection and a refrigerant outlet connection, a water coil that is also equipped with the necessary connections. The outer tank is made from steel. The end caps may either be welded on or bolted on, depending on the type of service that may be required. The refrigerant outlet tube is extended almost to the bottom of the tank so that it can pick up the liquid refrigerant when necessary without the tank being completely full. The water coil is equipped with fins on the tubing to increase the efficiency and reduce the size of the coil needed per unit of refrigeration capacity.

The water supplied to the condenser enters the coil from the water inlet connection, flows through the coil, and leaves through the outlet connection. Generally this water flow is in the opposite direction to the flow of refrigerant. To reduce waste and help keep the cost of operation to a minimum, the water leaves the coil by the outlet connection and is returned to a cooling tower where it is cooled and sent through the water circuit over and over again. See Figure 9–10.

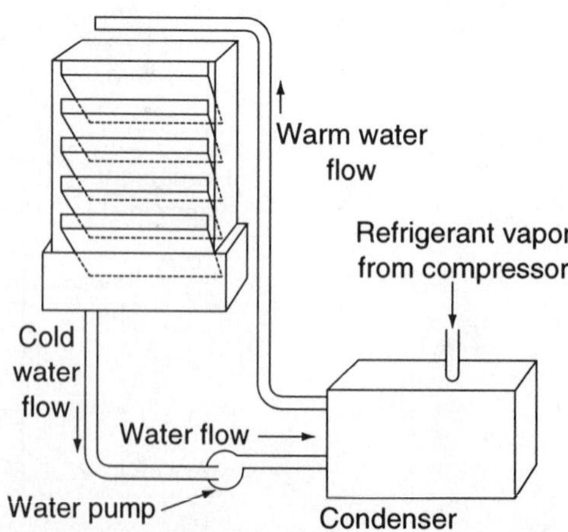

Figure 9–10 Cooling tower water circuit

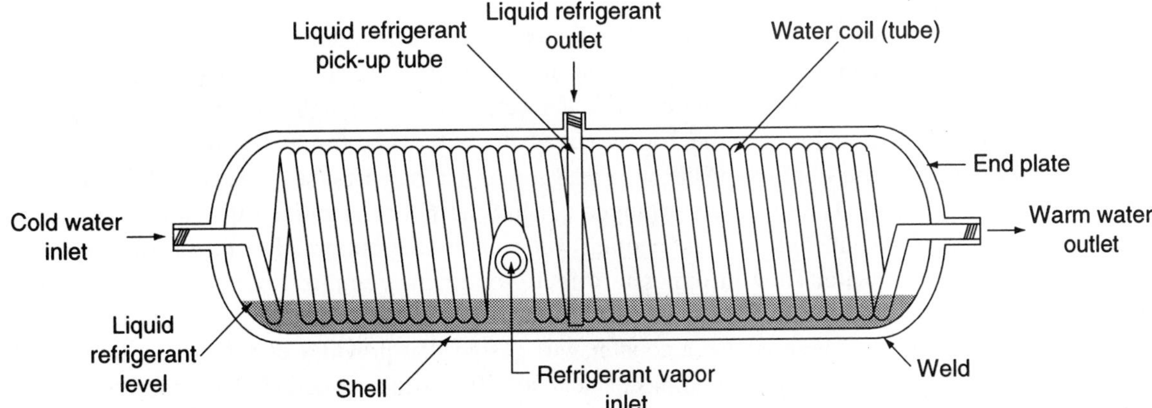

Figure 9–11 Liquid refrigerant collected in the bottom of shell and tube condenser

During operation, the hot, vaporous refrigerant leaves the compressor and enters the condenser inlet connection. It then goes into the condenser shell. At this point it comes into contact with the water coil, which is much cooler than the refrigerant vapor, and heat is given up to the coil and the water inside it. The heat is then carried away by the water and dissipated to the atmosphere in the cooling tower. Some of the heat is also conducted through the condenser shell walls to the surrounding atmosphere inside. When enough heat has been removed, the vapor begins to change to a liquid and settle to the bottom of the tank where it is stored until needed by the flow control device. See Figure 9–11.

The liquid refrigerant enters the outlet tube, which extends down into the bottom of the tank. It then flows into the liquid line, which directs the refrigerant to the flow control device and the evaporator.

Tube-in-Tube Condensers

Tube-in-tube water-cooled condensers are made with one tube inside another. See Figure 9–12.

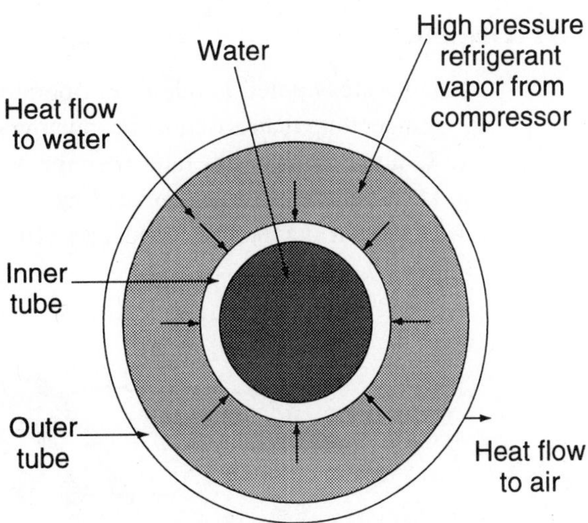

Figure 9–12 Cross section of a tube-in-tube condenser

Through special fittings on the ends of the tubing, the water is directed through the inside tube, and the refrigerant vapor is directed through the space between the inner and the outer tube. As the arrows in Figure 9–12 show, the heat is transferred through the wall of the inner tube to the water flowing through it and a small amount of heat is transferred through the wall of the outer tube to the atmosphere. This small heat transfer through the outer wall somewhat increases the capacity of the condenser.

The tubes are arranged so that the hot refrigerant vapor enters the top of the condenser and the cooling water enters the bottom. The refrigerant vapor and the water flow in opposite directions, or in a counterflow pattern, to increase the efficiency.

There is also a counterswirl design that provides excellent heat transfer characteristics and good mechanical stability of the condenser. The tubes are designed so that both the water and the refrigerant vapor are swirling inside their respective tubes. This greatly increases the contact of both the refrigerant and the water with the surface of each tube, which in turn increases the efficiency of the condenser. See Figure 9–13.

The swirling action of the water helps to keep scale and other debris from collecting on the surfaces of the tube, reducing the amount of required maintenance.

Tube-in-tube condensers are also used on cooling towers to reduce the amount of water required for effective and efficient operation.

SUMMARY 9–3

- When there is an adequate supply of water available, water-cooled condensers are sometimes preferred because they operate with a lower condensing pressure and temperature.
- When a cooling tower is used, the water can be recirculated continuously while keeping the water consumption to a minimum.
- Shell-in-tube water-cooled condensers look very much like a liquid receiver tank.
- Generally, the water flow is in the opposite direction to the refrigerant flow.
- Tube-in-tube water-cooled condensers are made with one tube inside the other.
- Through special fittings on the ends of the tubes, the water is directed through the inside tube and the refrigerant vapor is directed through the space between the inner and outer tube.

REVIEW QUESTIONS 9–3

1. Do water-cooled condensers operate with a lower condensing pressure and temperature than air-cooled condensers?
2. To almost what temperature can the water be cooled in a cooling tower?
3. Does counter-flow or regular-flow provide the best heat transfer?
4. Does swirling increase condenser efficiency?

9–13 Counterswirl design principle (Courtesy of Packless Industries)

9–4 EVAPORATIVE CONDENSERS

Evaporative condensers are popular where lower condensing temperatures and pressures than those possible with straight air-cooled condensers are required and where water usage must be kept to a minimum. Their main uses are in applications where it is desirable to have the cooling unit running during periods of low ambient temperatures, especially when the temperatures are below freezing. In this case the water can be drained from the tower and the unit operated as an air-cooled unit.

The coil in an evaporative condenser can be operated as an air-cooled condenser and/or a water-cooled condenser. See Figure 9–14.

There is a fan located on top of the tower that forces air through the coil. The water spray helps cool the refrigerant as well as cool the tower circulating water. The air flow is directed into the bottom of the tower, through the water spray, and into the condensing coil, then it is discharged out the top of the tower to the atmosphere. Through the evaporation process of the evaporative condenser, the circulating water temperature can be lowered almost to the wet-bulb temperature of the ambient air. By removing the latent heat, the sensible heat is also removed. The temperature of the water remaining in the tower sump is also lowered.

In practice, the refrigerant is piped from the compressor discharge to the refrigerant coil located in the evaporative condenser. The condensing coil is located in a spray chamber, where the refrigerant is cooled and condensed by both the water spray and the air flowing over and through the coil fins. The fins are spaced extra wide so that the formation of scale will not rapidly plug up the space between them. The water is stored in a sump in the bottom of the evaporative condenser until it is needed. A water pump moves the water from the sump up through the pipes to the spray chamber, where it is broken into a very fine spray. The spray is directed over the coil where it absorbs the latent heat of vaporization to cool the coil. The water cooled by the evaporation process then falls down into the sump where it is ready to be pumped again. There is only a very small percentage of the water evaporated.

SUMMARY 9–4

- Evaporative condensers are popular where lower condensing temperatures and pressures than those possible with straight air-cooled condensers are required and where water usage must be kept to a minimum.

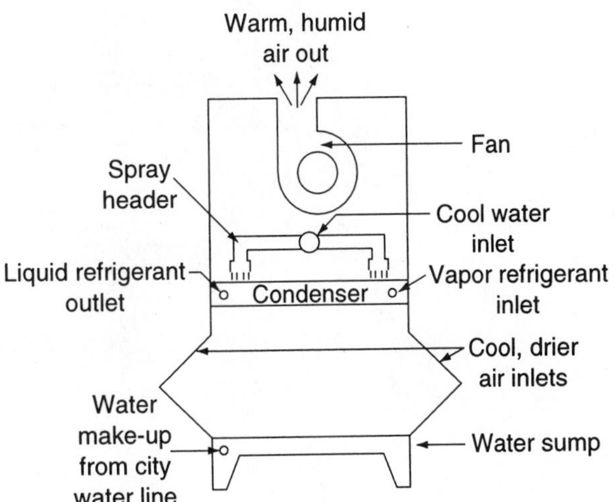

Figure 9–14 Evaporative condenser

- Evaporative condensers are mainly used in applications where it is desirable to have the cooling unit running during periods of low ambient temperatures, especially when the temperatures are below freezing.
- In an evaporative condenser, there is only a very small percentage of the water evaporated.

REVIEW QUESTIONS 9–4

1. What is the most popular application of evaporative condensers?
2. How can the coil in an evaporative condenser be operated?
3. Why are the fins on the coil of an evaporative condenser spaced wide?

9–5 COUNTERFLOW OF COOLING WATER

Water-cooled condensers are designed to cool the refrigerant to approximately the same temperature as the cooling water. This is possible only by using the counterflow principle. See Figure 9–15.

In the counterflow principle, the cooling water enters the condenser at one end and the refrigerant vapor enters at the other. Thus, the hottest refrigerant contacts the

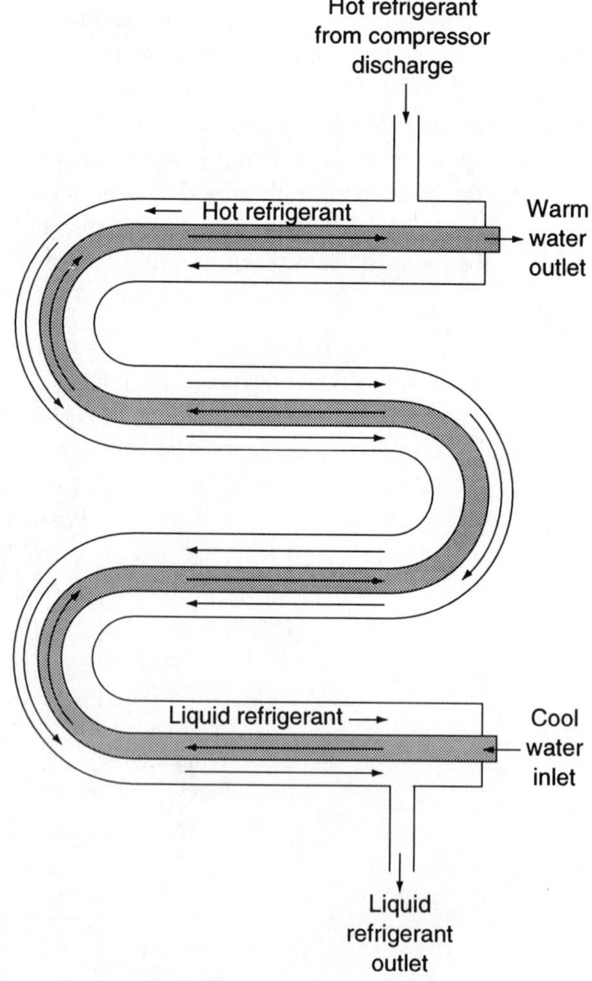

Figure 9–15 Counterflow principle

warmest water first and the coolest water last. Thus, the liquid refrigerant leaving the condenser is cooled to approximately the same temperature as the cool entering water. If this were reversed, the refrigerant and water would flow in the same direction and the hottest refrigerant would contact the coolest water first and remain in contact with the same water completely through the condenser. The liquid refrigerant would only be cooled to approximately the same temperature as the leaving water, resulting in a minimum amount of heat transferred from the refrigerant to the water.

SUMMARY 9–5

- Water-cooled condensers are designed to cool the refrigerant to approximately the same temperature as the cooling water. In the counterflow principle, the cooling water enters the condenser at one end and the refrigerant vapor enters at the other.
- The liquid refrigerant leaving the condenser is cooled to approximately the same temperature as the cool entering water.

REVIEW QUESTIONS 9–5

1. To what temperature are water-cooled condensers designed to cool the refrigerant?
2. To what temperature is the liquid refrigerant leaving a water-cooled condenser cooled?

9–6 CONDENSER CAPACITY

Several factors that determine the heat transfer capacity of a condenser are:

1. Condenser surface area.
2. The contact between the refrigerant and the internal surface of the condenser tube.
3. The difference in temperature between the cooling medium and the refrigerant.
4. The velocity of the refrigerant passing through the tubes. Generally, the greater the velocity, the greater the capacity.
5. The flow rate of the cooling medium used to cool the condenser. The amount of heat transferred increases with the velocity of the cooling medium. The capacity of air-cooled condensers also increases as the density of air increases.
6. The material from which the condenser coil is made.
7. The cleanliness of the heat transfer surfaces. Dirt, scale, or corrosion can reduce the heat transfer capabilities of the condenser.
8. The speed with which the condensed refrigerant is removed from the condenser and is replaced by uncooled refrigerant vapor.

When these factors are considered, the primary variable is the temperature difference between the refrigerant vapor and the condensing medium. Other variables are the cleanliness of the heat transfer surfaces and the amount of the cooling medium passing over, or through, the condenser. These are the factors with which the service technician is most concerned. He must be able to determine if the surfaces are clean and if the desired quantity of cooling medium is supplied to the condenser. If not, what can be done to bring these factors in line with what is required.

SUMMARY 9–6

- There are several factors that determine the heat transfer capacity of an air-cooled condenser. When these factors are considered, the primary variable is the temperature difference between the refrigerant vapor and the condensing medium.
- Other variables are the cleanliness of the heat transfer surfaces and the amount of cooling medium passing over, or through, the condenser.
- The service technician must be able to determine and be able to correct any of these problems.

REVIEW QUESTIONS 9–6

1. When an air-cooled condenser is used, why will the discharge pressure be lower during a rain storm?
2. How will a refrigerant overcharge reduce condenser capacity?
3. What two things must a service technician keep in mind when servicing a condenser?

9–7 CONDENSING TEMPERATURE

The condensing temperature is the temperature at which the refrigerant vapor changes to a liquid in the condenser. This temperature is not the same as the temperature of the cooling medium. The condensing temperature must always be higher than the cooling medium for heat to be transferred from the refrigerant.

For the refrigerant to condense in the condenser, heat must be removed from the condenser at the same rate the refrigerant vapor brings it in. The only way to increase the capacity of a condenser under a given set of conditions is to increase the temperature difference between the refrigerant and the cooling medium.

Because the compressor is designed to operate under pressure, the refrigerant pressure inside the condenser will continue to rise until the temperature difference between the cooling medium and the refrigerant is great enough to allow the necessary amount of heat to be transferred. The larger the condenser, the smaller the temperature difference must be. If the amount of cooling medium for a small condenser is also small, the temperature difference between the medium and the refrigerant must be larger. When something happens to upset this balance, the result will be a higher than desirable discharge pressure that may cause problems with the system. Thus, it is necessary that the condenser be operating as it was designed.

The condensing temperature and the resulting condensing pressure are determined by several factors, such as: the capacity of the condenser, the temperature of the cooling medium, and the heat content of the refrigerant discharged from the compressor. The heat content of the discharged refrigerant is dependent on its volume, density, and temperature.

Condensing Temperature Difference

A condenser is normally sized for a system with the capacity to carry the load at the desired operating temperature difference between the discharged refrigerant and the cooling medium. Normally, air-cooled condensers are selected to operate with temperature differences of between 20°F and 30°F at design conditions. There are both higher and lower temperature differences available for use on special applications. Most manufacturers use air-cooled condensers that are satisfactory for use over a wide range of applications. The condensing temperature difference at high suction pressures may range from 30°F to 40°F. At low evaporating temperatures, the temperature difference is usually not more than 4°F to 10°F. For water-cooled units, the design temperature difference is generally determined by the temperature of the condensing water and the flow rate. The condensing temperature difference for these applications may be as low as 90°F or as high as 120°F.

Condensers must have a capacity higher than the evaporator capacity. It must be equal to the evaporator capacity plus the heat of compression and the efficiency loss of the compressor-motor. Most condensers are rated according to the evaporator capacity. However, the manufacturer may provide factors that allow for these factors, in addition to the evaporator capacity, to aid in making the proper condenser selection.

SUMMARY 9–7

- The condensing temperature difference is the temperature at which the refrigerant vapor changes to a liquid in the condenser.
- This temperature is not the same as the temperature of the cooling medium.
- For the refrigerant to condense in the condenser, heat must be removed at the same rate the refrigerant vapor brings it in.
- Because the compressor is designed to operate under a pressure, the refrigerant pressure inside the condenser will continue to rise until the temperature difference between the cooling medium and the refrigerant is great enough to allow the necessary heat to be transferred.
- A condenser is normally sized for a system with the capacity to carry the load at the desired operating temperature difference between the discharged refrigerant and the cooling medium.
- Condensers must have a capacity greater than the evaporator capacity.

REVIEW QUESTIONS 9–7

1. What is the condensing temperature of a refrigerant?
2. What must occur in the condenser of a refrigeration unit?
3. What happens to the condenser capacity when the temperature difference between the refrigerant and the cooling medium is increased?
4. Upon what is the heat content of discharge refrigerant dependent?
5. With what temperature difference are air-cooled condensers selected to operate?
6. What is the condensing temperature range of most water-cooled condensers?
7. To what capacity must the condenser be equal?

9–8 NONCONDENSABLE GASES

Both of the elements contained in air, nitrogen, and oxygen are a gas at all temperatures encountered in refrigeration and air conditioning work. These gases may be liquefied under extremely high pressures accompanied by extremely low temperatures. They are considered to be noncondensables when inside a refrigeration system.

Dalton's Law, one of the basic laws of nature, is the fact that gases, when in combination, will act as though only one gas is present. Each will act completely independent of all the other gases in the combination. The total pressure inside a system is the sum total pressures of all the gases present. Charles' Law states that any gas within an enclosed space has the characteristic that if the space remains constant so that the gas cannot expand, its pressure will vary directly with the temperature of the gas.

It can then be seen that, if air is sealed in a system with refrigerant, the nitrogen and oxygen will each exert their own pressures in the system in addition to that exerted by the refrigerant. The pressure exerted will increase with an increase in temperature.

Air, because it is a noncondensable, will remain in the gaseous form and will collect in the top of the condenser and the receiver tank. When the system is operating, the discharge pressure will be a total of the refrigerant pressure and the pressures exerted by both the nitrogen and the oxygen. The amount of pressure increase will be determined, to a great extent, on the amount of air trapped inside the system. It can easily exceed pressures greater than 50 psig. It should be suspected any time that a system has a high discharge pressure that there are noncondensables present. The noncondensables must be removed before the system can function as it was designed.

SUMMARY 9–8

- Both of the elements contained in air, nitrogen, and oxygen are a gas at all temperatures encountered in refrigeration and air conditioning work.
- Dalton's Law, one of the basic laws of nature, is the fact that gases, when in combination, will act as though only one gas is present.
- It can be seen that if air is sealed in a refrigeration system with refrigerant, the nitrogen and oxygen will each exert their own pressures in the system in addition to that exerted by the refrigerant.
- Air, because it is a noncondensable, will remain in the gaseous form and will collect in the top of the condenser and the receiver tank.
- It should be suspected that any time a system has a high discharge pressure there are noncondensables present.

REVIEW QUESTIONS 9–8

1. In a refrigeration system, how will a combination of gases act?
2. When noncondensables are in a refrigeration system, what will determine the resultant discharge pressure?
3. What should be suspected any time a system is operating with a higher than normal discharge pressure?

9–9 CLEANING THE CONDENSER

The purpose of the condenser is to remove the heat from the refrigerant. Therefore, the heat transfer surfaces must be kept clean. Any accumulation of dirt, scale, or lint will effectively act as an insulator and prevent the necessary heat transfer. A lack of heat transfer will cause an increase in the discharge pressure. The surfaces of the condenser must be cleaned periodically to remove any debris that may have accumulated since it was last cleaned. On air-cooled condensers, this is relatively easy to do: just reverse-wash the condenser with a garden hose and a high pressure nozzle. Severe accumulations may require that a cleaning solvent may also be used. Water-cooled condensers present a different problem. An accumulation of scale on the inside surfaces of the condenser tubes is much more difficult to find and remove. The type of mineral present will cause a specific type of scale that is generally removed with some type of liquid cleaning agent. This is normally called descaling, or acidizing, the unit. The type of cleaning agent used will be determined by the type of scale. There are some general type scale removers on the market that usually do a respectable job of removing the scale. However, caution must be used to prevent damage to the system piping. In some cases, it is desirable to consult with a local chemical company to determine the type of scale remover best to use.

SUMMARY 9–9

- The heat transfer surfaces of a condenser must be kept clean.
- A lack of heat transfer will cause an increase in the discharge pressure.
- On water-cooled condensers, an accumulation of scale on the inside surfaces of the condenser tubes is difficult to find and remove.

REVIEW QUESTIONS 9–9

1. What maintenance procedure must be done to a condenser to help it do its job?
2. What is the normal procedure for cleaning a scaled-up water-cooled condenser?

9–10 CONDENSER LOCATION

Air-cooled condensers require a high volume of free-flowing air to cool and liquefy the refrigerant. Because of this requirement, where the condenser is located is very important. The air-cooled condenser should be located where there will be little out-of-door activity, but still where there is plenty of free air. If there is not enough air, the discharge pressure will be higher than normal, the system capacity will be reduced, and the cost of operation will increase.

When locating water-cooled condensers, be sure to choose a place where the ambient temperature will not drop below the freezing point while water is in the unit. If a water-cooled condenser should freeze, permanent damage to the coil could result. If the water tubes should freeze and burst, water will enter the refrigeration system and contaminate all the components. When this happens, a massive cleanup is required to remove all of the water and moisture from the refrigerant circuit. Most of the time all of the moisture cannot be removed without very expensive procedures and repairs.

If possible, a water-cooled condenser should be located where the ambient temperature never drops below 60°F. When the system is to be shut down for the winter months, the cooling water must be drained from the condenser, and the connecting tubing and all of the devices cleared of all water to protect them from freeze damage.

When horizontal water-cooled condensers are used, the tendency to only disconnect the water lines must be avoided. This procedure simply will not allow enough of the trapped water to drain from the system to prevent freeze damage. To properly drain a horizontal water-cooled condenser requires that the connecting lines be disconnected and the condenser, lines, and coils be cleared of all water by blowing through them with compressed air or some other form of pressure that will push the water out of the system. In some installations, it will be desirable to completely remove the condenser head to make certain that all of the water has been removed. When a vertical type water-cooled condenser is used, it can be satisfactorily drained by removing the drain plug in the bottom of the condenser shell or remove one of the connecting water lines. It is always wise to make certain that all of the water has drained from the condenser regardless of which type is used. To clean up a system that has had a frozen condenser is very expensive and time consuming.

SUMMARY 9–10

- Air-cooled condensers require a large volume of free-flowing air to cool and liquefy the refrigerant.
- The air-cooled condenser should be located where there will be little outdoor activity.
- When locating water-cooled condensers, be sure to choose a place where the ambient temperature will not drop below the freezing point while water is in the unit.
- In some installations, it will be a good idea to completely remove the condenser head to make certain that all of the water has been removed.

REVIEW QUESTIONS 9–10

1. Where should an air-cooled condenser be located?
2. Where should water-cooled condensers be located?
3. What must be done to water-cooled condensers during freezing temperatures?

9–11 WATER-FLOW CONTROL VALVES

A water-control valve is used on water-cooled condensers to reduce the amount of water used and to help control the compressor discharge pressure. These valves control the flow of water automatically by sensing the compressor discharge pressure. On some of the larger systems, the discharge pressure is controlled by cycling the tower fan.

These vales are generally of the bellows-type pressure operated valve. In this valve, the bellows are connected to the compressor discharge. When installed in this manner, the valve senses the discharge or high side pressure and responds accordingly. When the discharge pressure increases, the valve bellows senses the increase and gradually opens the valve seat to allow more water to flow through the condenser and lower the discharge pressure. When the discharge pressure drops, the valve bellows senses the drop and allows the valve to close off to reduce the water flow and increase the discharge pressure. In operation, the valve is never completely closed. It modulates somewhere around the half-open position to maintain the desired discharge pressure.

The water-flow control valve is installed in one of the water lines to the condenser. Some manufacturers recommend that the valve be installed in the water-inlet line, and others recommend that it be installed in the water-outlet line. Follow the manufacturer's recommendations. See Figure 9–16.

The valve must always be installed with the arrow on the valve indicating the correct direction of water flow. If it is installed in the water-inlet line the arrow

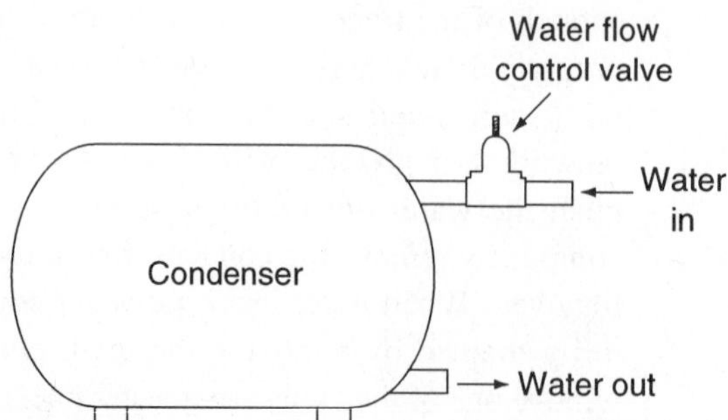

Figure 9–16 Water flow control valve in water inlet line

should point toward the condenser. If it is installed in the water-outlet line the arrow should point toward the tower, or away from the condenser.

SUMMARY 9–11

- A water valve is used on water-cooled condensers to reduce the amount of water used and to help control the compressor discharge pressure.
- When the discharge pressure increases, the valve bellows senses the increase and gradually opens the valve seat to allow more water to flow through the condenser and lower the discharge pressure.
- The water-flow control valve is installed in one of the water lines to the condenser.
- The valve must always be installed with the arrow on the valve indicating the correct direction of water flow.

REVIEW QUESTIONS 9–11

1. What is used on water-cooled condensers to help control the discharge pressure?
2. When controlling the discharge pressure on a water-cooled unit, where is the pressure sensed?
3. How must a water control valve be installed?

9–12 LIQUID RECEIVERS

Liquid receivers are tanks that are installed at the outlet of the condenser so that liquid will readily flow into them. Some systems have sufficient room in the condenser to contain the refrigerant and others do not. Those that do not need either a receiver or some other means of storing the refrigerant when repairs to the high side of the system are needed. In most instances, when normal system operation requires 8 pounds or more of refrigerant, a liquid receiver is used. They are sized by the total refrigerant charge required for proper system operation. The purpose of the receiver is to store refrigerant until the system needs it, or to store the complete charge when repairs are to be made to the high side of the system.

The outlet connection of the condenser is connected to the inlet connection of the receiver tank. See Figure 9–17.

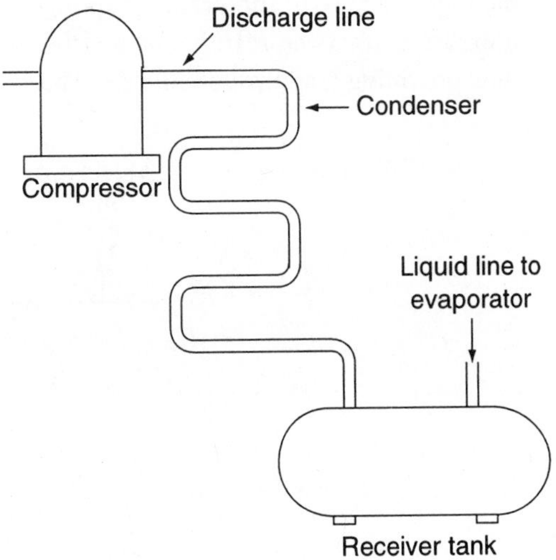

Figure 9–17 Location of liquid receiver tank

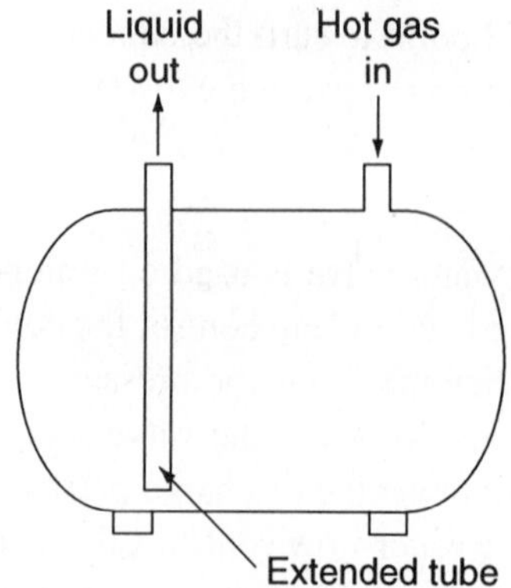

Figure 9–18 Receiver pickup tube

The outlet of the receiver is connected either to the liquid line or to a liquid line service (king) valve. The outlet connection has a tube that reaches almost to the bottom of the tank so that the liquid refrigerant can enter the liquid line when the receiver is not full. See Figure 9–18.

This is so that as small a charge as possible can be used and still allow liquid refrigerant to enter the liquid line. If the refrigerant charge drops to the point that the end of the pick-up tube is not covered by liquid refrigerant, gaseous refrigerant will enter the liquid line and cause poor operating conditions. Also, if the system has an overcharge of refrigerant, the liquid may fill the receiver and not allow room above it for gaseous refrigerant. In this case the compressor will have a higher than normal discharge pressure and excessive operating costs.

Liquid Receiver Safety Devices

The purpose of liquid receiver safety devices is to prevent the receiver shell from rupturing in case of excessive pressures. The safeties are generally piped outdoors so that the refrigerant will be purged from the building in case of fire or some other condition causing excessive receiver pressures. The safety device may be in the form of a fusible plug or a spring-loaded relief valve. The safety device is generally located in the place on the receiver that is most likely to rupture. See Figure 9–19.

The fusible plug is designed to melt and let the refrigerant out of the receiver when the pressure and temperature reach a predetermined point, rather than allowing the receiver tank to rupture. The spring-loaded relief valve is set to a predetermined pressure so that the refrigerant will be allowed to escape before the tank ruptures, thus providing a controlled release. The metal used in the fusible plug must never be

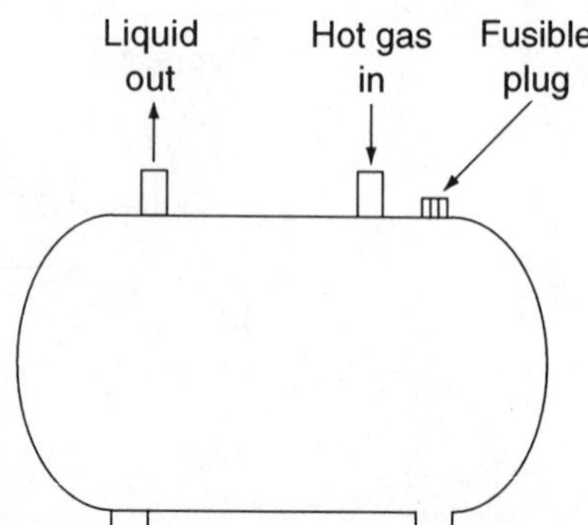

Figure 9–19 Location of fusible plug

replaced with regular soft solder because the melting temperature may be too high and allow the tank to rupture, or it may be too low and allow the refrigerant to escape without reason. The spring-loaded device must never be adjusted to a pressure different than that set by the manufacturer. To do so may cause the same problems encountered with replacing the melting type metal in the fusible plug devices. Never subject the receiver tank or a refrigerant cylinder to temperatures above 125°F. Domestic refrigerators and freezers hold a relatively small charge of refrigerant; therefore, they are not required to have safety devices.

When charging a system that is equipped with a receiver tank, the system is charged until the receiver tank is approximately one-third full of liquid. This is to make certain that the end of the pick-up tube is always immersed in liquid refrigerant and that there will still be enough room to store the complete system charge in the receiver tank during repairs.

SUMMARY 9–12

- Liquid receivers are tanks that are installed at the outlet of the condenser so that liquid will readily flow into them.
- Liquid receivers are sized by the total refrigerant charge required for proper operation of the system.
- The purpose of the receiver is to store refrigerant until the system needs it, or to store the complete charge when repairs are to be made to the high side of the system.
- If the refrigerant charge drops to the point that the end of the pick-up tube is not covered by liquid refrigerant, gaseous refrigerant will enter the liquid line and cause poor operating conditions.
- The purpose of liquid receiver safety devices is to prevent the receiver shell from rupturing in case of excessive pressures.
- The safety device is generally located in a place that is most likely to rupture.
- The fusible plug is designed to melt and let the refrigerant out of the receiver when the pressure and temperature reach a predetermined point.
- The spring-loaded relief valve is set to a predetermined pressure so that the refrigerant will be allowed to escape before the tank ruptures.
- When charging a system that is equipped with a receiver tank, the system is charged until the receiver tank is approximately one-third full of liquid.

REVIEW QUESTIONS 9–12

1. What is placed at the condenser outlet?
2. When is a liquid receiver most likely to be used?
3. How are liquid receivers sized?
4. What device is used at the outlet of the receiver tank?
5. To where is the refrigerant piped from the receiver safety devices?
6. Name two types of liquid receiver safety devices.
7. To what point is a system with a receiver tank charged?

CHAPTER

10

EVAPORATORS

OBJECTIVES:

Upon completion of this chapter, you should be able to:

- Know the purpose of the evaporator in a refrigeration system
- Know the different types of evaporators used in refrigeration systems
- Better understand the heat transfer through an evaporator
- Better understand how temperature difference affects the humidity in a refrigerated cabinet
- Be more familiar with evaporator defrosting
- Be more familiar with the methods used to defrost an evaporator

INTRODUCTION

The evaporator is the system component that actually does the cooling of the space or product. Therefore, it is a very important component. It may be defined as a device that is used for the purpose of absorbing heat into the refrigeration system. It is located in the space to be cooled and the heat is absorbed because of the latent heat of vaporization of the refrigerant inside it.

10–1 TYPES OF EVAPORATORS

Evaporators are generally divided into two distinct types: the flooded type and the dry or direct expansion type. A discussion of these types of evaporators follows.

Flooded Evaporators

Flooded evaporators, as their name implies, have a puddle of liquid refrigerant in the evaporator during the normal operating cycle. The liquid refrigerant is recirculated by the use of a separation chamber (surge drum) that is connected to the evaporator tubing. See Figure 10–1.

During operation, the liquid refrigerant enters the separation chamber through the flow-control device. It then falls down to the bottom of the chamber where the evaporator inlet is connected. The liquid then flows through the evaporator where it boils and absorbs heat from the space or product. It then leaves the evaporator as a vapor and returns to the top of the separation chamber, where it is drawn into the suction line and recirculated through the system. Any refrigerant not evaporated flows to the bottom of the separation chamber, where it again enters the evaporator.

Controlling the level of the liquid refrigerant and recirculating the unevaporated liquid assure that the complete inside of the evaporator tubing is virtually in contact with liquid refrigerant under almost any load conditions. The flooded type evaporator is very efficient in operation.

Dry or Direct-Expansion (DX) Evaporators

Direct-expansion evaporators are manufactured to provide a continuous path for the refrigerant to flow through. The liquid refrigerant enters the evaporator through the flow-control device in metered amounts to keep the evaporator fully refrigerated at all times. It passes through the evaporator, where it boils and absorbs heat. As the refrigerant boils off, it gradually changes from a liquid to a vapor before it leaves the evaporator. See Figure 10–2.

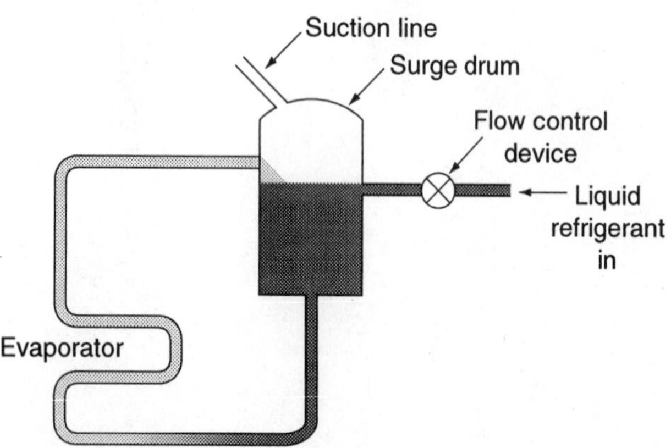

Figure 10–1 Flooded evaporator

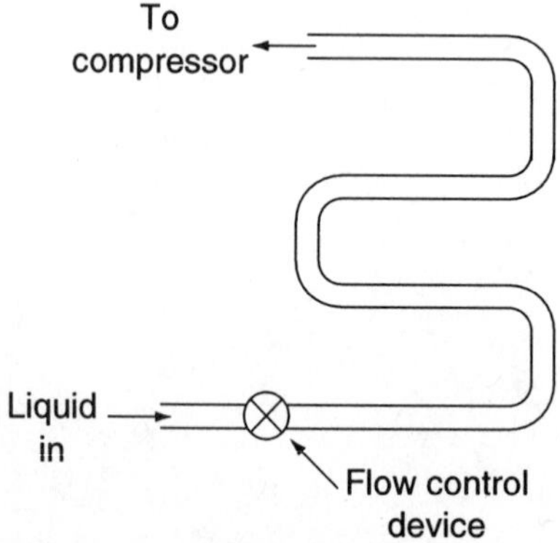

Figure 10–2 Dry or direct expansion evaporator

Notice that there is no puddle of liquid refrigerant in this type of evaporator and that there is no separation line between the liquid and the vapor in the evaporator. The vaporous refrigerant then leaves the evaporator and enters the suction line where it is returned to the compressor. The flow-control device on these types of evaporators generally operates with some degree of superheat gained in the evaporator during the boiling phase of the refrigerant. In most installations, there is little or no liquid refrigerant allowed to leave the evaporator. The dry or direct expansion evaporator is almost as efficient as the flooded type when operating with a thermostatic expansion valve.

SUMMARY 10–1

- Flooded evaporators have a puddle of liquid refrigerant in the evaporator during the normal operating cycle.
- Controlling the level of the liquid refrigerant and recirculating the unevaporated liquid assures that the complete inside of the evaporator tubing is virtually in contact with liquid refrigerant under almost any load condition.
- Direct expansion evaporators are manufactured to provide a continuous path for the refrigerant to flow through.
- Notice that there is no puddle of liquid in the DX evaporator and that there is no separation line between the liquid and the vapor in the evaporator.
- In most installations, there is little or no liquid refrigerant allowed to leave the evaporator.

REVIEW QUESTIONS 10–1

1. Name the two types of evaporators used in refrigeration systems.
2. In reference to the refrigerant, what is the difference between the two types of evaporators?
3. What is the purpose of an evaporator in a refrigeration system?
4. In a DX coil, what tends to keep the evaporator fully refrigerated?

10–2 EVAPORATOR STYLES

There is an evaporator made for just about every type of application that can be thought of. Until the particular unit is specified, it is difficult to determine which is the best type for the application.

The most popular type of evaporator is the blower coil unit. In this particular type of evaporator, the refrigerant is inside the tubes and the air passes through the fins and over the tubes to give up heat to the coil. The fins increase the heat transfer characteristics of the coil, allowing a smaller-sized coil to be used for a given application. See Figure 10–3.

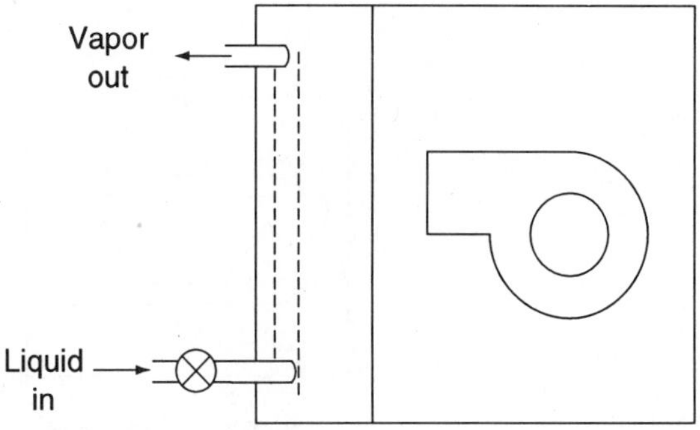

Figure 10–3 Blower coil type evaporator

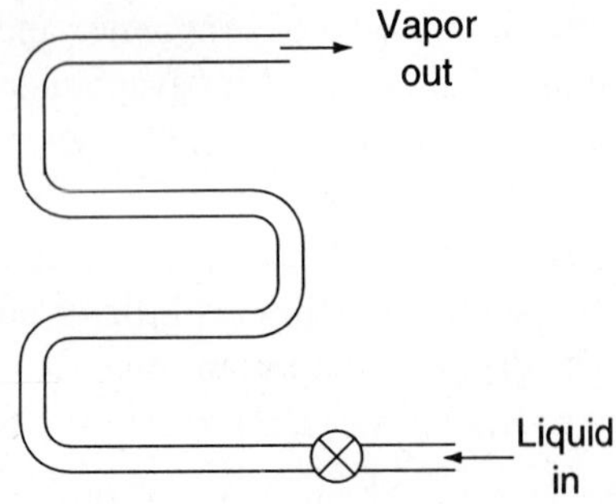

Figure 10–4 Bare tube type coil

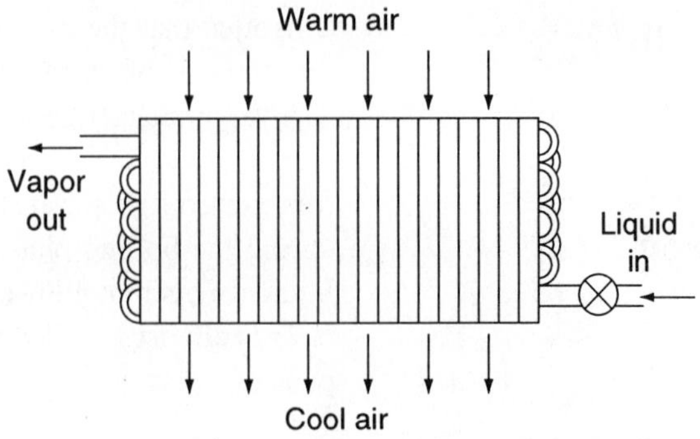

Figure 10–5 Gravity type evaporator

There are some applications that require that bare tube coils be used. These coils have no fins. See Figure 10–4.

Gravity Type Evaporators

Gravity type coils that use natural convection air flow are used in some commercial and industrial refrigeration applications. See Figure 10–5.

These types of coils sometimes have fins, but sometimes they are of the bare tube design.

Flat-Plate Type Evaporators

These types of evaporators are used in process refrigeration applications. The product to be cooled may be placed directly on the plate to aid in faster cooling. See Figure 10–6.

SUMMARY 10–2

- There is an evaporator made for just about every type of application that can be thought of.
- The most popular type of evaporator is the blower coil unit.
- Gravity type coils are used in some commercial and industrial refrigeration applications.
- Flat-plate evaporators are used in process refrigeration applications.

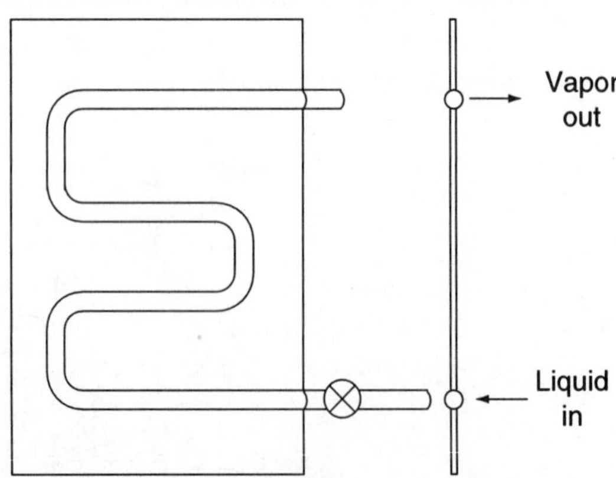

Figure 10–6 Flat-plate type evaporator

REVIEW QUESTIONS 10–2

1. What is the most popular type of evaporator coil?
2. What type of evaporator will probably be used in process refrigeration units?

10–3 HEAT TRANSFER IN EVAPORATORS

During the cooling process the heat must be transferred from the space or process being cooled to the fins, then to the evaporator tubes, and then to the refrigerant inside the tubes. During the cooling process there are two steps that occur. They are:

1. The heat from the space or material being cooled must first be absorbed by the evaporator fins and tubing.
2. The heat must then pass through the walls of the evaporator tubing and be absorbed by the refrigerant inside the evaporator.

There are several factors that control the transfer of heat from the space or material being cooled. They are:

1. The surface of the evaporator metal. (Is it smooth, dull, or rough?)
2. The temperature difference between the surface of the evaporator and the surrounding air. If this difference is small, the rate of heat transfer will also be small.
3. The velocity of the air flowing through the evaporator. The velocity of the air currents inside the conditioned space.
4. The heat transfer capabilities of the metal from which the evaporator was made.
5. The amount of frost on the evaporator.

The velocity of the air flowing over the evaporator is very important. As the air has been cooled, it must immediately be replaced with warm air. If this circulation is reduced, the amount of heat transferred will also be reduced and the space will not be cooled as desired. Properly circulated air will increase the heat transferred from four to six times that of still air. When forced air circulation is used, the heat transferred will sometimes increase to as much as twenty times that of natural-draft circulation.

The amount of heat transferred from the evaporator surface to the refrigerant inside is also very important in the cooling process. The factors that control this heat transfer are:

1. The total surface area of the evaporator. The cooling capacity of the evaporator is proportionate to the amount of surface exposed. That is, when the exposed area is small, the capacity of the evaporator will be small. Fins are added to help increase evaporator capacity.
2. The temperature difference of the refrigerant and the surrounding medium.
3. The refrigerant velocity as it passes through the evaporator tubes. (Normally, the higher the velocity, the greater the heat transfer.)
4. The amount of primary surface to secondary surface. (The area of the fins compared to the area of the tube surface.)
5. The condition of the refrigerant. (Is the evaporator flooded or dry?)
6. The thickness of the oil film on the inside of the tube. A thick oil film will reduce the heat transfer of the evaporator.

7. How fast is the evaporated refrigerant removed. The evaporated refrigerant must be removed so that more liquid refrigerant can enter the evaporator. If it is not removed, the pressure will increase, raising the evaporating temperature and reducing the refrigeration effect.
8. The type of medium being cooled. When a liquid is used to surround the evaporator, the heat transfer is about five times greater than when air is the medium.
9. The dew-point temperature of the air passing through the evaporator. When the entering air temperature is above the dew point temperature of the evaporator surface, both latent heat and sensible heat must be removed.

SUMMARY 10–3

- During the cooling process, the heat must be transferred from the space or process being cooled to the fins, then to the evaporator tubes, and then to the refrigerant inside the tubes.
- There are several factors that control the transfer of heat from the space or material being cooled.
- The velocity of the air flowing over the evaporator is very important.
- Properly circulated air will increase the heat transferred from four to six times.

REVIEW QUESTIONS 10–3

1. How will a small temperature difference between the evaporator and the air affect the capacity of the evaporator?
2. If the air flow over an evaporator is increased, how will the heat transfer be affected?
3. Will a flooded or a dry type evaporator be more efficient?
4. Will a compressor that pumps an excessive amount of oil affect the evaporator capacity?

10–4 CALCULATION OF HEAT TRANSFERRED

When any coil is operating with a steady temperature differential, the total amount of heat transferred can be calculated by using the formula:

$$Q = U \times A\,(T_1 - T_2)$$

Where: Q = Total amount of Btu transferred per hour of operation
U = Overall heat transfer coefficient
A = Area of surface in square feet through which the heat is being transferred
T_1 = Space temperature °F
T_2 = Refrigerant temperature °F

Example: We have an evaporator that will absorb 3 Btu per square foot per hour for each 1°F temperature difference. The total surface area of the evaporator is 5 square feet. The temperature of the refrigerated space is 45°F, and the refrigerant temperature is 5°F. What is the evaporator capacity?

$$\begin{aligned} Q &= 3 \times 5 \times (45 - 5) \\ &= 15 \times 40 \\ &= 600 \text{ Btu/hr} \end{aligned}$$

From this example it can be readily seen how much the capacity of a coil is affected by the heat conductivity, the exposed surface area, and the temperature difference between the space temperature and the refrigerant temperature. If we could increase any one of these factors the total capacity of the coil will also be increased. Likewise if any one of them is decreased the total capacity of the coil is decreased.

Rate of Heat Transfer

Flooded evaporators will deliver about 50% more heat than the direct expansion evaporator. This is because the refrigerant in the direct expansion evaporator is a mist rather than a liquid. The mist or vaporous refrigerant cannot make as good a contact with the inside evaporator surface as the liquid in the flooded evaporator. A flooded evaporator using gravity-type air circulation will absorb approximately 3 Btu per square foot per hour for every 1°F temperature difference between the space temperature and the refrigerant temperature. On the other hand, the direct expansion evaporator will absorb only about 1 1/2 to 2 Btu per square foot per hour for each 1°F temperature difference.

SUMMARY 10–4

- When any coil is operating with a steady temperature differential, the total amount of heat transferred can be calculated easily.
- Flooded evaporators will transfer about 50% more heat than direct expansion evaporators.

REVIEW QUESTIONS 10–4

1. Why is a flooded evaporator more efficient than a DX evaporator?
2. Approximately how many Btu/ft^2 will a DX coil transfer?

10–5 EVAPORATOR DESIGN FACTORS

The refrigerant, as it flows through the evaporator tubing, encounters some resistance to the flow. When this resistance exceeds certain limits, the capacity of the evaporator is decreased accordingly. This is because of the resulting lower refrigerant pressure at the evaporator outlet. This lower pressure reduces the specific volume of the refrigerant returning to the compressor, and the weight of the refrigerant vapor that the compressor pumps also decreases. Because of this single factor, the length of tubing used in an evaporator must be as short as possible. When large capacity evaporators are designed, the coil is divided into separate refrigerant circuits. See Figure 10–7.

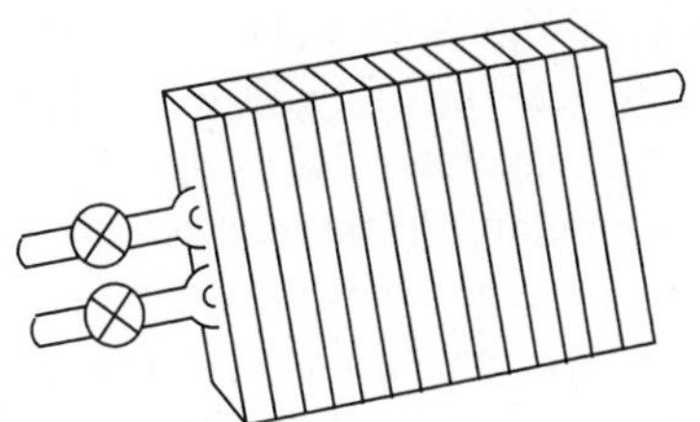

Figure 10–7 Large evaporator divided into circuits

Medium- and high-temperature installations will operate satisfactorily with 1 to 2 psi pressure drop through the evaporator. The pressure drop through the evaporator in low-temperature installations is commonly designed for about 1/2 to 2 psi pressure drop.

Some other things that must be kept in mind about evaporator design are: the tube in the evaporator must be sized to maintain sufficient refrigerant velocity to keep the lubricating oil moving through the evaporator and back to the compressor. With sufficient refrigerant velocities, the refrigerant will scrub the tubing walls to remove the oil film that tends to cling there. This increases the efficiency of the unit by keeping the heat transfer to a maximum. It should be noted that low pressure drop and high refrigerant velocities are in opposition to each other. The design of an evaporator is, at best, a compromise.

SUMMARY 10–5

- The refrigerant, as it flows through the evaporator tubing, encounters some resistance to the flow.
- When this resistance exceeds certain limits, the capacity of the evaporator is decreased accordingly.
- Medium- and high-temperature applications will operate satisfactorily with 1 to 2 psi pressure drop through the evaporator.
- It should be noted that low-pressure drop and high refrigerant velocities are in opposition to each other. The design of an evaporator is, at best, a compromise.

REVIEW QUESTIONS 10–5

1. Will a high or a low resistance in an evaporator reduce its capacity?
2. How will the pumping capacity of a compressor be affected by a higher than normal restriction in the evaporator?
3. What is considered to be a normal pressure drop through an air conditioning evaporator?

10–6 TEMPERATURE DIFFERENCE AND DEHUMIDIFICATION

In the operation of evaporators, the dew point temperature of air has considerable importance. The dew point temperature is the temperature at which moisture will start to condense out of the air.

In evaporator design, the physical characteristics of an evaporator are fixed. The primary variable is the temperature difference between the medium being cooled and the evaporating refrigerant inside the evaporator. When this temperature difference is increased, the capacity of the evaporator will decrease. For the best economy, the temperature difference should be kept as small as possible. This is because the smaller temperature differential is generally accompanied by a higher suction pressure. The higher suction pressure allows the compressor to operate more efficiently because more refrigerant can be drawn into the cylinder on each suction stroke. Temperature differences of 5°F to 20°F are the most common.

An evaporator that is operating with too great a temperature differential will cause low humidity in the refrigerated space. The amount of moisture taken out of the air during the cooling process is in direct relation to the evaporator temperature. When fresh vegetables, meats, fruits, and other such items are subjected to low-humidity conditions, they will tend to dry out, ruining the product. A very high relative humidity condition is required for these types of products, and a temperature difference of between 8°F to 12°F is generally recommended. When lower relative humidities are required, greater temperature differences can be used satisfactorily.

SUMMARY 10–6

- In the operation of evaporators, the dew point temperature is important.
- The primary variable in evaporator operation is the temperature difference between the medium being cooled and the evaporating refrigerant inside the evaporator.
- An evaporator that is operating with too great a temperature differential will cause low humidity in the refrigerated space.

REVIEW QUESTIONS 10–6

1. In an evaporator, what is the primary variable?
2. Why will a reduced air flow over the evaporator decrease its capacity?
3. What are the most common evaporator temperature differences?
4. What will cause low humidity in a refrigerated space?

10–7 EVAPORATOR FROSTING

Some frost accumulation on an evaporator is normal. However, any accumulation must be removed because the frost acts as an insulator. The frost forms because the air is cooled down to the dew point temperature and condensation accompanies any further drop in temperature of the air. If the moisture is then cooled down to the freezing temperature, the condensation will freeze, forming frost. This frost is then cooled down to evaporator temperature. If we calculated the amount of heat required to accomplish this frost formation, we would find that more than 1,200 Btu were required to make each pound of frost. The problem is that when the frost is melted, only a very small amount of this heat is recovered and used in the refrigerated space. Frost is an expensive necessity of refrigeration systems operating below the freezing temperature of the condensate. Thus, condensate must be kept to a minimum.

Defrosting Evaporators

When an evaporator operates at or below the freezing temperature of the condensate, frost will appear on it. If the frost is allowed to accumulate, air flow will gradually be decreased until it is completely stopped. The frost accumulation must be removed to keep the efficiency of the unit as high as possible, especially if continuous operation of the equipment is desired. To automatically remove the frost, defrost cycles are necessary.

In installations where the return air is 32°F or above, the defrosting will normally occur during the OFF cycle of the unit, especially when the fan is allowed to run continuously. This is known as off-cycle defrost. Fan operation can be set for any amount of time desired that will remove the frost.

In installations where the return air temperature is below 32°F, some source of heat must be used to melt the frost fast enough so that the coil is defrosted for the next "ON" cycle. The two most popular methods of defrosting low-temperature coils are electric and hot-gas defrost.

Electric Defrost. Some coils are equipped with an electric heating element embedded into the coil, or mounted onto it in some way to provide sufficient heat for a complete defrost. When the coil becomes frosted enough to decrease unit efficiency, a defrost control will stop the compressor and start the electric defrost automatically. The evaporator fans are stopped during defrost so that the heat will not be blown into the refrigerated space. When the defrost period is over, the defrost control stops heating the coil and starts the compressor and evaporator fans to resume normal operation.

Hot-Gas Defrost. Hot-gas defrost is also known as reverse-cycle defrost. When the defrost control determines that a defrost cycle is needed, the control will switch a reversing valve to change the direction of refrigerant flow through the system. There are other components required for this system to operate, but they will not be covered at this time. The evaporator fan will stop, but the compressor continues to operate. The hot gas from the compressor discharge is directed to the evaporator so that it can melt the frost. When the defrost control determines that the frost has all been removed, it will reverse the reversing valve piston, start the evaporator fan, and resume normal operation.

Most defrost systems also provide some means of keeping the condensation from refreezing in the evaporator drain pan. If the condensation were to refreeze, it would stop up the drain, causing the condensate to run inside the refrigerated space.

SUMMARY 10–7

- Any frost accumulation on the evaporator must be removed because the frost acts like an insulator.
- The problem is that when the frost is melted, only a very small amount of the heat is recovered and used in the refrigerated space.
- If frost is allowed to accumulate, air flow will gradually be decreased until it is completely stopped.
- The frost accumulation must be removed to keep the efficiency of the unit as high as possible.
- Some coils are equipped with an electric heating element embedded in the coil, or mounted on it in some way to provide sufficient heat for complete defrost.
- Hot-gas defrost is also known as reverse-cycle defrost.

REVIEW QUESTIONS 10–7

1. Why is frost on an evaporator not desirable?
2. Approximately how much heat is required to make a pound of frost on an evaporator?
3. When will OFF cycle defrost of an evaporator occur?
4. When is heat required to defrost an evaporator?

10–8 OIL CIRCULATION

In a refrigeration system, oil is continually circulating throughout the system. This is because the refrigerant vapor, as it passes through the compressor, collects some oil and carries it along. A small amount of oil will lubricate the moving components in the system. However, oil in large quantities will cause problems by reducing the heat transfer in the evaporator, and by a lack of compressor lubrication. The oil must be kept moving through the system to prevent these problems. The proper sizing of the refrigerant lines will help this problem.

Some types of compressors pump large quantities of oil during normal operation. When this problem is encountered, the system must be equipped with oil traps, and

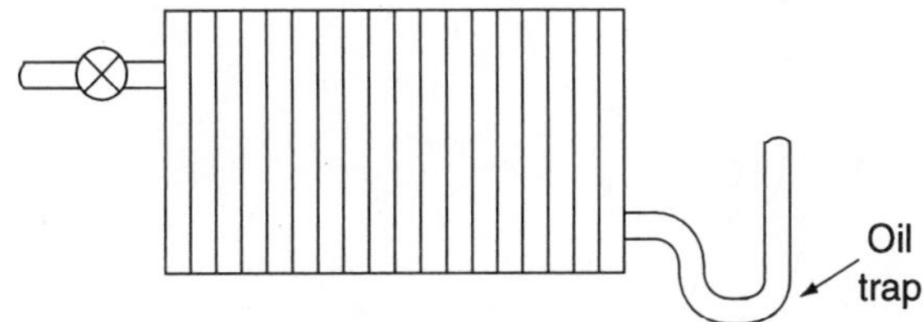

Figure 10–8 Oil trap installed at evaporator outlet

perhaps an oil separator, to keep the oil in the compressor crankcase and not in the system where it can cause problems.

When the evaporator is installed below the compressor, oil return is often a problem. This problem can be alleviated by the installation of an oil trap in the suction line as it leaves the evaporator. The oil will collect in the trap and, when the refrigerant flow restriction is sufficient, the oil will be forced out of the trap, up the suction line, and back to the compressor. See Figure 10–8.

SUMMARY 10–8

- In a refrigeration unit, oil is continually circulating throughout the system.
- The oil must be kept moving through the system to prevent many problems.
- When the evaporator is installed below the compressor, oil return is often a problem.
- Oil can be kept moving through the system by installing oil traps in the lines at the proper places.

REVIEW QUESTIONS 10–8

1. How will large amounts of oil circulating through a refrigeration system affect its operation?
2. What can be done to reduce large quantities of oil circulating through a system?
3. When a compressor is installed above the evaporator, how can oil problems be helped?

CHAPTER

11

FLOW-CONTROL DEVICES

OBJECTIVES

Upon completion of this chapter, you should be able to:

- Know the major types of flow-control devices in use today
- Know the purpose of flow-control devices
- Better understand the theory of operation of automatic expansion valves
- Better understand the theory of operation of thermostatic expansion valves
- Better understand the theory of operation of capillary tubes
- Know how to properly determine the operating superheat of a thermostatic expansion valve
- Better understand how to troubleshoot thermostatic expansion valves

INTRODUCTION

There are many types of flow-control devices used in refrigeration equipment. Their purpose is to feed the proper amount of refrigerant into the evaporator. Smaller systems usually use the most simple flow-control devices possible. However, most of the larger systems sometimes use very complex control systems. The use and purpose of the system will dictate the type of control system used.

Before any flow-control can be analyzed, repaired, or adjusted, it is necessary for the technician to understand the use, operation, and function of the particular control. It sometimes takes much study and experience to fully understand how a particular flow-control device operates.

11–1 PURPOSE

The basic function of a flow-control device is to control the flow of refrigerant into the evaporator and keep it refrigerated as designed. This is probably one of the most misunderstood components in a refrigeration system. It must be kept in mind that its purpose is to keep the evaporator fully refrigerated. There may be several different forces that operate the control, such as temperature, pressure, and/or a combination of these two.

The reasons that the flow of refrigerant into the evaporator is so important are:

1. For the evaporator to operate as it was designed, the flow-control device must feed the correct amount of refrigerant into it. The refrigerant is usually fed in a swirling pattern so that it will scrub the inside walls of the evaporator tubes. When either too much or too little refrigerant is allowed into the evaporator, the efficiency will drop. The best heat transfer is obtained when the inside of the evaporator tubes are completely wetted with liquid refrigerant, except in the very end of the evaporator, which is used to add superheat to the refrigerant vapor before it enters the suction line to the compressor.
2. All of the liquid refrigerant must be evaporated inside the evaporator; if not, it could possibly return to the compressor, causing damage from liquid flood-back. Liquid flood-back can cause damage to the compressor valves and bearings, rendering the compressor inoperative and perhaps ruined beyond repair.

SUMMARY 11–1

- The basic function of a flow-control device is to control the flow of refrigerant into the evaporator and keep it refrigerated as designed.
- The refrigerant is usually fed in a swirling pattern so that it will scrub the inside walls of the evaporator tubes.

REVIEW QUESTIONS 11–1

1. What will happen when too much refrigerant is fed into the evaporator?
2. What could happen to the compressor if the flow-control device feeds too much refrigerant?

11–2 TYPES OF FLOW-CONTROL DEVICES

The three popular types of flow-control devices in use today are: (1) automatic expansion valve (AXV), (2) thermostatic expansion valve (TXV), and (3) capillary tube or restrictor.

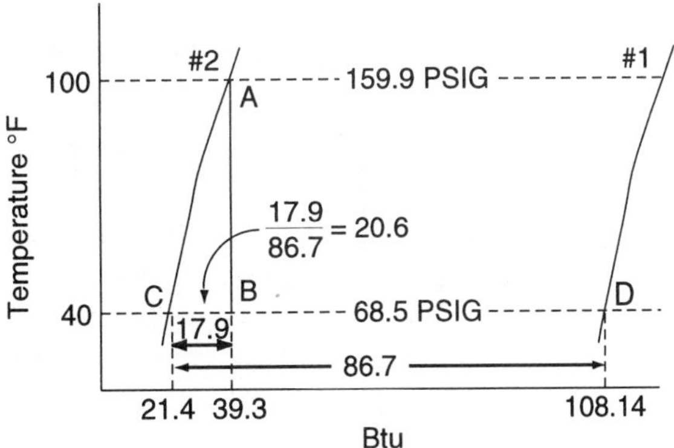

Figure 11–1 Operation of a flow-control device

Theory of Operation

A complete knowledge of the refrigerating process is required before an understanding of how the flow-control device performs as it was designed can be achieved. What happens as the refrigerant flows through the flow-control device is shown in Figure 11–1.

In this figure, the area to the left of line 2 represents the refrigerant in the liquid state. The area between lines 1 and 2 represents a mixture of liquid and vapor. The area to the right of line 2 represents the refrigerant in a completely vapor state.

In this illustration, liquid Refrigerant–22 enters the flow-control device at the point labeled A, with a pressure of 195.9 psi and a temperature of 100°F. After passing through the flow-control device, the pressure drops to 68.5 psi and the temperature drops to 40°F, point B. The refrigerant at this point is a mixture of liquid and vapor, point D. This vapor is present because some of the liquid refrigerant evaporated to absorb some of the sensible heat from the remaining liquid when cooling it from 120°F down to 40°F. This changing to gas is termed flash gas and occurs in all refrigeration equipment.

The amount of flash gas that occurs in this illustration is 20.6%. The illustration shows how flash gas is calculated. At point C on line 2, the saturated liquid line, all of the refrigerant at this point is a liquid at a temperature of 40°F. Each pound of Refrigerant–22 at these conditions would contain 21.4 Btu/lb. At point D on line 1, the saturated vapor line, all of the refrigerant has changed into a vapor at 40°F. At these conditions, the refrigerant contains 108.14 Btu/lb.

From this it can be seen that the refrigerant actually gains no heat as it passes through the flow-control device. One pound of liquid contains 39.3 Btu/lb when it enters the flow-control device, and it contains 39.3 Btu/lb when it leaves the flow-control device.

The area between points C and D represents 86.7 Btu, the total amount of heat that the refrigerant could absorb at these conditions. The line between points C and B represents 17.9 Btu that are flash gas and are used to cool the remaining refrigerant down to 40°F from 100°F. The amount of flash gas can be calculated by dividing 17.9 by 86.7 and is found to be 20.6%. This indicates that 20.6% of the liquid has been changed to flash gas.

In a practical application, there is a small amount of heat loss of the refrigerant because of the heat gained as it flows through the flow-control device, the refrigerant lines to the evaporator, distributor, and other system components through which the refrigerant flows. However, this amount of heat loss is too small for concern.

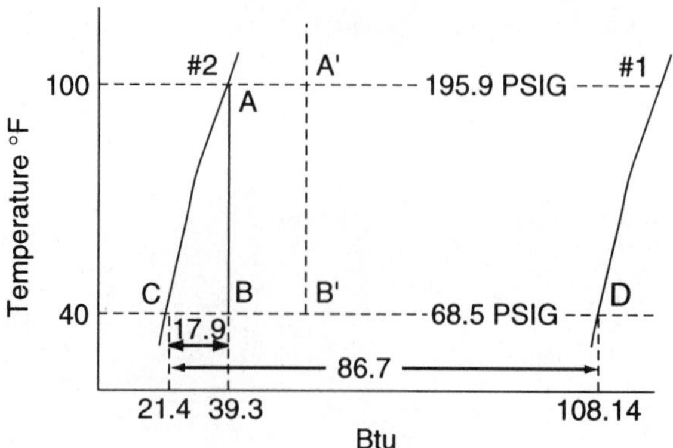

Figure 11–2 Flash gas

The refrigerant changing from a liquid to a vapor, or flash gas, can be caused by several factors. This may also be illustrated with a graph. See Figure 11–2.

This figure is the same as Figure 11–1, with the exception that line A' B' has been added. This line shows the effect that an increased compression ratio has on flash gas in a refrigeration system. For example, if the low-side pressure remains the same, but the compression ratio has been increased due to an increase in discharge pressure.

The line from points C to B is representative of the flash gas in Figure 11–1. The line from point C to B' represents the new quantity of flash gas caused by the higher compression ratio. It can be seen from this illustration that the compression ratio must be kept to a minimum.

The entire refrigeration cycle can be imposed on one temperature-Btu chart. See Figure 11–3.

In this chart, the area to the left of line 2 represents refrigerant in the liquid state. The area between lines 1 and 2 represents a mixture of liquid and vapor. The area to the right of line 1 represents the refrigerant in the vapor state.

To follow the refrigerant through a complete cycle we can consider that point A in Figure 11–3 is the refrigerant entering the suction valve of the compressor. Then, from point A to point B, the vapor is being compressed. Notice at this point that not only is the refrigerant temperature increased, but there is also an increase in the heat

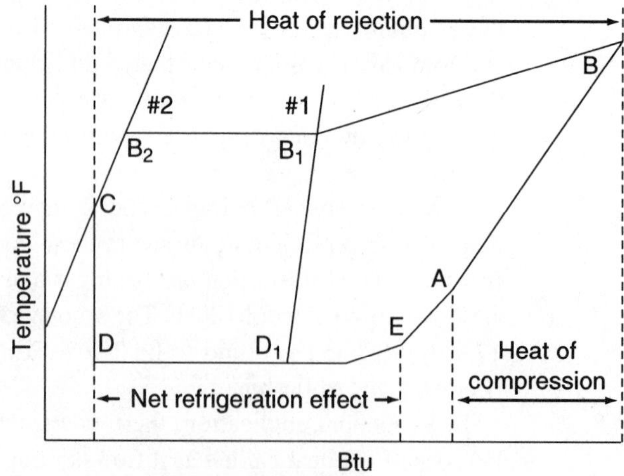

Figure 11–3 Refrigeration cycle

content of the vapor. This additional temperature and heat is the result of the compressing of vaporous refrigerant and is termed the heat of compression. From point B to B_1, the superheat is removed from the vapor and it is cooled to the saturated vapor line, or the condensing temperature of the refrigerant. Then from point B_1 to point B_2, the vapor is condensed and changed to a liquid. From point B_2 to point C, the liquid is subcooled. The amount of heat removed from the refrigerant during this process is known as the heat of rejection.

Then from point C to point D, the liquid passes through the flow-control device, where both the pressure and temperature are reduced. Because of this change in pressure and temperature, there is also a change in the refrigerant. Some of it changes into flash gas. The amount of heat remains the same, as we learned from the previous example in Figure 11–1.

As the refrigerant flows through the rest of the cycle, heat is absorbed into the refrigerant. From point D to point D_1, heat is gained because the remainder of the liquid has evaporated. This heat is latent heat, which results in a change of state of the refrigerant. Then from point D_1 to point E, superheat is absorbed into the system because all of the liquid refrigerant has evaporated. Point E is the exit of the evaporator. The heat gained from point D to point E is known as the net refrigerating effect. The net refrigerating effect is the amount of actual work done by the system. From point E to point A, a small amount of heat is absorbed in the form of superheat. This heat is absorbed by the suction line and simply serves to add superheat to the suction gas.

This cycle is the same for all compression refrigeration systems. If this cycle is understood, the system can be completely analyzed both easily and accurately.

SUMMARY 11–2

- A complete knowledge of the refrigerating process is required before an understanding of how the flow-control device performs as it was designed.
- The refrigerant actually gains in heat as it passes through the flow-control device.
- As the refrigerant is compressed, it gains additional heat and temperature, termed heat of compression.

REVIEW QUESTIONS 11–2

1. Name the three popular types of flow-control devices.
2. Define flash gas.
3. Theoretically, how much heat is gained by the refrigerant in the flow-control device?

11–3 AUTOMATIC EXPANSION VALVES (AXVS)

The automatic expansion valve, or constant pressure valve, was the first automatic expansion valve used on refrigeration systems. It is the basis on which the thermostatic expansion valve is based. The valve is so named because it opens and closes simply from pressure within the low side of the system. It does not have any external devices controlling its operation. Through its operation, the refrigerant pressure in the evaporator and low side of the system remains almost constant. The automatic opening and closing of the valve is in direct response to the low-side refrigerant pressure. As the pressure increases, the valve closes down to reduce the amount of refrigerant entering the evaporator. When the pressure drops, the valve opens to allow more refrigerant into the low side to keep the low-side pressure at the valve

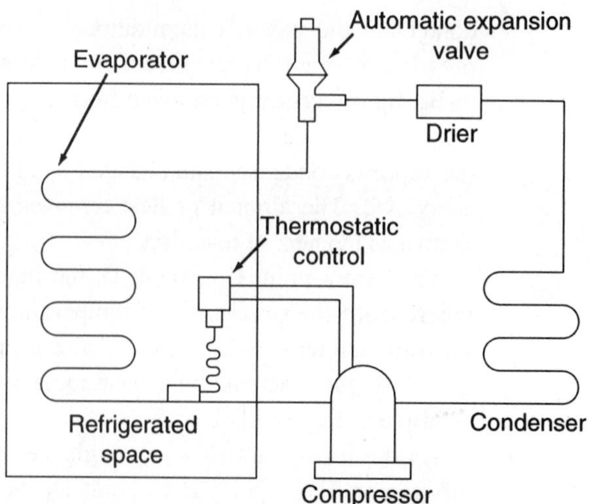

Figure 11–4 Location of an automatic expansion valve

setting. The automatic expansion valve will not compensate for varying load conditions in the system.

Function

Automatic expansion valves are basically pressure regulating valves. They react only to pressure changes at the valve outlet. They are mounted on the evaporator inlet to control the flow of refrigerant into the low side of the system. See Figure 11–4.

Through their operation, a constant low-side pressure is maintained when the unit is in operation.

The components of these valves are a diaphragm, a control spring, and the valve needle (ball) and seat. The control spring, located above the diaphragm, exerts pressure, causing the diaphragm to move downward, causing the valve to open. The opposing force on the opposite side of the diaphragm comes from the evaporator pressure, which acts as a valve-closing force.

When the unit cycles off, the valve will close because refrigerant pressure in the low side of the system increases. When the unit starts operating, the compressor quickly reduces the pressure in the low side to match the control spring force. At this point the expansion valve is ready to open. It will open when the evaporator pressure just drops below the control-spring setting.

This point is known as the valve opening pressure. As the compressor continues to operate, the valve must open further to satisfy its demands. As the evaporator pressure continues to drop, the valve continues to open-feed more refrigerant into the evaporator. The valve opens until enough refrigerant is being fed into the evaporator, where it evaporates to exactly match the pumping capacity of the compressor. At this time the valve will maintain the evaporator pressure at that setting while the unit is operating.

Adjustment

Automatic expansion valves are adjusted manually. Turning the adjustment screw either increases or decreases the tension on the control spring, which changes the valve opening pressure. The valve may be adjusted to open at any predetermined pressure within the range of the control spring. The operating pressure point of the valve will be just a few pounds below the opening point. The exact pressure is determined by the compressor pumping capacity. When a low-side gauge is connected to the system, the opening point is the pressure registered on the gauge.

When an automatic expansion valve has been replaced, or on a new system, the unit should be allowed to operate for about 24 hours before making any adjustments. This is to allow the refrigerant and oil to be properly distributed and the system to reach its normal operating temperatures. Do not turn the adjusting stem more than 1/4 turn at one time. Then wait about 15 minutes to let the valve settle out before making another adjustment. Be sure to keep the unit running continuously when checking the expansion valve setting.

Features

Because automatic expansion valves are pressure regulating devices, they offer a number of features that are used on a variety of applications. Following are the operating characteristics of an automatic expansion valve on a properly charged system:

1. Protection against evaporator icing: Automatic expansion valves eliminate the problem of frost accumulation on the evaporator or a refrigeration unit and prevent the freezing of the product load. Air conditioning systems, operating during periods of low load conditions, will accumulate frost on the evaporator or a refrigeration unit and reduce the capacity of the system unless some type of protection is provided.

   A constant low-side pressure is maintained and, as a result, the evaporator temperature is maintained at that corresponding temperature. When the automatic expansion valve is adjusted to provide an evaporator temperature just above the freezing point of water, frost accumulation on the evaporator will be eliminated, regardless of the ambient temperature, heat load on the unit, or the amount of operating time.

   Automatic expansion valves are popular on applications such as drinking fountains, soda-fountain water coolers, photo developing tanks, and other industrial fluid and liquid chiller applications.
2. Control of relative humidity: The automatic expansion valve, when adjusted to maintain an evaporator pressure and corresponding temperature just above the freezing point of water, will not only prevent the accumulation of frost but the low-side pressure will be maintained to allow for maximum water removal, causing the desired humidity.
3. Motor overload protection: The close control of the low-side pressure eliminates the possibility of excessive current draw by the compressor motor because of a high suction pressure. This is a desired protection when the unit is operating under high load conditions. The automatic expansion valve is set to constantly maintain the desired suction pressure, which does not fluctuate when the load fluctuates. Because of this, there is no variation in the motor current draw requirements. The current draw of the condensing unit is automatically maintained within safe operating limits.

   This feature allows for lower cost electric wiring because motor horsepower is established on the basis of the total system load at the design operating temperatures. Higher ambients will not place an extra load on the unit because the evaporator pressure is under constant control. There are no reserve motor requirements, eliminating this extra cost. Also, the required surface area of the condenser will be smaller.

4. Simplified field service: Units equipped with automatic expansion valves are easier to service. Air conditioning manufacturers generally use fixed adjustment valves that are set at some point above the freezing temperature of water for the refrigerant and unit design features.

   Charging the system with the correct amount of refrigerant is fairly simple. When an expansion valve is overcharged with refrigerant, the valve automatically adjusts the flow during the ON cycle to properly feed the evaporator. Any excess refrigerant is kept either in the bottom of the condenser or in the receiver tank. Also, the ambient temperature has little effect on the operation of an expansion valve system.
5. The right valve for water coolers: When automatic expansion valves are used on water-cooler applications, they are adjusted to provide minimum water temperatures. The automatic expansion valve will also provide absolute insurance that the unit will not freeze up and damage the water cooler because of constant suction pressure control. The valve must always be adjusted to provide water cooler evaporator refrigerant temperatures above 32°F. Moisture in the refrigerant system will not freeze at the valve orifice at these temperatures.
6. An ideal valve for low-starting torque motors: When the automatic expansion valve is equipped with a bleed orifice, the low-side pressure equalizes during the OFF cycle. This allows the compressor motor to start unloaded, eliminating the need for a more expensive high-starting torque motor.
7. A low-capacity bypass valve: The automatic expansion valve functions as a high- to low-side pressure regulator. When it is installed as a bypass valve responding to the outlet pressure, the valve will open when the low-side pressure drops to the opening point of the valve. In this manner, the low-side system pressure will be maintained under almost any operating conditions.

Constant Low-Side Pressure

The automatic expansion valve provides a constant flow of refrigerant to the evaporator. The flow rate is equal to the pumping capacity of the compressor at that given time. During operation, the automatic expansion valve will open to feed refrigerant into the evaporator in an amount that exactly matches that being removed by the compressor. Thus, the low-side pressure and the resulting evaporator temperature are maintained at the desired level.

We can thus see that the refrigeration system is balanced between the operation of the automatic expansion valve and the pumping capacity of the compressor. Automatic expansion valves are differential-type controls. That is, the opening point of the valve, and the operating point of the valve, are automatically determined by the running time of the unit. The differential provides the required amount of valve movement to feed the liquid refrigerant into the evaporator at the same rate it is being pumped out by the compressor.

These types of flow controls are ideal for use in refrigeration systems where the evaporator temperature must be maintained at a constant value.

Bleed-Type Valves for Off-Cycle Unloading

Systems using low-starting torque motors, such as the split-phase type used in compressors, use expansion valves that provide compressor unloading during the OFF cycle.

Bleed-type automatic expansion valves (also known as the slotted orifice valve) allow the refrigerant in the high-side to pass over into the low-side during the unit OFF cycle. This pressure change allows the compressor to start in an unloaded condition, reducing the required starting torque. Because the pressures are basically equal on both sides of the system, the compressor starts under practically no load.

The bleed-type expansion valve is made from a standard valve with a small slot machined into the valve orifice to prevent the valve from completely closing during the unit OFF cycle. This slot allows the refrigerant to escape from the high-side to the low-side, permitting an unloaded compressor condition for the next start-up.

This slot also helps to increase the total valve capacity because of the additional refrigerant flowing through the slot. These bleed slots provide a fixed amount of capacity depending on (1) the bleed port size, (2) the density of the liquid refrigerant, and (3) the amount of pressure drop across the valve.

Selecting the Properly Sized Bleed. Selecting a valve with the properly sized bleed port requires a certain amount of care for any particular application. The selection process will usually feature the smallest bleed-slot available that will permit the desired unloading during the shortest possible OFF cycle. This step is necessary so that the bleed slot will not interfere with normal valve operation during the unit ON cycle. A large bleed slot, when used with a small capacity unit, will cause problems at low-suction pressures because the slot may allow enough refrigerant to flow through and prevent the valve from actually opening or opening only slightly. The pumping capacity of the compressor could possibly be reached without the valve opening. In some instances, the suction pressure may not be allowed to drop low enough to reach the desired evaporator temperature and pressure.

To check and determine if the bleed slot is the proper size, turn the adjusting stem to a setting that will cause a lower pressure than normally desired. Allow the unit to operate while checking the low-side pressure. If the suction pressure drops to the valve setting, or at least to some point below the normal operating pressure, the bleed slot will not interfere with proper unit operation.

Effect of Elevation Change on Valve Setting

In the automatic expansion valve, the control-spring side of the diaphragm is exposed to the surrounding atmospheric pressure. The control-spring, plus the atmospheric pressure, tends to cause the valve to move in the opening direction. The spring pressure can be changed by turning the adjustment screw. After adjustment, the valve will remain at that pressure setting. However, if the atmospheric pressure changes greatly, the valve must be readjusted to provide the evaporator pressure and temperature desired. Thus, if the system is moved to a different altitude, the valve may need to be readjusted for that atmospheric pressure.

Factors Affecting Valve Capacity

The capacity of automatic expansion valves are affected by the following factors:

1. Size of the orifice
2. Amount of needle movement
3. The pressure drop across the valve
4. The type of refrigerant used
5. The condensing temperature or pressure
6. The size of the bleed slot
7. Evaporating pressure or temperature
8. The amount of liquid subcooling

SUMMARY 11–3

- The automatic expansion valve was the first automatic expansion valve used on refrigeration systems.
- The automatic expansion valve is so named because it opens and closes simply from pressure within the low-side of the system.
- The components of an automatic expansion valve are a diaphragm, a control spring, and the valve needle (ball) and seat.
- Turning the adjustment either increases or decreases the tension on the control spring, which changes the valve opening pressure.
- The valve may be adjusted to open at any predetermined pressure within the range of the control spring.
- A new valve should be allowed to operate for about 24 hours before making any adjustments.
- Any excess refrigerant charge is kept either in the bottom of the condenser or in the receiver tank.
- Automatic expansion valves automatically adjust the flow rate to equal the pumping capacity of the compressor.

REVIEW QUESTIONS 11–3

1. What causes an automatic expansion valve to open?
2. When a refrigerated cabinet using an automatic expansion valve is replenished with products, will the valve adjust for this change?
3. What determines the adjustment range of automatic expansion valves?
4. How does an automatic expansion valve provide motor overload protection?
5. What determines the suction pressure on an automatic expansion valve system?

11–4 THERMOSTATIC EXPANSION VALVES (TXVS)

The most common flow-control device used is the thermostatic expansion valve. See Figure 11–5.

An orifice located in the valve seat regulates the flow of refrigerant into the evaporator. The flow rate is metered by a needle-type plunger and seat that controls the amount of orifice opening.

These are precision devices that are designed to control the flow of refrigerant into the evaporator in an amount that exactly matches the evaporation rate of the refrigerant in the evaporator. By matching this flow, the possibility of liquid refrigerant returning to the compressor is greatly reduced. TXVs act in response to the temperature of the suction gas as it leaves the evaporator, along with the refrigerant pressure inside the evaporator. By responding to both of these factors, the TXV can control the amount of refrigerant exiting the evaporator by maintaining a predetermined superheat setting.

Recall that a superheated vapor is at a temperature that is higher than the corresponding saturation temperature for that pressure. That is, the amount of superheat is the increase in temperature above the saturation temperature at that pressure.

Consider an evaporator operating with the refrigerant R-22 at a suction pressure of 68.5 psi. The saturation temperature at 68.5 psi is 40°F. See Figure 11–6.

As long as there is any liquid refrigerant at this pressure, the refrigerant temperature will remain at 40°F.

As the refrigerant passes through the evaporator, the liquid boils off into a vapor. The refrigerant becomes less liquid and more vapor toward the end of the evaporator.

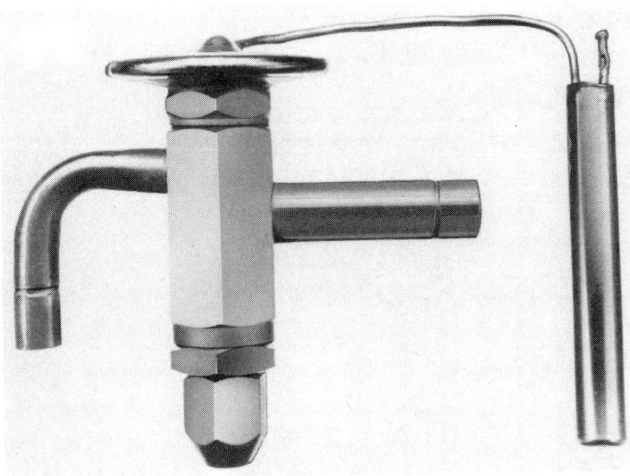

Figure 11–5 Thermostatic expansion valve (Courtesy of Alco Controls Div., Emerson Electric Co.)

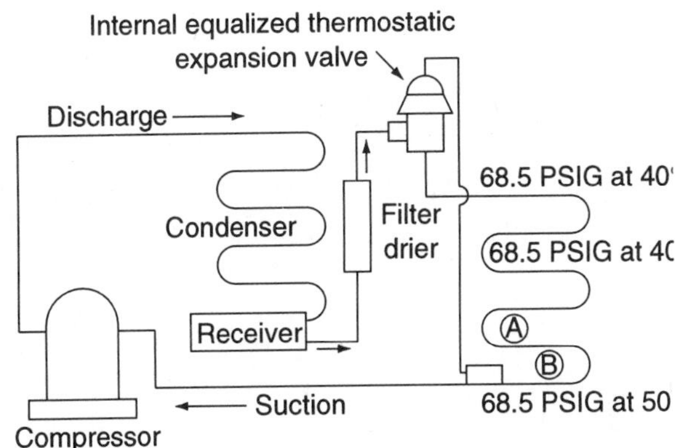

Figure 11–6 Basic refrigeration schematic

When the refrigerant has reached point A in Figure 11–6, it has absorbed sufficient heat for it all to be changed from a liquid to a vapor. As the vapor continues to travel through the remainder of the evaporator at 68.5 psi pressure, its temperature increases because more heat is absorbed from the refrigerated space. When the refrigerant vapor has reached point B, it has a temperature of 50°F. The refrigerant has gained more superheat. The amount of superheat is 50°F – 40°F = 10°F. The amount of superheat that the refrigerant can gain when passing through an evaporator is determined by two things. First, by the amount of refrigerant being metered into the evaporator and second, by the amount of heat being absorbed by the evaporator.

Operation

There are three forces controlling the operation of a thermostatic expansion valve. See Figure 11–7.

The forces are: (1) the pressure inside the remote bulb and power assembly (P_1), (2) the pressure inside the evaporator (P_2), and (3) the pressure of the superheat spring (P_3).

The remote bulb and power assembly is a sealed assembly. In the following discussion, the remote bulb and power assembly is assumed to contain the same type of refrigerant as that used in the refrigeration system.

The pressure (P_1) developed within the remote bulb and power assembly corresponds to the saturation pressure and temperature of the refrigerant exiting the evaporator. This force causes the pin to move the valve pin in the opening direction. This force is opposed by the evaporator pressure (P_2) on the underside of the diaphragm, which tends to close the valve. This closing force is assisted by force (P_3), the valve superheat spring. In operation, the valve will take a stable control position when all three of these forces are in equilibrium. Equilibrium occurs when $P_1 = P_2 + P_3$. As the refrigerant inside the evaporator becomes superheated, a pressure is caused in the remote bulb assembly that is greater than the combined pressures created by the refrigerant in the evaporator and the superheat adjustment spring. This increase in pressure causes the valve pin to open further. When the temperature of the refrigerant vapor leaving the evaporator drops, the pressure in the remote bulb also decreases, dropping below the combined pressures of the evaporator and the superheat adjustment spring, causing the valve pin to partially close the valve.

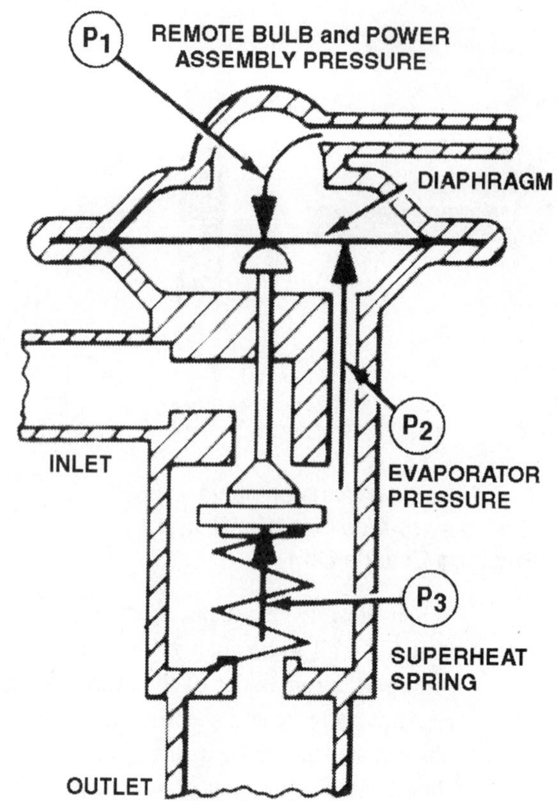

Figure 11–7 Thermostatic expansion valve basic forces (Courtesy of Alco Controls Div., Emerson Electric Co.)

The manufacturer sets the superheat of a TXV so that the pin is just starting to move away from the seat at the desired setting. TXVs are designed so that an increase of 4°F superheat will cause the valve to be completely open.

When the superheat setting is increased, the evaporator capacity is decreased, because more of the evaporator surface is required to add the additional superheat to the refrigerant vapor. See Figure 11–8.

It is most important that the superheat setting be made according to the manufacturer's recommendations. It is necessary that a minimum change in the superheat be required to cause the valve pin to move to the full open position. The correct superheat setting allows a smaller evaporator to be used, reducing the initial cost of the system and the operating costs. To provide maximum evaporator efficiency and capacity under all load conditions, it is necessary that an accurate control of liquid refrigerant into the evaporator be maintained.

Adjustment

The superheat setting is set at the factory before shipment of the valve. For most installations, this setting will not need to be changed. However, they can be adjusted to provide fine-tuning or to meet the requirements of unusual applications.

Some equipment manufacturers use nonadjustable TXVs on their original equipment design. See Figure 11–9.

These types of valves are set at a predetermined superheat setting and cannot be readjusted.

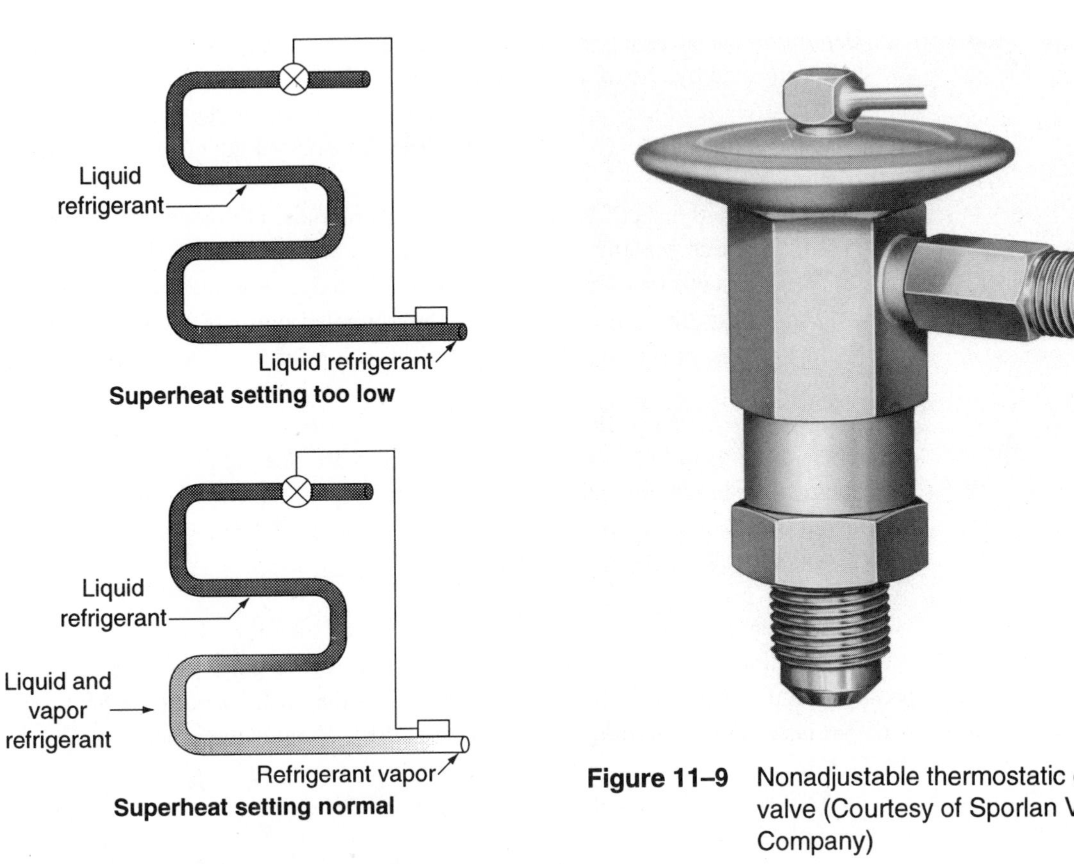

Figure 11–8 Relationship of superheat and evaporator capacity

Figure 11–9 Nonadjustable thermostatic expansion valve (Courtesy of Sporlan Valve Company)

In most cases, the nonadjustable valve is a modification of the adjustable type. The adjusting components can be added in the field when necessary.

Determining the Superheat Setting

Observing the suction pressure or the position of the frost line will not help in determining the performance of a thermostatic expansion valve. The first thing that should be done when checking the operation of the TXV, after making certain that the system is properly charged, is to properly measure the operating superheat setting. There are four steps that are used to determine the superheat setting of a TXV. They are as follows:

1. Measure the suction line temperature at the point where the remote bulb is clamped.

2. Determine the suction pressure in the suction line at the point where the remote bulb is located by one of the two following methods:
   a. The preferred method is to install a compound gauge in the suction line where the external equalizer is connected or at the external equalizer connection on the TXV.
   b. Determine the suction pressure at the compressor suction service valve. Add to the gauge reading the estimated pressure drop through the suction line between the remote bulb location and the compressor suction service valve. The sum of the gauge reading and the estimated pressure drop in the suction line will approximately equal the suction line pressure at the remote bulb location.
3. Convert the pressure calculated in Step 2 (a) or (b) to the saturated evaporator temperature by use of a pressure-temperature chart. See Table 11–1.
4. Subtract the two temperatures determined in Step 1 and Step 3. The superheat is the difference between them.

   Example: Refer to Figure 11–10.

The temperature of the suction line at the bulb location is 52°F, the suction pressure at the compressor suction service valve is 66 psig, and the estimated pressure drop is 2 psi. The total suction pressure is then calculated to be 66 + 2= 68 psi. This is equal to a saturation temperature of 40°F. What is the superheat setting of the TXV?

The 52°F suction line temperature, minus the 40°F saturation temperature, equals 12°F superheat setting.

BOLD FIGURES = INCHES MERCURY VACUUM   LIGHT FIGURES = PSIG

°F	R-12	R-13	R-22	R-500	R-502	R-717 Ammonia	°F	R-12	R-13	R-22	R-500	R-502	R-717 Ammonia
−100	**27.0**	7.5	**25.0**	–	**23.3**	**27.4**	16	18.4	211.9	38.7	24.2	47.8	29.4
−95	**26.4**	10.9	**24.1**	–	**22.1**	**26.8**	18	19.7	218.8	40.9	25.7	50.1	31.4
−90	**25.7**	14.2	**23.0**	–	**20.7**	**26.1**	20	21.0	225.7	43.0	27.3	52.5	33.5
−85	**25.0**	18.2	**21.7**	–	**19.0**	**25.3**	22	22.4	233.0	45.3	29.0	55.0	35.7
−80	**24.1**	22.2	**20.2**	–	**17.1**	**24.3**	24	23.9	240.3	47.6	30.7	57.5	37.9
−75	**23.0**	27.1	**18.5**	–	**15.0**	**23.2**	26	25.4	247.8	49.9	32.5	60.1	40.2
−70	**21.8**	32.0	**16.6**	–	**12.6**	**21.9**	28	26.9	255.5	52.4	34.3	62.8	42.6
−65	**20.5**	37.7	**14.4**	–	**10.0**	**20.4**	30	28.5	263.2	54.9	36.1	65.4	45.0
−60	**19.0**	43.5	**12.0**	–	**7.0**	**18.6**	32	30.1	271.3	57.5	38.0	68.3	47.6
−55	**17.3**	50.0	**9.2**	–	**3.6**	**16.6**	34	31.7	279.5	60.1	40.0	71.2	50.2
−50	**15.4**	57.0	**6.2**	–	0.0	**14.3**	36	33.4	287.8	62.8	42.0	74.1	52.9
−45	**13.3**	64.6	**2.7**	–	2.1	**11.7**	38	35.2	296.3	65.6	44.1	77.2	55.7
−40	**11.0**	72.7	0.5	**7.9**	4.3	**8.7**	40	37.0	304.9	68.5	46.2	80.2	58.6
−35	**8.4**	81.5	2.6	**4.8**	6.7	**5.4**	45	41.7	327.5	76.0	51.9	88.3	66.3
−30	**5.5**	91.0	4.9	**1.4**	9.4	**1.6**	50	46.7	351.2	84.0	57.8	96.9	74.5
−28	**4.3**	94.9	5.9	0.0	10.6	0.0	55	52.0	376.1	92.6	64.2	106.0	83.4
−26	**3.0**	98.9	6.9	0.7	11.7	0.8	60	57.7	402.3	101.6	71.0	115.6	92.9
−24	**1.6**	103.0	7.9	1.5	13.0	1.7	65	63.8	429.8	111.2	78.2	125.8	103.1
−22	**0.3**	107.3	9.0	2.3	14.2	2.6	70	70.2	458.7	121.4	85.8	136.6	114.1
−20	0.6	111.7	10.1	3.1	15.5	3.6	75	77.0	489.0	132.2	93.9	148.0	125.8
−18	1.3	116.2	11.3	4.0	16.9	4.6	80	84.2	520.8	143.6	102.5	159.9	138.3
−16	2.1	120.8	12.5	4.9	18.3	5.6	85	91.8	–	155.7	111.5	172.5	151.7
−14	2.8	125.7	13.8	5.8	19.7	6.7	90	99.8	–	168.4	121.2	185.8	165.9
−12	3.7	130.5	15.1	6.8	21.3	7.9	95	108.3	–	181.8	131.2	199.7	181.1
−10	4.5	135.4	16.5	7.8	22.8	9.0	100	117.2	–	195.9	141.9	214.4	197.2
−8	5.4	140.5	17.9	8.8	24.4	10.3	105	126.6	–	210.8	153.1	229.7	214.2
−6	6.3	145.7	19.3	9.9	26.0	11.6	110	136.4	–	226.4	164.9	245.8	232.3
−4	7.2	151.1	20.8	11.0	27.7	12.9	115	146.8	–	242.7	177.3	262.6	251.5
−2	8.2	156.5	22.4	12.1	29.5	14.3	120	157.7	–	259.9	190.3	280.3	271.7
0	9.1	162.1	24.0	13.3	31.2	15.7	125	169.1	–	277.9	203.9	298.7	293.1
2	10.2	167.9	25.6	14.5	33.1	17.2	130	181.0	–	296.8	218.2	318.0	315.0
4	11.2	173.7	27.3	15.7	35.0	18.8	135	193.5	–	316.6	233.2	338.1	335.0
6	12.3	179.8	29.1	17.0	37.0	20.4	140	206.6	–	337.3	248.8	359.1	365.0
8	13.5	185.9	30.9	18.4	39.1	22.1	145	220.6	–	358.9	265.2	381.1	390.0
10	14.6	192.1	32.8	19.8	41.1	23.8	150	234.6	–	381.5	282.3	403.9	420.0
12	15.8	198.6	34.7	21.2	43.3	25.0	155	249.9	–	405.2	300.1	427.8	450.0
14	17.1	205.2	36.7	22.7	45.5	27.5	160	265.12	–	429.8	318.7	452.6	490.0

Table 11–1 Temperature-pressure chart (Courtesy of Alco Controls Div., Emerson Electric Co.)

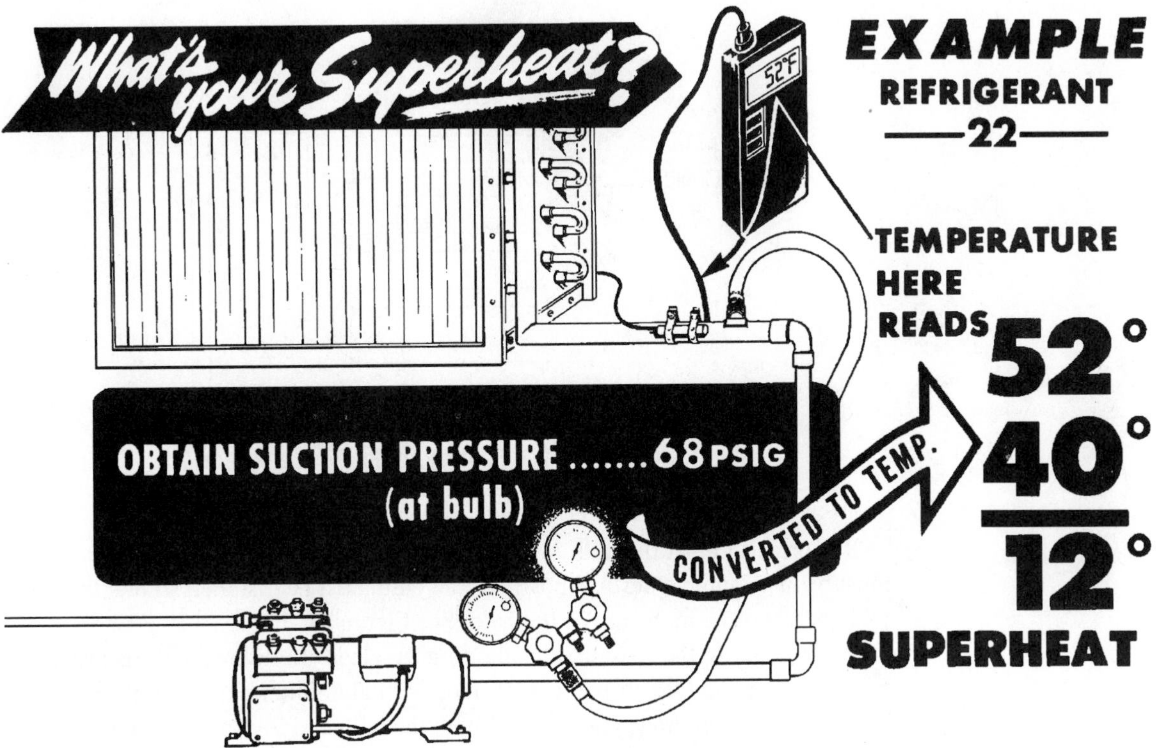

Figure 11–10 Determination of superheat (Courtesy of Sporlan Valve Company)

This is in contrast to measuring the temperature at the evaporator inlet and at the evaporator outlet. When the superheat is measured in this manner, any pressure drop in the evaporator tubing will not be taken into consideration and will not indicate an accurate superheat setting.

Changing the Superheat Setting

The superheat setting is adjusted by turning the adjustment stem. Turning the stem counterclockwise reduces the superheat setting. When the superheat setting is to be increased, the adjusting stem is turned clockwise. Do not make more than one turn on the stem at a time. Then observe the suction pressure for about 30 minutes before making another adjustment. This procedure is recommended to prevent overshooting the desired setting.

To determine the correct superheat setting for a given unit, consult the manufacturer's specifications. Generally, the proper superheat setting is determined by the temperature difference (TD) between the refrigerant and the product being cooled. When air conditioning is used, the superheat setting may be as high as 15°F without any noticeable evaporator capacity loss. When low TDs are used, such as in low-temperature blower coils, the superheat setting may be 10°F or lower for the recommended evaporator capacity rating.

Remote Bulb Location and Installation

Location of the bulb is an extremely important step in the installation of a thermostatic expansion valve. It will sometimes determine the success or failure of the entire installation. There must be good thermal contact between the bulb and the suction line. The bulb must be securely fastened to the suction line with two bulb clamps. It should be installed on a straight section of the line.

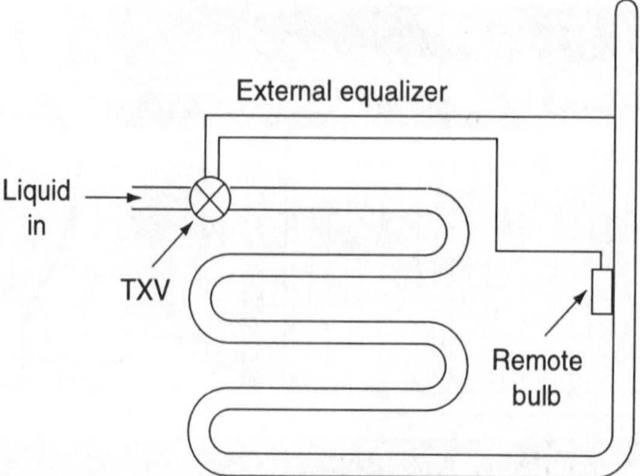

Figure 11–11 Remote bulb installation on vertical tubing

It is preferred that the bulb be installed on a horizontal run of the suction line. If it must be installed in a vertical position, the bulb must be installed so that the capillary tubing extends from the top of the bulb. See Figure 11–11.

When making the installation, thoroughly clean the tubing before fastening the bulb. It is a good idea to paint steel lines with aluminum paint before installing the bulb. This will minimize corrosion of both the bulb and line at this point.

On suction lines smaller than 7/8 inch OD, the bulb may be installed on top of the line. On lines larger than 7/8 inches OD, the bulb should be mounted at about the 4 or 8 o'clock position. See Figure 11–12.

It is recommended that the remote bulb be protected from any air stream. On installations where the temperature is above freezing, use a material such as sponge rubber that will not absorb water to cover the bulb. When the temperature will be below freezing, use cork or a similar material that is sealed against moisture to insulate the bulb.

Good suction line piping practices should be used so that a proper bulb location can provide the best possible valve control. The bulb must never be installed in a line trap. Liquid refrigerant or a mixture of liquid refrigerant and oil boiling out of a trap

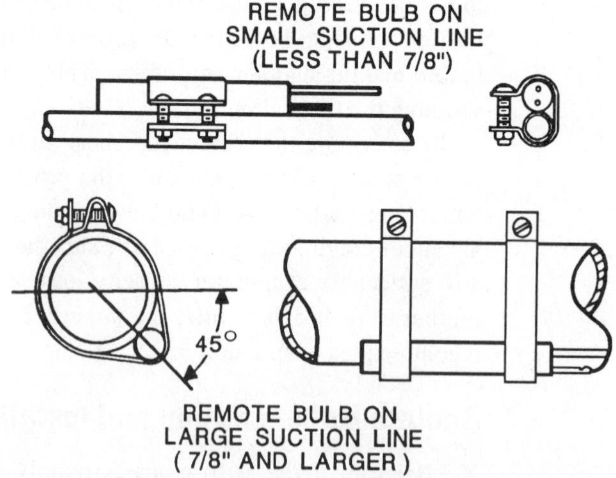

Figure 11–12 Remote bulb installation on horizontal tubing (Courtesy of Alco Controls Div., Emerson Electric Co.)

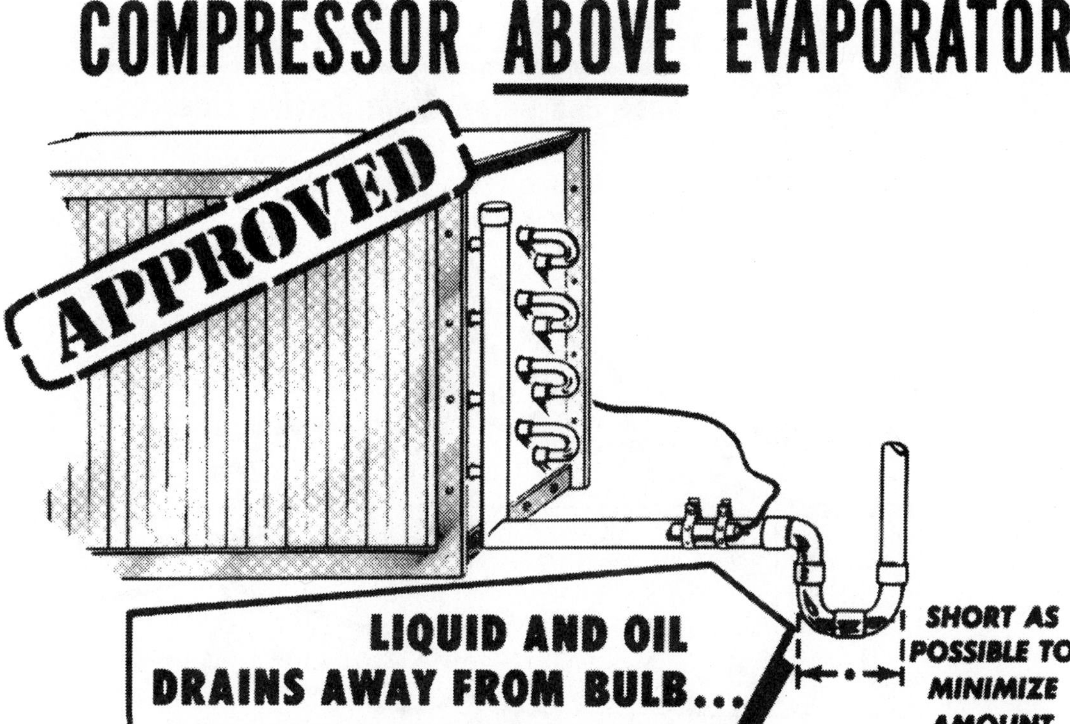

Figure 11–13 Evaporator suction piping (Courtesy of Sporlan Valve Company)

could cause a colder than normal temperature resulting in poor valve control and poor system performance.

The recommended suction line piping uses a horizontal line exiting the evaporator. The remote bulb is mounted on this piece of line. The suction line should be slightly pitched downward toward the compressor. When a vertical riser is used, a trap should be installed immediately ahead of the vertical line. See Figure 11–13.

When a trap is used, it will collect the liquid refrigerant and oil flowing from the evaporator and prevent them from influencing the remote bulb operation.

When installing a multiple evaporator system, the suction piping should be so that the refrigerant flow from one evaporator cannot affect the operation of the valve for another evaporator. Approved piping practices include the use of traps in the proper position so that each valve can be properly controlled. See Figure 11–14.

When the evaporator is located above the compressor, the suction line should be extended to the top of the evaporator so that liquid refrigerant and oil will not drain by gravity to the compressor during the OFF cycle. See Figure 11–15.

If a unit pump-down cycle is used, the suction line may be turned down immediately after leaving the evaporator without a trap being installed.

TXV Equalizer

Thermostatic expansion valves depend on three pressures for proper operation. The three pressures are: the bulb pressure on top of the diaphragm, which must always be equal to the sum of the evaporator pressure (suction pressure) and the adjustment spring pressure, both of which are applied to the bottom or evaporator side of the diaphragm.

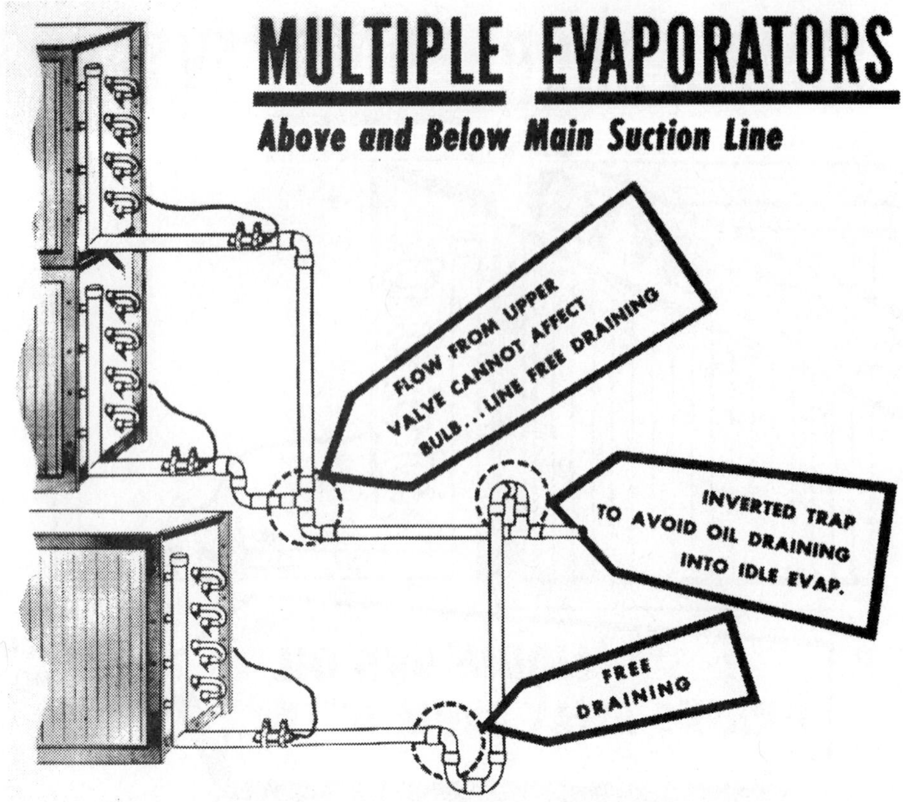

Figure 11–14 Evaporator suction piping on multiple evaporators (Courtesy of Sporlan Valve Company)

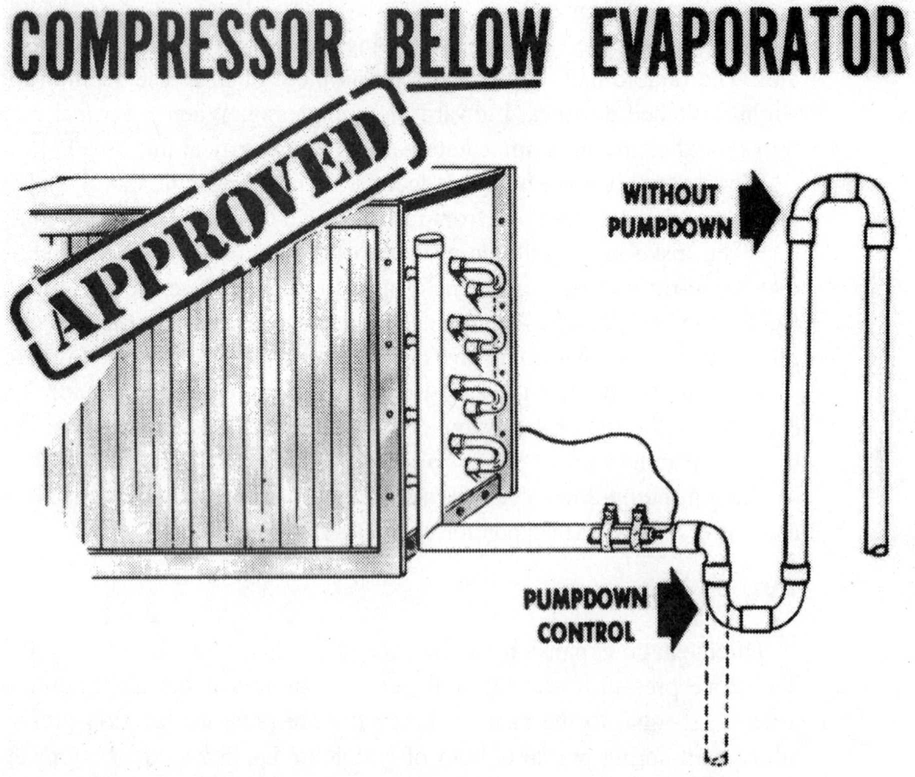

Figure 11–15 Evaporator suction piping with compressor below evaporator (Courtesy of Sporlan Valve Company)

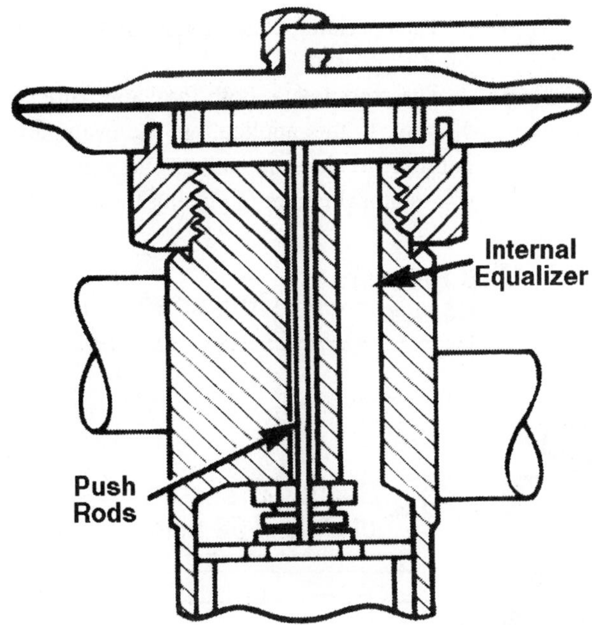

Figure 11–16 Internally equalized expansion valve (Courtesy of Sporlan Valve Company)

When an internally equalized valve is used, the pressure at the valve outlet (evaporator inlet) is applied to the evaporator side of the diaphragm through a passageway inside the valve. See Figure 11–16.

When an externally equalized expansion valve is used, the evaporator side of the diaphragm is not affected by the outlet pressure of the valve. The suction line pressure is piped from the outlet of the evaporator to the bottom of the diaphragm through an equalizer connection on the valve. This equalizer line is installed in the suction line downstream of the bulb location. See Figure 11–17.

Internally equalized expansion valves are used on evaporator coils having a low internal pressure drop. External equalized expansion valves are used on evaporator coils having a high internal pressure drop or use a refrigerant distributor.

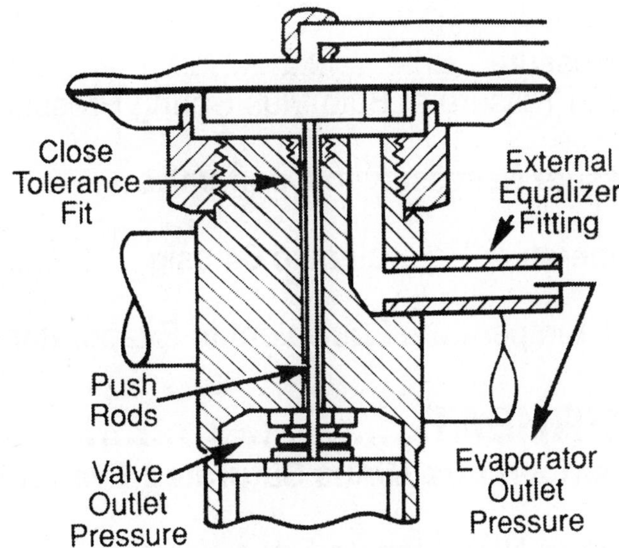

Figure 11–17 Externally equalized expansion valve (Courtesy of Sporlan Valve Company)

The operating characteristics of an internally equalized valve are shown in Figure 11–18.

The pressures at both the valve outlet and the remote bulb location are the same, 52 psig. In this application, the evaporator side of the diaphragm senses 52 psi and attempts to close the valve. The adjusting spring pressure, which is 12 psi, helps the evaporator pressure in closing the valve. As a result, the valve adjusts the flow rate of

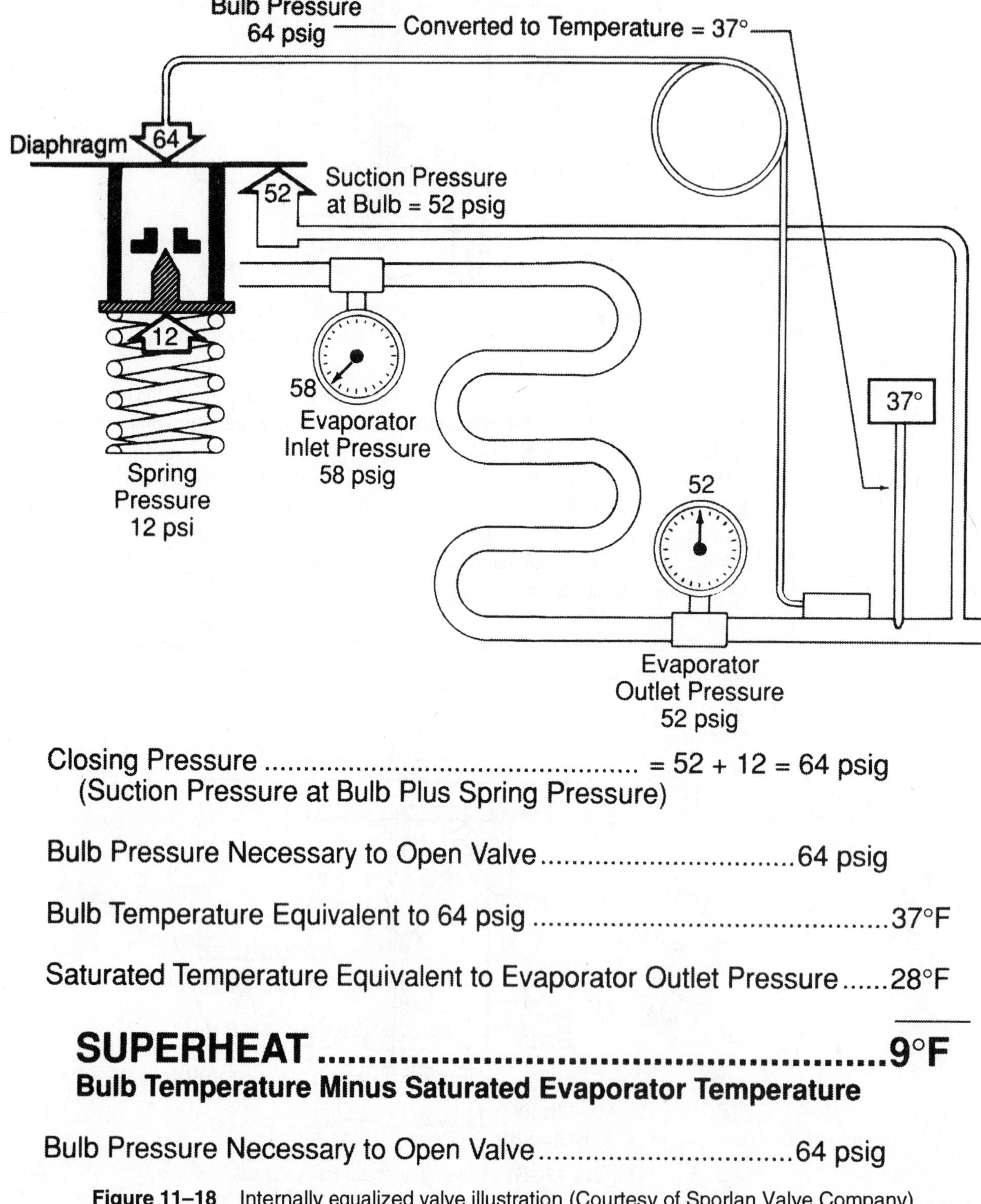

Figure 11–18 Internally equalized valve illustration (Courtesy of Sporlan Valve Company)

the refrigerant until the suction vapor gains enough superheat to heat the remote bulb to a temperature of 37°F. The two forces are now equal to the evaporator spring pressure, which is causing a superheat of 9°F.

Now, if we installed this same valve, without making any adjustments on an evaporator having a 6 psi internal pressure drop, the operating superheat would equal 13°F. See Figure 11–19.

The higher superheat occurs because the valve is sensing a 58 psig pressure at the valve outlet (or evaporator inlet). This makes a total closing force of 58 plus 12,

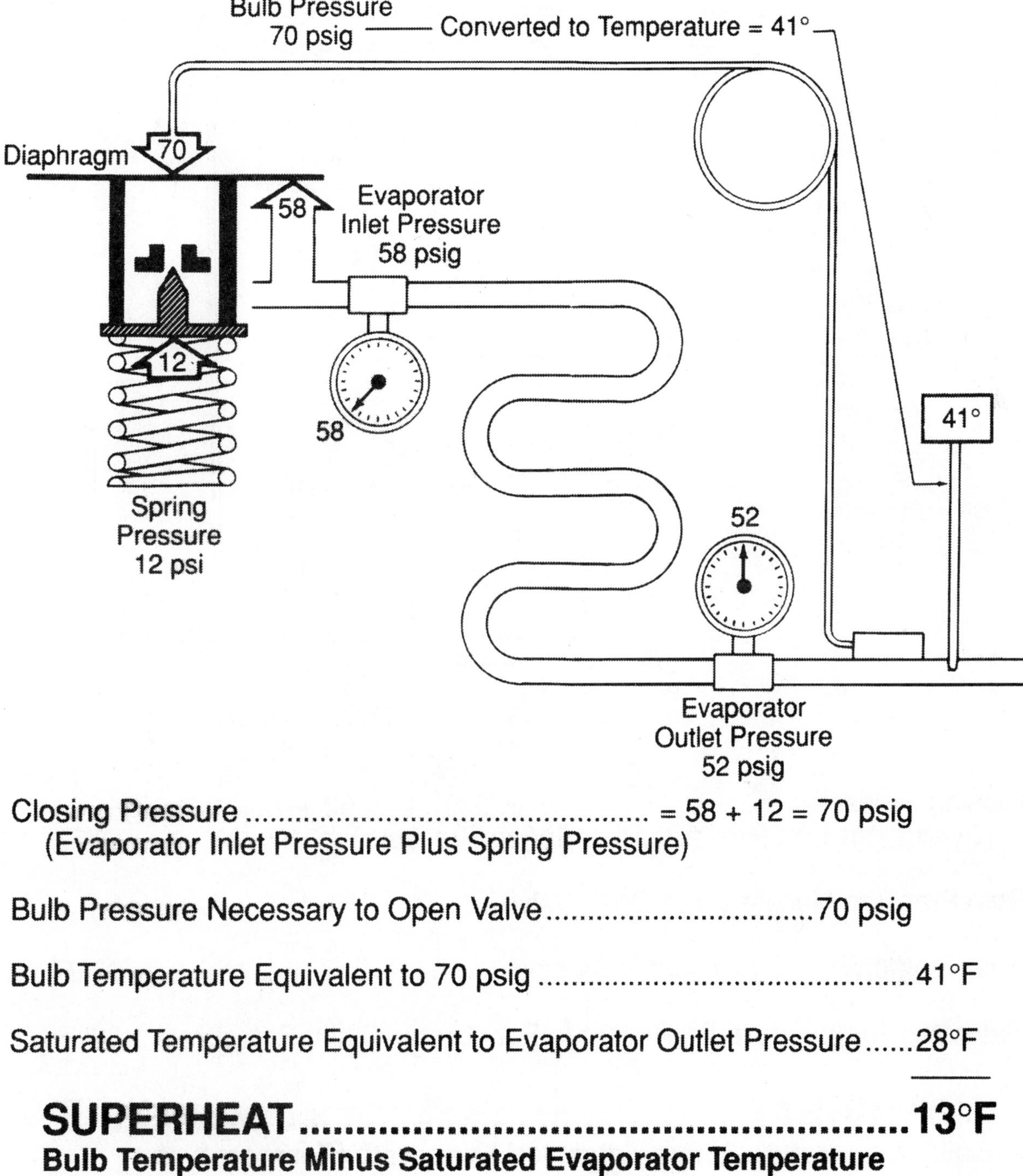

Figure 11–19 Thermostatic expansion valve with coil pressure drop (Courtesy of Sporlan Valve Company)

which equals 70 psig. The valve is designed to operate with equalized pressures above and below the diaphragm. This higher pressure below the diaphragm causes the valve to close down and create the needed superheat to cause the needed pressure in the remote bulb. This example illustrates how a high pressure drop in an evaporator, using an internally equalized expansion valve, will create an unusually high superheat, resulting in a large reduction in evaporator capacity.

This problem may be remedied through the use of an externally equalized expansion valve. See Figure 11–20.

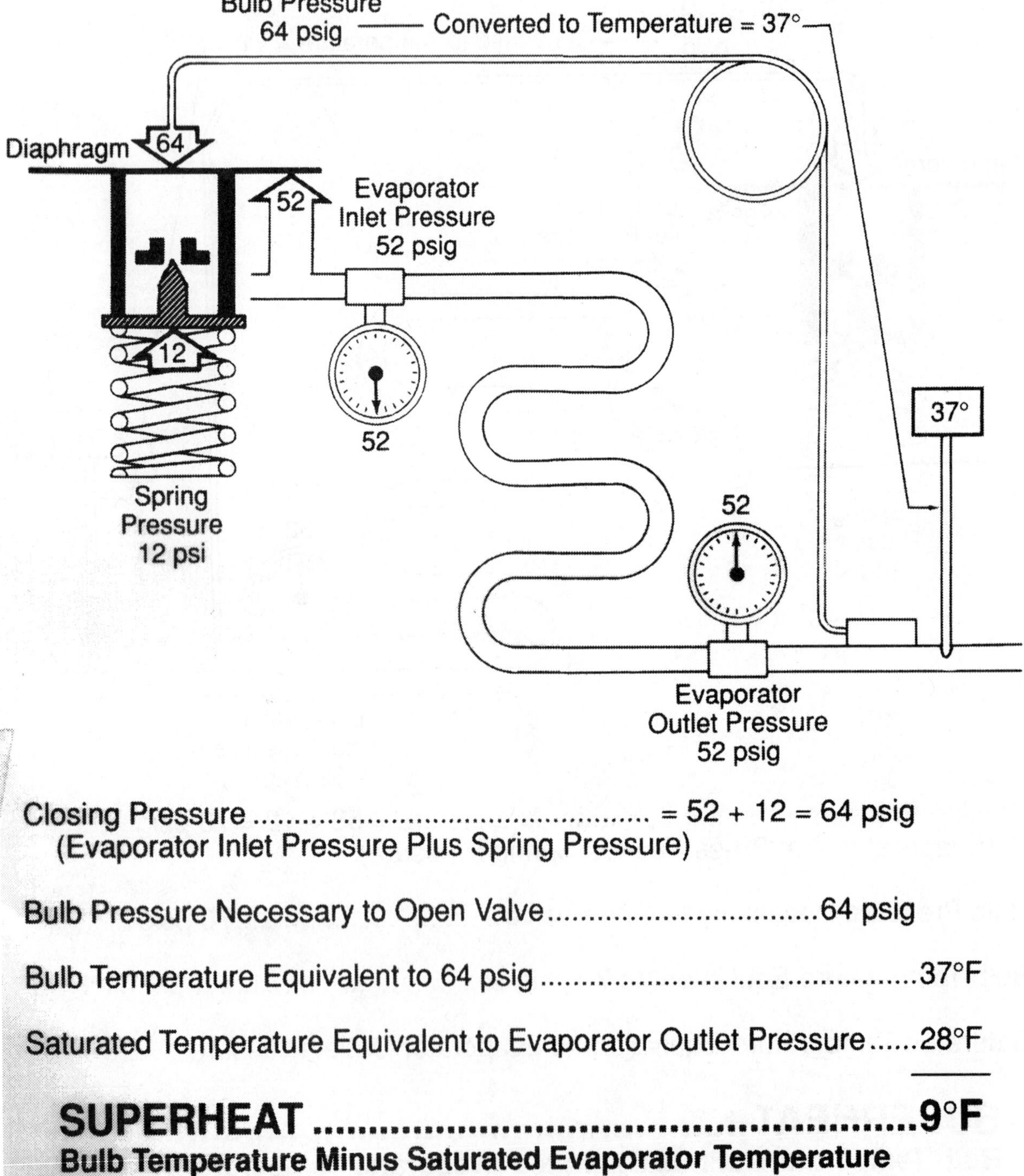

Figure 11–20 Externally equalized expansion valve installation (Courtesy of Sporlan Valve Company)

This illustration is the same as that used in Figure 11–19, but with an externally equalized expansion valve. In this application, the suction pressure at the remote bulb location is directed to the evaporator side of the diaphragm through the external equalizer line. The system is operating exactly like the one in Figure 11–19. The superheat is now 9°F.

Distributors

When very large evaporators are needed, they are generally split up into multiple refrigerant circuits to reduce the pressure drop. This usually requires some means of feeding the refrigerant evenly through all of the circuits. These multiple circuits are fed by a device called a distributor. See Figure 11–21.

The refrigerant distributor is mounted on the evaporator to feed the refrigerant through the required tubing. See Figure 11–22.

When the liquid refrigerant passes through the expansion valve, a part of the liquid changes to flash gas to cool the remaining liquid refrigerant down to the evaporator temperature. This refrigerant is then fed into the distributor from the expansion valve. The distributor then evenly distributes it to all the circuits through the small feeder tubes. Each circuit requires a feeder tube and connection on the distributor.

If a distributor was not used, the refrigerant would be divided into separate liquid and vapor layers, resulting in some of the circuits being starved and all over poor refrigeration. To prevent this, the feeder tubes must also be exactly the same size and length.

Figure 11–21 Refrigerant distributors (Courtesy of Sporlan Valve Copany)

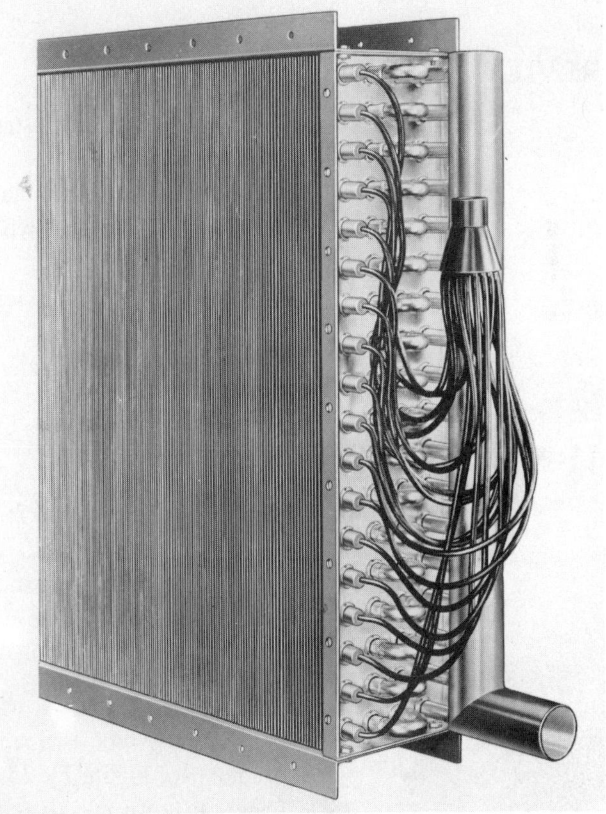

Figure 11–22 Coil prepared for a distributor (Courtesy of Sporlan Valve Company)

There are two different designs used in distributors. If one is to be installed, be sure to use the one recommended by the equipment manufacturer. There is a high-pressure distributor that uses the turbulence created by the orifice to produce good refrigeration efficiency. The low-pressure drop distributor depends on a contour-flow to create the desired distribution of the refrigerant. Both types produce good results when properly applied. For thermostatic expansion valve service hints, see Table 11–2.

SUMMARY 11–4

- The most common flow-control device is the thermostatic expansion valve.
- Thermostatic expansion valves control the flow of refrigerant into the evaporator in an amount that exactly matches the evaporation rate of the refrigerant in the evaporator.
- Thermostatic expansion valves act in response to the temperature of the suction gas as it leaves the evaporator, along with the refrigerant pressure inside the evaporator.
- The three forces that operate a thermostatic expansion valve are: (1) the pressure inside the remote bulb and power element (P_1), (2) the pressure inside the evaporator (P_2), and (3) the pressure of the superheat spring (P_3).
- Thermostatic expansion valves are designed so that an increase of 4°F superheat will cause the valve to completely open.
- The first thing that should be done when checking the operation of a thermostatic expansion valve is to make certain that the system is properly charged.
- The remote bulb must be properly installed for proper operation.
- Large evaporators with a high-pressure drop require that separate circuits be used that are fed with a distributor.

REVIEW QUESTIONS 11–4

1. To what does a thermostatic expansion valve control the flow of refrigerant?
2. Define superheat.
3. Name the three forces that operate a thermostatic expansion valve.
4. What is the first step when checking the operation of a thermostatic expansion valve?
5. When installing or servicing a system with multiple evaporators, what should be done with the suction line?
6. From where does an external equalizer line receive its pressure?

11–5 CAPILLARY TUBES

Capillary tubes are the most simple of all the refrigerant flow-control devices. There are no moving parts and no adjustments to be made. Capillary tubes are limited to single cabinet installations having their own condensing unit. They are not suitable for use on multiple evaporator systems or multiple temperature applications.

Function

The capillary tube is a small diameter tube used to feed the refrigerant into the evaporator. See Figure 11–23.

They are not true valves because they are nonadjustable and cannot be easily regulated. These devices are generally used only on flooded type systems. The

LOW SUCTION PRESSURE . . . HIGH SUPERHEAT	
Probable Cause	**Remedy**
A. Expansion Valve Limiting Flow	
Inlet pressure too low from excessive vertical lift, undersize liquid line or excessive low condensing temperature. Resulting pressure difference across valve too small.	Increase head pressure. If liquid line is too small, replace with proper size. See Table 3.
Gas in liquid line . . . due to pressure drop in the line or insufficient refrigerant charge. If there is no sight glass in the liquid line, a characteristic whistling noise will be heard at the expansion valve.	Locate cause of liquid line flash gas and correct by use of any or all of the following methods: 1. Add charge. 2. Clean strainers, replace filter driers. 3. Check for proper line size (see Table 3). 4. Increase head pressure or decrease temperature to insure solid liquid refrigerant at valve inlet.
Valve restricted by pressure drop through evaporator.	Change to an expansion valve having an external equalizer.
External equalizer line plugged, or external equalizer connector capped without providing a new valve cage or body with internal equalizer.	If external equalizer is plugged, repair or replace. Otherwise, replace with valve having correct equalizer.
Moisture, wax, oil or dirt plugging valve orifice. Ice formation or wax at valve seat may be indicated by sudden rise in suction pressure after shutdown and system has warmed up.	Wax and oil indicate wrong type oil is being used. Purge and recharge system, using proper oil. Install an ALCO ADK filter-drier to prevent moisture and dirt from plugging valve orifice.
Valve orifice too small.	Replace with proper valve.
Superheat adjustment too high.	See "Measuring and Adjusting Operating Superheat".
Power assembly failure or partial loss of charge.	Replace power assembly (if possible) or replace valve.
Gas charged remote bulb of valve has lost control due to remote bulb tubing or power head being colder than the remote bulb.	Replace with a "W" cross ambient power assembly. See "Remote Bulb and Power Assembly Charges".
Filter screen clogged.	Clean all filter screens.
Wrong type oil.	Purge and recharge system and use proper oil.
B. Restriction in system other than expansion valve. (Usually, but not necessarily, indicated by frost or lower than normal temperatures at point of restriction.)	
Strainers clogged or too small.	Remove and clean strainers. Check manufacturers catalog to make sure that correct strainer was installed. Add an ALCO ADK filter-drier to system.
A solenoid valve not operating properly or is undersized.	Refer to ALCO "Solenoid Valve Service Hints" pamphlet. If valve is undersized, check manufacturers catalog for proper size and conditions which would cause malfunction.
King valve at liquid receiver too small or not fully opened. Hand valve stem failure or valve too small or not fully opened. Discharge or suction service valve on compressor restricted or not fully opened.	Repair or replace faulty valve if it cannot be fully opened. Replace any undersized valve with one of correct size.
Plugged lines.	Clean, repair or replace lines.
Liquid line too small.	Install proper size liquid line. See Table 3.
Suction line too small.	Install proper size suction line. See Table 2.
Wrong type oil in system, blocking refrigerant flow.	Purge and recharge system and use proper oil.

LOW SUCTION PRESSURE . . . LOW SUPERHEAT	
Probable Cause	**Remedy**
Poor distribution in evaporator, causing liquid to short circuit through favored passes and throttling valve before all passes receive sufficient refrigerant.	Clamp power assembly remote bulb to free draining suction line. Clean suction line thoroughly before clamping bulb in place. See Figure 3. Install an ALCO Venturi-Flo distributor. Balance evaporator load distribution.
Compressor oversize or running too fast due to wrong size pulley.	Reduce speed of compressor by installing proper size pulley, or provide compressor capacity control.
Uneven or inadequate evaporator loading due to poor air distribution or brine flow.	Balance evaporator load distribution by providing correct air or brine distribution.
Evaporator too small . . . often indicated by excessive ice formation.	Replace with proper size evaporator.
Excessive accumulation of oil in evaporator.	Alter suction piping to provide proper oil return or install oil separator, if required.

Table 11–2 Thermostatic expansion valve service hints

HIGH SUCTION PRESSURE . . . HIGH SUPERHEAT	
Probable Cause	**Remedy**
Unbalanced system having an oversized evaporator, and undersized compressor and a high load on the evaporator. Load in excess of design conditions.	Balance system components for load requirements.
Compressor undersized.	Replace with proper size compressor.
Evaporator too large.	Replace with proper size evaporator.
Compressor discharge valves leaking.	Repair or replace valves.
HIGH SUCTION PRESSURE . . . LOW SUPERHEAT	
Probable Cause	**Remedy**
Compressor undersized.	Replace with proper size compressor.
Valve superheat setting too low.	See "Measuring and Adjusting Operating Superheat".
Gas in liquid line with oversized expansion valve.	Replace with proper size expansion valve. Correct cause of flash gas.
Compressor discharge valves leaking.	Repair or replace discharge valves.
Pin and seat of expansion valve wire drawn, eroded, or held open by foreign material, resulting in liquid flood back.	Clean or replace damaged parts or replace valve. Install an ALCO ADK filter-drier to remove foreign material from system.
Ruptured diaphragm or bellows in a constant pressure (automatic) expansion valve, resulting in liquid flood back.	Replace valve power assembly.
External equalizer line plugged, or external equalizer connection capped without providing a new valve cage or body with internal equalizer.	If external equalizer is plugged, repair or replace. Otherwise, replace with valve having correct equalizer.
Moisture freezing valve in open position.	Apply hot rags to valve to melt ice. Install an ALCO ADK filter-drier to insure a moisture-free system.
FLUCTUATING SUCTION PRESSURE	
Probable Cause	**Remedy**
Incorrect superheat adjustment.	See "Measuring and Adjusting Operating Superheat".
Trapped suction line.	Install "P" trap to provide a free draining suction line.
Improper remote bulb location or installation.	Clamp remote bulb to free draining suction line. Clean suction line thoroughly before clamping bulb in place. See Figures 3, 4 or 5.
"Flood back" of liquid refrigerant caused by poorly designed liquid distribution device or uneven evaporator loading. Improperly mounted evaporator.	Replace faulty distributor with an ALCO Venturi-Flo distributor. If evaporator loading is uneven, install proper load distribution devices to balance air velocity evenly over evaporator coils. Remount evaporator lines to provide proper angle.
External equalizer lines tapped at a common point although there is more than one expansion valve on same system.	Each valve must have its own separate equalizer line going directly to an appropriate location on evaporator outlet to insure proper operational response of each individual valve. For one example, see Figure 5.
Faulty condensing water regulator, causing change in pressure drop across valve.	Replace condensing water regulator.
Evaporative condenser cycling, causing radical change in pressure difference across expansion valve. Cycling of blowers or brine pumps.	Check spray nozzles, coil surface, control circuits, thermostat overloads, etc. Repair or replace any defective equipment. Clean clogged nozzles, coil surface, etc.
Restricted external equalizer line.	Repair or replace with correct size.
FLUCTUATING DISCHARGE PRESSURE	
Probable Cause	**Remedy**
Faulty condensing water regulating valve.	Replace condensing water regulating valve.
Insufficient charge . . . usually accompanied by corresponding fluctuation in suction pressure.	Add charge to system.
Cycling of evaporative condenser.	Check spray nozzles, coil surface, control circuits, thermostat overloads, etc. Repair or replace any defective equipment. Clean clogged nozzles, coil surface, etc.
Inadequate and fluctuating supply of cooling water to condenser.	Check water regulating valve and repair or replace if defective. Check water circuit for restrictions.
Cooling fan for condenser cycling.	Determine cause for cycling fan, and correct.
Fluctuating discharge pressure controls on low ambient air cooled condenser.	Adjust, repair, or replace controls.

Table 11–2 (Continued) Thermostatic expansion valve service hints

HIGH DISCHARGE PRESSURE	
Probable Cause	**Remedy**
Insufficient cooling water due to inadequate supply or faulty water valve.	Start pump and open water valves. Adjust, repair or replace any defective equipment.
Condenser or liquid receiver too small.	Replace with correct size condenser or liquid receiver.
Cooling water above design temperature.	Increase supply of water by adjusting water valve, replacing with a larger valve, etc.
Air or non-condensable gases in condenser.	Purge and recharge system.
Overcharge of refrigerant.	Bleed to proper charge.
Condenser dirty.	Clean condenser.
Insufficient cooling air circulation over air cooled condenser.	Properly locate condenser to freely dispel hot discharge air. Tighten or replace slipping belts or pulleys and be sure blower motor is of proper size.

Table 11–2 (continued) Thermostatic expansion valve service hints

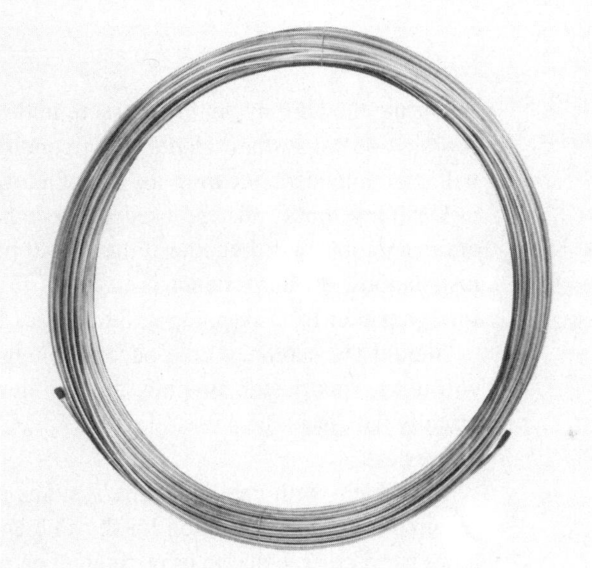

Figure 11–23 Capillary tubing (Courtesy of Robinair)

refrigerant is fed into the evaporator at a predetermined rate. The rate is determined by the unit size, size of the capillary tube, and the heat load for the unit. Capillary tubes act much like a small-diameter water pipe that hold back the water until the pressure forces the water out with a small flow rate. The same theory is used in capillary tubes. They hold back the refrigerant until the discharge pressure has built up enough to force the liquid refrigerant into the evaporator.

Because the opening in the capillary tube is fixed, the rate of flow also cannot be varied to a great extent. During conditions when the load and the discharge pressure do not vary, capillary tubes perform very well. However, when the discharge pressure or the load varies considerably, the capillary tube will either underfeed or overfeed the evaporator.

Figure 11–24 Capillary tube and strainer (Courtesy of Parker Hannifin Corporation)

During the OFF cycle, the pressure in the high-side is gradually bled over into the low-side of the system. Under these conditions, when the compressor next starts, it will start unloaded, allowing the use of a low-starting torque motor.

Capillary tubes must be used on extremely clean systems. Otherwise, the small diameter tube may become plugged, stopping the refrigeration process. In most installations, a filter-strainer is installed just ahead of the capillary tube to prevent foreign matter from stopping it. See Figure 11–24.

Should the capillary tube become plugged, the evaporator will defrost, the unit will not normally stop running, and the thermal overload may stop the compressor. Usually the discharge pressure will become very high, causing the compressor motor to overload.

A problem with capillary tube systems is the adjusting of the amount of charge required for proper operation for the ambient temperature at charging time. Capillary tubes use a critical charge of refrigerant because there is no receiver in which to store any excess. An overcharge of refrigerant will cause high discharge pressures, compressor motor overloading, and possible liquid floodback to the compressor during the OFF cycle. A refrigerant undercharge will allow refrigerant vapor to enter the capillary tube, resulting in a reduction in refrigeration capacity.

The simplicity of the capillary tube eliminates the need for a receiver tank and the need for a high-starting torque motor. Because of these factors, the capillary tube is a very inexpensive flow-control device.

Accurately sizing capillary tubes is sometimes a very difficult task. The final selection is usually done by an actual test for system operation. When the properly sized tube has been determined for a given size unit, it can be used on identical systems with very few problems. There are tables and charts to aid in sizing of capillary tubes. Figures 11–25, 11–26, 11–27, 11–28, and 11–29 can be used to tentatively select capillary tubes for various size units.

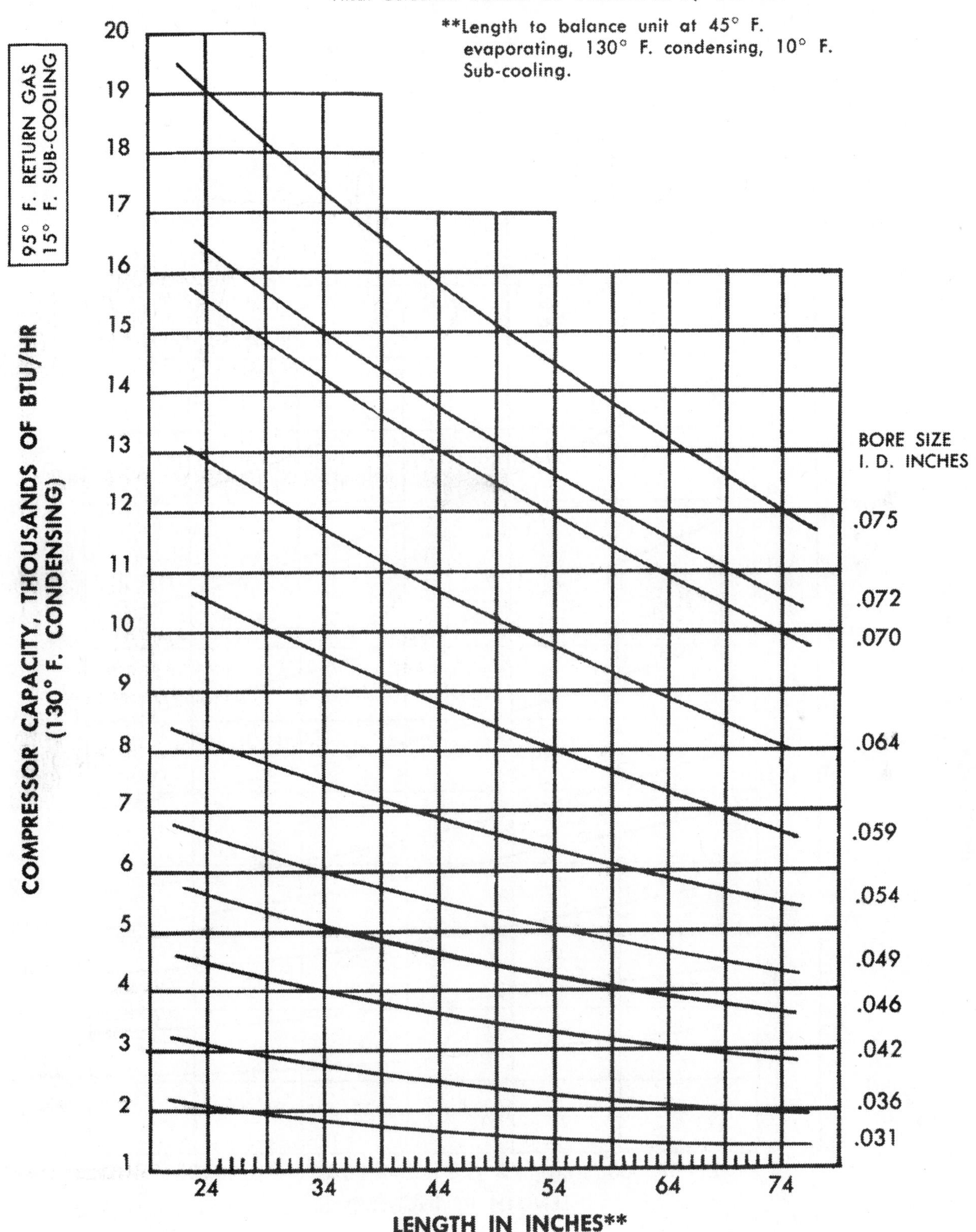

Figure 11–25 Capillary tube selection—R–22 high temperature (Courtesy of Copeland Corp.)

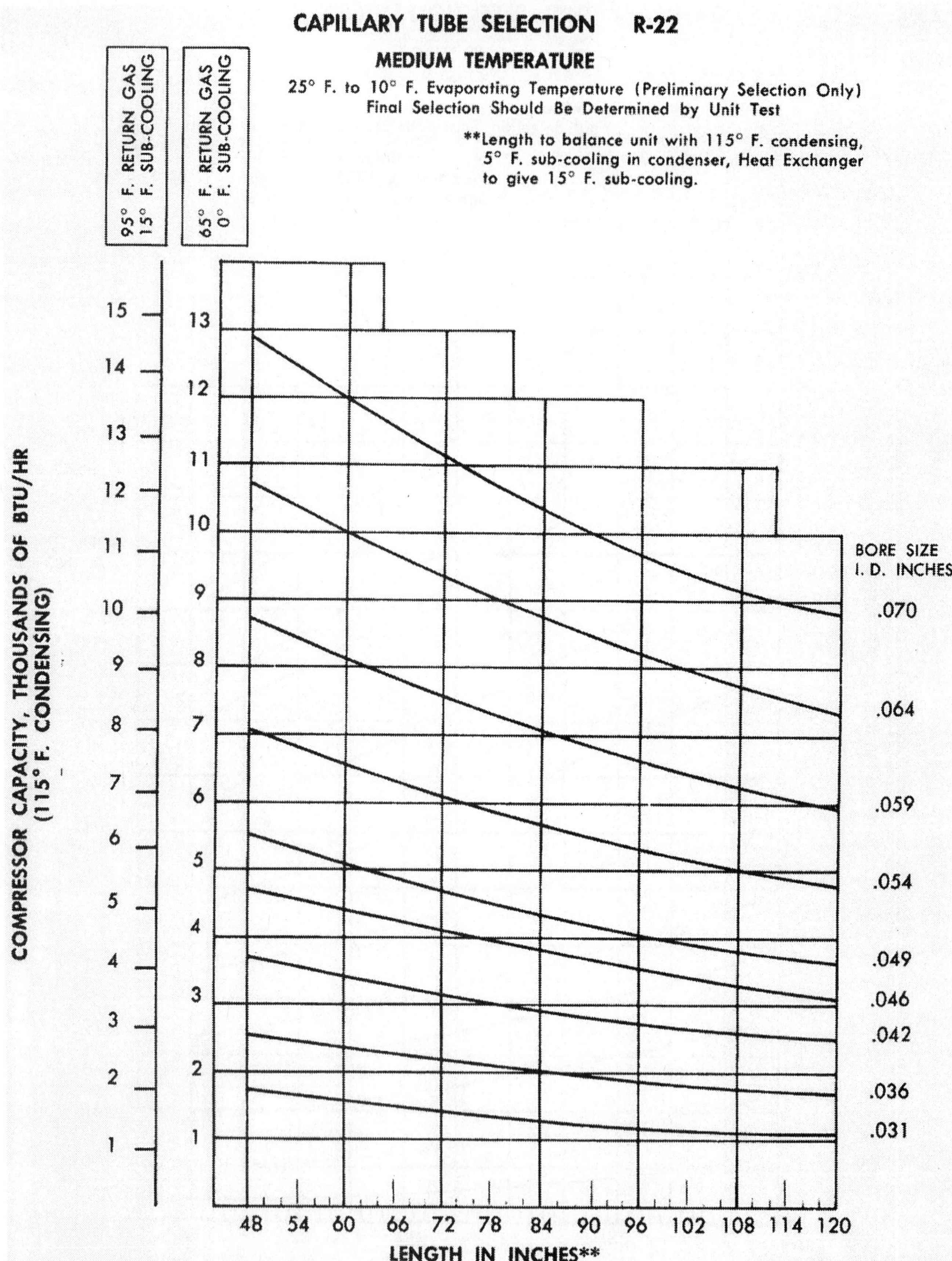

Figure 11–26 Capillary tube selection—R–22 medium temperature (Courtesy of Copeland Corp.)

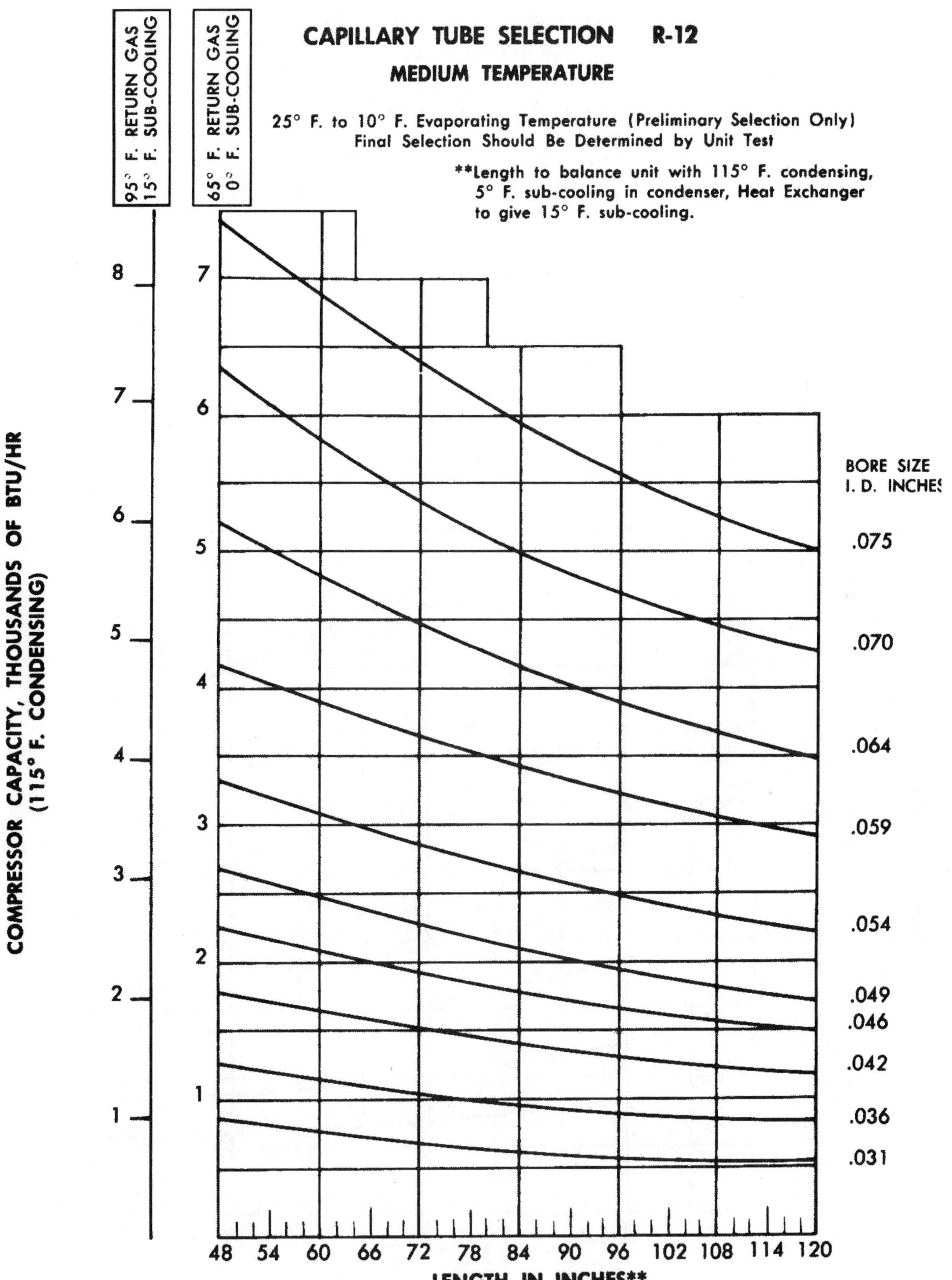

Figure 11–27 Capillary tube selection—R–12 medium temperature (Courtesy of Copeland Corp.)

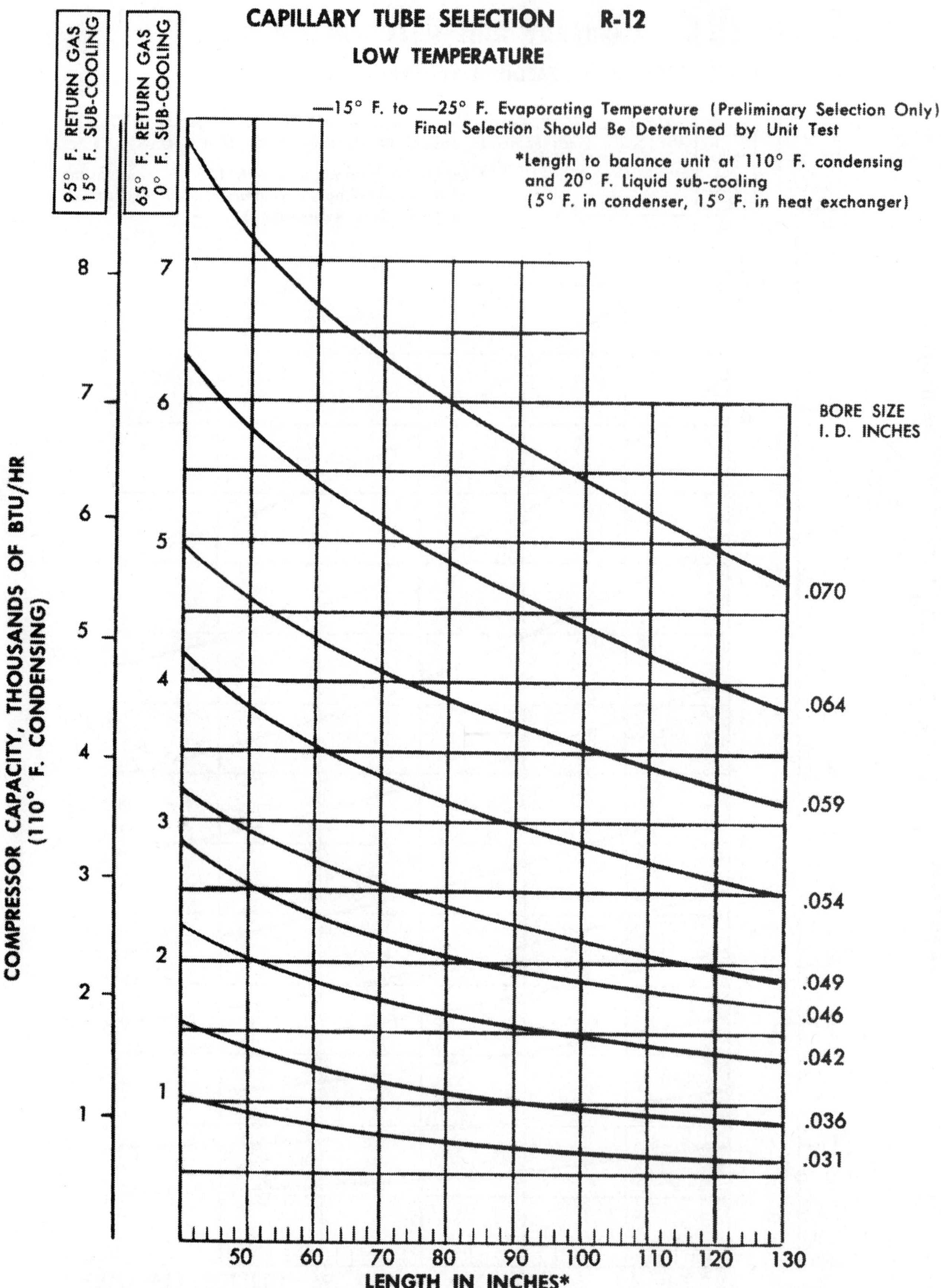

Figure 11–28 Capillary tube selection—R–12 low temperature (Courtesy of Copeland Corp.)

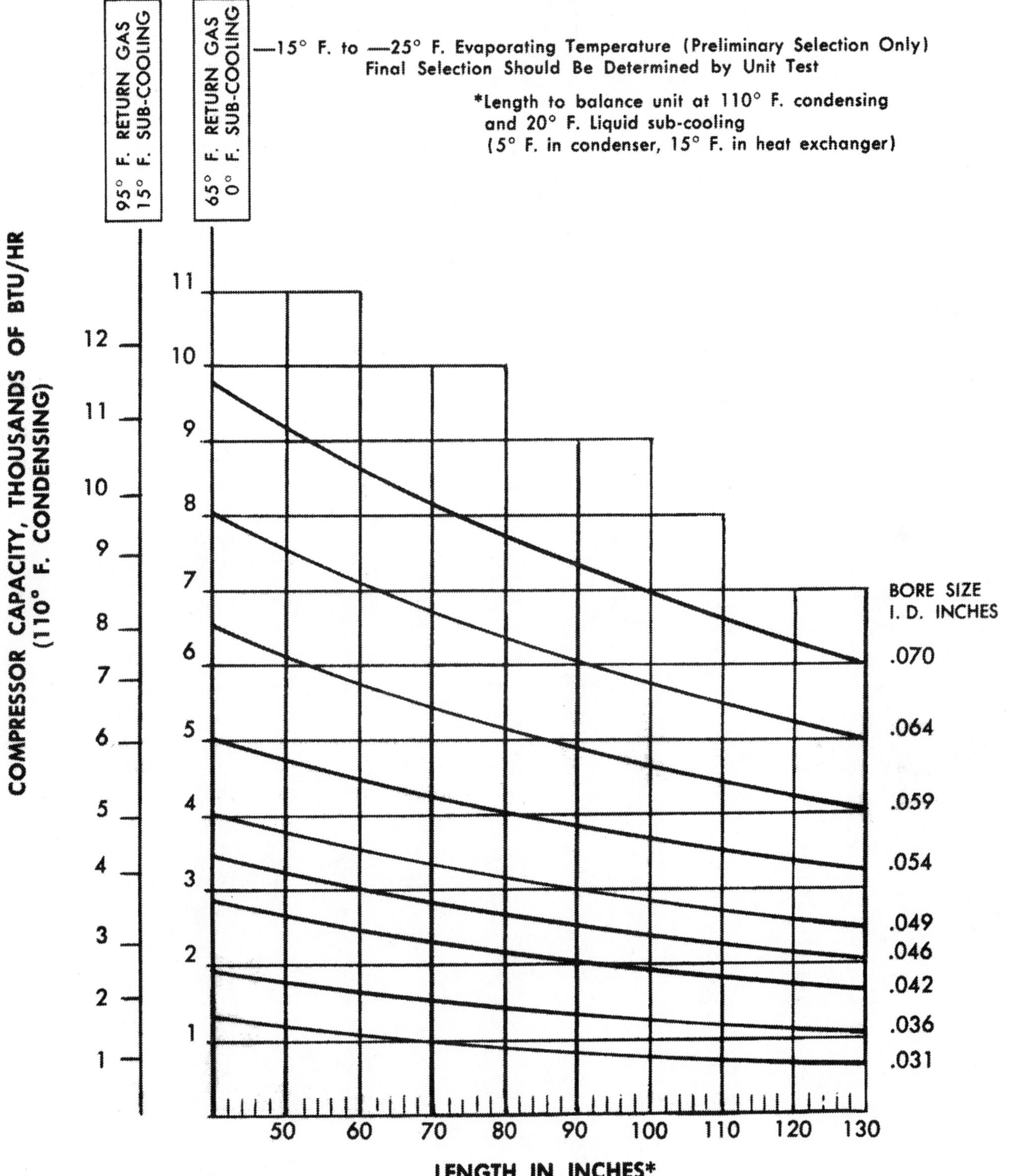

Figure 11–29 Capillary tube selection—R–502 low temperature (Courtesy of Copeland Corp.)

SUMMARY 11–5

- Capillary tubes are the most simple of all the refrigerant flow-control devices.
- Capillary tubes are generally used on flooded type systems.
- Capillary tube systems are critically charged.
- Accurate sizing of capillary tubes is sometimes a very difficult task.

REVIEW QUESTIONS 11–5

1. Why are capillary tube systems critically charged?
2. What happens to the suction pressure on an increase in the discharge pressure on a capillary tube system?
3. Why are strainers important on capillary tube systems?

CHAPTER

12

ACCESSORIES

OBJECTIVES

Upon completion of this chapter, you should be able to:

- Better understand the purpose of the accessories used on refrigeration systems
- Know the purpose of filter-driers used in a refrigerant circuit
- Know the purpose of suction line filter-driers
- Understand the use of crankcase heaters

INTRODUCTION

An accessory can be defined as an article or device that adds to the convenience or the effectiveness of the total system, but it is not a required component. There are many accessories available for use on refrigeration and air conditioning systems. Their use will depend on the desired operation of the total unit and its application. Accessories are used to aid the refrigeration or air conditioning system to operate more efficiently, economically, or to certain specifications.

12–1 ACCUMULATORS

Accumulators are designed to prevent liquid refrigerant and oil from returning to the compressor where it could possibly cause severe damage. Some types of refrigeration systems are more susceptible than others to the return of liquid refrigerant to the compressor in quantities large enough to cause damage. Any liquid refrigerant allowed to return to the compressor has the capability to break compressor valves, dilute the lubricating oil, wash out the bearings, and cause complete loss of oil in the crankcase due to foaming. When liquid returns to the compressor, it causes a condition known as liquid slugging. When the oil foams and leaves the crankcase, it can possibly break valves, pistons, rods, crankshafts, and other compressor components. When the oil is pumped from the crankcase, it causes a condition known as oil pumping or slugging.

The purpose of an accumulator is to act as a reservoir to hold any liquid refrigerant or oil that may be returning to the compressor through the suction line. The liquid refrigerant is boiled off and returned to the compressor as a vapor. The oil is returned to the compressor crankcase through a metering hole located in the accumulator tubing.

Accumulators are installed in the suction line immediately before the compressor suction service valve. See Figure 12–1.

When the accumulator is properly sized, it can regulate relatively large quantities of liquid refrigerant and oil. Liquid refrigerant collects in the bottom of the accumulator shell where it is metered back to the compressor, along with any oil that has accumulated at a rate that will neither damage the compressor nor cause a starved condition. See Figure 12–2.

The refrigerant vapor is fed back to the compressor through the opening in the suction line near the top of the accumulator. Accumulators will almost always sweat (condense water) on their outer surface as well as gain some heat. Some manufacturers recommend insulating the accumulator shell to prevent moisture damage to the surrounding area and to reduce the amount of heat absorbed into the system at this point.

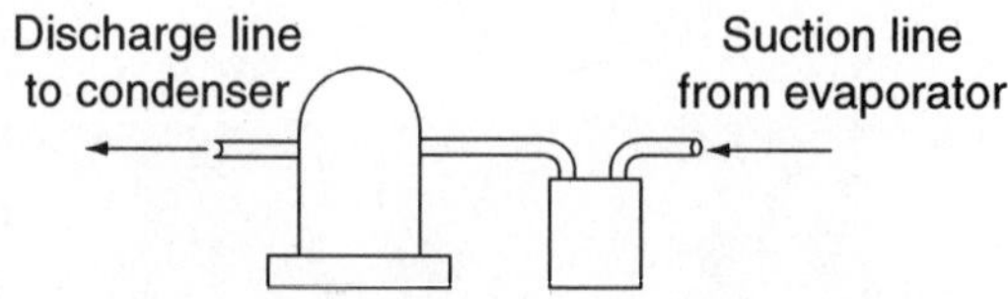

Figure 12–1 Location of accumulator

Figure 12–2 Internal view of suction accumulator (Courtesy of Refrigeration Research, Inc.)

Accumulator Selection

A properly selected accumulator will not produce any appreciable pressure drop. It will provide proper oil return to the compressor crankcase. It will also have enough capacity to allow at least 50% of the total system refrigerant charge to be stored in it. The sizes of the inlet and outlet connections do not necessarily need to match those of the suction line or other components.

SUMMARY 12–1

- Accumulators are designed to prevent liquid refrigerant and oil from returning to the compressor.
- The purpose of an accumulator is to act as a reservoir to hold any liquid refrigerant and oil that may be returning to the compressor.
- A properly selected accumulator will not produce any appreciable pressure drop.

REVIEW QUESTIONS 12–1

1. Why are accumulators needed on some refrigeration systems?
2. What will a properly sized accumulator do?
3. With what capacity should an accumulator be designed?

Figure 12–3 Liquid line drier (Courtesy of Alco Controls Div., Emerson Electric Co.)

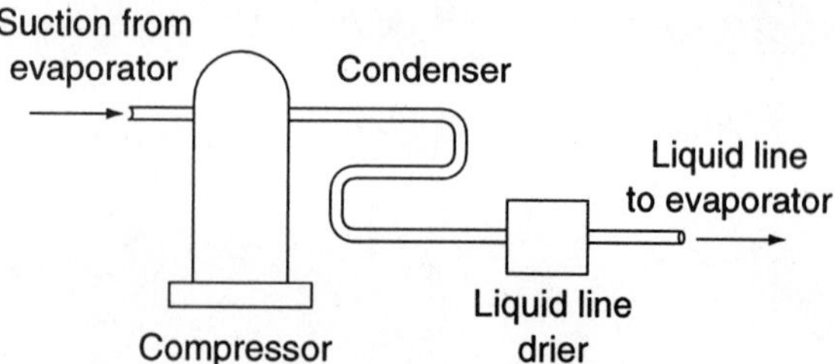

Figure 12–4 Location of liquid line drier

12–2 FILTER-DRIERS

Filter-driers are used in refrigeration systems to remove moisture, dirt, and other debris from the system. Moisture is probably the most harmful contaminant that can enter a refrigeration system. Filter-driers are installed on almost every installation, whether it is factory assembled or a field built-up system. See Figure 12–3.

Liquid line driers have been used for many years as moisture-removing devices. They are installed in the liquid line where both the refrigerant and moisture are concentrated. See Figure 12–4.

It is necessary that moisture be removed from the system because it is the single most contributing factor in the formation of acids, sludge, and corrosion. Just a small amount of moisture mixed with hydrocarbon refrigerants form hydrochloric and hydrofluoric acids. Moisture is most noticeable when it freezes in the orifice of the flow-control device, causing the refrigeration unit to stop cooling. When a large amount of moisture is present in the system, major mechanical breakdowns are possible. It is necessary that any moisture, acid, or corrosion be removed from the system as soon as possible to prevent costly damage to the system components.

Selection of Liquid Line Driers

There are several factors that must be considered when selecting driers. The following is a list of the most important:

1. Type and amount of refrigerant. This is especially important with some of the HFC refrigerants.
2. Refrigeration system capacity
3. Line size
4. Allowable pressure drop through the drier

In most instances, the refrigerant line size, the type of refrigerant, and the application are known. It then becomes a matter of selecting the drier on the drying capacity and refrigerant-flow capacity. This information is published by the filter-drier manufacturer in tables. See Table 12–1.

FILTER-DRIER SELECTION SERIES "H" "DRI-CORE"®										CAPACITY RATINGS**						
CATALOG NUMBER		SIZE	CORE FILTER AREA	DIMENSIONS INCHES			WT.	RECOMMENDED TONNAGE*		DROPS OF WATER				FLOW CAP. @ 2 P.S.I. IN TONS		
				SHELL	OVERALL LGT.			REFRIGERANTS		R-12(15PPM)		R-22 (60 PPM)				
										LIQUID LINE TEMP. °F						
FLARE	O.D.S.	CONN.	SQ. IN.	DIA	FLARE	O.D.S.	LBS.	R-12	R-22	7C°	125°	75°	126°	R-12	R-22	R-502
H032	H032-S	1/4	11	1 5/8	4 3/16	3 1/2	1/2	3/4	3/4	46	33	31	20	2.3	3.0	2.0
H033	H033-S	3/8			4 7/16	3 9/16										
H052	H052-S	1/4	17	2 1/2	5	4 5/16	1	1	1	92	66	62	40	2.3	3.0	2.0
H053	H053-S	3/8			5 5/16	4 7/16								4.0	5.2	3.5
H082	H082-S	1/4	24	2 1/2	5 13/16	5 1/8	1 1/8	1	1	156	112	107	68	2.7	3.5	2.4
H083	H083-S	3/8			6 1/4	5 1/4		2	2					5.3	6.8	4.7
H084	H084-S	1/2			6 1/2	5 3/4	1 1/4							8.2	10.6	7.2
H162		1/4	36	3	6 7/16		2	2	2	282	202	192	122	2.7	3.5	2.4
H163	H163-S	3/8			6 3/4	5 7/8		3	3					5.5	7.1	4.8
H164	H164-S	1/2			7 1/8	6 1/4	2 1/4	4	4					8.7	11.2	7.7
H165	H165-S	5/8			7 3/8	6 1/2		5	5					11.0	14.2	9.7
	H167-S	7/8				6 7/8	2 1/2	7 1/2	7 1/2					13.2	17.0	11.6
H303		3/8	57	3	9 11/16		3 1/2	4	5	490	352	335	212	5.8	7.5	5.1
H304	H304-S	1/2			10	9 1/4		7 1/2	7 1/2					11.8	15.2	10.4
H305	H305-S	5/8			10 1/4	9 1/2	3 3/4	10	10					15.3	19.7	13.5
	H307-S	7/8				9 7/8			15					24.9	32.1	21.9
	H309-S	1 1/8				10 1/8	4							30.0	38.7	26.4
H413		3/8	71	3 1/2	9 3/4		5 3/8	5	71/2	710	506	482	305	5.8	7.5	5.1
H414	H414-S	1/2			9 15/16	9 1/8		10	10					12.1	15.6	10.6
H415	H415-S	5/8			10 5/16	9 3/8	5 1/2		15					16.0	20.6	14.1
	H417-S	7/8				9 13/16		15	20					25.9	33.4	22.8
	H419-S	1 1/8				10 1/8								31.0	40.0	27.3
	H607-S	7/8	106	3		16 1/8	6	20	25	1158	579	562	432	31.0	40.0	27.3
	H609-S	1 1/8				16 3/8	6 1/4	25	30					34.0	45.0	30.0
H755	H755-S	5/8	123	3 1/2	9 3/4	14 7/8	8 1/4	15	20	1320	950	905	507	17.5	22.6	15.4
	H757-S	7/8				15	8 3/4	25	30					28.2	36.4	24.8
	H759-S	1 1/8				15 1/4								33.8	43.6	29.7

Table 12–1 Filter-drier selection (Courtesy of Henry Valve Company)

Types of Liquid Line Driers

The two types of liquid line driers are the straight-through sealed type and the angle-replaceable core type. See Figures 12–5 and 12–6.

The sealed-type driers are for one-time use only. The angle-replaceable core types are for multiple use applications. They are especially useful when a system must be cleaned up and more than one filter will be needed to satisfactorily complete the job.

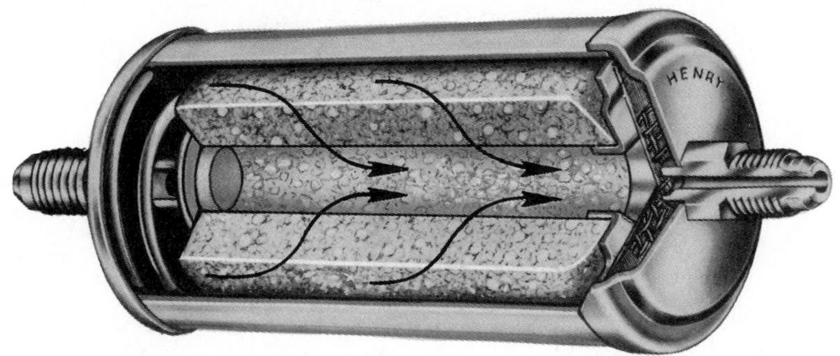

Figure 12–5 Straight-through liquid line drier (Courtesy of Henry Valve Company)

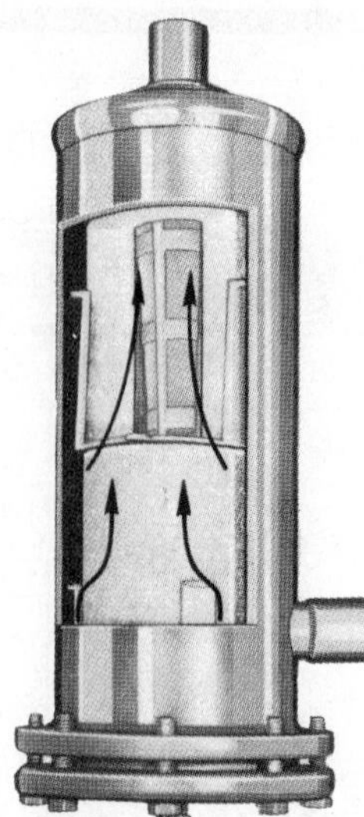

Figure 12–6 Angle-replaceable core drier (Courtesy of Henry Valve Company)

Suction-Line Filter

A properly selected and installed suction-line filter will provide protection for the compressor by collecting any foreign matter that travels through the suction line. They are installed in the suction line directly upstream of the compressor suction line connection. See Figure 12–7.

Should any solid particles enter the compressor and circulate with the oil, they could cause serious damage to the bearings and other machined surfaces.

Suction-line filters should be installed just before the system is started up, or during service procedures that require one be used.

Selection of Suction-Line Filter Driers. Following is a list of factors that must be considered when selecting suction-line filters.

1. The type of refrigerant used
2. The size of the suction line
3. The allowable pressure drop through the filter
4. The application (air conditioning, commercial refrigeration, or low-temperature application)

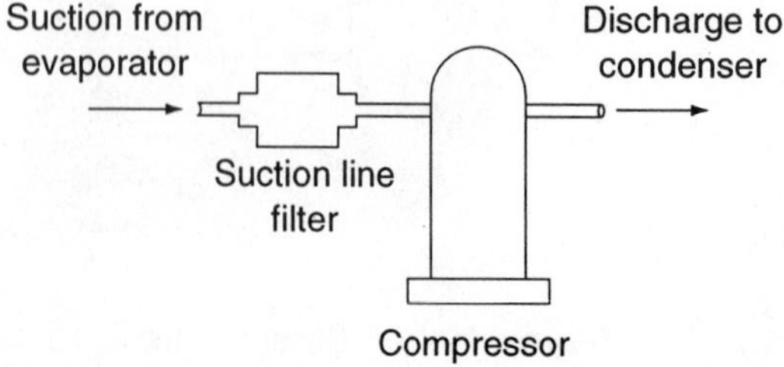

Figure 12–7 Location of suction-line filter

CATALOG NUMBER O.D.S.	SIZE CONN.	CORE FILTER AREA SQ. IN.	DIMENSIONS–INCHES DESSICANT CUBIC VOLUME	SHELL DIA.	OVER-ALL LENGTH	WT. LBS.
HS164–S	1/2	33	16	3	6 3/8	2 1/4
HS165–S	5/8				6 5/8	2 3/8
HS166–S	3/4				6 5/8	2 3/8
HS307–S	7/8	53	30	3	9 7/8	3 1/4
HS419–S	1 1/8	64	41	3 1/2	10 1/8	4 1/2
HS4311–S	1 3/8	67	48	4 9/16	10 1/2	7 1/2
HS4313–S	1 5/8				10 1/2	7 1/2

Table 12–2 Suction-line filter selection (Courtesy of Henry Valve Company)

In most cases the type of refrigerant, the line size, and the application are known. The filter is then selected on the amount of refrigerant flow expressed in tons of refrigeration capacity. The flow-capacity is published in the filter manufacturer's tables for all refrigerants for which the filter was designed. See Table 12–2.

Types of Suction-Line Filters. There are two types of suction-line filters available. They are the straight-through sealed type and the angle-replaceable core type. See Figures 12–8 and 12–9.

Suction-Line Filter-Drier. The suction-line filter-drier removes foreign particles from the system as well as moisture and acid from the refrigerant. They are manufactured using a type of material that has a high affinity for moisture and for the particular type of refrigerant used. With some of the HFC refrigerants, a specified type of drying material must be used. They are designed to remove moisture, acid, and foreign particles from the refrigerant in the vapor form. See Figure 12–10.

They have a large filtering area and a traverse-flow passageway to allow the refrigerant vapor to flow through them with a very small refrigerant pressure drop. When filter-driers are properly installed, they will permit the return of clean, dry

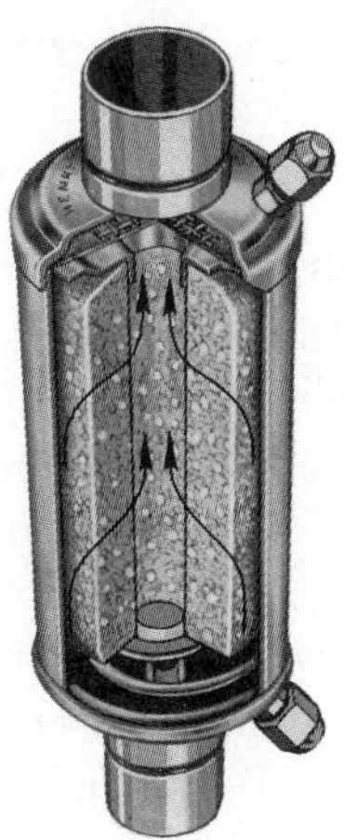

Figure 12–8 Straight-through suction-line filter (Courtesy of Henry Valve Company)

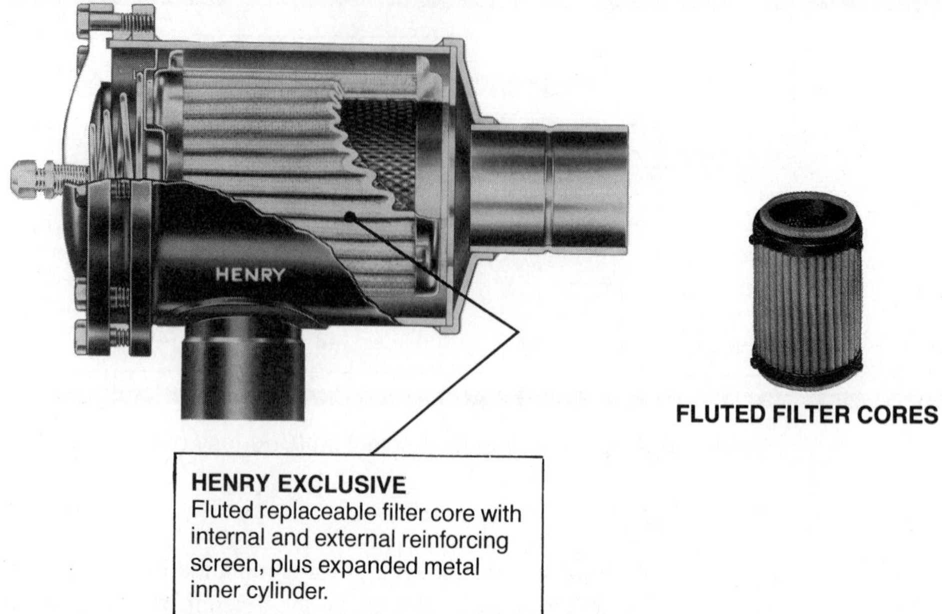
HENRY
FLUTED FILTER CORES
HENRY EXCLUSIVE
Fluted replaceable filter core with internal and external reinforcing screen, plus expanded metal inner cylinder.

Figure 12–9 Angle-replaceable core suction-line filter (Courtesy of Henry Valve Company)

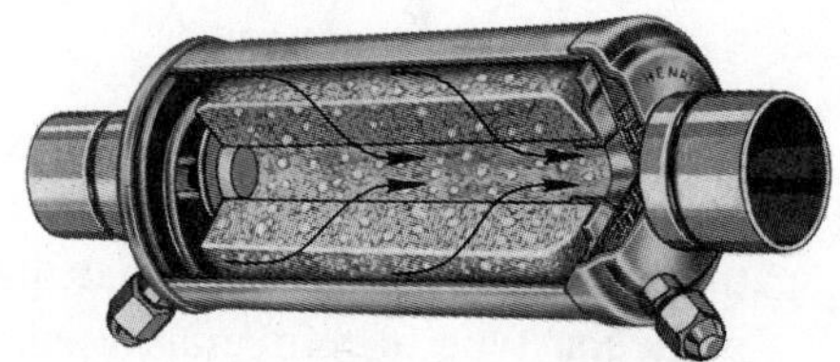

Figure 12–10 Suction-line filter-drier (Courtesy of Henry Valve Company)

refrigerant to the compressor. Suction-line filter-driers must be carefully monitored for several days and if a refrigerant pressure drop of 5 psi occurs, the filter-drier must be replaced with either a new drier or a new core. This procedure is followed until it is safe to leave the filter-drier in the system without monitoring.

SUMMARY 12–2

- Filter-driers are used in refrigeration systems to remove moisture, acid, dirt, and other debris from the system.
- Liquid line driers have been used for many years as moisture removing devices.
- A properly selected and installed suction-line filter will provide protection for the compressor by collecting any foreign matter that travels through the suction line.
- Suction-line filters are installed in the suction line directly upstream of the compressor suction line connection.
- Suction-line filters are selected on the amount of refrigerant flow expressed in tons of refrigeration capacity.
- Suction-line filter-driers must be carefully monitored for several days and if a refrigerant pressure drop of 5 psi occurs, the filter-drier must be replaced.

REVIEW QUESTIONS 12–2

1. What is probably the most harmful contaminant in a refrigeration system?
2. Why are liquid-line driers installed in the liquid line?
3. How are liquid-line driers sized in the field?
4. What is the purpose of suction-line filters?

12–3 STRAINERS

Strainers are shell-like devices placed in the refrigerant lines to remove foreign particles. They are installed directly ahead of the system component they are to protect. If the foreign particles were left in the system they could possibly clog the orifices and other moving parts, preventing proper operation of the component and the system as a whole.

Types of Strainers. There are three general types of strainers used. They are the straight-through sealed type (see Figure 12–11), the cleanable angle type (see Figure 12–12), and the cleanable "Y" type (see Figure 12–13).

Figure 12–11 Straight-through sealed strainer (Courtesy of Henry Valve Company)

Figure 12–12 Cleanable angle strainer (Courtesy of Henry Valve Company)

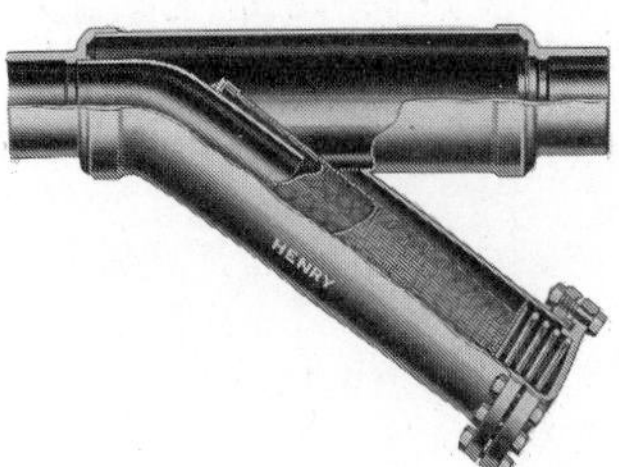

Figure 12–13 Cleanable "Y" strainer (Courtesy of Henry Valve Company)

SUMMARY 12–3

- Strainers are shell-like devices placed in the refrigerant lines to remove any foreign particles.
- Strainers are installed directly ahead of the system component they are to protect.

REVIEW QUESTIONS 12–3

1. Where are strainers installed?
2. Why must foreign particles be removed from a refrigeration system?

12–4 MOISTURE-LIQUID INDICATORS

Moisture-liquid indicators are glass eyes that are placed in the liquid line between the condenser or receiver and the flow-control device. Their purpose is to allow the technician to easily determine if the system has a sufficient refrigerant charge or if there is moisture in the system. An overcharge of refrigerant will not be indicated by these devices. See Figure 12–14.

If the system is short of refrigerant, there is a restriction in the liquid line or there is improper liquid subcooling. The sight glass will show bubbles moving in the direction of refrigerant flow. The system must be operating before the condition of the refrigerant charge can be determined. It is usually better if the system has operated long enough to reach all the operating temperatures. When the system is found low on refrigerant charge, the leak must be found and repaired before recharging the system.

Purpose of the Moisture Indicator. The moisture indicator will show a green dot in the center when the system is sufficiently dry. As the level of moisture increases,

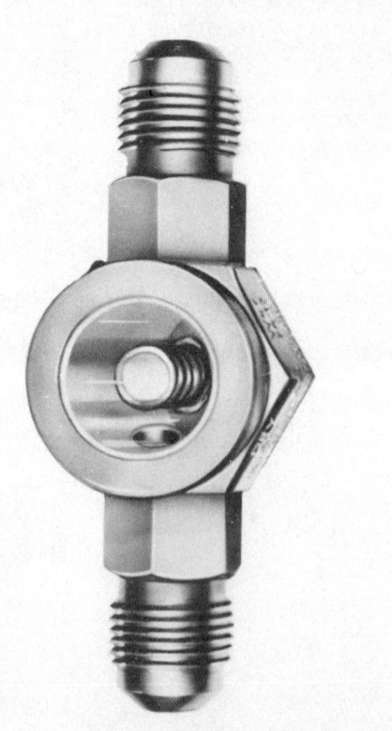

Figure 12–14 Moisture-liquid indicator (Courtesy of Sporlan Valve Company)

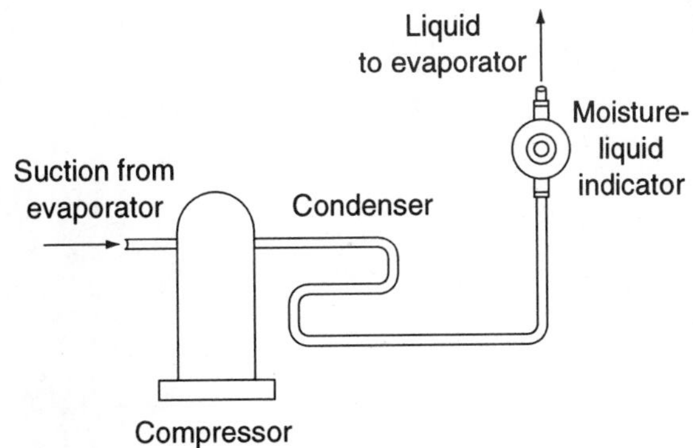

Figure 12–15 Location of moisture-liquid indicator

the dot will gradually change from green toward yellow. When the dot shows yellow, there is too much moisture in the system. The system must be cleaned immediately to prevent certain damage. The moisture indicator must always be installed in the liquid line for complete accuracy. See Figure 12–15.

The system must be in operation for the sensitized element to correctly indicate the condition of the refrigerant. An excessive amount of moisture and heat will damage the sensing element and cause it to remain yellow or discolored, depending on the amount of damage. The liquid-moisture indicator must then be replaced.

SUMMARY 12–4

- Moisture-liquid indicators are glass eyes that are placed in the liquid line between the condenser or receiver and the flow control device.
- The purpose of moisture-liquid indicators is to allow the technician to easily determine if the system has a sufficient refrigerant charge or if there is moisture in the system.
- The moisture indicator will show a green dot in the center when the system is dry.
- The system must be in operation for the sensitized element to correctly indicate the condition of the refrigerant.

REVIEW QUESTIONS 12–4

1. Is the purpose of liquid indicators to indicate a refrigerant overcharge?
2. Can a liquid indicator show bubbles if a system has a restriction in it?
3. What must be done to a system before a moisture indicator will operate properly?

12–5 OIL SEPARATOR

There is always oil circulating through any refrigeration system. The oil is taken from the compressor crankcase by the refrigerant as it passes through the compressor. Each type of refrigerant has a different affinity for the oil. Therefore, the amount of oil traveling through the system will depend on the type of refrigerant used, the evaporation temperature, and the refrigerant pressure. Regardless of the refrigerant type used, there will always be some oil present in the system.

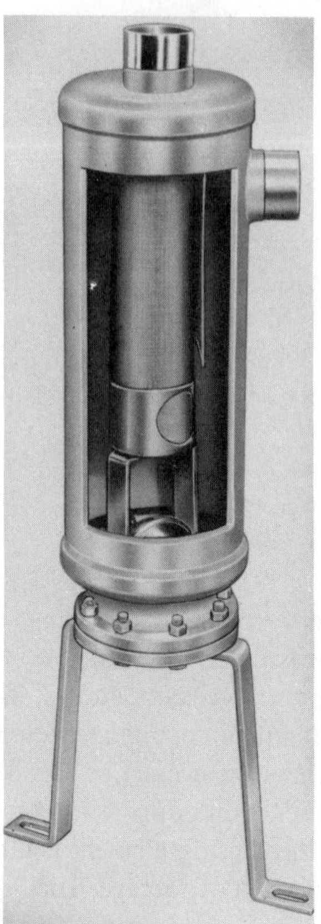

Figure 12–16 Oil separator (Courtesy of AC&R Components, Inc.)

When oil loss becomes a problem, an oil separator can be installed in the system to reduce the amount of oil circulating with the refrigerant. See Figure 12–16.

Some design engineers recommend the installation of an oil separator on every low-temperature and large air conditioning job, up to about 150 tons capacity. The use of an oil separator can greatly increase the operating efficiency of almost any system. This is especially true when open-top display cases are used in commercial refrigeration installations where the evaporating temperatures can easily reach –30°F to –40°F. Also, most two-stage compressor equipment manufacturers require the installation of an oil separator.

Oil Separator Purpose

The purpose of an oil separator is to maintain the desired oil level in the compressor crankcase during operation. However, there are more benefits realized by preventing the free circulation of oil. The oil can settle in the evaporator or condenser and reduce the system effectiveness and overall performance and efficiency.

These devices are designed to separate the oil from the refrigerant and return the oil to the compressor crankcase to prevent lubrication problems. To be effective, this oil must be separated from the refrigerant before it can reach the other system components. Oil is designed to stay in the compressor crankcase and not circulate through the system.

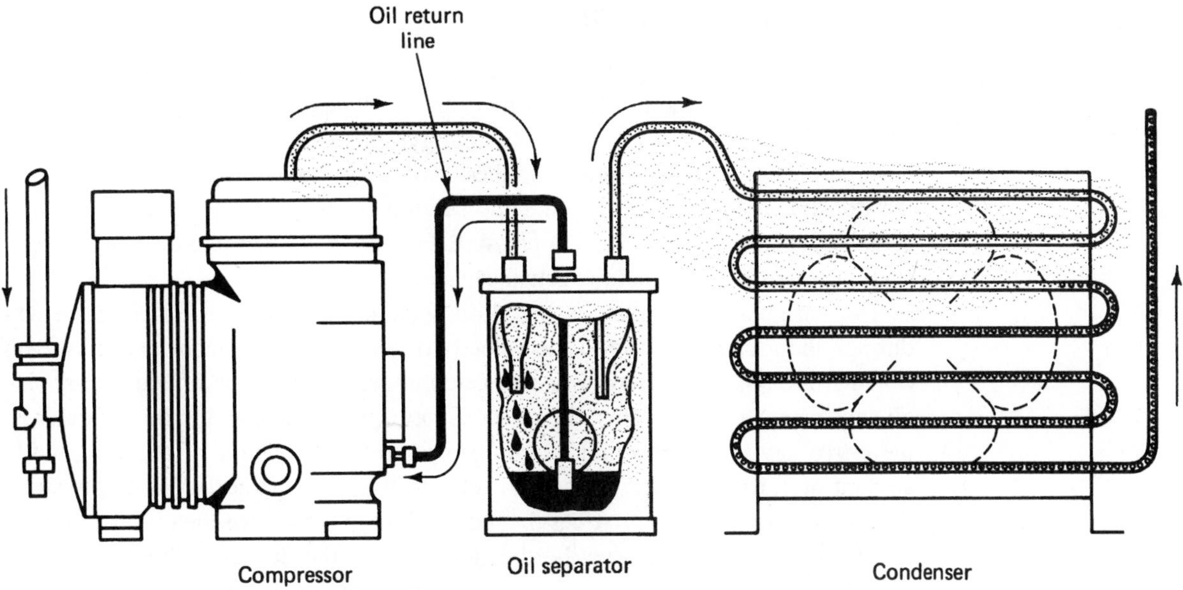

Figure 12–17 Oil separator location (Courtesy of AC&R Components, Inc.)

How the Oil Separator Functions

The oil separator is installed in the discharge line between the compressor and the condenser. See Figure 12–17.

The hot discharge gas from the compressor enters the inlet of the separator in a foglike vapor and flows through the inlet baffling. At this point, the vapor changes direction several times as it strikes the baffles. The oil, at this point, is in the form of very fine droplets, or atomized. These droplets fall, or separate, from the refrigerant by striking against each other and becoming larger, heavier droplets of oil. This separation of the oil is caused by a reduction in the refrigerant velocity inside the oil separator. The shell of the oil separator is designed to cause this reduction in velocity. When the velocity of the refrigerant is reduced, the oil droplets are heavier than the refrigerant vapor and have more momentum. They collide with each other as well as strike the surfaces of the separator shell. The relatively oil-free refrigerant vapor then passes out of the separator, through the outlet screens, and the refrigerant velocity is allowed to increase to the original velocity. The refrigerant vapor then goes to the condenser. The separated oil drops to the bottom of the separator, where it is stored until sufficient quantity has accumulated. It is then returned to the compressor crankcase. When enough has accumulated, a float valve will open, allowing the oil to return to the compressor crankcase. Also, any foreign particles that may have entered the separator can settle and be kept out of circulation. The float valve is located high enough in the separator to allow only clean oil to be returned. To keep the compressor from operating with insufficient oil, it takes only a small amount in the separator to open the float valve.

Sizing an Oil Separator

Oil separator manufacturers generally rate their oil separators in tons capacity or by the horsepower. Caution should be exercised when using the horsepower rating because the actual Btu rating of the compressor may be quite different than the horsepower rating. The difference in these two are because of the compressor speed, the density of the return gas, and the suction pressure. Compressor capacity is actually dependent on the suction pressure. For any given compressor, the higher the suction pressure, the higher the actual capacity.

The refrigerant line sizing is based on obtaining the desired refrigerant velocity through the system. A drop in the velocity of the refrigerant vapor and oil mixture flowing through the oil separator is a determining factor in the effectiveness of the oil separator. Because of this, the relationship of the volume of the oil separator and the refrigerant line connections, when compared to the volume of the shell and the amount of baffling, is very important for effective oil separation. When the size of the compressor increases, so does the requirement in the size of the oil separator. The line connections must always be the same size as the discharge line to avoid excessive pressure drop. It is generally recommended that the line connections be at least as large as the discharge line.

SUMMARY 12–5

- The amount of oil traveling through a system will depend on the type of refrigerant used, the evaporation temperature, and the refrigerant pressure.
- Some design engineers recommend the installation of an oil separator on every low-temperature and large air conditioning job, up to about 150 tons capacity.
- The purpose of an oil separator is to maintain the desired oil level in the compressor crankcase during operation.
- The oil separator is installed in the discharge line between the compressor and the condenser.
- Oil separator manufacturers generally rate their oil separators in tons capacity or by the horsepower.
- It is generally recommended that the line connections be at least as large as the discharge line.

REVIEW QUESTIONS 12–5

1. Why does oil circulate through a refrigeration system?
2. What type of refrigerant prevents oil circulation through the system?
3. What causes the separation of oil from the refrigerant in an oil separator?
4. How is oil returned from the separator to the compressor?
5. What caution should be taken when sizing an oil separator?

12–6 VIBRATION ELIMINATORS

When the refrigeration unit is in operation, there is always some noise created. This noise will be transmitted through the discharge line to the structure and cause an annoying noise inside the building. A vibration eliminator is a flexible metallic hose that effectively separates the compressor from the refrigerant line. See Figure 12–18.

In most installations, the vibration eliminator is installed in the discharge line. See Figure 12–19.

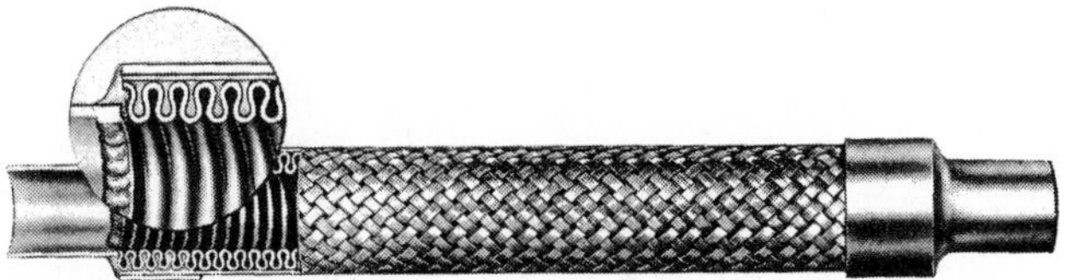

Figure 12–18 Metallic vibration eliminator (Courtesy of Packless Industries)

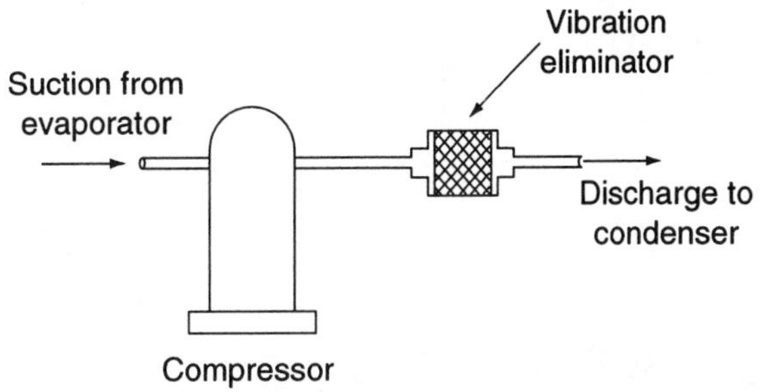

Figure 12–19 Vibration eliminator in discharge line

When extreme vibration is encountered, vibration eliminators are installed in both the discharge and the suction line. See Figure 12–20.

When compressor noise is experienced on smaller units in which soft copper tubing is used for the refrigerant lines, the tubing may be coiled once or twice to provide adequate vibration protection.

Vibration eliminators must never be bent or installed in a strain. Either of these conditions will cause the eliminator to fail in a short period of time.

SUMMARY 12–6

- Vibration eliminators are used to prevent noise from traveling through the lines to the structure.
- When extreme vibration is encountered, vibration eliminators are installed in both the discharge and the suction line.
- Vibration eliminators must never be bent or installed in a strain.

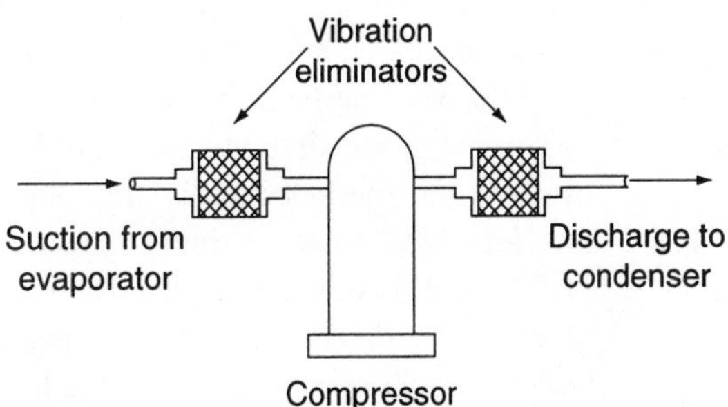

Figure 12–20 Vibration eliminators in both discharge and suction lines

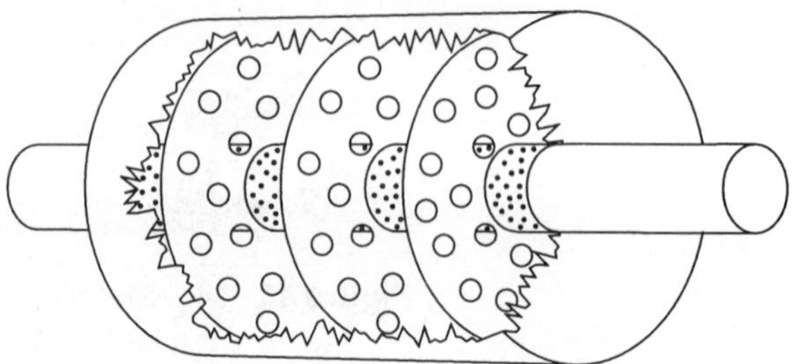

Figure 12–21 Discharge muffler

REVIEW QUESTIONS 12–6

1. What is the purpose of vibration eliminators?
2. When would a vibration eliminator be needed in both the suction and discharge lines?
3. What will result if a vibration eliminator is installed in a strain?

12–7 DISCHARGE MUFFLERS

Some compressors cause a pulsating noise to travel through the refrigerant lines. On smaller systems this problem can sometimes be helped by installing a discharge muffler. The discharge muffler is installed in the discharge line between the compressor and condenser. It is basically a shell with baffle plates inside. The internal volume of the shell depends on the displacement of the compressor and condenser. When the muffler is designed, the sound waves are also taken into account so that adequate sound prevention is obtained. See Figure 12–21.

Discharge Muffler Purpose

Discharge mufflers are used to dampen the pulsating noise created by the compressor pumping. These pulsations are sometimes transmitted through the lines and into the building.

All reciprocating compressors create some hot-vapor pulsations. The two closely related problems that are possible when the pulsations are severe are: (1) the pulsating noise, which is not damaging to anything but the user's ears, and (2) the vibration can cause the refrigerant lines to break. In many systems these two problems occur very close together.

Location of the Discharge Muffler

The discharge muffler is usually installed as close to the compressor discharge as possible. Most hermetic compressors are designed with the muffler inside the shell.

Discharge mufflers are basically shells that provide a natural oil trap. They will trap oil and could possibly trap liquid refrigerant. When installing a discharge muffler, there are some things that must be considered. When installed in a vertical line, it should be in the downward direction. See Figure 12–22.

When the discharge muffler is installed in a horizontal line, it must be installed with the larger portion on top of the line. See Figure 12–23.

Should the muffler be installed with the larger portion on the bottom or side it will act as an oil and refrigerant trap, starving the system of these two fluids.

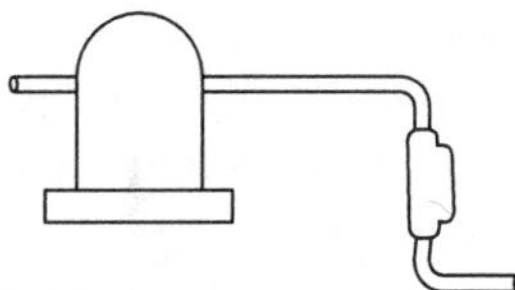

Figure 12–22 Discharge muffler installed in a vertical position

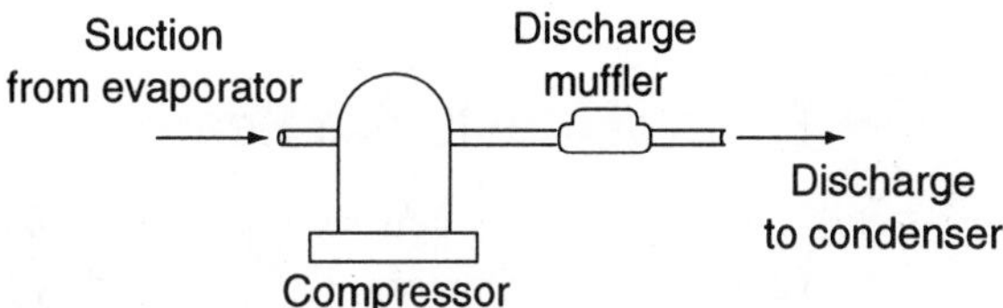

Figure 12–23 Discharge muffler installed in a horizontal position

Selection of Discharge Mufflers

Selecting discharge mufflers is an engineering problem. Much time and effort go into making this selection. Any field selection should be avoided because sometimes the problems can be increased rather than decreased when the wrong muffler is chosen.

SUMMARY 12–7

- Discharge mufflers are used to dampen the pulsating noise created by the compressor pumping.
- The discharge muffler is installed in the discharge line between the compressor and condenser, usually as close to the compressor discharge connection as possible.
- Discharge mufflers are basically a shell that provide a natural oil trap.
- If a discharge muffler is installed with the large portion on the bottom or the side, it will act as an oil and refrigerant trap, starving the system of these two fluids.
- Any field selection of a discharge muffler should be avoided because sometimes the problems can be increased rather than decreased when the wrong muffler is chosen.

REVIEW QUESTIONS 12–7

1. What type of compressor is most likely to cause pulsating noises?
2. Why must discharge mufflers be installed with the large part on the top?
3. Why do hermetic compressors not require a discharge muffler?

12–8 CRANKCASE HEATERS

Compressors installed in places where the possibility of refrigerant migration is probable, a crankcase heater should be used. Migration is most probable when the evaporator pressure will be greater than the compressor crankcase pressure. The crankcase heater is used to keep the oil in the compressor crankcase warm enough to prevent refrigerant migrating to it during the OFF cycle. Refrigerant will not migrate to the crankcase while the compressor is running because of the circulation created by the compressor.

When liquid refrigerant migrates to the crankcase it will rapidly evaporate when the compressor starts, causing the oil to foam. This foaming action will cause the oil to flow from the compressor, resulting in a lack of oil in the crankcase to provide sufficient lubrication. Also, there is a great possibility that, when the oil leaves the compressor, it will damage the valves and other compressor components. The liquid refrigerant will also act as a lubricant dilution substance allowing possible damage to the moving compressor parts.

Types of Crankcase Heaters

There are three general types of crankcase heaters: (1) the insert type (see Figure 12–24), (2) the externally mounted type (see Figure 12–25), and (3) the motor winding type (see Figure 12–26).

Crankcase Heater Selection

Crankcase heaters are low-wattage resistance elements placed in strategic places to keep the compressor crankcase warmer than the surrounding ambient temperature. They are always energized when the electricity to the condensing unit is turned on.

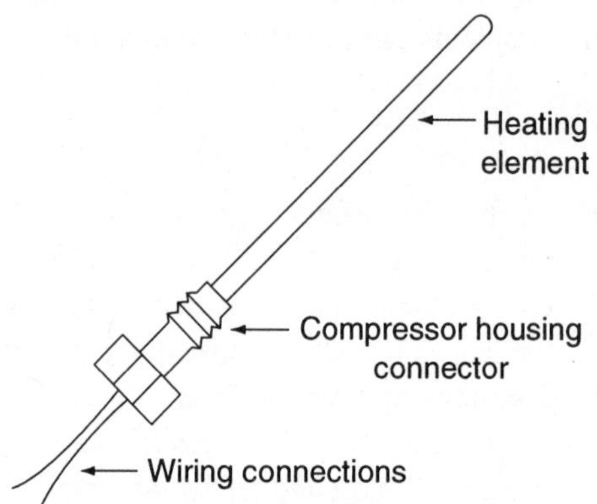

Figure 12–24 Insert type crankcase heater

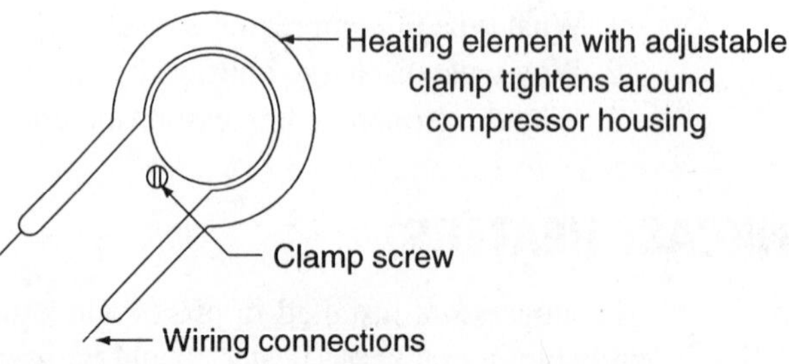

Figure 12–25 External type crankcase heater

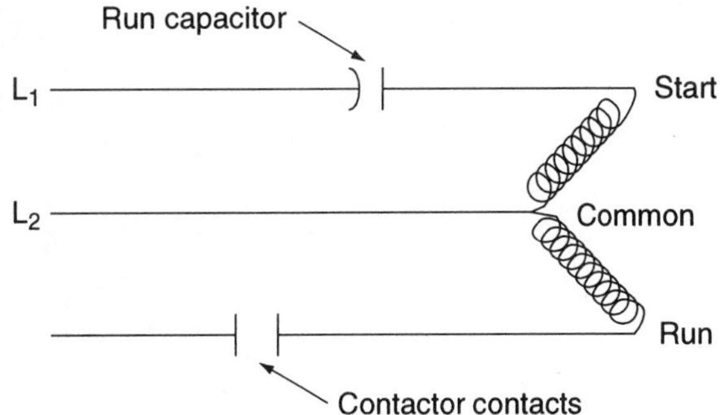

Figure 12–26 Motor winding as a crankcase heater

Sometimes they may be energized during operation; however, this is a rare occasion. Crankcase heaters are carefully selected to prevent overheating the compressor oil. Check with the unit manufacturer for recommended heaters for their unit.

SUMMARY 12–8

- The crankcase heater is used to keep the oil in the compressor crankcase warm enough to prevent refrigerant migrating to it during the OFF cycle.
- When liquid refrigerant migrates to the crankcase, it will rapidly evaporate when the compressor starts, causing the oil to foam.
- The liquid refrigerant will also act as a lubricant dilution substance, allowing possible damage to the moving compressor parts.
- Crankcase heaters are low-wattage resistance elements placed in strategic places to keep the compressor crankcase warmer than the surrounding ambient temperature.

REVIEW QUESTIONS 12–8

1. What will cause refrigerant to migrate to the compressor crankcase?
2. What causes the foaming action in the compressor crankcase on start-up?
3. What precaution must be used when selecting a crankcase heater?

12–9 COMPRESSOR SERVICE VALVES

Open-type and serviceable hermetic and semi-hermetic compressors are usually equipped with service valves. Generally one is used for the suction and one for the discharge ports on the compressor so that these pressures can be checked, as well as other service operations performed. They have no operating function in the system; however, when service operations are required they are indispensable.

Compressor service valves are forged from either brass or cast iron. The valve connections are designed to provide a full, unrestricted refrigerant flow. They are machined to fit a standard fitting. There are also valves available for use on flanged compressor fittings.

These valves are usually of the back-seating type. They are designed so that the valve seat will cause a positive seal, whether the valve stem is screwed in (front-seated) or screwed out (back-seated). Some service valves use a valve packing around

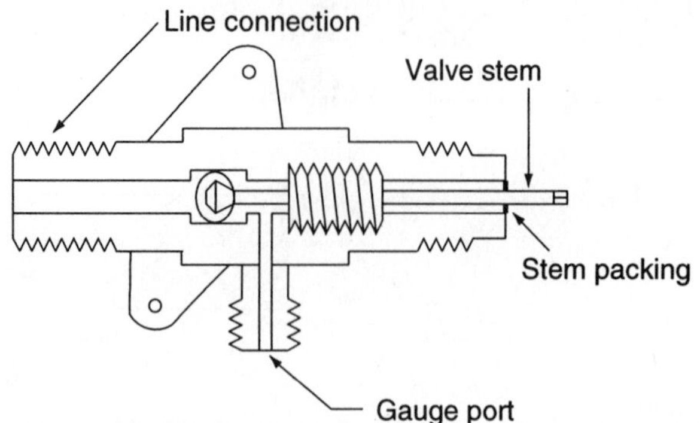

Figure 12–27 Double port compressor service valve

the stem to prevent refrigerant leakage at that point. Other types use a bellows-type seal to prevent refrigerant leakage at this point. See Figure 12–27.

During use, when the valve is back-seated, the valve is fully open to the flow of refrigerant to the compressor and the gauge port is closed off. In this position, the gauge port cover can be removed and the gauge hose screwed on the valve without any loss of refrigerant.

SUMMARY 12–9

- Service valves are usually of the back-seating type. They are designed to provide a positive seal when the valve stem is front-seated or back-seated.
- During use, when the valve stem is back-seated, the valve is fully open to the flow of refrigerant to the compressor and the gauge port is closed off.
- Service valves have no operating function in the system; however, when service operations are required they are indispensable.

REVIEW QUESTIONS 12–9

1. During normal system operation, in what position is the service valve?
2. When referring to a service valve, what does the term front-seated indicate?

12–10 CHECK VALVES

Check valves are used in design applications when the refrigerant must not be allowed to flow in the opposite direction in a given line. See Figure 12–28.

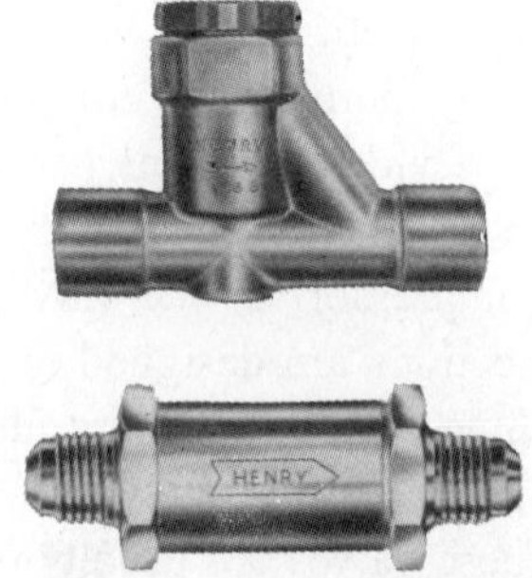

Figure 12–28 Check valves (Courtesy of Henry Valve Company)

Purpose of Check Valves

A check valve checks or prevents the flow of refrigerant in one direction, while allowing full flow in the opposite direction. An application for a check valve would be when two evaporators of different temperatures are fed by only one condensing unit. A check valve should be installed in the suction line from the lower temperature evaporator to prevent the suction vapor from the higher-temperature evaporator from backing-up into it. Check valves must be installed with the flow in the direction indicated by an arrow on the body.

SUMMARY 12–10

- Check valves are used in design applications when the refrigerant must not be allowed to flow in the opposite direction in a given line
- A check valve checks or prevents the flow of refrigerant in one direction while allowing it full flow in the opposite direction.
- Check valves must be installed with the flow in the direction indicated by the arrow on the body.

REVIEW QUESTIONS 12–10

1. What is the purpose of check valves?
2. On a two-temperature type of refrigeration system, where would a check valve most likely be installed?

12–11 WATER-REGULATING VALVES

When water-cooled condensing units are used, there must be some method of controlling the discharge pressure and reduce the use of water. It is the purpose of the water-regulating valve to accomplish these two functions. See Figure 12–29.

Preferably, these valves are installed in the water line from the condenser so that the condenser is always completely full of water to prevent unnecessary scaling of the tubes. See Figure 12–30.

If the valve is installed in the inlet line and it closes down, a turbulence in the water can be created, preventing all of the tubes from being completely covered with water, allowing unnecessary scaling.

Purpose of Water-Regulating Valves

These valves may be actuated by either pressure or temperature from the compressor discharge. They act to either allow more water to flow through the condenser or to throttle the amount of water flowing through the condenser, depending on the discharge pressure. These valves should be properly sized. A valve that is too large will allow the discharge pressure to fluctuate and one that is too small will cause the discharge pressure to be higher than desired due to a lack of cooling water to the condenser.

SUMMARY 12–11

- It is the purpose of the water-regulating valve to control the discharge pressure and reduce the use of water.
- Water-regulating valves may be actuated by either pressure or temperature from the compressor discharge.

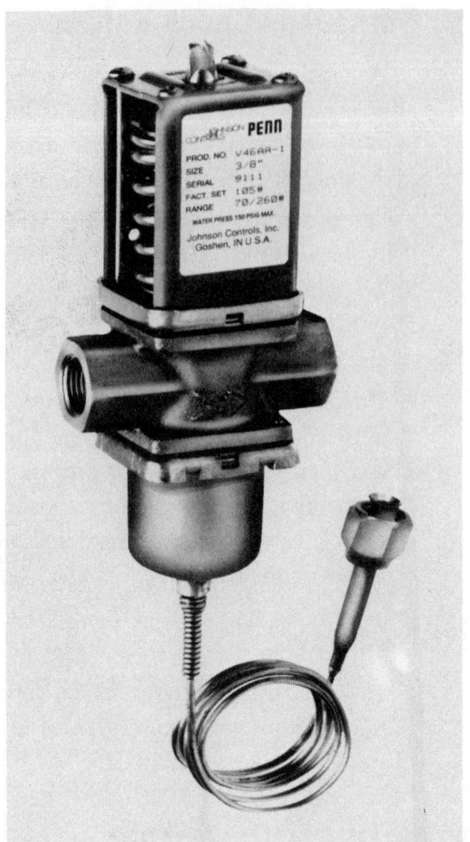

Figure 12–29 Pressure-actuated water regulating valve (Courtesy of Penn Controls, Inc.)

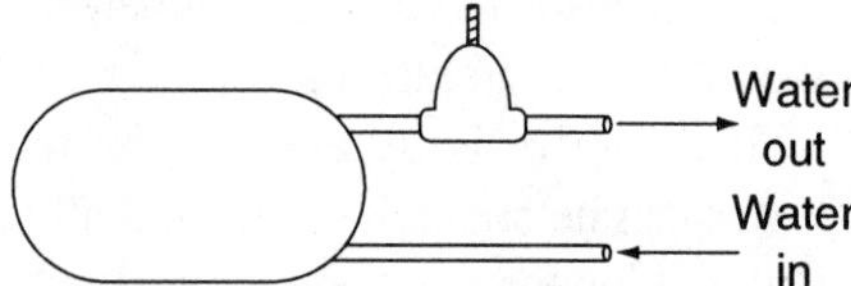

Figure 12–30 Location of water valve

- These valves should be properly sized.
- A valve that is too large will allow the discharge pressure to fluctuate and one that is too small will cause the discharge pressure to be higher than desired due to the lack of cooling water to the condenser.

REVIEW QUESTIONS 12–11

1. What would be the result of installing a water-regulating valve in the condenser water inlet line?
2. What could possibly cause a fluctuating discharge pressure?

CHAPTER

13

REFRIGERANTS

OBJECTIVES

Upon completion of this chapter, you should be able to:

- Understand the purpose of refrigerants
- Better understand refrigerant characteristics
- Better understand the effects that pressure has on the boiling point of refrigerants
- Understand refrigerant condensing pressure
- Understand refrigerant vaporizing pressure
- Know the most popular types of refrigerants used in refrigeration systems
- Know more about refrigerant and oil relationships
- Better understand how to handle refrigerant cylinders
- Better understand how to use the P-H diagram

INTRODUCTION

The substance that absorbs the heat from the conditioned space in a refrigeration system is known as the refrigerant. The compressor compresses the refrigerant, the condenser changes the refrigerant from a vapor to a liquid, the flow-control device controls the flow of refrigerant into the evaporator, and the lines are the path that direct the refrigerant from one component to another. With the EPA Clean Air Act came many new refrigerants. Some of these refrigerants will be discussed here. All of the older refrigerants will be covered because they will be with us for some time to come, even with the phasing out of their use.

The generally accepted definition of a refrigerant is a fluid that absorbs heat by evaporating at a low temperature and pressure and gives up that heat by condensing at a higher temperature and pressure.

13–1 REFRIGERANT CHARACTERISTICS

Most of the refrigerants in use today are a vapor at atmospheric pressures and temperatures. For these refrigerants to be useful, we must be able to change them from a vapor form to a liquid. To make this change they must be compressed and cooled. This is the purpose of the condensing unit on a refrigeration system. For our definition, a fluid may be one of two states: a vapor or gas, and a liquid. For practical purposes, the words *gas* and *vapor* are interchangeable. However, to be completely technical, when a gas is close to its condensation point it is called a vapor. All fluids may exist as either a liquid or a gas. Fluids that have a high boiling point exist as a gas only when heated to a high enough temperature or are subjected to a vacuum. The fluids that have low boiling points exist as a vapor at ordinary room temperatures and pressures. The fluorocarbon-type refrigerants are members of this category. For the vapors of these types of refrigerants to change to a liquid they must be compressed and cooled or condensed.

For example, if we think of water as being a refrigerant we can use it to follow the boiling process of a refrigerant. At atmospheric pressure and temperature, water exists as a liquid. When the water is heated to 212°F at atmospheric pressure, at sea level it will begin to boil. It will continue to boil until all the water is evaporated. The water has changed to the gaseous form known as steam or water vapor. If the water is heated in an open container, its temperature will not exceed 212°F. All of the heat applied to the water is used to cause it to boil or evaporate.

Likewise, if we place a refrigerant in an open container, it will immediately begin to boil and evaporate. The heat applied will come from the surrounding material or air. The liquid refrigerant will not boil at 212°F. It will boil at some lower temperature because refrigerents have a lower boiling temperature than water. For example, liquid Refrigerant–22, when exposed to atmospheric pressure, will boil at –41.4°F. With this boiling temperature, it will absorb sufficient heat from the surroundings to boil very rapidly and change to a vapor. No external source of heat is required.

A refrigerant, when boiling, will absorb an exact amount of heat equal to its latent heat of vaporization. That is the amount of heat required to change it from a liquid to a vapor.

SUMMARY 13–1

- A refrigerant is a fluid that absorbs heat by evaporating at a low temperature and pressure and gives up that heat by condensing at a higher temperature and pressure.

- Most of the refrigerants in use today are vapor at atmospheric pressures and temperatures.
- A refrigerant, when boiling, will absorb an exact amount of heat equal to its latent heat of vaporization.

REVIEW QUESTIONS 13–1

1. What is required of a refrigerant to make it useful?
2. What is the purpose of the condensing unit on a refrigeration system?
3. From where does a refrigerant get the heat that causes it to boil?
4. How much heat will a refrigerant absorb when boiling?

13–2 EFFECTS OF PRESSURE ON THE BOILING POINT

It should be known by now that the boiling point of a liquid can be changed by changing the pressure applied to it. Thus, the greater the pressure, the higher the boiling point. Conversely, when the pressure is lowered, the boiling point will also be lowered. By using pressure, the boiling point of a liquid can be changed to meet some particular requirements. For example, if we place a liquid under a partial vacuum, the boiling point will be lower.

Some types of refrigerants require a high pressure when used in a refrigeration system. For example, R–112 has a boiling point of 199°F. See Table 13–1.

PRODUCT	FORMULA	MOLECULAR WEIGHT	BOILING POINT °F
"FREON–14"	CF_4	88.0	–198.4
"FREON–23"	CHF_3	70.0	–115.7
"FREON–13"	$CClF_3$	104.5	–114.6
"FREON–116"	CF_3–CF_3	138.0	–108.8
"FREON–13B1"	$CBrF_3$	148.9	–72.0
"FREON–502"	$CHClF_2$/$CClF_2$–CF_3 (48.8/51.2% by weight)	121.2	–50.1
"FREON–22"	$CHClF_3$	86.5	–41.4
"FREON–115"	$CClF_2$–CF_3	154.5	–37.7
"FREON–12"	CCl_2F_2	120.9	–21.6
"FREON–C318"	C_4F_8(cyclic)	200.0	21.5
"FREON–114"	$CClF_2$–$CClF_2$	170.9	38.4
"FREON–21"	$CHCl_2F$	102.9	48.1
"FREON–11"	CCl_3F	137.4	74.8
"FREON–114B2"	$CBrF_2$–$CBrF_2$	259.9	117.5
"FREON–113"	CCl_2F–CClF2	187.4	117.6
"FREON–112"	CCl_2F–CCl_2F	203.9	199.0

Table 13–1 Boiling point of "Freon" fluorocarbon refrigerants
(Courtesy of Freon Products Div., E. I. DuPont de Nemours & Co., Inc.)

The type of system that uses this refrigerant requires more energy than does one that uses a refrigerant having a lower boiling point. The major portion of the operating power is required to compress the refrigerant so that it can be cooled and condensed. The refrigerants R–12 and R–502 do not require such high pressures. R–12, R–22, R–500, and R–502 are mostly used in air conditioning and commercial refrigeration systems.

SUMMARY 13–2

- The boiling point of a liquid can be changed by changing the pressure applied to it.
- By using pressure, the boiling point of a liquid can be changed to meet some particular set of requirements.
- The major part of the operating power used by a refrigeration system is required to compress the refrigerant so that it can be cooled and condensed.

REVIEW QUESTIONS 13–2

1. How is pressure used in a refrigeration system?
2. Why is the majority of the power in a refrigeration system used to compress the refrigerant?

13–3 CRITICAL TEMPERATURE

A practical definition of critical temperature is the temperature of a vapor above which the vapor cannot be liquified regardless of the amount of pressure applied to it. It cannot be condensed because the molecules in the vapor are vibrating so rapidly that pressure cannot compress them close enough together to cause them to become a liquid when cooled.

In the refrigeration cycle, the refrigerant changes from a liquid to a vapor and back to a liquid as it passes through the system. Because of these changes, a refrigerant is required that will meet the system requirements and remain below its critical temperature. See Table 13–2.

When the temperature of a vapor is lowered, the pressure required to cause it to change to a liquid is also lowered. From this we can see that for each and every degree below the critical temperature, there is a pressure that will cause it to change to a liquid.

Refrigerant manufacturers have developed graphs that indicate, for each temperature below the critical temperature, the pressure required to condense the refrigerant. See Figure 13–1, page 244-245.

REFRIGERANT	CRITICAL TEMPERATURE °F
R–11	388.4
R–12	233.6
R–22	204.8
R–502	194.1

Table 13–2 Critical temperatures of fluorocarbon refrigerants

The temperature in degrees Fahrenheit is along the bottom line of the graph. The degrees Centigrade is along the top line. The absolute pressure (psia) is located on the left side of the graph and the inches of mercury-vacuum are located on the right.

To use the graph: if we have a quantity of R–22 at 80°F, what is the saturation pressure both psig and absolute? (Locate the 80°F temperature on the bottom line. Follow this line vertically, until the curve for R–22 is intersected. Read the gauge pressure on the right side and the absolute pressure on the left side of the graph. R–22 at this temperature has an absolute pressure of 158.3 psia and a gauge pressure of 143.6 psig. It will take this pressure at 80°F to cause the refrigerant vapor to change to a liquid (condense).

SUMMARY 13–3

- A practical definition of critical temperature is the temperature above which the vapor cannot be liquified regardless of the amount of pressure applied to it.
- Refrigerant manufacturers have developed graphs that indicate the pressure required to condense the refrigerant for each temperature below the critical temperature.

REVIEW QUESTIONS 13–3

1. When a refrigerant is above the critical temperature, why can it not be liquified?
2. At 0°F, what is the saturated pressure of R–22 refrigerant?

13–4 STANDARD CONDITIONS

When the temperatures of the refrigerant inside a system change there are several other conditions that also change in direct response to these changes. The capacity of the system changes, the condensing pressure of the refrigerant, the latent heat of the refrigerant, and the vaporizing pressure also vary with a change in refrigerant temperature. Because of these changes, certain standards have been developed to compare different refrigerants. These are known as the standard conditions. These conditions require that certain temperatures be maintained at various points throughout the system; for example, a refrigerant temperature of 5°F in the evaporator; a refrigerant temperature in the saturated portion of the condenser of 86°F; a liquid temperature at the flow-control device of 77°F; and a suction-vapor temperature of 14°F.

SUMMARY 13–4

- Because of certain changes that happen to the refrigerant as it travels through a system, certain standards have been developed to be used when comparing different refrigerants.
- Standard conditions require that certain temperatures be maintained at various points throughout the system.

REVIEW QUESTIONS 13–4

1. With what evaporator temperature are refrigerants rated?
2. What device has been established for use when comparing different refrigerants?

13–5 CONDENSING PRESSURE

The pressure at which the refrigerant will condense will depend on the temperature at which it will change from a vapor to a liquid. The condensing temperature

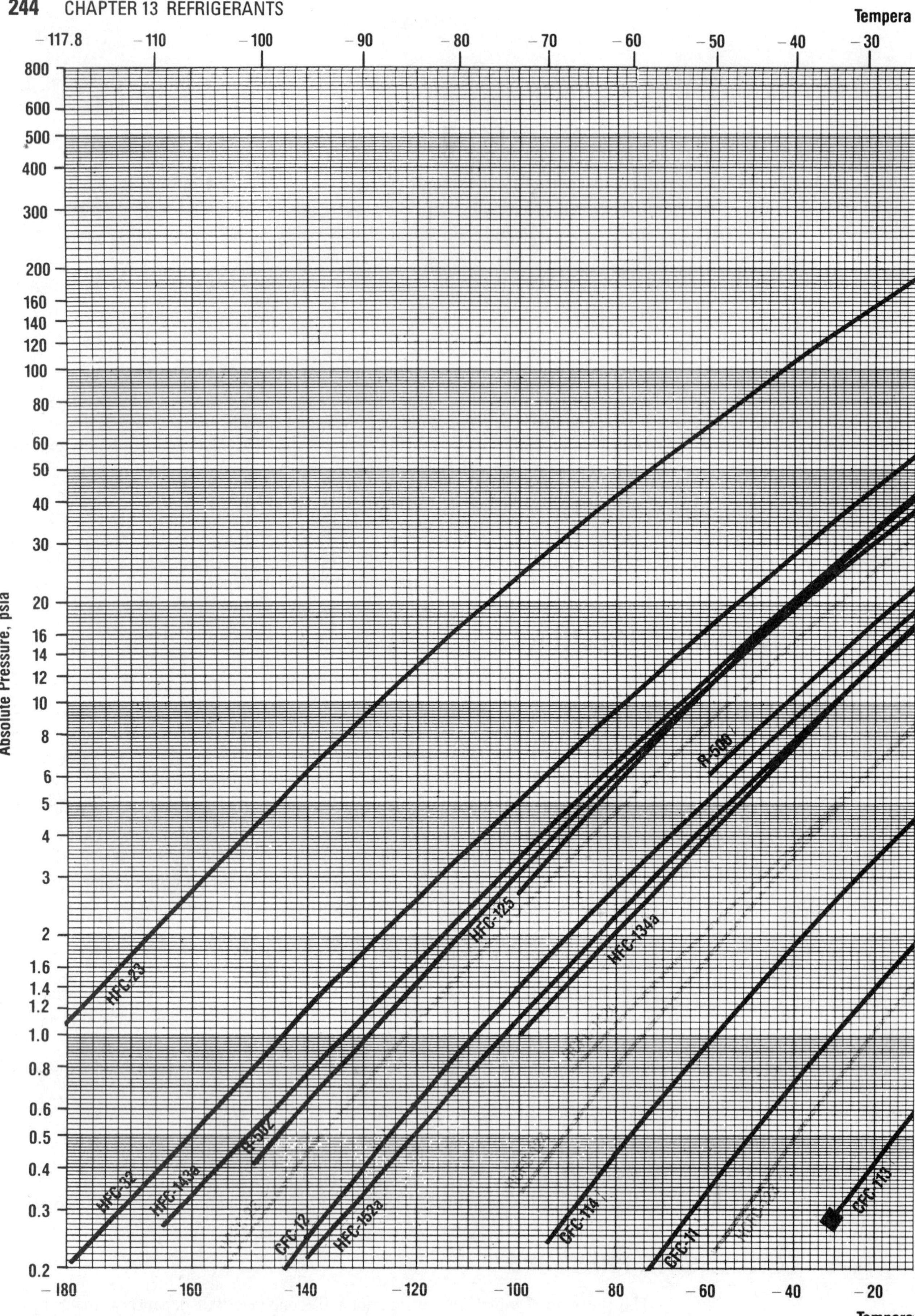

Figure 13–1 Pressure-temperature relationships of freon compounds (Courtesy of E.I. Dupont de Numours & Co., Inc.)
● Critical Point ■ Freezing Point

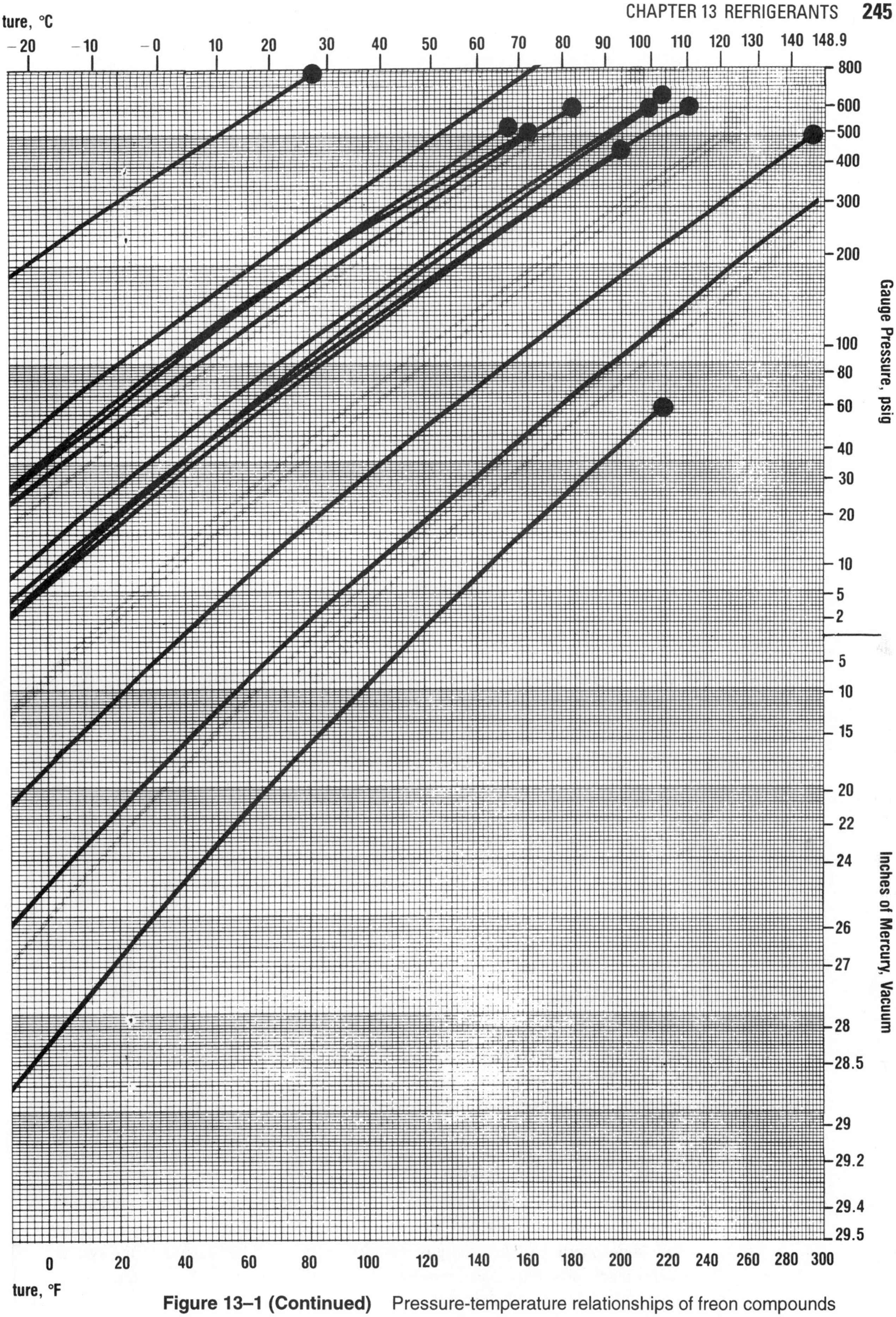

ture, °C

−20 −10 −0 10 20 30 40 50 60 70 80 90 100 110 120 130 140 148.9

800 600 500 400 300 200 100 80 60 40 30 20 10 5 2

Gauge Pressure, psig

5 10 15 20 22 24 26 27 28 28.5 29 29.2 29.4 29.5

Inches of Mercury, Vacuum

0 20 40 60 80 100 120 140 160 180 200 220 240 260 280 300

ture, °F

Figure 13–1 (Continued) Pressure-temperature relationships of freon compounds (Courtesy of E.I. Dupont de Numours & Co., Inc.)

should be as low as possible so that a smaller condenser can be used. This requires that the cooling medium be cool enough to liquify the vapor at a low pressure. In most cases, when the same type refrigerant is used for the same type of application, a water-cooled condenser will operate with a lower discharge pressure than an air-cooled condenser. It is generally assumed that an air-cooled condenser will have a condensing temperature of about 25°F to 35°F higher than the ambient temperature. There are some conditions that will cause this to change. For example, the efficiency of the condenser, the location of the condenser, condenser cleanliness, and amount of air passing over the condenser. Water-cooled condensers generally operate with condensing temperatures that are lower than the surrounding ambient temperature.

It must be remembered that the refrigerant saturation pressures shown in Figure 13–1 are not the same as the operating discharge pressures of a refrigeration unit. Because of the heat of compression, the operating discharge pressures will be somewhat higher than shown here. Also, if the refrigerant is to be changed from a vapor to a liquid in the condenser, the temperature of the vapor must be higher than the temperature of the cooling medium.

SUMMARY 13–5

- The pressure at which a refrigerant will condense depends on the temperature that it will change from a vapor to a liquid.
- The condensing temperature should be as low as possible so that a smaller condenser can be used.
- It is generally assumed that an air-cooled condenser will have a condensing temperature of about 25°F to 35°F higher than the ambient temperature.
- Water-cooled condensers generally operate with a condensing temperature that is lower than the surrounding ambient temperature.
- It must be remembered that the refrigerant saturation pressures are not the same as the operating discharge and suction pressures of a refrigeration unit.

REVIEW QUESTIONS 13–5

1. What is required to liquify vapor refrigerant at a low pressure?
2. In the condenser, what is required to condense the refrigerant?
3. Are the pressures shown on a pressure-temperature chart the same as the discharge pressure?

13–6 VAPORIZING PRESSURE

The refrigerant chosen must evaporate at a temperature that will not require a lower-than-necessary suction pressure. Low suction pressures cause reduced compressor efficiency, increased operating costs, high compression ratios, and many other undesirable conditions. A refrigerant temperature in the evaporator of about 5°F is equal to the evaporating pressure for most domestic refrigerators. Generally, the desired refrigerant will have an evaporating temperature of 5°F at about atmospheric pressure. See Table 13–3.

Refrigerants that require a vacuum to cause the desired evaporating temperature for the application are not considered to be practical. When the system operates below atmospheric pressure, there is a tendency for air to be drawn into the system, causing many problems. Refrigeration systems that operate with pressures above atmospheric will not pull air into them if a leak occurs.

REFRIGERANT	PRESSURE AT 5°F
R–11	–24" HG
R–12	11.8
R–22	28.1
R–500	16.4
R–502	36

Table 13–3 Vaporizing pressures of refrigerants at 5°F

The pressure in the low-side of the system will essentially be the same all the way back to the compressor. The major difference will be caused by pressure drop in the evaporator and suction line because of a restriction. The refrigerant evaporating temperature will be the same as the temperature indicated on the pressure-temperature chart, by the evaporator pressure, and in the low side of the system. See Table 13–4.

SUMMARY 13–6

- The refrigerant chosen for a particular application must evaporate at a temperature that will not require a lower than necessary suction pressure.
- Refrigerants that require a vacuum to cause the desired evaporating temperature for the application are not considered to be practical.
- The pressure in the low-side of the system will essentially be the same all the way back to the compressor.
- The refrigerant evaporating temperature will be the same as the temperature indicated on the pressure-temperature chart.

REVIEW QUESTIONS 13–6

1. Will a low-suction pressure cause high compression ratios?
2. Why are refrigerants used that require the low-side pressure to be below zero psig not desirable?
3. What will cause the refrigerant pressure to be lower at the compressor than in the evaporator?

13–7 LATENT HEAT OF VAPORIZATION

The latent heat of vaporization is the amount of heat in Btu required to change a liquid into a vapor, the change taking place at a constant temperature. We can now apply this definition to one pound of refrigerant vaporizing at atmospheric pressure and at a liquid temperature equal to its boiling point when the operation starts.

When any refrigerant is evaporating, it must absorb the required heat from inside the refrigerated space and its contents. The amount of heat absorbed will be exactly equal to the latent heat of vaporization of the refrigerant. A refrigerant having a high latent heat of vaporization will absorb more heat when vaporizing than one having a lower latent heat value. Thus, a refrigerant having a high latent heat of vaporization

TEMPER-ATURE	REFRIGERANT					TEMPER-ATURE	REFRIGERANT				
°F	12	22	500	502	717	°F	12	22	500	502	717
–60	*19.0**	*12.0*	-------	*7.0*	*18.6*	27	26.1	51.2	33.3	61.4	41.4
–55	*17.3*	*9.2*	-------	*3.6*	*16.6*	28	26.9	52.4	34.3	62.7	42.6
–50	*15.4*	*6.2*	-------	0.0	*14.3*	29	27.7	53.6	35.2	64.1	43.8
–45	*13.3*	*2.7*	-------	2.1	*11.7*	30	28.4	54.9	36.1	65.4	45.0
–40	*11.0*	0.5	*7.9*	4.3	*8.7*	31	29.2	56.2	37.0	66.8	46.3
–35	*8.4*	2.6	*4.8*	6.7	*5.4*	32	30.1	57.5	38.0	68.2	47.6
–30	*5.5*	4.9	*1.4*	9.4	*1.6*	33	30.9	58.8	39.0	69.7	48.9
–25	*2.3*	7.4	1.1	12.3	1.3	34	31.7	60.1	40.0	71.1	50.2
–20	0.6	10.1	3.1	15.5	3.6	35	32.6	61.5	41.0	72.6	51.6
–18	1.3	11.3	4.0	16.9	4.6	36	33.4	62.8	42.0	74.1	52.9
–16	2.0	12.5	4.9	18.3	5.6	37	34.3	64.2	43.1	75.6	54.3
–14	2.8	13.8	5.8	19.7	6.7	38	35.2	65.6	44.1	77.1	55.7
–12	3.6	15.1	6.8	21.2	7.9	39	36.1	67.1	45.2	78.6	57.2
–10	4.5	16.5	7.8	22.8	9.0	40	37.0	68.5	46.2	80.2	58.6
–8	5.4	17.9	8.9	24.4	10.3	41	37.9	70.0	47.2	81.8	60.1
–6	6.3	19.3	9.8	26.0	11.6	42	38.8	71.4	48.4	83.4	61.6
–4	7.2	20.8	11.0	27.7	12.9	43	39.8	73.0	49.6	85.0	63.1
–2	8.2	22.4	12.1	29.4	14.3	44	40.7	74.5	50.7	86.6	64.7
0	9.2	24.0	13.3	31.2	15.7	45	41.7	76.0	51.8	88.3	66.3
1	9.7	24.8	13.9	32.2	16.5	46	42.6	77.6	53.0	90.0	67.9
2	10.2	25.6	14.5	33.1	17.2	47	43.6	79.2	54.2	91.7	69.5
3	10.7	26.4	15.1	34.1	18.0	48	44.6	80.8	55.4	93.4	71.1
4	11.2	27.3	15.7	35.0	18.8	49	45.7	82.4	56.6	95.2	72.8
5	11.8	28.2	16.4	36.0	19.6	50	46.7	84.0	57.8	96.9	74.5
6	12.3	29.1	17.0	37.0	20.4	55	52.0	92.6	64.1	106.0	83.4
7	12.9	30.0	17.7	38.0	21.2	60	57.7	101.6	71.0	115.6	92.9
8	13.5	30.9	18.4	39.0	22.1	65	63.8	111.2	78.1	125.8	103.1
9	14.0	31.8	19.0	40.0	22.9	70	70.2	121.4	85.8	136.6	114.1
10	14.6	32.8	19.8	41.1	23.8	75	77.0	132.2	93.9	148.0	125.8
11	15.2	33.7	20.5	42.2	24.7	80	84.2	143.6	102.5	159.9	138.3
12	15.8	34.7	21.2	43.2	25.6	85	91.8	155.7	111.5	172.5	151.7
13	16.4	35.7	21.9	44.3	26.5	90	99.8	168.4	121.2	185.8	165.9
14	17.1	36.7	22.6	45.4	27.5	95	108.2	181.8	131.3	199.8	181.1
15	17.7	37.7	23.4	46.6	28.4	100	117.2	195.9	141.9	214.4	197.2
16	18.4	38.7	24.2	47.7	29.4	105	126.6	210.8	153.1	229.8	214.2
17	19.0	39.8	24.9	48.9	30.4	110	136.4	226.4	164.9	245.8	232.3
18	19.7	40.8	25.7	50.1	31.4	115	146.8	242.7	177.4	262.7	251.5
19	20.4	41.9	26.5	51.2	32.5	120	157.6	259.9	190.3	280.3	271.7
20	21.0	43.0	27.3	52.4	33.5	125	169.1	277.9	204.0	298.7	293.1
21	21.7	44.1	28.2	53.7	34.6	130	181.0	296.8	218.2	318.0	-------
22	22.4	45.3	29.0	54.9	35.7	135	193.5	316.6	233.2	338.1	-------
23	23.2	46.4	29.8	56.2	36.8	140	206.6	337.2	248.8	359.1	-------
24	23.9	47.6	30.7	57.4	37.9	145	220.3	358.9	265.2	381.1	-------
25	24.6	48.8	31.6	58.7	39.0	150	234.6	381.5	282.3	403.9	-------
26	25.4	49.9	32.4	60.0	40.2	155	249.5	405.1	300.2	427.8	-------

*Vacuum—*Italic figures*

Table 13–4 Temperature-pressure chart—gauge pressures

REFRIGERANT	LATENT HEAT Btu/lb
R–11	83.459
R–12	68.204
R–22	93.206
R–500	82.45
R–502 at 40°F	63.1

Table 13–5 Latent heat of vaporization at 5°F in Btu/lb

will require smaller system components because a smaller amount of refrigerant must be circulated to produce the same refrigerating effect. See Table 13–5.

The refrigerant pressures are very important in the vaporizing of a refrigerant because the latent heat of vaporization will vary with the operating pressures and temperatures. When lower refrigerant temperatures and pressures are used, the latent heat of vaporization increases.

SUMMARY 13–7

- The latent heat of vaporization is the amount of heat in Btu required to change a liquid into a vapor, the change taking place at a constant temperature.
- When any refrigerant is evaporating, it must absorb the required heat from inside the refrigerated space and its contents.
- A refrigerant having a high latent heat of vaporization will require smaller system components because a smaller amount of the refrigerant must be circulated to produce the same refrigerating effect.
- The refrigerant pressures are very important in the evaporation of a refrigerant because the latent heat of vaporization will vary with the operating pressure and temperatures.

REVIEW QUESTIONS 13–7

1. When a liquid refrigerant boils, how much heat does it absorb?
2. What type of refrigerant requires smaller system components?
3. What two things cause the latent heat of vaporization of a refrigerant to change?

13–8 TYPES OF REFRIGERANTS

There are many types of refrigerants that are available. Some of the more popular types will soon be obsolete because of the Clean Air Act. Probably the oldest refrigerant still in use is ammonia, used in very large capacity refrigeration equipment. There is a refrigerant developed for virtually every use. The CFC issue will cause many of the old stand-by refrigerants to be phased out in the near future. For this reason, it is desirable that every effort be made to conserve these refrigerants so that the equipment that is designed to use them will be in service longer. The following is a description of the most popular refrigerants used in refrigeration and air conditioning systems today.

R–11 Trichlorofluoromethane (CCl_3F). R–11 is a synthetic chemical refrigerant. It is stable, nonflammable, and has a very low toxicity rating. This refrigerant is a

Evaporator Pressure at 5°F, psia	2.9373
Condenser Pressure at 86°F, psia	18.186
Compression Ratio (86°F/5°F)	6.19
Latent Heat of Vaporization at 5°F, Btu/lb	83.459
Net Refrigerating Effect, Btu/lb	66.796
Refrigerant Circulated per Ton of Refrigeration, lb/min	2.9942
Saturated Liquid Volume at 86°F, cu ft/lb	0.010942
Liquid Circulated per Ton of Refrigeration, cu in/min	56.614
Saturated Vapor Density at 5°F, lb/cu ft	0.081933
Saturated Vapor Density at 86°F, lb/cu ft	0.44668
Compressor Displacement per Ton of Refrigeration, cu ft/min	36.544
Refrigeration per Cubic Foot of Compressor Displacement, Btu	5.473
Heat of Compression, Btu/lb	13.292
Temperature of Compressor Discharge, °F	110.9
Coefficient of Performance	5.025
Horsepower per Ton	0.9383

Table 13–6 R–11 refrigerant standard ton characteristics (Courtesy of Freon Products Div., E. I. DuPont de Nemours & Co., Inc.)

CFC and is controlled by the Clean Air Act of 1990. Its availability will be very limited. There are many large tonnage units using this refrigerant still in operation, but their existence depends on conserving it for future use. During operation, with an evaporator temperature of 5°F, the low-side pressure is 24 inches of mercury (Hg) vacuum. The high-side pressure is 18.3 psia with a condensing temperature of 86°F. At 5°F its latent heat value is 83.459 Btu/lb. See Table 13–6.

R–11 leak detection is accomplished through the use of several types of leak detectors, such as soap bubble solution, halide torch, electronic leak detectors, fluorescent dyes, and ultrasonic leak detectors.

The cylinder color code for R–11 is orange.

R–12 Dichlorodifluoromethane ($CCl_2 F_2$). This refrigerant, widely used in many applications, is a CFC and is controlled by the Clean Air Act of 1990. Its boiling point is –21.6°F at atmospheric pressure. It is a vapor at normal room temperatures and pressures. It will liquify when exposed to a pressure of 76 psig and a temperature of 75°F. See Table 13–7.

This refrigerant has very little odor; however, in large concentrations a faint, sweet odor may be detected. It has a critical temperature of 233.6°F. It is considered to be colorless in both the vapor and liquid forms. R–12 is nontoxic, nonflammable, and nonirritating. It is noncorrosive to any of the ordinary types of metal used in refrigeration systems, even in the presence of water. R–12 is stable under all the conditions and temperatures encountered in refrigeration systems. The liquid will dissolve lubricating oils in all proportions. The oils used in refrigeration will rapidly absorb R–12 vapor. Normally, there is not enough separation of the oil and refrigerant in the evaporator to interfere with the normal evaporating process.

This refrigerant is only slightly soluble in water. Because of this, corrosion is prevented in the refrigeration system, using R–12 as the refrigerant.

Leak detection of R–12 is accomplished through several methods, such as soap bubbles, halide torch, electronic leak detectors, fluorescent dyes, and ultrasonic leak detectors.

The cylinder color code for R–12 is white.

Evaporator Pressure at 5°F, psia	26.483
Condenser Pressure at 86°F, psia	108.04
Compression Ratio (86°F/5°F)	4.08
Latent Heat of Vaporization at 5°F, Btu/lb	68.204
Net Refrigerating Effect, Btu/lb	50.035
Refrigerant Circulated per Ton of Refrigeration, lb/min	3.9972
Saturated Liquid Volume at 86°F, cu ft/lb	0.012396
Liquid Circulated per Ton of Refrigeration, cu in/min	85.621
Saturated Vapor Density at 5°F, lb/cu ft	0.68588
Saturated Vapor Density at 86°F, lb/cu ft	2.6556
Compressor Displacement per Ton of Refrigeration, cu ft/min	5.8279
Refrigeration per Cubic Foot of Compressor Displacement, Btu	34.318
Heat of Compression, Btu/lb	10.636
Temperature of Compressor Discharge, °F	100.84
Coefficient of Performance	4.704
Horsepower per Ton	1.0023

Table 13–7 R–12 refrigerant standard ton characteristics
(Courtesy of Freon Products Div., E. I. DuPont de Nemours & Co., Inc.)

R–22 Chlorodifluoromethane ($CHClF_2$). R–22 is an HCFC type refrigerant. It is controlled by the Clean Air Act of 1990. It was developed for installations operating with a low evaporating temperature. R–22 and R–12 have many of the same characteristics. The major differences are that the operating pressures for R–12 are higher. It has a latent heat of vaporization of 93.21 Btu/lb at 5°F and a greater refrigeration capacity for a given volume of the saturated vapor because of its lower specific volume than R–12. It has a boiling point of –41.36°F at atmospheric pressure. The normal operating discharge pressure is 172.8 psia at an 86°F condensing temperature. See Table 13–8.

The normal evaporator pressure at 5°F is 42.88 psia. This is a nontoxic, nonflammable, noncorrosive, and nonirritating refrigerant. R–22 is stable under the normal operating pressures of a refrigeration system. It has high compression ratios at low evaporating temperatures. Because of this, the temperature of compressed R–22 vapor

Evaporator Pressure at 5°F, psia	42.888
Condenser Pressure at 86°F, psia	172.87
Compression Ratio (86°F/5°F)	4.03
Latent Heat of Vaporization at 5°F, Btu/lb	93.206
Net Refrigerating Effect, Btu/lb	70.027
Refrigerant Circulated per Ton of Refrigeration, lb/min	2.8560
Saturated Liquid Volume at 86°F, cu ft/lb	0.013647
Liquid Circulated per Ton of Refrigeration, cu in/min	67.351
Saturated Vapor Density at 5°F, lb/cu ft	0.80422
Saturated Vapor Density at 86°F, lb/cu ft	3.1622
Compressor Displacement per Ton of Refrigeration, cu ft/min	3.5512
Refrigeration per Cubic Foot of Compressor Displacement, Btu	56.32
Heat of Compression, Btu/lb	15.022
Temperature of Compressor Discharge, °F	128.4
Coefficient of Performance	4.662
Horsepower per Ton	1.0114

Table 13–8 R–22 refrigerant standard ton characteristics
(Courtesy of Freon Products Div., E. I. DuPont de Nemours & Co., Inc.)

may get so high that the compressor could possibly be damaged when operating in ultra-low temperature applications.

R–22 is commonly chosen as the refrigerant in self-contained units and other applications where there is limited space. When R–22 is used in applications where the evaporating temperature will reach –40°F, the lubricating oil will start separating from the refrigerant and a film will be formed on the evaporator surface. This film will interfere with the normal evaporating process.

Leak detection of R–22 is accomplished with the normal leak-detection methods of soap bubble solution, halide torch, electronic leak detectors, fluorescent dyes, and ultrasonic leak detectors.

The cylinder color code for R–22 is green.

HCFC–123 ($CHCl_2CF_3$). This is one of the new refrigerants that have been developed as an interim replacement for some of the CFC refrigerants. It is being used as a replacement for R–11 and other refrigerants in chillers and low-temperature brine applications.

HCFC–123 has a boiling point of 82°F at atmospheric pressure. Therefore, it is a liquid at room temperatures and pressures. It has an evaporating pressure of 2.3 psia and a latent heat value of 78.9 Btu/lb at 5°F. The discharge pressure is 15.6 psia at 86°F. See Figure 13–2.

HCFC–123 may require a different type of compressor lubricating oil than that normally used for other refrigerants. Be sure to check with the equipment manufacturer for their recommendations. Replacing R–11 with HCFC–123 may require extensive retrofit procedures for satisfactory operation. Some localities require complete equipment room refrigerant monitoring systems be installed when the changeover is completed.

Leak detection is accomplished by use of soap bubbles, fluorescent dyes, or ultrasonic leak detectors. The equipment room is usually equipped with halogen-sensitive or oxygen-sensitive monitors.

The cylinder color code for HCFC–123 is light blue-gray.

HCFC–124 (CHC_lFCF_3). This is a ternary blend of refrigerant that has been developed to replace R–11. It is one of the refrigerants that is being used as an interim replacement for some of the CFC refrigerants being phased-out because of the Clean Air Act of 1990. It is a replacement for R–11 and other refrigerants in chillers and medium temperature applications.

HCFC–124 has a boiling point of 10.25°F at atmospheric pressure. At 5°F it has an evaporating pressure of 12.99 psia and a latent heat value of 70.7 Btu/lb. At 86°F the condensing pressure is 64.58 psia. See Figure 13–3.

Leak detection is accomplished by use of soap bubbles, fluorescent dyes, or ultrasonic leak detectors. The equipment room is usually equipped with halogen-sensitive or oxygen-sensitive monitors.

The cylinder color code for HCFC–124 is not permanently assigned as yet, but the interim color is light green-gray.

HFC–134a (CH_2FCF_3). HFC–134a has been developed as a replacement for R–12 in almost all applications. For the same amount of subcooling, HFC–134a produces a greater refrigerating effect than R–12. It is generally compatible with all the components used in R–12 systems. HFC–134a has a boiling point of –14.9°F at atmospheric pressure. At a 5°F evaporating temperature it has a pressure of 23.77 psia and a latent heat value of 90.2 Btu/lb. At 86°F it has a condensing pressure of 111.83 psia. See Figure 13–4.

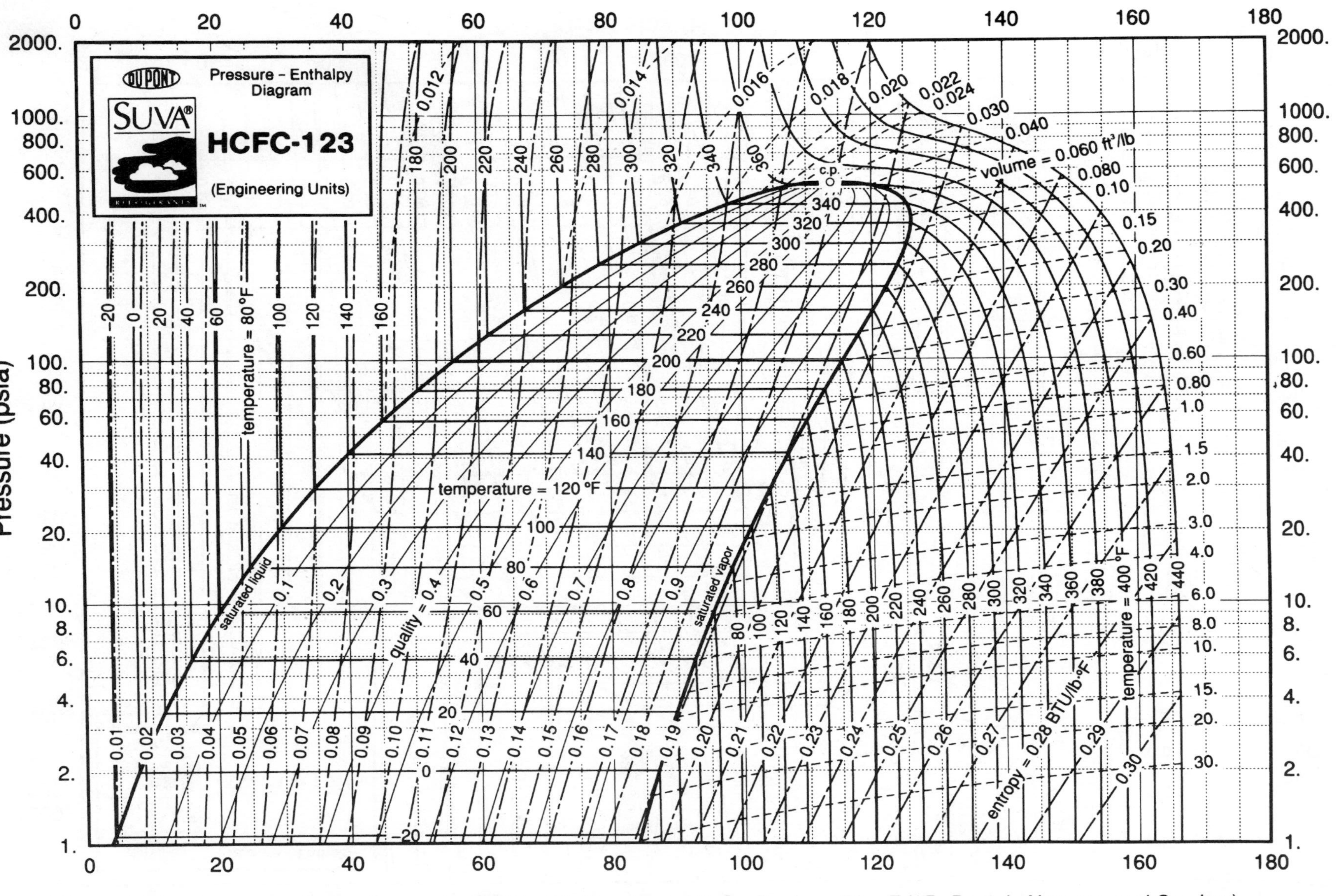

Figure 13–2 Pressure-enthalpy diagram for HCFC–123 (Courtesy of Freon Products Div., E.I. DuPont de Numours and Co., Inc.)

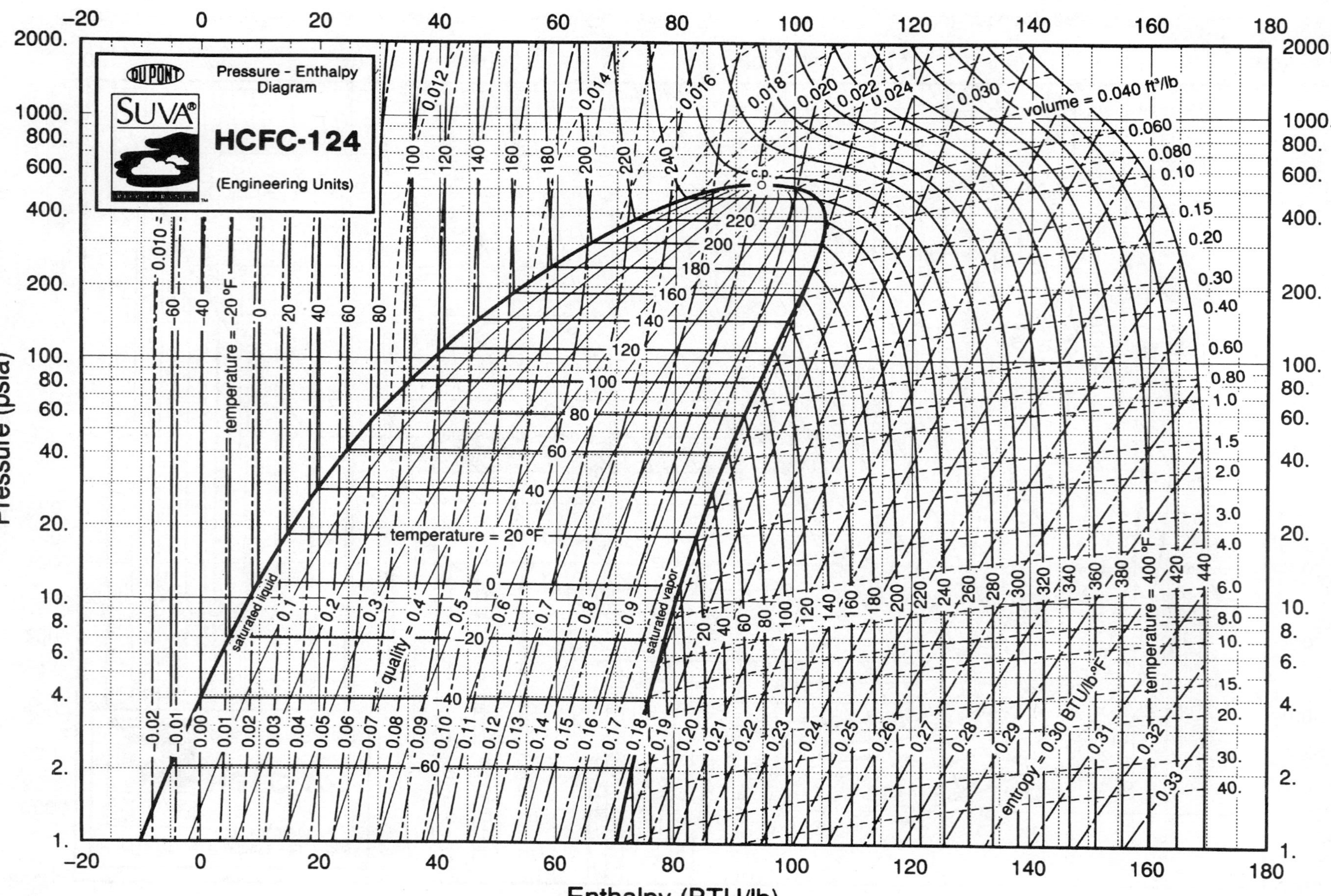

Figure 13–3 Pressure-enthalpy diagram for HCFC–124 (Courtesy of Freon Products Div., E.I. DuPont de Numors and Co., Inc.)

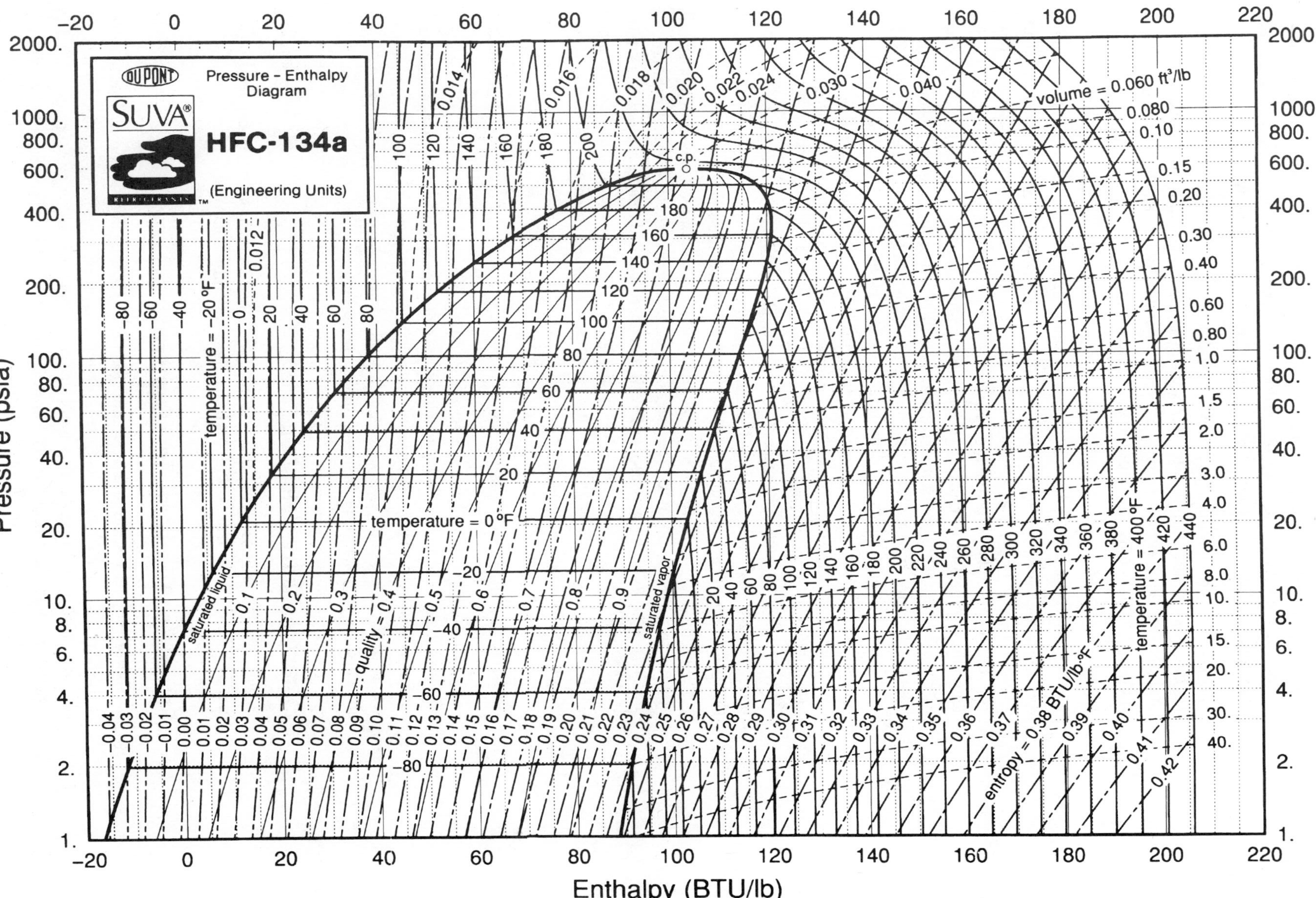

Figure 13–4 Pressure-enthalpy diagram for HCFC–134a (Courtesy of Freon Products Div., E.I. DuPont de Numors and Co., Inc.)

Water solubility: Liquid HFC–134a has a high affinity for water. Therefore, there is less possibility of moisture freezing at the flow-control device than with either R–12 or R–22. However, this does not reduce the need for extreme caution to prevent moisture entering the system. (See the "Moisture" section following.)

Lubricants. Mineral oils and HFC–134a are not miscible. HFC–134a requires that special oils be used. Polyol ester oils are being recommended by some manufacturers and poly alkylene glycols (PAGs) are recommended by others. Be sure to check with the equipment manufacturer for the recommendations.

Moisture. Polyol ester oils, while not as hygroscopic (the ability to absorb moisture) as the PAGs, are 100 times more hygroscopic than mineral oils. This moisture is extremely difficult to remove from the system even with a vacuum and heat. Thus, extreme caution must be exercised when working with this refrigerant and these oils.

Utmost care must be taken to prevent moisture from getting into the system. Do not leave the compressor or the system open to the atmosphere for more than 15 minutes. The maximum system moisture capacity is 80 PPM. After running the system with the appropriate drier installed, the system moisture level should be no more than 10 PPM.

Always use an appropriate drier in any system using HFC–134a. The polyol ester oils used with HFC–134a are prone to hydrolyze with moisture, resulting in the formation of acids in the system. Most manufacturers require that an appropriate drier be used in every application.

The molecular sieve-type driers that are presently compatible with R–22 should be used. The XH–6 (bonded core), XH–7, and XH–9 types are recommended. The XH–6 (loose-fill) type is not recommended because of its short life span.

Solid core driers, if made with bauxite, would have the tendency to absorb both polyol ester oil and moisture. The ester could hydrolyze and form acidic materials. If the drier were to become overloaded with moisture, it could release the acidic materials back into the system. Clearly this would not be good for the compressor and other parts. Thus, most manufacturers do not recommend solid-core driers made with bauxite for systems using polyol ester oils.

Capillary Tube Selection

In general, HFC–134a has a greater refrigerating effect than R–12, thus reducing the required mass flow for a given capacity. Consequently, a nonheat exchange capillary tube may require a change, either more restrictive or less restrictive, depending on the actual application. A heat exchange capillary tube may not require any change. As with any capillary tube selection, system testing is necessary.

Expansion Valve Selection

The expansion valve manufacturers have designed expansion valves especially for use with HFC–134a. Check with them for their recommendations.

Return Gas/Discharge Temperatures

The theoretical discharge temperature for HFC–134a is slightly lower than that of R–12 at similar conditions. Therefore, existing compressor guidelines regarding return gas and discharge temperatures should apply to HFC–134a compressors as well. In general, keeping the return gas cool without flooding the compressor is beneficial in keeping the compressor discharge and motor temperatures to acceptable levels.

Refrigerant Quantity

The refrigerant quantity will depend on the system components used. In general, based on limited application data, 5% to 30% less HFC–134a will be required when compared to R–12.

Refrigerant Charging

In general, refrigerant charging equipment such as charging boards, valves, and hoses that are compatible with R–22 (considered more aggressive with gaskets and plastics than R–12), should be compatible with HFC–134a. This equipment, once designated for HFC–134a use, should be used specifically for HFC–134a only. Converted R–12 equipment should be clean of all residual R–12. Pulling a deep vacuum (25 to 50 microns) and repeated flushing with HFC–134a should be sufficient. Consult your equipment and component manufacturer for specific recommendations for converting R–12 equipment for use with HFC–134a.

HFC–134a can be charged in either the liquid or vapor state. If refrigerant charging is done in the liquid state, it should be done into the liquid line. Vapor charging can be done into the suction line while the compressor is running. CAUTION: Always break the vacuum with refrigerant vapor before applying electrical power to the compressor.

Leak testing is accomplished by using leak detection equipment specifically designed for HFC–134a. Some of the standard leak detection methods can be used, such as soap bubbles, special electronic leak detectors, fluorescent dyes, and ultrasonic leak detectors.

The cylinder color code for HFC–134a is light blue.

R–500 Refrigerant (CCl_2F_2/CH_3CHF_2). R–500 is an azeotropic refrigerant mixture of 26.2% R–152a and 73.8% of R–12. It is considered to be a CFC type refrigerant and is controlled by the Clean Air Act of 1990. Azeotrope is a name that scientists have given to a specific mixture of different compounds, which result in a mixture having different characteristics from either of its original components. It can be evaporated and condensed without any change in its composition. The pressure-temperature curve is relatively constant and is different from either of its original components. It has an evaporator pressure of 31.21 psia at 5°F. The condensing pressure at 86°F is 127.6 psia. The boiling point of R–500 is –28°F at atmospheric pressure. It has a latent heat value of 82.45 Btu/lb at 5°F.

This refrigerant is very popular in both commercial and industrial refrigeration applications. It is used primarily with reciprocating compressors.

R–500 is very soluble with water and requires that proper evacuation techniques be used, along with adequate refrigerant driers. It is fairly soluble with oil, which is a plus in low-temperature applications.

Leak detection is accomplished by the usual means, such as soap bubbles, halide leak detector, electronic leak detector, fluorescent dyes, and ultrasonic leak detectors.

The cylinder color code for R–500 is yellow.

R–502 Refrigerant ($CHClF_2/CClF_2CF_3$). This is an azeotropic refrigerant mixture of 48.8% of R–22 and 51.2% of R–115. It is much like R–12 and R–22 in most of its physical characteristics. It has an evaporator pressure of 30.01 psia at –20°F. The latent heat of vaporization is 72.5 Btu/lb at –20°F. It has a boiling point of –49.76°F at atmospheric pressure. The evaporator pressure is 15 psia at –20°F. The condensing pressure is 297.4 psia at 120°F. See Table 13–9.

EVAPORATING TEMPERATURE = –20°F RETURN GAS TEMPERATURE = 65°F CONDENSING TEMPERATURE = 120°F	
Evaporator Pressure, psia	30.01
Condenser Pressure, psia	297.4
Compression Ratio	9.91
Net Refrigerating Effect*, Btu/lb	45.26
Refrigerant Circulated per Ton, lb/min	4.418
Saturated Liquid Volume at 120°F, cu ft/lb	0.01472
Liquid Circulated per Ton, cu in/min	112.4
Vapor Density at 65°F†, lb/cu ft	0.6165
Compressor Displacement per Ton, cu ft/min	7.167
Refrigeration Capacity, Btu/cu ft	27.90
Heat of Compression, Btu/lb	22.52
Temperature of Compressor Discharge, °F	226.0
Coefficient of Performance	2.010
Horsepower per Ton	2.346

Table 13–9 R–502 Refrigerant standard ton characteristics (Courtesy of Freon Products Div., E. I. DuPont de Nemours & Co., Inc.)

* Enthalpy of vapor 65°F and evaporator pressure-enthalpy of liquid at 120°F.

† It is assumed that vapor enters the compressor cylinder at a temperature of 65°F.

When R–502 is chosen, a smaller compressor may be used to obtain the same refrigerating effect as when R–12 is used. R–502 is very well suited for applications that require a low-evaporating temperature. It is generally used in single-stage systems operating with an evaporating temperature of 0°F or below. It may be used in ultra-low temperature, two-stage applications with good success. Some medium-temperature applications operate very well using R–502 as the refrigerant. R–502 is nonflammable, nontoxic, noncorrosive, and is very stable in normal applications. Oil separators are usually recommended when the evaporating temperature is below –40°F because below this temperature the oil has a tendency to separate from the refrigerant.

Leak detection is accomplished by the usual means, such as soap bubbles, halide torch, electronic leak detectors, fluorescent dyes, and ultrasonic leak detectors.

The cylinder color code for R–502 is orchid.

SUMMARY 13–8

- R–11 is used in many large tonnage units.
- R–12 has very little odor; however, in large concentrations a faint, sweet odor may be detected.
- R–22 is a nontoxic, nonflammable, noncorrosive, and nonirritating refrigerant.
- HCFC–123 is being used as a replacement for R–11 and other refrigerants in chillers and low-temperature brine applications. HCFC–124 is a ternary blend refrigerant that has been developed to replace R–11.
- HFC–134a has been developed as a replacement for R–12 in almost all applications. HFC–134a has a high affinity for water; therefore, there is less probability of moisture freezing at the flow-control device than with either R–12 or R–22.
- Special oils must be used with HFC–134a.

- The molecular sieve-type driers that are presently compatible with R–22 should not be used with HFC–134a. If the drier were to become overloaded with moisture, it could release the acidic materials back into the system.
- Existing compressor guidelines regarding return gas temperatures should apply to HFC–134a compressors as well.
- Equipment that is once used for HFC–134a should be used specifically for HFC–134a only.
- R–500 is an azeotropic refrigerant mixture. It is popular in both commercial and industrial refrigeration applications.
- R–502 is an azeotropic refrigerant mixture. It is very well suited for applications that require a low-evaporating temperature.

REVIEW QUESTIONS 13–8

1. What is the pressure of saturated R–22 refrigerant at 40°F?
2. Why can R–22 not be used on ultra-low temperature applications?
3. What is the boiling temperature of HCFC–123 at 0 psig?
4. When retrofitting a system to use HCFC–123, what is usually required?
5. Why is moisture a big problem in HFC–134a systems?
6. When retrofitting an R–12 system to HFC–134a, how much refrigerant will be needed?
7. In what type of compressor is R–500 usually used?
8. In what type of application is R–502 well suited?

13–9 REFRIGERANT-OIL RELATIONSHIPS

The compressors used in refrigeration systems require oil. The refrigerant and the oil are mixed continuously under many different conditions. The oils used in these compressors are soluble in the refrigerant, and, under some conditions, will mix completely. The term used to describe this condition is miscibility.

The oil that travels with the refrigerant through the system is constantly exposed to high temperatures in the compressor and condenser and low temperatures in the evaporator and low side of the system. Because of these conditions, wax or any other impurity cannot be allowed in the system. Therefore, only highly refined oils designed specifically for use in refrigeration systems should be used. The type of oil is generally recommended by the equipment manufacturer. With the new refrigerants that are being introduced to the industry, the proper oil is even more important.

Because a small amount of oil is required to lubricate the cylinders, oil circulates with the refrigerant through the complete system. It has been determined that oils and refrigerant vapor do not readily mix. Because of this, it is important that the refrigerant velocities be kept high enough to cause the oil to return to the compressor crankcase.

Liquid refrigerant is attracted to the lubricating oil in the system. The liquid will boil and migrate through the system and eventually reach the compressor crankcase. This action occurs even when there is no pressure differential between the components. In the crankcase, the vapor will condense back into a liquid. This migration continues until the oil is completely saturated with liquid refrigerant.

When the compressor next starts, the liquid will rapidly boil off, causing the oil to foam violently. This foaming action causes the oil to leave the crankcase, resulting in compressor lubrication problems. This migration should be stopped to prevent damage to the compressor, either through slugging or lack of lubrication. This problem is generally solved by installing a crankcase heater.

SUMMARY 13–9

- The oil that travels through the system with the refrigerant is constantly exposed to high and low temperatures; therefore, impurities such as wax cannot be allowed in the system.
- Liquid refrigerant is attracted to the lubricating oil in the system. It will collect in the compressor crankcase during the OFF cycle and will rapidly boil off when the compressor starts, causing a foaming action in the oil.

REVIEW QUESTIONS 13–9

1. Why does oil travel through the system with the refrigerant?
2. What is the best way to determine what type of oil to use in a refrigeration compressor?
3. What can be done to prevent compressor damage due to liquid refrigerant?

13–10 REFRIGERANT TABLES

Refrigerant tables are compiled listing all the characteristics of each individual refrigerant. Very precise information can be obtained from them. When specific data are required, it is easily obtained from these tables. The data for HFC–134a is shown in Table 13–10.

This table lists the saturation properties of HFC–134a by pressure, in absolute; volume; density; enthalpy; and entropy at each temperature listed on the side of the table.

Enthalpy

This is the term used in thermodynamics to describe the heat content of fluid. It is expressed in Btu/lb of the refrigerant. The generally accepted zero point for the enthalpy of a refrigerant is –40°F. That is, at –40°F the refrigerant has no enthalpy. Liquid refrigerant with a temperature below –40°F is considered to have a negative enthalpy and any liquid above –40°F is considered to have a positive enthalpy.

The enthalpy in different sections of the refrigeration system is used to determine the performance of the system at those operating temperatures and pressures. If the refrigerant flow rate through a coil and the heat content per pound are known, the cooling capacity of the unit can be determined.

Entropy

Entropy is the mathematical ratio used in thermodynamics. It is used in solving engineering design problems and is not very useful in the field. Therefore, we will not discuss it in this text.

SUMMARY 13–10

- Refrigerant tables list the saturation properties of refrigerants. The information is very precise.
- The pressure, volume, density, enthalpy, and entropy for each refrigerant at each temperature is listed.
- Entropy describes the heat content of a fluid. It is expressed in Btu/lb of the refrigerant.
- The generally accepted zero point for the enthalpy of a refrigerant is –40°F.

TEMP.	PRESSURE	VOLUME ft^3/lb		DENSITY lb/ft^3		ENTHALPY Btu/lb			ENTROPY Btu/(lb)(°R)		TEMP.
°F	psia	LIQUID v_f	VAPOR v_g	LIQUID $1/v_f$	VAPOR $1/v_g$	LIQUID h_f	LATENT h_{fg}	VAPOR h_g	LIQUID s_f	VAPOR s_g	°F
30	40.800	0.0124	1.1538	80.96	0.8667	21.6	85.9	107.4	0.0473	0.2227	30
31	41.636	0.0124	1.1315	80.85	0.8838	21.9	85.7	107.6	0.0480	0.2226	31
32	42.486	0.0124	1.1098	80.74	0.9011	22.2	85.5	107.7	0.0486	0.2226	32
33	43.349	0.0124	1.0884	80.62	0.9188	22.5	85.3	107.9	0.0492	0.2225	33
34	44.225	0.0124	1.0676	80.51	0.9367	22.8	85.2	108.0	0.0499	0.2224	34
35	45.115	0.0124	1.0472	80.40	0.9549	23.2	85.0	108.1	0.0505	0.2223	35
36	46.018	0.0125	1.0274	80.28	0.9733	23.5	84.8	108.3	0.0512	0.2223	36
37	46.935	0.0125	1.0080	80.17	0.9921	23.8	84.6	108.4	0.0518	0.2222	37
38	47.866	0.0125	0.9890	80.05	1.0111	24.1	84.4	108.6	0.0525	0.2221	38
39	48.812	0.0125	0.9705	79.94	1.0304	24.5	84.2	108.7	0.0531	0.2221	39
40	49.771	0.0125	0.9523	79.82	1.0501	24.8	84.1	108.8	0.0538	0.2220	40
41	50.745	0.0125	0.9346	79.70	1.0700	25.1	83.9	109.0	0.0544	0.2219	41
42	51.733	0.0126	0.9173	79.59	1.0902	25.4	83.7	109.1	0.0551	0.2219	42
43	52.736	0.0126	0.9003	79.47	1.1107	25.8	83.5	109.2	0.0557	0.2218	43
44	53.754	0.0126	0.8837	79.35	1.1316	26.1	83.3	109.4	0.0564	0.2217	44
45	54.787	0.0126	0.8675	79.24	1.1527	26.4	83.1	109.5	0.0570	0.2217	45
46	55.835	0.0126	0.8516	79.12	1.1742	26.7	82.9	109.7	0.0576	0.2216	46
47	56.898	0.0127	0.8361	79.00	1.1960	27.1	82.7	109.8	0.0583	0.2216	47
48	57.976	0.0127	0.8210	78.88	1.2181	27.4	82.5	109.9	0.0589	0.2215	48
49	59.070	0.0127	0.8061	78.76	1.2405	27.7	82.3	110.1	0.0596	0.2214	49
50	60.180	0.0127	0.7916	78.64	1.2633	28.0	82.1	110.2	0.0602	0.2214	50
51	61.305	0.0127	0.7774	78.53	1.2864	28.4	81.9	110.3	0.0608	0.2213	51
52	62.447	0.0128	0.7634	78.41	1.3099	28.7	81.8	110.5	0.0615	0.2213	52
53	63.604	0.0128	0.7498	78.29	1.3337	29.0	81.6	110.6	0.0621	0.2212	53
54	64.778	0.0128	0.7365	78.16	1.3578	29.4	81.4	110.7	0.0628	0.2211	54
55	65.963	0.0128	0.7234	78.04	1.3823	29.7	81.2	110.9	0.0634	0.2211	55
56	67.170	0.0128	0.7106	77.92	1.4072	30.0	81.0	111.0	0.0640	0.2210	56
57	68.394	0.0129	0.6981	77.80	1.4324	30.4	80.8	111.1	0.0647	0.2210	57
58	69.635	0.0129	0.6859	77.68	1.4579	30.7	80.6	111.3	0.0653	0.2209	58
59	70.892	0.0129	0.6739	77.56	1.4839	31.0	80.4	111.4	0.0659	0.2209	59
60	72.167	0.0129	0.6622	77.43	1.5102	31.4	80.2	111.5	0.0666	0.2208	60
61	73.459	0.0129	0.6507	77.31	1.5369	31.7	80.0	111.6	0.0672	0.2208	61
62	74.769	0.0130	0.6394	77.19	1.5640	32.0	79.7	111.8	0.0678	0.2207	62
63	76.096	0.0130	0.6283	77.06	1.5915	32.4	79.5	111.9	0.0685	0.2207	63
64	77.440	0.0130	0.6175	76.94	1.6194	32.7	79.3	112.0	0.0691	0.2206	64
65	78.803	0.0130	0.6069	76.81	1.6477	33.0	79.1	112.2	0.0698	0.2206	65
66	80.184	0.0130	0.5965	76.69	1.6764	33.4	78.9	112.3	0.0704	0.2205	66
67	81.582	0.0131	0.5863	76.56	1.7055	33.7	78.7	112.4	0.0710	0.2205	67
68	83.000	0.0131	0.5764	76.44	1.7350	34.0	78.5	112.5	0.0717	0.2204	68
69	84.435	0.0131	0.5666	76.31	1.7649	34.4	78.3	112.7	0.0723	0.2204	69
70	85.890	0.0131	0.5570	76.18	1.7952	34.7	78.1	112.8	0.0729	0.2203	70
71	87.363	0.0131	0.5476	76.05	1.8260	35.1	77.9	112.9	0.0735	0.2203	71
72	88.855	0.0132	0.5384	75.93	1.8573	35.4	77.6	113.0	0.0742	0.2202	72
73	90.366	0.0132	0.5294	75.80	1.8889	35.7	77.4	113.2	0.0748	0.2202	73
74	91.897	0.0132	0.5206	75.67	1.9210	36.1	77.2	113.3	0.0754	0.2201	74
75	93.447	0.0132	0.5119	75.54	1.9536	36.4	77.0	113.4	0.0761	0.2201	75
76	95.016	0.0133	0.5034	75.41	1.9866	36.8	76.8	113.5	0.0767	0.2200	76
77	96.606	0.0133	0.4950	75.28	2.0201	37.1	76.6	113.7	0.0773	0.2200	77
78	98.215	0.0133	0.4868	75.15	2.0541	37.4	76.3	113.8	0.0780	0.2200	78
79	99.844	0.0133	0.4788	75.02	2.0885	37.8	76.1	113.9	0.0786	0.2199	79
80	101.494	0.0134	0.4709	74.89	2.1234	38.1	75.9	114.0	0.0792	0.2199	80
81	103.164	0.0134	0.4632	74.75	2.1589	38.5	75.7	114.1	0.0799	0.2198	81
82	104.855	0.0134	0.4556	74.62	2.1948	38.8	75.4	114.3	0.0805	0.2198	82
83	106.566	0.0134	0.4482	74.49	2.2312	39.2	75.2	114.4	0.0811	0.2197	83
84	108.290	0.0134	0.4409	74.35	2.2681	39.5	75.0	114.5	0.0817	0.2197	84
85	110.050	0.0135	0.4337	74.22	2.3056	39.9	74.8	114.6	0.0824	0.2196	85
86	111.828	0.0135	0.4267	74.08	2.3436	40.2	74.5	114.7	0.0830	0.2196	86
87	113.626	0.0135	0.4198	73.95	2.3821	40.5	74.3	114.9	0.0836	0.2196	87
88	115.444	0.0135	0.4130	73.81	2.4211	40.9	74.1	115.0	0.0843	0.2195	88
89	117.281	0.0136	0.4064	73.67	2.4607	41.2	73.8	115.1	0.0849	0.2195	89

Table 13–10 HFC–134a saturation properties—temperature table (Courtesy of Freon Products Div., E.I. DuPont de Numours and Co., Inc.)

- The enthalpy in different sections of the refrigeration system is used to determine the performance of the system at those operating temperatures and pressures.

REVIEW QUESTIONS 13–10

1. What is the generally accepted zero point for the enthalpy of a refrigerant?
2. At what temperature does a liquid have a negative enthalpy?
3. Where is specific data for a particular refrigerant obtained?
4. What is used to determine the performance of a refrigeration system?

13–11 POCKET PRESSURE-TEMPERATURE CHARTS

These are small tables that may be carried in the pocket for use when servicing refrigeration and air conditioning equipment. They have the most common temperatures and the related saturation pressures of various refrigerants listed for easy use and referral. See Table 13–11.

When the temperature is known, the saturation pressure for any listed type of refrigerant can easily be determined. If the pressure is known, the temperature at which the refrigerant is operating can be determined. The manufacturers of most refrigeration gauges have these pressures and temperatures listed in a table on the gauge face for easy reference.

SUMMARY 13–11

- Pocket pressure-temperature charts have the most common temperatures and the related saturation pressures of the various refrigerants listed for easy use and referral.
- For any known temperature, the related saturated pressure can be determined.

REVIEW QUESTIONS 13–11

1. At a 40°F temperature, what is the gauge pressure of R–22?
2. We have an air-cooled condensing unit using R–22 refrigerant and operating in an ambient temperature of 100°F. What should the discharge pressure be?

13–12 THE P-H DIAGRAM

One diagnostic tool that is available to everyone but is seldom used is the pressure-enthalpy diagram, sometimes called the Mollier Diagram, but generally called the P-H diagram. This tool is used by most design engineers but is unfortunately overlooked by service technicians.

Remember that the important thing about learning theP-H diagram is that it will not be used to plot every system that is serviced. However, just learning the chart will provide a new and fresh insight on what is happening inside the system.

The P-H diagram is a simple, graphic way to plot a system cycle and observe its characteristics. No two systems are exactly alike. Each one has its own personality. When using the diagram, a single pound of refrigerant will be followed completely around the system and when and where it changes states and what that change means will be learned. Since only one pound of refrigerant is followed, the diagram may be applied to any size system. The pounds per minute circulated would determine the total system capacity. To better understand this process, some basics must be reviewed, starting with diagram construction.

	Vapor Pressure, Psig							
°F	"Freon–113"	"Freon–114"	"Freon–13"	R–500**	"Freon–11"	"Freon–12"	"Freon–507"	"Freon–22"
–50		27.2*	57.0			15.4*	0.0	6.0"
–48		27.0*	60.0			14.6*	0.8	4.7"
–46		26.8*	63.0			13.8*	1.6	3.3"
–44		26.6*	66.2			12.9*	2.5	1.8"
–42		26.3*	69.4			11.9*	3.4	0.3"
–40		26.1*	72.7	7.9*	28.4*	11.0*	4.3	0.6
–38		25.9*	76.2	6.7*	28.3*	10.0*	5.2	1.4
–36		25.6*	79.7	5.4*	28.2*	8.9*	6.2	2.3
–34		25.3*	83.3	4.2*	28.1*	7.8*	7.2	3.2
–32		25.0*	87.1	2.8*	28.0"	6.7*	8.3	4.1
–30	29.3*	24.7*	90.9	1.4*	27.8*	5.5*	9.4	5.0
–28	29.3*	24.4*	94.9	0.0	27.7*	4.3*	10.5	6.0
–26	29.2*	24.0*	98.9	0.8	27.5*	3.0*	11.7	7.0
–24	29.2*	23.7*	103.0	1.5	27.4*	1.6*	12.9	8.1
–22	29.1*	23.3*	107.3	2.3	27.2*	0.3*	14.7	9.2
–20	29.1*	22.9*	111.7	3.1	27.0*	0.6	15.5	10.3
–18	29.0*	22.5*	116.2	4.0	26.9*	1.3	16.9	11.5
–16	28.9*	22.1*	120.8	4.9	26.7*	2.1	18.3	12.7
–14	28.9*	21.6*	125.7	5.8	26.5*	2.8	19.7	13.9
–12	28.9*	21.1*	130.5	6.8	26.2*	3.7	20.2	15.2
–10	28.7*	20.6*	135.4	7.8	26.0*	4.5	22.4	16.6
–8	28.6*	20.1*	140.5	8.8	25.8*	5.4	24.4	18.0
–6	28.5*	19.6*	145.7	9.9	25.5*	6.3	26.0	19.4
–4	28.4*	19.0*	151.1	11.0	25.3*	7.2	27.7	20.9
–2	28.3*	18.4*	156.5	12.1	25.0*	8.2	29.4	22.5
0	28.2*	17.8*	162.2	13.3	24.7*	9.2	31.2	24.1
2	28.1*	17.2*	167.9	14.5	24.4*	10.2	33.1	25.7
4	28.0*	16.5*	173.7	15.7	24.1*	11.2	35.0	27.4
6	27.9*	15.8*	179.8	17.0	23.8*	12.3	37.0	29.2
8	27.7*	15.1*	185.9	18.4	23.5*	13.5	39.0	31.0
10	27.6*	14.3*	192.2	19.8	23.1*	14.6	41.1	32.9
12	27.5*	13.5*	198.6	21.2	22.7*	15.8	43.2	34.9
14	27.3*	12.7*	205.2	22.7	22.3*	17.1	45.4	36.9
16	27.1*	11.9*	211.9	24.2	21.9*	18.4	47.7	39.0
18	27.0*	11.0*	218.8	25.7	21.5*	19.7	50.1	41.1
20	26.8*	10.1*	225.8	27.3	21.1*	21.0	52.4	43.3
22	26.6*	9.1*	233.0	29.0	20.6*	22.4	54.9	45.5
24	26.4*	8.1*	240.3	30.7	20.2*	23.9	57.4	47.9
26	26.2*	7.1*	247.8	32.5	19.7*	25.4	60.0	50.2
28	26.0*	6.1*	255.5	34.3	19.1*	26.9	62.7	52.7
30	25.8*	5.0*	263.3	36.1	18.6*	28.5	65.4	55.2
32	25.6*	3.9*	271.3	38.0	18.1*	30.1	68.2	57.8
34	25.3*	2.7*	279.5	40.0	17.5*	31.7	71.1	60.5
36	25.1*	1.5*	287.8	42.0	16.9*	33.4	74.1	63.3
38	24.8*	0.2*	295.3	44.1	16.3*	35.2	77.1	66.1
40	24.5*	0.5	305.0	46.2	15.6*	37.0	80.2	69.0
42	24.2*	1.2	313.9	48.4	14.9*	38.8	83.4	72.0
44	23.9*	1.9	322.9	50.7	14.2*	40.7	86.6	75.0
46	23.6*	2.6	332.2	53.0	13.5*	42.7	90.0	78.2
48	23.3*	3.3	341.6	55.4	12.8*	44.7	93.4	81.4
50	22.9*	4.0	351.2	57.8	12.0*	46.7	96.9	84.7
52	22.6*	4.8	361.1	60.3	11.2*	48.8	100.5	88.1
54	22.2*	5.6	371.1	62.9	10.4*	51.0	104.1	91.5
56	21.8*	6.4	381.3	65.5	9.5*	53.2	107.9	95.1
58	21.4*	7.3	391.7	68.2	8.7*	55.4	111.7	98.8

Table 13–11 Condensed pressure-temperature chart
(Courtesy of Freon Products Div., E. I. DuPont de Nemours & Co., Inc.)

* Inches mercury below one atmosphere ** Patented by Carrier Corporation

	Vapor Pressure, Psig							
°F	"Freon–113"	"Freon–114"	"Freon–13"	R–500**	"Freon–11"	"Freon–12"	"Freon–507"	"Freon–22"
60	21.0*	8.1	402.4	71.0	7.7*	57.7	115.6	102.5
62	20.6*	9.0	413.3	73.8	6.8*	60.1	119.6	106.3
64	20.1*	9.9	424.2	76.7	5.8*	62.5	123.7	110.2
66	19.7*	10.9	435.6	79.7	4.8*	65.0	127.9	114.2
68	19.2*	11.9	447.0	82.8	3.7*	67.6	132.2	118.3
70	18.7*	12.9	458.8	85.8	2.6*	70.2	136.6	122.5
72	18.2*	13.9	470.7	89.0	1.5*	72.9	141.1	126.8
74	17.6*	15.0	482.9	92.3	0.4*	75.6	145.6	131.2
76	17.1*	16.1	495.3	95.6	0.4	78.4	150.3	135.7
78	16.5*	17.7	508.1	99.0	1.0	81.3	155.1	140.3
80	15.9*	18.3	521.0	102.5	1.6	84.2	159.9	145.0
82	15.3*	19.5	534.1	106.1	2.2	87.2	164.9	149.8
84	14.6*	20.7	547.5	109.7	2.9	90.2	170.0	154.7
86	13.9*	22.0		113.4	3.6	93.3	175.1	159.8
88	13.2*	23.3		117.3	4.3	96.3	180.4	164.9
90	12.5*	24.6		121.2	5.0	99.8	185.8	170.1
92	11.8*	25.9		125.1	5.7	103.1	193.3	175.4
94	11.0*	27.3		129.2	6.5	106.5	196.9	180.9
96	10.2*	28.7		133.3	7.3	110.0	202.6	186.5
98	9.4*	30.2		137.6	8.1	113.5	208.4	192.1
100	8.6*	31.7		141.9	8.9	117.7	214.4	197.5
102	7.7*	33.2		146.3	9.8	120.9	220.4	203.8
104	6.8*	34.8		150.9	10.6	124.6	226.6	209.9
106	5.9*	36.4		155.4	11.5	128.5	232.9	216.0
108	4.9*	38.0		160.1	12.5	132.4	239.3	222.3
110	4.0*	39.7		164.9	13.4	136.4	245.8	228.7
112	3.0*	41.4		169.8	14.4	140.5	252.6	235.2
114	1.9*	43.2		174.8	15.3	144.7	259.7	241.9
116	0.8*	45.0		179.9	16.4	148.5	266.1	248.7
118	0.1	46.9		185.0	17.4	153.2	273.1	255.6
120	0.7	48.7	ABOVE CRITICAL TEMPERATURE	190.3	18.5	157.7	280.3	262.6
122	1.3	50.7		195.7	19.6	162.2	287.6	269.7
124	1.9	52.7		201.2	20.7	166.7	295.0	277.0
126	2.5	54.7		206.7	21.9	171.4	302.5	284.4
128	3.1	56.7		212.4	23.0	176.7	310.7	291.8
130	3.7	58.8		218.2	24.3	181.0	318.0	299.3
132	4.4	61.0		224.1	25.5	185.9	326.0	307.1
134	5.1	63.2		230.1	26.8	191.0	334.1	315.2
136	5.8	65.5		236.3	28.1	196.1	342.3	323.6
138	6.5	67.7		242.5	29.4	201.3	350.7	332.3
140	7.2	70.1		248.8	30.8	206.6	359.2	341.3
142	8.0	72.5			32.2	212.0	367.8	350.3
144	8.8	74.9			33.7	217.5	376.7	359.4
146	9.6	77.4			35.1	223.1	385.6	368.6
148	10.4	80.0			36.6	228.8	394.7	377.9
150	11.2	82.6			38.2	234.6	404.0	387.2
152	12.1	85.2			39.7	240.5	413.4	396.6
154	13.0	87.9			41.3	246.5	423.0	406.1
156	13.9	90.7			43.0	252.6	432.7	415.6
158	14.8	93.5			44.6	258.8	442.6	425.1
160	15.7	96.4			46.3	265.1	452.6	434.6

Table 13–11 Condensed pressure-temperature chart (continued)
(Courtesy of Freon Products Div., E. I. DuPont de Nemours & Co., Inc.)

* Inches mercury below one atmosphere ** Patented by Carrier Corporation

Saturation, Subcooling, and Superheat

To become a proficient technician, you must learn about the three conditions that a refrigerant exists in while it is traveling around the system. They are: saturation, subcooling, and superheat. To pinpoint which condition the refrigerant is in at any given point, both a temperature reading and a pressure reading must be taken. Taking only one of these readings will not be sufficient.

A good method for learning about the three conditions is to study water, with which most people are familiar. Water and refrigerants are similar except that at a-tmospheric pressure (0 psig) it will boil at 212°F rather than, for example, R–22 which boils at –41°F at that same pressure. The boiling point, incidentally, is called the *saturation temperature.*

The diagram for water in Figure 13–5, plots the heat content in Btu/lb along the horizontal axis versus the temperature along the vertical axis. See Figure 13–5.

Two types of heat will be considered in this process. They are *sensible heat* and *latent heat.* The sensible heat exchange involves a change in temperature and latent heat involves a change of state (at a constant temperature). The diagram is plotted at 0 psig (atmospheric) pressure. As the water is warmed from 52°F to 212°F, sensible heat is used because only a change in temperature is experienced. Since subcooling is defined as the number of degrees the liquid temperature is below its saturation temperature, the liquid then is in the subcooling state. At 52°F, the subcooling is 160°F, and at 212°F it is zero degrees because it has reached the saturation point.

When the water reaches its saturation point of 212°F at atmospheric pressure, boiling will start at a constant temperature as long as the pressure remains the constant. During this change of state, from a liquid to a vapor, the water absorbs approximately 1,000 Btu for each pound of water evaporated. This is called the *latent heat of evaporation.*

When all of the liquid has boiled off and more heat is added to the vapor, the vapor temperature will rise above the saturation temperature of 212°F. Since the temperature of the vapor changes, sensible heat is again experienced. The water now exists as a vapor and is in the superheated state. Superheat is defined as the amount

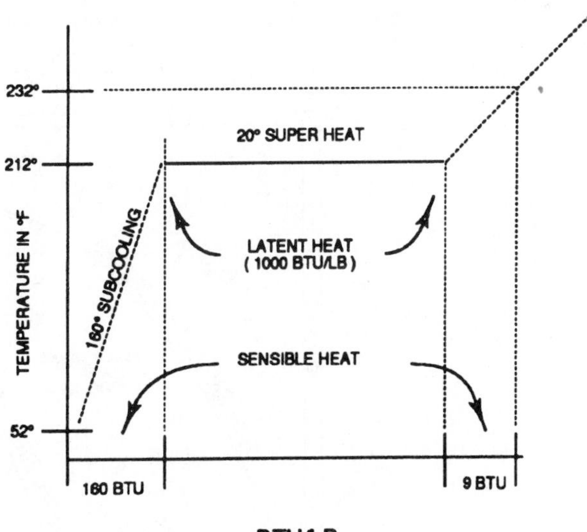

Figure 13–5 Btu/lb (Courtesy of John A. Hogan, *Get Smart Learn Your P-H Chart,* © 1989, by permission)

of degrees the vapor temperature is above the saturation temperature at a given pressure. If the temperature of the superheated vapor were measured at 232°F, then the degrees of superheat would be 20 because it is 20° above the saturation temperature of 212°F.

Note the tremendous amount of heat required to boil a fluid, or to cause it to go through a change of state (latent heat of vaporization), when compared to the amount it takes to simply change its temperature. For this reason, the evaporator should be kept as full of liquid refrigerant as possible without flooding back to the compressor in order to get the maximum amount of heat transfer.

Since all of the conditions shown on the diagram occur at one pressure (0 psig), it can be seen that just by taking a pressure reading does not tell what the system is doing. In order to determine if the refrigerant is subcooled, saturated, or superheated, the temperature at a given point must also be known. Whenever the pressure reading is taken, immediately check a pressure-temperature chart to see what the saturation temperature is for that refrigerant at that pressure. Now, when the temperature at that point is known, it can be determined if the refrigerant is subcooled, saturated, or superheated. If the temperature is found to be above the saturation temperature, then it is superheated. If it is found to be below the saturation temperature, then it is subcooled. If the temperature is the same as the pressure-temperature chart indicates, then there is both liquid and vapor in the saturated state present.

The Three Zones. Notice in Figure 13–6 that the P-H diagram is divided into three zones, *saturated*, *subcooled*, and *superheated*.

As the refrigerant travels around the system, it will exist in one of the three zones indicated on the diagram. It is always changing conditions and will never exist in any two at the same time. The refrigerant is either saturated, subcooled, or superheated. If a technician does not have an understanding of these basic principles, he will never thoroughly understand how the system operates.

Saturated Zone. Any time the refrigerant is inside the envelope or dome, it is in the saturated zone. When a refrigerant is saturated, it will contain both liquid and vapor. The right-hand curve of the saturated zone is the *100% saturated vapor line* and the curve on the left-hand side of the zone is the *100% saturated liquid line.*

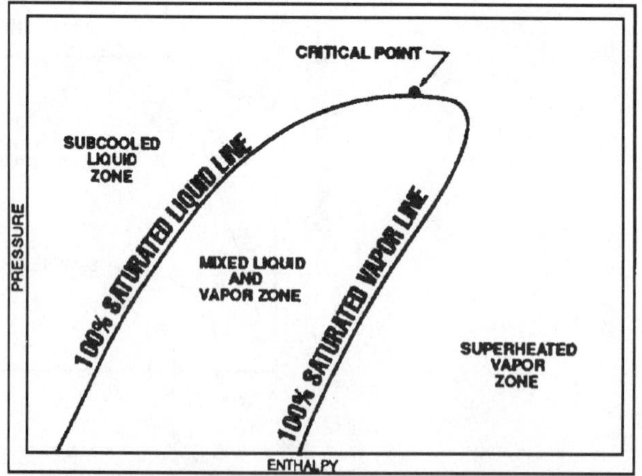

Figure 13–6 P-H diagram zones (Courtesy of John A. Hogan, *Get Smart Learn Your P-H Chart*, © 1989, by permission)

Some charts (E.I. Du Pont, for example) will have a series of vertical lines within the saturated zone. These lines indicate the percentage of vapor along each line. Saturation occurs in the evaporator where the refrigerant is changing state from a liquid to a vapor (boiling) and in the condenser where the reverse is happening (condensing). As previously noted, the temperatures and pressures shown on the pressure-temperature chart are all in the saturated state.

Superheated Zone. To the right of the 100% saturated vapor line, the refrigerant is above the saturation temperature. The superheated condition should exist from within the outlet of the evaporator to within the inlet portion of the condenser. The superheat measurement is the temperature of the vapor minus the saturation temperature at a given pressure.

Subcooled Zone. To the left of the saturated line is the *subcooled liquid zone*. The refrigerant is in the liquid state, where its temperature is below the saturated temperature at a given pressure. The subcooled condition should exist from the outlet of the condenser to the inlet of the expansion device. The degrees of subcooling are found by subtracting the temperature of the liquid from the condensing temperature at that pressure.

Critical Point. The critical point on the P-H diagram is located where the saturated liquid curve and the saturated vapor curve converge. At any temperature above this point, the refrigerant may exist in the vapor phase only.

Five Refrigerant Properties. Most refrigeration gauges read in pounds per square inch above atmospheric pressure (psig). At sea level, 0 pounds gauge, the pressure would be 14.7 psia. Pressures below 0 psig would be considered a partial vacuum and read in inches of mercury. A perfect vacuum would be defined as 0 pounds per square inch absolute (psia). The P-H diagram is scaled for *absolute pressures*. See Figure 13–7.

Most computations use psia. Consequently, with a gauge pressure reading of 10 psig, to obtain absolute pressure simply add 14.7 psi to the gauge reading. This would give an absolute pressure reading of 24.7 psia. Many people forget to make this conversion when reading the P-H diagram. Always convert pressures to absolute for this use.

Figure 13–7 Absolute (psia) (Courtesy of John A. Hogan, *Get Smart Learn Your P-H Chart*, © 1989, by permission)

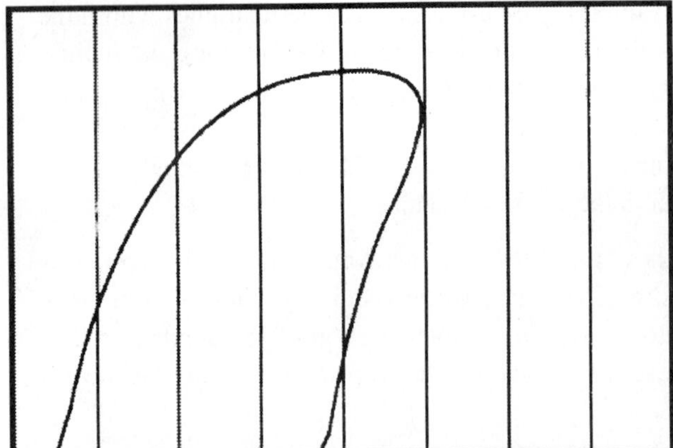

Figure 13–8 Enthalpy (Btu/lb)(Courtesy of John A. Hogan, *Get Smart Learn Your P-H Chart*, © 1989, by permission)

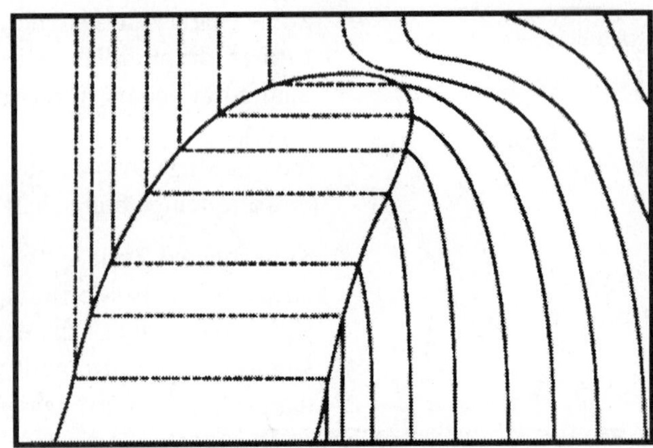

Figure 13–9 Temperature (F) (Courtesy of John A. Hogan, *Get Smart Learn Your P-H Chart*, © 1989, by permission)

The *enthalpy scale* shown in Figure 13–8 is usually shown at both the top and bottom of the diagram, and the lines of constant enthalpy run vertically. Enthalpy represents the total heat content in Btu per pound.

In Figure 13–9, please note that the lines of *constant temperature* run almost vertically in the superheated zone, horizontally in the saturated zone, and then go vertical in the subcooled zone.

Entropy is defined as the heat available measured in Btu per pound per degree of change for a substance. The lines of constant entropy extend diagonally up to the right from the saturated vapor line. See Figure 13–10.

The main concern of the service technician is that when the refrigerant is compressed, it follows up the line of constant entropy.

The lines of *constant volume* extend out from the saturated vapor line into the superheated zone. See Figure 13–11.

These values indicate how much space (cu ft) is taken up by each pound of refrigerant vapor.

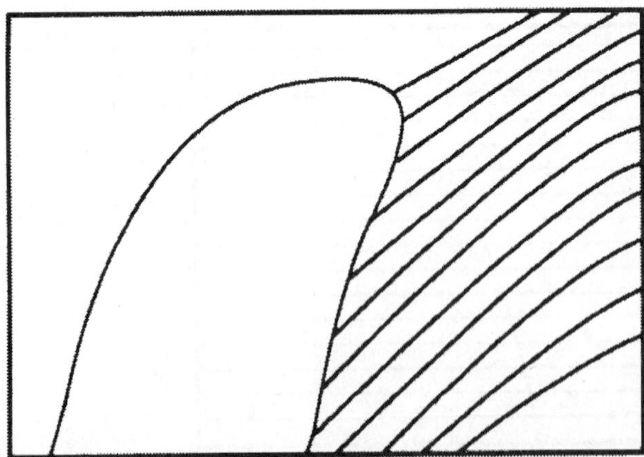

Figure 13–10 Entropy (Btu/lb/R) (Courtesy of John A. Hogan, *Get Smart Learn Your P-H Chart*, © 1989, by permission)

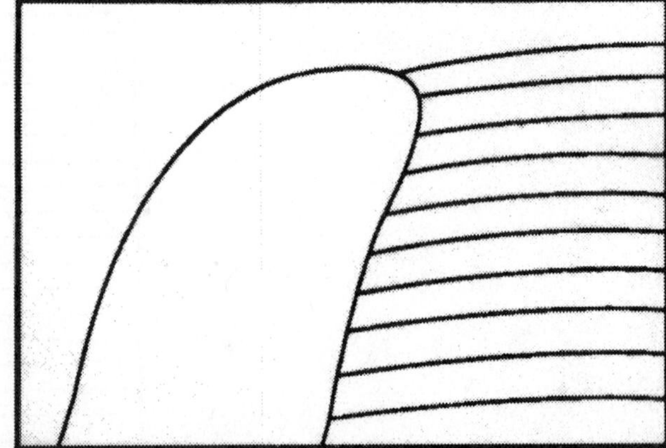

Figure 13–11 Constant volume (cu ft/lb) (Courtesy of John A. Hogan, *Get Smart Learn Your P-H Chart*, © 1989, by permission)

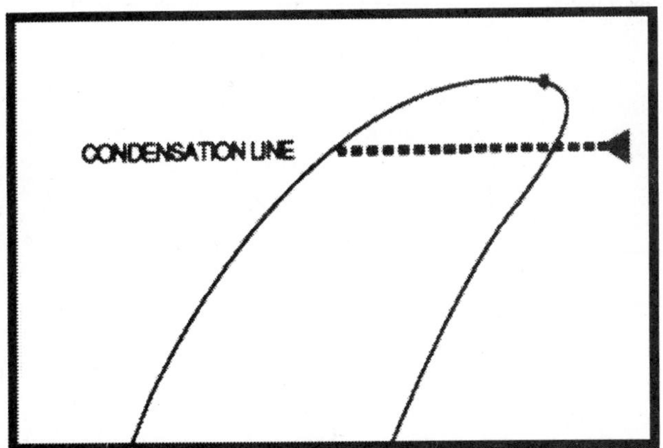

Figure 13–12 Condensation line (Courtesy of John A. Hogan, *Get Smart Learn Your P-H Chart*, © 1989, by permission)

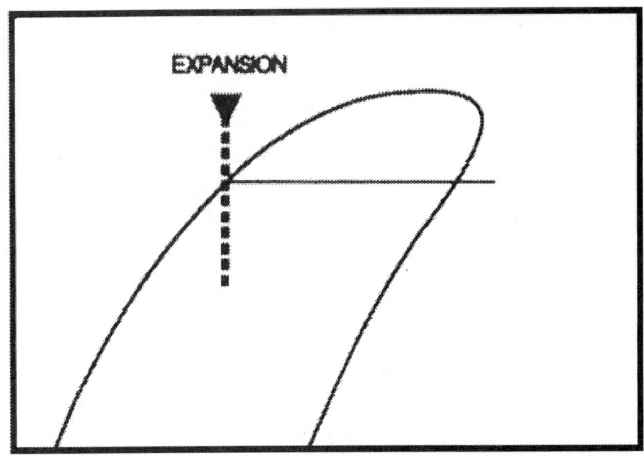

Figure 13–13 Expansion line (Courtesy of John A. Hogan, *Get Smart Learn Your P-H Chart*, © 1989, by permission)

The Refrigeration Cycle. Usually, the refrigeration cycle is shown as an ideal cycle. This is generally an illustration of a cycle but it does not show superheating in the evaporator, suction line, heat exchangers, vapor-cooled compressors, etc. Also, subcooling of the liquid and pressure losses are not shown. These things are important to the service technician, but to understand the basic cycle, the ideal cycle will be discussed first.

The four functions of the refrigeration cycle are: compression, condensing, expansion, and evaporating. Following is a discussion of these functions.

To draw a simple, ideal cycle, only two pressures are required: the evaporation pressure and the condensing pressure. They are also called the high- and low-side pressures. Do not forget to add 14.7 psi to the gauge readings to obtain the absolute pressure required for the diagram.

Draw the condensing and evaporating horizontal pressure lines. See Figure 13–12.

Next, drop the expansion line down from the intersection of the condensation line and the saturated liquid line to intersect with the horizontal evaporation line. See Figures 13–13 and 13–14.

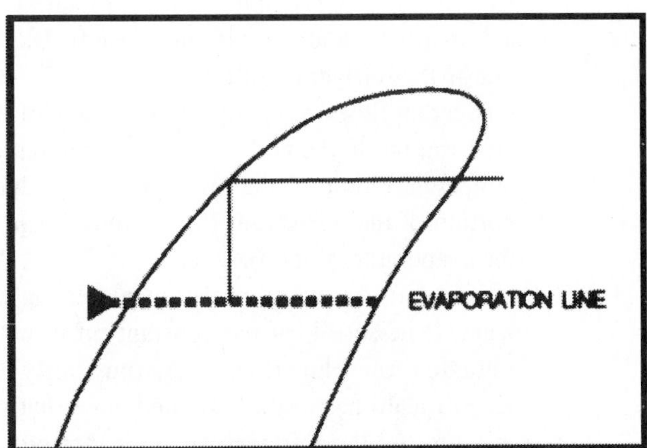

Figure 13–14 Evaporation line (Courtesy of John A. Hogan, *Get Smart Learn Your P-H Chart*, © 1989, by permission)

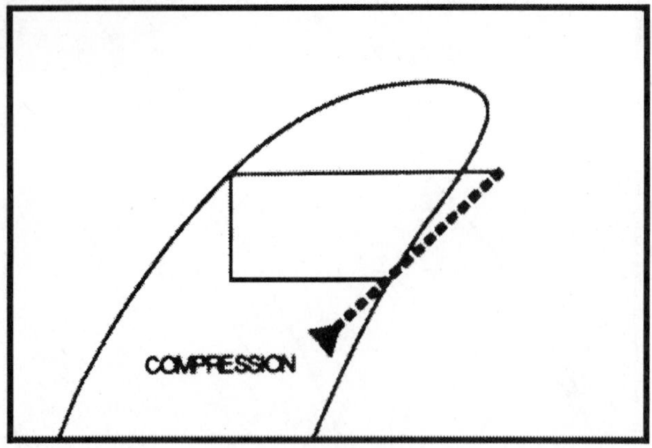

Figure 13–15 Compression line (Courtesy of John A. Hogan, *Get Smart Learn Your P-H Chart*, © 1989, by permission)

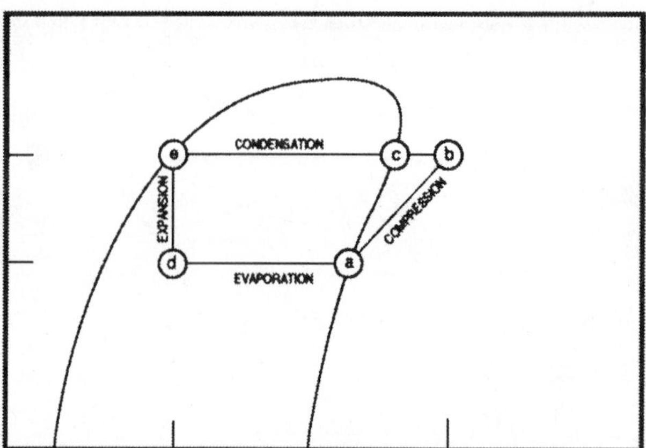

Figure 13–16 A pound of refrigerant around the cycle (Courtesy of John A. Hogan, *Get Smart Learn Your P-H Chart*, © 1989, by permission)

Finally, complete the cycle by extending the compression line up to the superheated zone, parallel with the constant entropy lines, until it intersects the condensation line. See Figure 13–15.

Let's follow a pound of refrigerant around the cycle as shown in Figure 13–16.

It leaves the compressor as high pressure vapor in a highly superheated condition (line a to b). By compressing the refrigerant vapor, we can condense it into a liquid with the cooling mediums available to us, either air or water, and reuse it for another cycle. As the pound of refrigerant moves left horizontally along the condensing pressure line, the condenser must remove the superheat before it reaches the 100% saturated vapor line (line b to c).

Condensation starts when the refrigerant crosses the saturated vapor line (point c) into the saturated zone. Note that as the refrigerant moves along the constant pressure line in the saturated zone, the temperature remains constant. When the pound of refrigerant reaches the 100% saturated liquid line (point e), all of the vapor has condensed to a liquid.

The next step is the expansion process where the pressure is dropped suddenly from the high pressure to the low pressure within the expansion device (and refrigerant distributor and tubes if one is used). There is no transfer of heat externally into or out of the refrigerant in this process under the ideal cycle discussion. Therefore the refrigerant flows vertically down the line of constant enthalpy (line e to d). Since it is crossing the horizontal lines of temperature while in the saturated zone, there is a temperature drop of the mixture. This is due to the latent heat of vaporization as a portion of the refrigerant flashes into a vapor. The refrigerant is now ready to enter the evaporator (point d).

The refrigerant enters the evaporator at the evaporator pressure in the saturated zone. It passes along the constant pressure line in the saturated zone to the 100% saturated vapor line (line a). As previously mentioned, on some P-H diagrams there are vertical lines in the saturated zone that represent the percentage of vapor in the mixture. As the refrigerant passes through the evaporator, it gains heat as it boils from a liquid to a vapor. When the refrigerant reaches the 100% vapor line, all the heat absorbed is contained in the vapor. It is now ready to be compressed again.

Do not forget, this is just the ideal cycle. A more real cycle will be discussed later.

Performance of the Cycle

With the help of the P-H diagram, certain quantities of the cycle may be obtained. The engineer who wants more exact figures would use the refrigerant tables. The P-H diagram, remember, is made up from the values in the tables. It is much simpler to use the P-H diagram.

The quantities are the heat of rejection, the refrigerating effect, the circulation rate, the compression ratio, the work of compression, the coefficient of performance, the volume rate of flow per ton, and the power per ton. Since the purpose of this presentation is to help the technician to better understand the refrigeration cycle through the study of the P-H diagram and not to confuse things with a lot of mathematical formulas, we will just discuss the first four quantities mentioned above.

We will consider an R–22 system that develops 10 tons of refrigeration while operating with a 90°F condensing temperature and 20°F evaporating temperature.

Heat of Rejection. Refer to Figure 13–17. This is the heat given up by the condenser. If the one pound of refrigerant has an enthalpy (total heat content) entering the condenser of 118 Btu/lb, and an enthalpy of 38 Btu/lb leaving the condenser, the difference then is the heat given up by the condenser. Note that the condenser is rejecting the heat picked up by the evaporator plus the heat of compression. The heat of compression is the heat added to the refrigerant by the work done by the compressor. Using the letter h for enthalpy, the formula becomes:

$$\text{Heat of Rejection} = h3 - h1$$
$$= 118 - 38 = \text{Btu/lb}$$

Refrigerating Effect. Refer to Figure 13–18. This is the heat absorbed by the evaporator as indicated by h2 and h1 on the diagram. Since the total heat content is 38 Btu/lb entering and 107 Btu/lb leaving, then the heat picked up in the evaporator is

$$\text{Refrigerating Effect} = h2 - h1$$
$$= 107 - 38 = 69 \text{ Btu/lb}$$

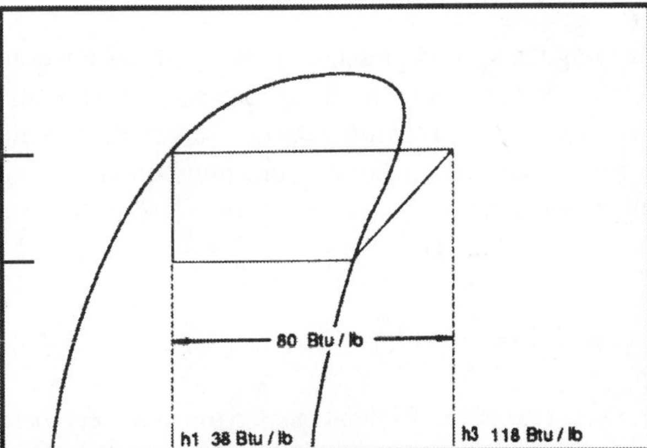

Figure 13–17 Heat of rejection (Courtesy of John A. Hogan, *Get Smart Learn Your P-H Chart,* © 1989, by permission)

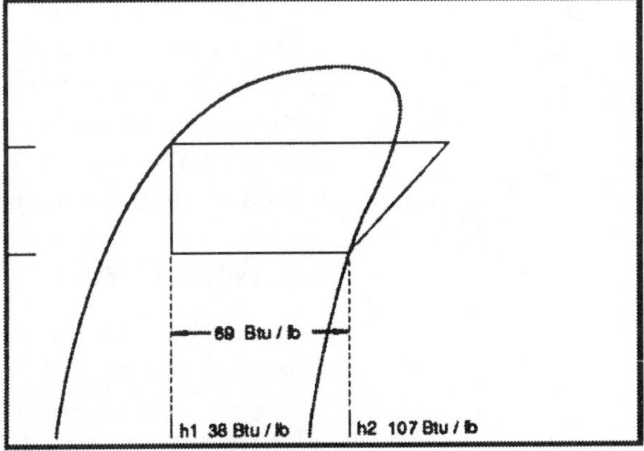

Figure 13–18 Refrigerating effect (Courtesy of John A. Hogan, *Get Smart Learn Your P-H Chart,* © 1989, by permission)

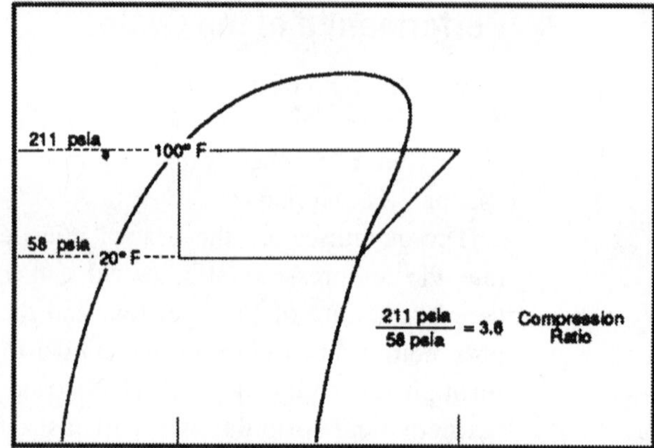

Figure 13–19 Compression ratio (Courtesy of John A. Hogan, *Get Smart Learn Your P-H Chart*, © 1989, by permission)

Circulation Rate. If the refrigerating capacity is known, then we can determine the total circulation rate in lb/min by dividing the capacity in Btu/min by the refrigerating effect of 69 Btu/lb.

$$\text{Refrigerant Flow} = \frac{(10 \text{ tons})(200 \text{ Btu/min/ton})}{69 \text{ Btu/min}}$$
$$= 29.98 \text{ Btu/min}$$

Compression Ratio. Refer to Figure 13–19. The compression ratio is simply stated as the absolute discharge pressure divided by the absolute suction pressure. Consequently, with a given discharge pressure, the higher the suction pressure the lower the compression ratio. Also, the lower the discharge pressure the lower the compression ratio.

While maintaining operating temperatures to satisfy the product requirements, we try to keep the compression ratio as low as possible. The lower the compression ratio the higher the volumetric efficiency. The higher the volumetric efficiency the more mass flow of refrigerant pumped by the compressor. The more mass flow pumped by the compressor the greater the system capacity, and less running time is required. The less running time the more money is saved.

As strange as it may seem, keeping the suction pressure up has a greater effect in lowering the compression ratio than lowering the discharge pressure by an equal amount. An added benefit of lower compression ratio is lower discharge temperatures resulting in less refrigerant-oil breakdown or contaminants being formed because of high operating temperatures.

The Actual Cycle

Remember that the ideal cycle did not have pressure drops, subcooling, or evaporator superheat.

Such a system does not exist, except on paper. Without evaporator superheat there would probably be liquid slugging the compressor, and without pressure drops, there would be no refrigerant flow. The ideal cycle, however, is excellent for use in understanding the basic diagram and, therefore, the cycle.

First, let's look at a simple system with just the basic components like the evaporator, compressor, and an expansion device. The other components can be added later. Also, we will illustrate some of the horrors that can happen in the field

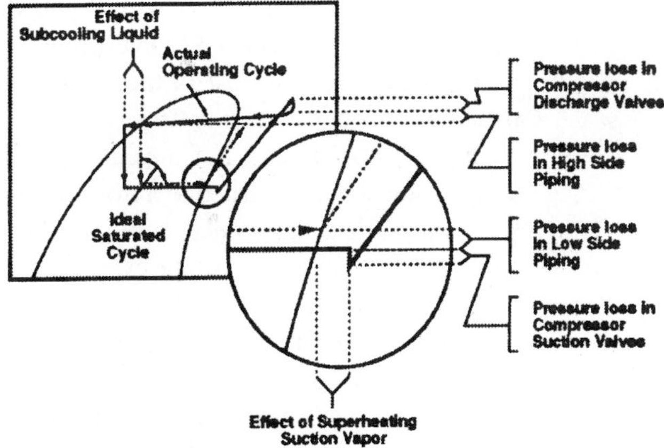

Figure 13–20 Plotting an actual cycle (Courtesy of John A. Hogan, *Get Smart Learn Your P-H Chart*, © 1989, by permission)

and show how your knowledge of the diagram can be used to provide a solution to the problems.

Let's start with the low pressure line. See Figure 13–20. Notice that it is not called the evaporator line now because it involves more than just the evaporator. There will be a pressure drop through the evaporator tubes and the suction line, superheat will be developed in the evaporator and additional superheat in the suction line will occur. Also, assume that this is a semi-hermetic, vapor-cooled type compressor. There would be additional superheat in the compressor body before the compression begins. Certain assumptions have been made, but they were to illustrate the difference between the ideal cycle and the actual cycle.

As can be seen, there is quite a difference in the discharge temperature between the two plots. Also notice the "spike" in the discharge temperature at the peak of its plot. This illustrates the high temperature experienced at the discharge valve of the compressor. The compressor engineers say that this temperature is approximately 75°F higher than what can be measured about 6 inches down the discharge line from the discharge service valve. Consequently, if the oil-refrigerant mixture starts to break down at about 300°F, then the temperature of the discharge vapor should be limited to about 225°F.

By observing the diagram, it can be seen that by moving the point where compression starts toward the left, the chances of running excessively high discharge temperatures that result in refrigerant-oil breakdown can be decreased. Following are some ideas to reduce the discharge temperature:

1. Check to see if the evaporator superheat is too high.
2. Reroute the suction line, if possible, to a cooler location.
3. Is a liquid-suction heat exchanger used unnecessarily?
4. Liquid injection into the suction line (consult the TXV manufacturer).

The idea of taking these steps is to move the compression line further to the left by decreasing the superheat and the discharge gas temperature. Also, if the superheat in the evaporator is too high, by lowering it to the recommended amount the evaporator capacity will also be increased. This is because it contains more refrigerant. This will also raise the suction pressure, which results in a lower compression ratio, which increases system capacity.

Some people equate frost with liquid refrigerant presence. DO NOT ADJUST THE SUPERHEAT TO A FROST LINE. Admittedly, there could possibly be some liquid present when frost is there, but not always. It depends on the operating pressures and temperatures of the system.

Suppose that we had an ice-cream case operating at a –40°F evaporator saturation temperature and we adjusted to a frost line at the suction service valve. This means that the temperature at the service valve would be about 32°F. Since superheat is defined as the vapor temperature minus the evaporator saturation temperature, this would result in a 72°F superheat at the suction service valve. This is excessive. Generally, case manufacturers recommend a 7°F superheat in the evaporator. With the valve adjusted properly and assuming a 20°F rise in the suction line, then the temperature at the suction service valve would be –13°F. There would certainly be frost at this temperature. However, there would be no liquid refrigerant present. This type of adjustment will also make quite a difference in the operating efficiency of the system.

The High Pressure Line. The high pressure line consists of the discharge line, condenser, receiver (if used), and the liquid line. Since the refrigerant is still superheated when it enters the condenser and is subcooled when it leaves the condenser, the condenser is the only component in which all three states exist simultaneously; superheat, saturation, and subcooling. There are some systems that have the liquid line subcooled below the evaporator saturation temperature. This instance would produce all three states in the evaporator, but those systems are the exception rather than the rule.

If a liquid-to-suction heat exchanger is used, about 10°F of subcooling is possible and the vertical expansion line will move to the left. Now there is a smaller value of enthalpy entering the evaporator that will give a greater capacity. It should be remembered that while adding a heat exchanger will increase the capacity on the low side, it will cause the opposite effect on the high side of the system. About 15°F can be added to the suction gas temperature entering the compressor, which as we discussed, will result in a higher discharge temperature. Before arbitrarily adding a heat exchanger it should be determined that this is the step that actually needs to be made. Sometimes a heat exchanger is required to subcool the liquid to insure 100% liquid at the expansion device inlet. This would justify the use of a heat exchanger.

Suppose that the subcooling at the receiver was found to be only 2°F and there is a 10 ft lift in the liquid line to the expansion device. There would be a minimum of 5 psi pressure drop because there is a 1/2 lb drop for every foot of lift. Add to this the line loss drop due to friction, plus the drop through the accessories. As can be seen on the plot on the P-H diagram, the pressure drop could carry the refrigerant down into the saturation zone where vapor would exist. The challenge is to move the vertical line to the left by subcooling, or to elevate the horizontal pressure line so that even with the pressure drops there will still be subcooled liquid in the subcooled zone.

The following are some suggested methods to increase the subcooling and move that line to the left:

1. Liquid-to-suction line heat exchanger.
2. Liquid-to-expansion device outlet heat exchanger.
3. Mechanical subcooling.
4. Water-cooled subcooler.
5. Surge receiver.

The Expansion Line. Most P-H diagrams show the expansion line as the pressure drop only through the expansion device. This is true in some cases but over the years there has been a tremendous increase in the use of refrigerant distributors. For this reason, a distributor has been added to the example. The drop from the TXV inlet to the evaporator inlet includes both the TXV and distributor. If a system were operating under full load, 200 psig at the TXV inlet and 70 psig at the evaporator inlet, the total pressure drop would be 130 psig. However, the drop across the TXV would be about 100 psig and the distributor tubes would be approximately 30 psig. This must be considered when selecting the TXV.

Today, mechanical subcooling is a design concept used by many engineers. As can be seen, with an increase in subcooling, the pressure drops lower before the expansion line crosses into the saturated zone. This means that not only is system capacity increased because of lowering the entering heat content, or enthalpy, but a greater mass flow through the expansion device is experienced because the vapor is formed during the expansion process. This is important to know when sizing the expansion device. Many systems are operating with oversized TXVs because of a high degree of subcooling. Sometimes this results in liquid floodback to the compressor. Expansion valve manufacturers provide subcooling capacity factors in their catalogs. Remember that subcooling also increases distributor and tube capacities and should be checked for proper sizing. Quite often, extensive mechanical subcooling is done on a system and the TXV and distributor sizing have not been considered, resulting in poor system performance and usually poor refrigerant distribution and floodback.

Plotting the EPR Valve. The EPR (evaporator pressure regulating valve) can be easily plotted on the P-H diagram. If the system has a single evaporator and a single compressor then there probably would not be much change in the plot because they are selected at a low pressure drop of 2 psi or less. This small amount would not show much on the diagram. However, if there is a supermarket system that has a high temperature evaporator in it then there could be as much as 20 psi, or even more, pressure drop across the valve. The reason for mentioning the EPR is because many people expect a high temperature drop to accompany the pressure drop. We will plot it.

If the EPR valve is located in the machine room and there was a superheat of about 40°F at the EPR location, then the pressure would be dropped down 20 psi from this point. Notice that since the plot is in the superheated zone, the temperature lines run almost vertical and thus only a slight temperature change would be detected, if any. If the operation was in the saturation zone then there would be a dramatic temperature drop.

SUMMARY 13–12

- Refrigerant tables are compiled listing all the characteristics of each individual refrigerant.
- Enthalpy is the term used in thermodynamics to describe the heat content of a fluid.
- The generally accepted zero enthalpy point of a refrigerant is –40°F.

REVIEW QUESTIONS 13–12

1. What is the saturation pressure of HFC–134a at 40°F?
2. What is the heat content of a refrigerant known as?
3. What can be determined in different sections of a refrigeration system and can be used to determine the system capacity?

13–13 HANDLING OF REFRIGERANT CYLINDERS

Cylinders used for the shipment of refrigerants are designed to meet certain specifications. They are to be filled only to about 75% liquid full at 80°F. The cylinder shall not be full of liquid when heated to a temperature of 131°F. This is required because when liquid refrigerant is compressed, the pressure builds at a much faster rate than when there is room for it to expand. If the cylinder is overfilled, or if heat is evenly and gradually applied to the cylinder, the liquid will expand until the cylinder becomes liquid full. When liquid full, the static pressure inside the cylinder increases very rapidly. The true pressure-temperature relationship exists up until the point that there is no more room for expansion. Then extremely high pressures will result.

When these pressures are reached, the cylinder cannot withstand the stress placed on it. It will then usually explode. This cylinder was subjected to approximately 1,300 psi.

If a cylinder is heated with a welding torch or some other type of spot-heating device, the wall of the cylinder will be weakened. It will rupture at that place rather than gradually building pressure to fill the inside of the cylinder. Fusible plugs are designed to protect the refrigerant cylinder in case of fire. They will not, however, protect the cylinder from a gradual uniform overheating. The fusible plug begins to melt at 157°F, but the hydrostatic pressure created by this temperature exceeds the cylinder test pressure.

A list of the recommended safety rules that should be observed when handling refrigerant cylinders follows:

1. Never allow a cylinder to be heated to 125°F.
2. Never store refrigerant cylinders in the direct sunlight.
3. Never apply a flame directly to a cylinder.
4. Never place an electric resistance heater in direct contact with a refrigerant cylinder.
5. Do not drop, dent, or otherwise abuse refrigerant cylinders.
6. When refilling small refrigerant cylinders, never exceed the weight stamped on them.
7. Always keep the valve cap and head cap in place when the refrigerant cylinder is not in use.
8. Always open all cylinder valves slowly.
9. Secure all refrigerant cylinders in the upright position to a stationary object with a strap or a chain when they are not mounted in a suitable stand.

SUMMARY 13–13

- Refrigerant cylinders are to be filled only to about 74% liquid full at 80°F.
- The cylinder shall not be full of liquid when heated to a temperature of 131°F.
- If a cylinder is heated with a welding torch or some other type of spot-heating device, the wall of the cylinder will be weakened.

REVIEW QUESTIONS 13–13

1. At 80°F, how full can a refrigerant cylinder be safely filled with liquid?
2. Why should a refrigerant cylinder never be heated with a welding torch?
3. What is the maximum temperature to which a refrigerant cylinder can be heated?

CHAPTER

14

REFRIGERANT RECOVERY, RECYCLING, AND RECLAIM

OBJECTIVES

Upon completion of this chapter, you should be able to:

- Better understand the theory behind ozone depletion
- Know some of the requirements of the Clean Air Act of 1990
- Better understand the National Recycling and Emission-Reduction Program
- Better understand the requirements of recovery and recycling equipment
- Know the different methods used for refrigerant recovery

INTRODUCTION

Many of the refrigerants in use today have been blamed for destruction of the ozone layer. There are two kinds of ozone that we come in contact with every day. The ozone close to the earth is a toxic pollutant and is responsible for smog and all kinds of health problems that are becoming more and more common. The stratospheric ozone layer, located from 6 to 30 miles above us, is considered to be a good ozone. Our objective should be to completely eliminate the atmospheric ozone and keep the stratospheric ozone.

14–1 STRATOSPHERIC OZONE

The stratospheric ozone layer is the shield that protects us from the ultraviolet rays of the sun. A reduction in the stratospheric ozone will allow more ultraviolet rays to strike the earth. Radiation, scientists tell us, will cause an increase in skin cancer and cataracts (the leading cause of blindness in the United States), and also cause some suppression of the immune system. When our immune systems are damaged, all of us will be more susceptible to more types of diseases. Damage to the ozone layer also presents a major threat to our food supply by reducing the farmers' crop yields. The stratospheric ozone layer protects everything and everyone on earth. Without it, all forms of plant and animal life will be threatened or eliminated.

Scientists have determined that CFCs and similar compounds that are extremely stable rise up into the atmosphere intact until they reach the stratosphere. Once in the stratosphere, the radiation from the sun causes them to break down, releasing the chlorine component of the chemicals. It is the chlorine that attacks and damages the ozone layer.

Drs. Sherwood Rowland and Marino Molina from the University of California are credited with doing the first research on destruction of the ozone layer. This study was published in 1974. At that time, there were no accurate measurements of the actual amount of ozone loss. It was just a scientific theory. However, it was because of this theory that, in 1978, the United States placed a ban on the use of CFCs in aerosol applications.

At this time the aerosol industry started searching for safe substitutes for this use. Progress was being made until about 1980, when the government reduced its demands in this area. This action caused a drop in the interest for further research for substitutes and the use of CFCs continued to grow.

Later, around 1985, scientists discovered there was a significant loss in the ozone layer over the southern hemisphere. They found a hole about the size of North America. When more measurements were made it was determined that there was a 50% loss in the total column and a loss greater than 95% at an altitude of 9 to 12 miles above the earth. This discovery brought new emphasis on international efforts to protect the ozone layer.

The Montreal Protocol on Substances that Deplete the Ozone Layer was negotiated in September of 1987. This original Protocol was signed by more than two dozen nations, including the United States. In January, 1989, the Protocol was placed into force. At that time, there were 68 nations that were parties to the signing. Since that time, others have joined the effort.

Not long after the Protocol was signed into being, scientists measured the ozone and found there were additional losses worldwide. The destruction was not limited to only remote areas, but had extended to cover the United States as well.

The actual loss of ozone was found to be much more than the computer models had indicated. This additional loss brought some new questions about whether or not the control measures set forth in the Montreal Protocol and EPA Regulations were strong enough to bring about the desired effect. In June, 1990, the parties to the Protocol met in London, England, and amended the original Protocol in hopes of providing adequate protection to the ozone layer. However, these new regulations were found to be lacking. There are four areas that require further attention by the parties to the Montreal Protocol; they are: (1) acceleration of the phase-out of CFCs and methyl chloroform schedules, (2) controlling, and ultimately eliminating, the production and use of hydrofluorocarbons (HCFCs), (3) eliminating, the emissions of ozone-destroying compounds, and (4) implementing effective trade sanctions. These areas are all approached in the Clean Air Act of 1990.

The concentration of natural chlorine in the stratosphere is about 0.6 parts per billion. In 1985, when the hole over Antarctica was discovered, the natural chlorine concentration was about 2.5 parts per billion. Most of this concentration was due to the release of CFCs, methyl chloroform, and similar chemicals into the atmosphere. Currently, the concentration of chlorine is at about the 3.0 parts per billion level, a record-high concentration that is still rising.

The largest producer and user of ozone-depleting chemicals in the entire world is the United States. CFCs are used as refrigerants in domestic refrigerators and freezers, automotive air conditioning units, and commercial refrigeration units. CFCs are used as blowing agents during the manufacture of furniture cushions and all kinds of packaging materials. Methyl chloroform is used as a degreasing solvent in metal cleaning procedures, as well as in manufacturing adhesives.

It has been estimated that in addition to the health benefits discussed above, the economic and environmental benefits of phasing out of CFCs will be approximately $58 billion through the year 2075. This includes $41 billion from the reduced damage to food crops.

In addition to these benefits, there is still the benefit of reducing the climate change that is predicted to occur because of an intensified greenhouse effect, a result of the reduction in the stratospheric ozone layer by the CFCs.

This global climate change is due to the accelerated accumulation of the greenhouse gases in the atmosphere, primarily carbon dioxide, methane, and CFCs. These gases act as a thermal blanket to trap the heat in the atmosphere surrounding the Earth, causing the Earth's surface temperature to increase.

About 50% of the predicted global warming is attributed to the emissions of carbon dioxide. CFCs are responsible for about 15% to 20% of the gases.

Each molecule of CFC gases has approximately 20,000 times more impact on the global climate than carbon dioxide. This makes the control of CFCs more important than controlling carbon dioxide in the prevention of global warming.

There are two major reasons for controlling the release of CFCs. First, this control is essential if the destruction of the ozone layer is to be stopped. Second, control of CFCs is the most important single step possible in controlling the greenhouse effect.

SUMMARY 14–1

- The stratospheric ozone layer is the shield that protects us from the ultraviolet rays of the sun.
- Scientists have determined that CFCs and similar compounds that are extremely stable rise up into the atmosphere intact until they reach the stratosphere.

- In the stratosphere, the radiation from the sun causes them to break down, releasing the chlorine component of the chemicals.
- The Montreal Protocol on Substances that Deplete the Ozone Layer was negotiated in September of 1987.
- The concentration of natural chlorine in the stratosphere is about 0.6 parts per billion. In 1985, when the hole over the Antarctic was discovered, the natural chlorine concentration was about 2.5 parts per billion. Most of this concentration was due to the release of CFCs, methyl chloroform, and similar chemicals into the atmosphere.
- The largest producer and user of ozone-depleting chemicals in the world is the United States.
- The global climate change is due to the accelerated accumulation of the greenhouse gases in the atmosphere, primarily carbon dioxide, methane, and CFCs.
- About 50% of the predicted global warming is attributed to the emissions of carbon dioxide. CFCs are responsible for about 15% to 20% of the gases.
- There are two major reasons for controlling the release of CFCs. First, this control is essential if the destruction of the ozone layer is to be stopped. Second, control of CFCs is the most important single step possible in controlling the greenhouse effect.

REVIEW QUESTIONS 14–1

1. What part in the make-up of the CFCs causes damage to the ozone layer?
2. Who were the first persons to be controlled in the release of CFCs into the atmosphere?
3. When was the original Montreal Protocol made a law?
4. To what is the concentration of chlorine in the stratosphere attributed?
5. What country produces and uses the most ozone-depleting chemicals?
6. What gas is the most responsible for global warming?
7. Which gas, when broken down into its molecules, has the most effect on the global warming climate?
8. What is the most important single step in controlling the greenhouse effect on the earth?

14–2 THE CLEAN AIR ACT

Title VI of the Clean Air Act "Protecting the Stratospheric Ozone," is just one more significant step toward protecting the Earth from the harmful rays from the sun.

Much effort has been expended in just a few years to protect the ozone layer from damage. A full decade was spent with much debate and little action. Then the international community began to take notice of the problem in about 1986. At this time it was a general consensus that something should be done to reduce the amount of chlorofluorocarbon (CFC) refrigerants and other ozone-depleting substances being released into the atmosphere. It was at this time that the problem was considered to be of a global nature, requiring global action. Because of the discussions that followed under the auspices of the United Nations Environment Program, a breakthrough was reached in September 1987. This led to the signing of the original Montreal Protocol by 23 nations.

The year 1990 was significant in the protection of the ozone layer. During this year the Clean Air Act was enacted and significant progress was made on a global level to protect it. Also, some of the regulations were strengthened and agreed to during the second meeting of the signatories of the Montreal Protocol. Thus, the year 1990 was the turning point in the political response to the threat of ozone depletion.

Since that time many things have transpired. There has been new scientific evidence establishing the fact that CFCs are the major cause of the hole in the ozone over the Antarctic and that the ozone levels in the northern mid-latitudes (above the United States) had dropped more than had been projected.

With this news, the international political community increased efforts to get widespread participation from all nations for participation in the Montreal Protocol and to strengthen the requirements set forth in it. Because of this increased effort, 68 nations had joined in the effort by January, 1991. At the June, 1991, meeting in London, England, amendments were adopted that would completely phase out CFCs and halons (used in fire extinguishers). Also, other ozone-depleting chemicals were placed on the list to be phased out. At this same time a fund was set up to support developing countries that wanted to participate in meeting and enforcing the regulations in the Montreal Protocol.

With the availability of new information, Congress started showing more interest in the problem. There was increased interest from the general population, causing Congress to include in the Clean Air Act major provisions that were greatly expanded from the original legislation and directed toward domestic efforts to protect the ozone layer from further damage.

Title VI of the Clean Air Act is very comprehensive and places production limits on ozone-depleting substances as well as restricting their use, emissions, and disposal. It includes specific instructions for companies about what is required for compliance in respect to what ozone-depleting chemicals they must stop manufacturing. There is also a program for reviewing any proposed replacements for the chemicals being phased out. This part of Title VI includes a section on the phase-out schedule for CFCs and HCFC (hydrofluorocarbons). HCFCs are chemicals that will serve as interim chemicals to be used as replacement chemicals for the CFCs. However, the HCFCs will be phased out in time because they also have chlorine in their composition.

The industrial areas of the world, and the consumer, will be greatly affected by the regulations of the Clean Air Act. There is mandatory recycling of CFCs and HCFCs, and warning labels are required on many of the products that can be purchased by the consumer.

Phase-Out Requirements

The following are some of the major provisions that are mandated by Title VI:

The legislation includes scheduled reductions that lead to the complete stoppage of production of CFCs, halons, and carbon tetrachloride by the year 2000. This has already been stepped up to the year 1995 for CFCs. It also calls for the freeze on production of HCFCs by the year 2015, and to be completely phased out by the year 2030. These restrictions have brought focus on production limits to the extent that these chemicals must be recovered, recycled, and reused if they are to continue to be available after these dates.

SUMMARY 14–2

- Title VI of the Clean Air Act "Protecting the Stratospheric Ozone" is just one more significant step toward protecting the Earth from the harmful rays of the sun.

- The year 1990 was significant in the protection of the ozone layer. During this year the Clean Air Act was enacted and significant progress was made on a global level to protect it.
- Title VI of the Clean Air Act is very comprehensive and places production limits on ozone-depleting substances, as well as restricting their use, emissions, and disposal.
- Title VI includes a section on the phase-out schedule of CFCs and HCFCs.
- The Clean Air Act includes scheduled reductions that lead to the complete stoppage of production of CFCs, halons, and carbon tetrachloride by the year 2000.
- These restrictions have brought focus on production limits to the extent that the chemicals must be recovered, recycled, and reused if they are to continue to be available after these dates.

REVIEW QUESTIONS 14–2

1. What was a significant time in regard to the protection of the ozone layer?
2. What part of the Clean Air Act is dedicated to protecting the ozone layer?
3. What are CFCs?
4. By what year are HCFCs to be completely phased out?
5. What must be done to make CFCs available for a longer period of time?

14–3 NATIONAL RECYCLING AND EMISSION-REDUCTION PROGRAM

In this section of Title VI, the EPA sets forth regulations that require that all intentional venting of any refrigerant into the atmosphere be stopped by July 1, 1992. These regulations also require that all other emissions of CFCs, HCFCs, halons, and methyl chloroform are to be stopped by November, 1995. They prohibit any person from knowingly venting any of the controlled substances, including HCFCs, during the servicing of refrigeration or air conditioning equipment (except for cars) beginning July 1, 1992, and require the safe disposal of these compounds by that date.

Warning Labels

The Clean Air Act establishes mandatory labeling requirements on all containers of CFCs, other ozone-depleting chemicals, and all products containing any of these chemicals (refrigerators, foam insulation, etc.), to aid consumers in choosing products and to help service technicians in deciding when to recycle and when to dispose of a refrigerant. There are certain conditions when warning labels may be required on products that are manufactured with, but do not contain, ozone-depleting substances (e.g., many electronics products and flexible foams) and eventually products that are manufactured using HCFCs.

Safe Alternatives

The EPA must be notified of, and evaluate, the overall environmental risks involved in using an alternative chemical. This is required so that the chemicals that are to be used as substitutes do not create any environmental problems.

SUMMARY 14–3

- Title VI of the Clean Air Act prohibits the intentional venting of any refrigerant by July 1, 1992

- All emissions of CFCs, HCFCs, halons, and methyl chloroform are to be stopped by November, 1995.
- The Clean Air Act establishes mandatory labeling requirements on all containers of CFCs and other ozone-depleting chemicals, and all products containing any of these chemicals.

REVIEW QUESTIONS 14–3

1. By what date must all venting of refrigerants stop?
2. Why are containers of ozone-depleting chemicals required to be labeled?

14–4 FEDERAL TAX

A federal tax will be charged on each pound of ozone-depleting chemical used by a manufacturer, producer, or importer. The amount of tax will be equal to (a) the base tax amount applied, (b) the ozone-depleting potential for each chemical as stated in the Montreal Protocol, and (c) the total pounds sold. As follows:

(a) Calendar Year	Base Tax Amount
1990 or 1991	$1.37
1992	1.67
1993 or 1994	2.65

The base tax shall be increased by $0.45 per pound each year after 1994.

(b) Chemical	ODP
CFC–11	1.00
CFC–12	1.00
CFC–113	0.8
CFC–114	1.0
CFC–115	0.6

Exceptions: No tax shall be imposed on the production, use of, or sale of any ozone-depleting chemicals:

- that have been recycled for resale,
- that are entirely consumed or transformed in the manufacture of any other chemical (companies can purchase these chemicals for transformation free of tax and need not apply for a refund or tax credit),
- that are exported (however, an exemption will not exceed a company's percentage of the 1986 production that is exported, multiplied by the total tax),
- that are produced with additional production allowances granted by EPA under CFR 52 (Protection of Stratospheric Ozone).

SUMMARY 14–4

- Federal tax will be charged on each pound of ozone-depleting chemical used by a manufacturer, producer, or importer.
- There will be no tax on refrigerants that have been recycled or reclaimed.

REVIEW QUESTIONS 14–4

1. How will refrigerants be taxed?
2. What is the ozone-depletion potential?
3. Which refrigerants will not be taxed?

14–5 SECTION 608 OF THE CLEAN AIR ACT

This section of the act requires that the U.S. Environmental Protection Agency (EPA) develop specific regulations prohibiting emissions of ozone-depleting chemicals during their use and disposal to the "lowest achievable level" and to maximize recycling of the chemicals. The act also prohibits the release of refrigerants into the atmosphere during the servicing, maintenance, and disposal of refrigeration and air conditioning equipment starting July 1, 1992.

The EPA has developed regulations requiring the recycling of ozone-depleting chemicals (both CFCs and HCFCs) during the servicing, repair, or disposal of refrigeration and air conditioning equipment. These regulations would also require that refrigerant in appliances, machines, and other goods be recovered from these items prior to their disposal.

Technician Certification

Technicians who service, install, and maintain air conditioning and refrigeration equipment must be certified by the EPA. The recovery and recycling equipment used during service operations must also be certified by the EPA. The technician certification is divided into 4 types, as follows: Type I (small appliances), Type II (high pressure), Type III (low pressure), and Type IV (universal certification). On the certification exam there will be 25 questions that are common to all 4 types of certification. These questions will focus on the environmental impact of CFCs and HCFCs, the regulations, and changing the outlook of the industry. There will also be questions concerning the filling and handling of refrigerant cylinders. There will also be questions that are related to the exposure levels allowed in equipment rooms. There will be 25 questions for each Type I, Type II, and Type II certification, which will also be specific for each sector covered by the particular certification. Type IV certification will have 75 questions, including 25 from each of the Types I, II, and III, including safety, cylinder shipping, and refrigerant disposal.

Type I (Small appliances) Certification. This test may be taken either on-site or by mail. The required passing grade is 84%. The exam will include questions concerning recovery devices that are unique to the small appliance sector of the industry. Also included will be system-dependent recovery techniques.

Type II (High-pressure) Certification. The questions for this certification will deal with vacuum levels required, proper use of recovery equipment for removing both liquid and vapor refrigerant from the system, the purpose of system receivers, and the use of refrigerant monitors in equipment rooms. This is a closed-book, proctored test. The required passing grade is 70%.

Type III (Low-pressure) Certification. This certification covers any type of equipment that contains a low-pressure refrigerant, such as CFC–11. The questions for this type of certification will include evaporator leak testing methods and proper procedures for deep evacuating the system. This is a closed-book, proctored test. The required passing grade is 70%.

Type IV (Universal) Certification. The exam for this type of certification will have 75 questions. There will be 25 questions from each of the Types I, II, and III, in addition to the 25 questions that are common to all 4 types of certification. The questions will also include shipping procedures, disposal, and proper refrigerant cylinder handling. This is a closed-book, proctored test. The required passing grade is 70%.

EPA has no regulations specifically for intermediate pressure and high-pressure systems because they are so uncommon. However, for both Type II and Type III certification exams, there will be some general questions covering these types of systems.

Technicians have until November, 1994, to become certified. Then they can only work on the types of systems for which they are certified. When a technician is certified there will be no recertification required. However, EPA may require that technicians demonstrate proper techniques for recovery-recycle unit operation. If proper procedures are not followed, the technician may loose certification.

Contractor Self-Certification

This is another type of certification that is required by EPA. Those wishing this type of certification can obtain the proper form from either the EPA or the OMB. The form is then filled out and mailed to the appropriate EPA regional office.

Those who service air conditioning and refrigeration systems (with the exception of automotive air conditioning systems) and those who dispose of appliances (with the exception of small appliances, room air conditioners, and automotive air conditioners) must self-certify that they either own or have leased certified recovery-recycle units. The deadline for self-certification is August 6, 1993.

Those persons who recover refrigerant from small appliances, room air conditioners, and automotive air conditioning systems before disposing of them must also be self-certified that the recovery-recycle units are EPA approved.

Self-certifications are not transferable to a new owner. When a business changes ownership, the new owner has 30 days to self-certify the equipment used.

Evacuation Standards

For systems containing a refrigerant charge of 200 lb or more of a high-pressure refrigerant such as HCFC–22, the system must be evacuated to 10 in Hg during the recovery process and before the system is opened to the atmosphere for repairs, with the exceptions listed next. This requirement is because of the difficulty in reaching lower vacuums with these types of refrigerants.

Systems containing a charge of more than 200 lb of CFC–12 or CFC–502 must be evacuated to 10 in HG during the recovery process and before the system is opened to the atmosphere for repairs, with the exceptions as listed below.

Systems containing a charge of less than 200 lb of CFC–12 or CFC–502 must be evacuated to 10 in Hg during the recovery process, with the exceptions listed next.

Low-pressure equipment must be evacuated to 29 in Hg during the recovery process and before the system is opened to the atmosphere. This evacuation is because the equipment for recovering this type of refrigerant is capable of pumping this low pressure on these types of systems.

Evacuation Standard Exceptions. There are two exceptions that permit the service technician to not meet these evacuation standards. They are:

1. "Nonmajor" Repairs. These are repairs that, after they are completed, there is no evacuation of the refrigerant to the atmosphere; and
2. Leaks that will not allow the required evacuation to be reached.

It was decided that both of these instances would actually cause greater emissions when the required evacuation level was attempted than when less, or none at all, was used.

Nonmajor Repairs. Nonmajor repairs are defined as those requiring only a small opening for a short period of time, only a few minutes. This type of repair would allow only a small amount of refrigerant to escape and the amount of air and moisture that could enter the system is minimal. Also included in this category is the replacement of components such as safety and pressure switches and filter-driers.

When making repairs of this nature, the system can be evacuated to 0 psig for high-pressure equipment and pressurization to 0 psig for low-pressure systems when performing service operations that do not require the evacuation of refrigerant to the atmosphere before being placed back in operation.

Major Repairs. This type of repair involves making large openings in the system. This includes such procedures as compressor replacement, condenser removal, evaporator removal, or an auxiliary heat exchanger removal. This category can be categorized by the need to evacuate the system after repairs and before charging the system for operation. This is when any refrigerant left in the system is purged through the vacuum pump to the atmosphere during the evacuation process.

Leak Repair Requirements

Systems that contain a refrigerant charge of 50 lb or more must have any leak repaired within 30 days. Systems containing less than 50 lb are not covered by this requirement. This regulation was effective by June 21, 1993. It is the responsibility of the equipment owner to have the repair completed if the leak exceeds 15% or 35% of the total charge (explained later). The technician should inform the owner of the leak and of the requirements by EPA. The owner cannot intentionally ignore any information that reveals that a leak exists. This is to prevent topping-off the charge in a leaking system.

Flexibility. There are two types of system leaks as mentioned earlier—15% and 35% leaks. There is a certain amount of flexibility in the requirements concerning leak repairs. These are as follows:

- Annual Leak Rate of 15% or Higher: Air conditioning systems such as those used in commercial buildings and hotels must have the leaks repaired.
- Annual Leak Rate of 35% or Higher: This area covers equipment that is used in an industrial process, commercial refrigeration, pharmaceutical systems, petrochemical systems, chemical systems, industrial ice machines, and ice rinks must have all leaks repaired.

The technician can estimate the size of the leak by checking past invoices for service procedures requiring refrigerant to be charged into the system. There are probably other means of estimating the leak size if the unit is maintained by the service company or a maintenance department.

There is one exception to the above requirements; that is, if the owner decided to replace or retrofit a leaking system but must wait a couple of months before making the necessary changes. When this situation occurs, the owner must develop a detailed plan within 30 days of being informed, showing his intentions to replace or retrofit the equipment. This plan must be dated and kept on-site and subject to EPA inspection. The repair of retrofit must be completed within one year from the date at which the plan was initiated.

Compliance Dates

The following are the compliance dates and what must be done to meet EPA guidelines:

- June 14, 1993: On systems containing a refrigerant charge of 50 lb or more, such as commercial refrigeration, or an industrial processing unit and a leak of 35% or more per year is found, the system owner must be notified. The owner must then repair the unit within 30 days.

When a leak of 15% per year is found on any type of system containing a refrigerant charge of 50 lb or more, the owner is responsible for having it repaired within 30 days.

Exception. If the owner developed a replacement or retrofit plan within 30 days of notification, there is a one-year exemption from making the repairs. The plan must be kept on-site and available for EPA inspection.

Purging. Only *de minimus* releases of refrigerant during the service procedure or equipment disposal are legal. Major releases are prohibited by law.

Recovery-Recycle Unit

To alter the design of a recovery-recycle unit in such a way as to affect its ability to meet EPA certification standards is illegal.

After July 13, 1993, to legally open a refrigeration system the service company must have at least one self-contained, EPA certified recovery unit at the business location.

Before opening any high-pressure system with a refrigerant charge of 50 lb or more, it must be pumped down to a vacuum of 10 in Hg before the system is opened if a major repair is to be performed. If a nonmajor repair is performed or if evacuation would substantially contaminate the recovered refrigerant, the system can be evacuated to 0 psig. The evacuation to 0 psig is also applicable if the system is not to be evacuated after the repair and before recharging the unit.

Before opening a low-pressure system, the pressure inside the system must be increased to 0 psig before opening it. This is to prevent pulling moisture and air into the system and contaminating the system and refrigerant.

When evacuating small appliances, such as refrigerators, freezers, room units, packaged terminal air conditioning units, packaged terminal heat pump systems, dehumidifiers, vending machines, drinking water coolers, or units containing a refrigerant charge of 5 lb or less, the level of evacuation depends on whether or not the system compressor is operational.

If the system compressor is operational, both the self-contained and the system-dependent recovery units must produce a 90% evacuation level. When the system compressor is not operational, the recovery procedure must remove 80% of the charge.

When disposing of a system containing refrigerant of any amount, an EPA-approved recovery-recycle unit must be used to evacuate the system to the levels discussed above.

Any refrigerant removed from a system or a part of a system to be serviced must be removed with an EPA-approved recovery-recycle unit.

When disposing of a small appliance, the last person to handle the appliance must recover any refrigerant or verify in writing that it has been recovered.

After August 12, 1993, it is illegal to open any refrigeration system for repair, maintenance, or disposal unless the technician has certified to EPA the ownership of the EPA certified recovery-recycle equipment. The certification form is available from EPA or the OMB.

It is also illegal to sell any refrigerant to a new owner unless it has been reclaimed and certified to meet ARI Standard 700 purity. The used refrigerant can be used in another system with the same owner without prohibitation.

Reclaimers must certify that the refrigerant has been cleaned to meet ARI 700 specifications and that not more than 1.5% of it will be vented during the reclamation process.

Reclaimers must report to EPA within 45 days from the end of the calendar year regarding the quantity of refrigerants received and reclaimed. They must also furnish the names of the persons who supplied the refrigerant to them.

November 15, 1993, all recovery-recycle units must be certified by EPA to meet the approved evacuation levels.

After this date, it is also illegal to sell anything but small appliances that are not equipped with a servicing aperture and a process tube so that the refrigerant can be recovered from the system.

The manufacturers of recovery-recycle units must have their units certified by an approved testing organization. The equipment must be able to meet the evacuation levels stated.

After November 14, 1994, a technician cannot legally open or dispose of refrigeration equipment unless proper technician certification has been attained.

Only training-testing organizations that are EPA approved can train and certify technicians after this date.

It is also illegal to sell refrigerant to anyone except EPA certified technicians.

The recovery and recycling of refrigerants represents a critical step in protecting the stratospheric ozone layer. As an encouragement for owners and users of air conditioning and refrigeration equipment to start recovering and/or recycling CFCs as soon as possible, EPA will grandfather recycling and recovery equipment. However, these provisions require that the equipment certified before January 1, 1994, must pump a vacuum of 4 inches mercury for high-pressure systems and a vacuum of 25 inches mercury for low-pressure systems.

All individuals who service air conditioning and refrigeration systems must be certified. There are four classifications of certification: (1) for servicing household appliances, (2) for servicing high-pressure equipment with a refrigerant charge below 50 pounds, (3) for servicing high-pressure systems with a refrigerant charge above 50 pounds, and (4) for servicing low-pressure equipment. Certification will depend on the passing of a proctored, EPA-approved test. The Agency authorizes organizations to administer the test.

There is also a provision requiring that contractors be certified. The contractors will be required to have sufficient recovery and recycling equipment to perform on-site recovery and recycling and will employ only certified technicians to service the equipment.

SUMMARY 14–5

- Section 608 of the Clean Air Act requires that the EPA develop specific regulations prohibiting the emissions of ozone-depleting chemicals during their disposal to the "lowest achievable level" and to maximize recycling of the chemicals.

- These regulations also require that refrigerant in appliances, machines, and other goods be recovered from these items prior to their disposal.
- Technicians had until November, 1994, to becomecertified.
- There are four types of technician certification.

REVIEW QUESTIONS 14–5

1. To what level must the emissions of ozone-depleting chemicals be prohibited?
2. What must be done before the disposal of any refrigerating equipment?
3. What type of technician certification is required to work on all refrigerating machines?
4. By what date must recovery equipment be certified?

14–6 RECOVERY AND RECYCLING EQUIPMENT

It must always be remembered that the hermetic or semihermetic compressor motor must never be used for recovering refrigerant. There are several reasons for this. First, the oil could be pumped from the compressor crankcase, causing problems from the lack of lubrication and also due to slugging of the compressor valves. Second, the insulating effect of the compressor-motor winding insulation is tremendously less when subjected to pressures below atmospheric, especially when there is no refrigerant surrounding it. Therefore, when the internal system pressure is reduced to the vacuum required by EPA, the compressor-motor winding could short out, requiring that the motor be rewound or completely replaced.

Recovery Equipment

Basically there are three types of refrigerant recovery units. The type of work most often done should dictate which one to buy. In any case it must be certified by EPA.

Depending on the purpose, the recovery unit may be designed with an internal system to be used for various purposes. Included may be an oil separator, an oil separator with multiple pass filter-drying system, or an oil separator with single pass filter-driers. See Figures 14–1 through 14–3.

Recovery Methods

The two basic methods of recovering refrigerant from a system are liquid and vapor.

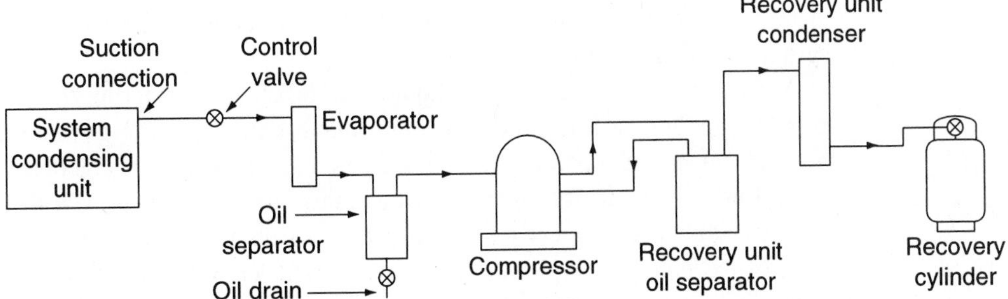

Figure 14–1 Recovery/recycle unit schematic using an oil separator

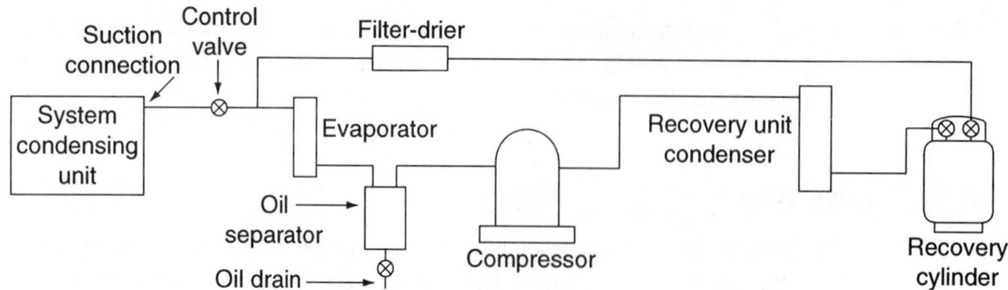

Figure 14–2 Recovery/recycle unit schematic using multiple pass filter-driers

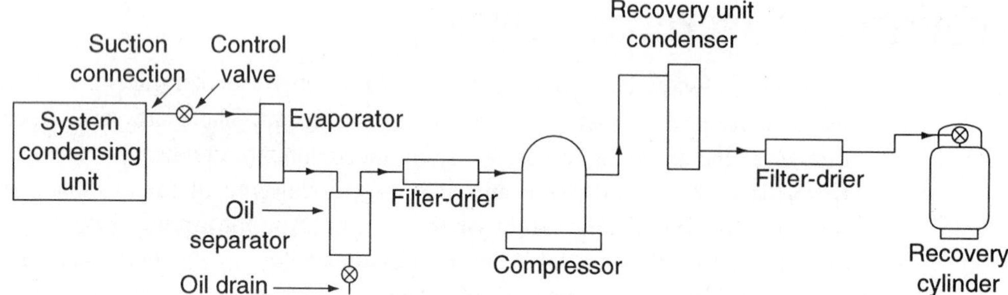

Figure 14–3 Recovery/recycle unit schematic using single pass filter-drier

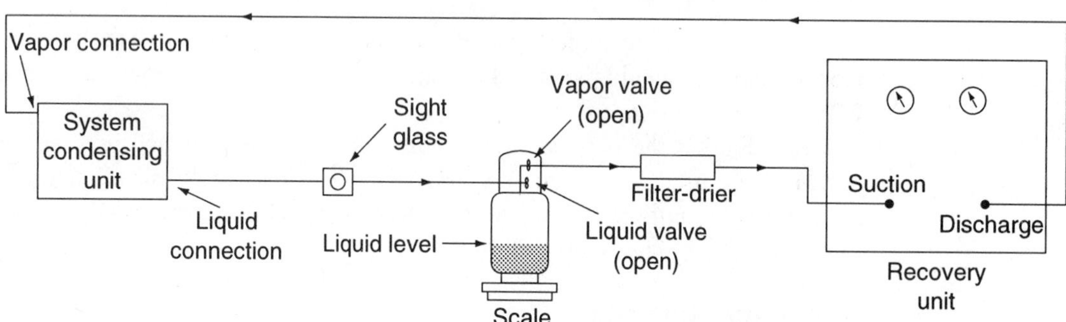

Figure 14–4 Liquid recovery connections

Liquid Recovery. Recovery of the refrigerant can be completed in much less time if most of the refrigerant can be removed in the liquid form. Therefore, use the liquid recovery procedure when possible so that the job can be completed as economically as possible. When using this recovery procedure, connect the recovery unit to the system as indicated in Figure 14–4.

Always use caution to prevent liquid refrigerant from entering the recovery unit compressor during this procedure. During this method, the pressure difference between the cylinder and the refrigeration system causes the refrigerant to flow into the cylinder. The suction of the recovery unit pulls the vapor from the top of the cylinder while the recovery unit discharge builds up a pressure over the liquid in the system, forcing it to flow into the recovery cylinder. When a drier is placed in the vapor line to the recovery unit, the refrigerant will be filtered and dried during the recovery process.

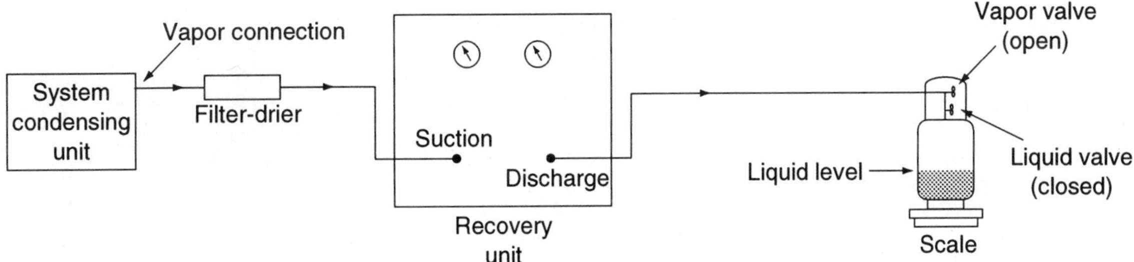

Figure 14–5 Vapor recovery connections

Vapor Recovery. To use the vapor method of refrigerant recovery, connect the unit as shown in Figure 14–5.

The refrigerant vapor is pumped from the system by the recovery unit compressor. It is then compressed, condensed, and forced into the recovery cylinder. This is the best method to use when only small amounts of refrigerant are to be removed or on systems that do not have liquid connections. A drier-filter placed in the vapor line to the recovery unit will clean and dry the refrigerant during the recovery process. This method is more time consuming than the liquid recovery method. However, it is the only method that will work on all systems.

SUMMARY 14–6

- It must be remembered that hermetic or semi-hermetic compressor motors must not be used for recovering refrigerant.
- Basically there are three types of refrigerant recovery units. The type of work most often done should dictate which one to buy. In any case, it must be certified by EPA.
- The two basic methods of recovering refrigerant are liquid and vapor.

REVIEW QUESTIONS 14–6

1. Why must a hermetic or semi-hermetic compressor never be used for refrigerant recovery?
2. What is the quickest method of refrigerant recovery?
3. Which recovery method is best when only a small amount of refrigerant is involved?

CHAPTER

15

INTRODUCTION TO ELECTRICITY

OBJECTIVES

Upon completion of this chapter, you should be able to:

- Better understand the nature of matter
- Better understand the structure of an atom
- Know more about the laws of electrical charges
- Know more about electrostatic fields
- Better understand the electron theory of electricity
- Better understand Ohm's Law
- Know more about series, parallel, and series-parallel circuits
- Better understand magnetism
- Know more about capacitance in an electrical circuit

INTRODUCTION

When we consider the living standards that we have gained over the years, we will more than likely conclude that electricity makes most of our conveniences possible. Most of us have always had electricity available and we take it for granted. However, if it was taken from us, there is little that we could do without some difficulty. There has been much discussion of the theories surrounding electricity but most will agree that it is one of the most important aspects of our daily living. Most of us actually know so little about it that it seems a mystery that surrounds the universe.

With the new discoveries, applications, and developments in the electrical field, an almost unlimited number of opportunities open up for anyone willing to take the time to learn and remove the mysteries surrounding electricity. It has not been long since very few people could begin to explain the electrical theories. Even though there was much known about it, no one actually knew what it was. However, today we know that it is simply the movement of small atomic particles known as electrons.

15–1 THE NATURE OF MATTER

Everything on earth is made up of matter. We can define matter as anything that occupies space or has mass. It has been determined by observation that matter has many different characteristics or properties. Some of these differences may be hardness, color, or taste.

All matter, as indicated by the accepted atomic theory, has an electrical nature. Our bodies are made up largely of a combination of positive and negative electrical charges. Everything is an electrical conductor; that is, it will allow electric current to pass through it. Some substances, however, more readily allow electricity to pass through them than others. From this it can be seen that a study of the structure of matter is necessary to gain an understanding of electricity.

The Elements

Elements are considered to be the building blocks that make up all matter. We can define an element as a substance containing only one kind of atom.

There are just over 100 known elements; however, there are only 92 natural elements. Those remaining are synthetic and are man-made through atomic research. This research continually allows new synthetic elements to be developed.

All matter is made from the various different elements; however, the elements cannot be produced by division or combining of the different elements. See Table 15–1.

The Compound

Many more materials exist than there are elements. This is due to the fact that certain atoms, when combined in certain combinations, will form a new chemical. The new chemical forms a new material that is completely different from either of the other elements. When the elements form the new chemical, it then becomes known as a compound. A compound that we are all familiar with is water. It is formed by combining two parts of hydrogen with one part of oxygen (H_2O). These are both gaseous elements, but when they are combined, a liquid is formed. See Figure 15–1.

ATOMIC NUMBER	NAME	SYMBOL	ATOMIC NUMBER	NAME	SYMBOL	ATOMIC NUMBER	NAME	SYMBOL
1	Hydrogen	H	32	Germanium	Ge	62	Samarium	Sm
2	Helium	He	33	Arsenic	As	63	Europium	Eu
3	Lithium	Li	34	Selenium	Se	64	Gadolinium	Gd
4	Beryllium	Be	35	Bromine	Br	65	Terbium	Tb
5	Boron	B	36	Krypton	Kr	66	Dysprosium	Dy
6	Carbon	C	37	Rubidium	Rb	67	Holmium	Ho
7	Nitrogen	N	38	Strontium	Sr	68	Erbium	Er
8	Oxygen	O	39	Yttrium	Y	69	Thulium	Tm
9	Fluorine	F	40	Zirconium	Zr	70	Ytterbium	Yb
10	Neon	Ne	41	Niobium	Nb	71	Lutetium	Lu
11	Sodium	Na		(Columbium)		72	Hafnium	Hf
12	Magnesium	Mg	42	Molybdenum	Mo	73	Tantalum	Ta
13	Aluminum	Al	43	Technetium	Tc	74	Tungsten	W
14	Silicon	Si	44	Ruthenium	Ru	75	Rhenium	Re
15	Phosphorus	P	45	Rhodium	Rh	76	Osmium	Os
16	Sulfur	S	46	Palladium	Pd	77	Iridium	Ir
17	Chlorine	Cl	47	Silver	Ag	78	Platinum	Pt
18	Argon	A	48	Cadmium	Cd	79	Gold	Au
19	Potassium	K	49	Indium	In	80	Mercury	Hg
20	Calcium	Ca	50	Tin	Sn	81	Thallium	Tl
21	Scandium	Sc	51	Antimony	Sb	82	Lead	Pb
22	Titanium	Ti	52	Tellurium	Te	83	Bismuth	Bi
23	Vanadium	V	53	Iodine	I	84	Polonium	Po
24	Chromium	Cr	54	Xenon	Xe	85	Astatine	At
25	Manganese	Mn	55	Cesium	Cs	86	Radon	Rn
26	Iron	Fe	56	Barium	Ba	87	Francium	Fr
27	Cobalt	Co	57	Lanthanum	La	88	Radium	Ra
28	Nickel	Ni	58	Cerium	Ce	89	Actinium	Ac
29	Copper	Cu	59	Praseodymium	Pr	90	Thorium	Th
30	Zinc	Zn	60	Neodymium	Nd	91	Protactinium	Pa
31	Gallium	Ga	61	Promethium	Pm	92	Uranium	U
93	Neptunium	Np	97	Berkelium	Bk	101	Mendelevium	Mv
94	Plutonium	Pu	98	Californium	Cf	102	Nobelium	No
95	Americium	Am	99	Einsteinium	E	103	Lawrencium	Lw
96	Curium	Cm	100	Fermium	Fm			

Table 15–1 The natural elements

The Molecule

The molecule is the smallest particle into which a compound can be reduced before being broken down into its basic elements. An example is water, being made up of two atoms of hydrogen and one of oxygen. Common table salt is another example. When one grain of table salt is broken in half until it is as small as possible without changing its chemical structure, there would be one molecule of salt remaining. See Figure 15–2.

If the grain of salt was further divided, it would return to its original elements, sodium and chlorine.

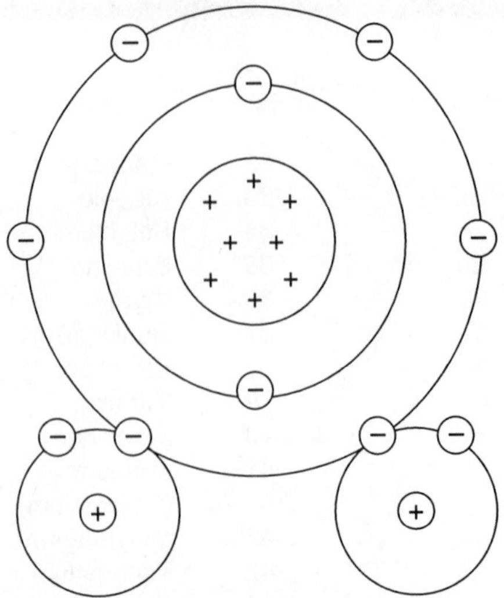

Figure 15–1 A molecule of water

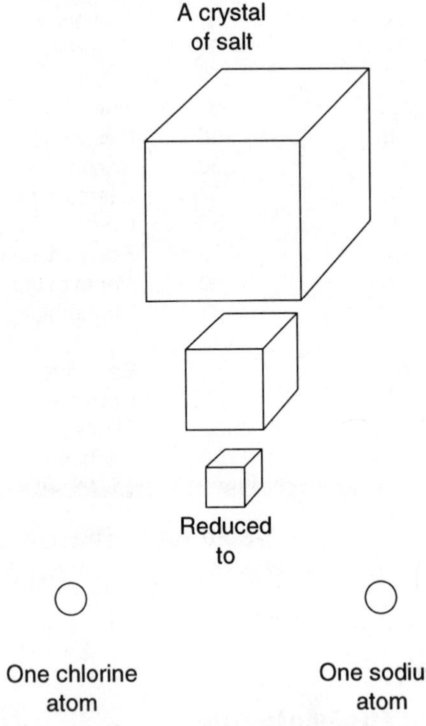

Figure 15–2 A molecule of salt reduced

The Atom

The atom is the smallest particle into which an element can be reduced and still retain its original properties. If a drop of water were reduced to its smallest size, a molecule of water would remain. If the drop was further reduced, hydrogen and oxygen atoms would be produced. See Figure 15–3.

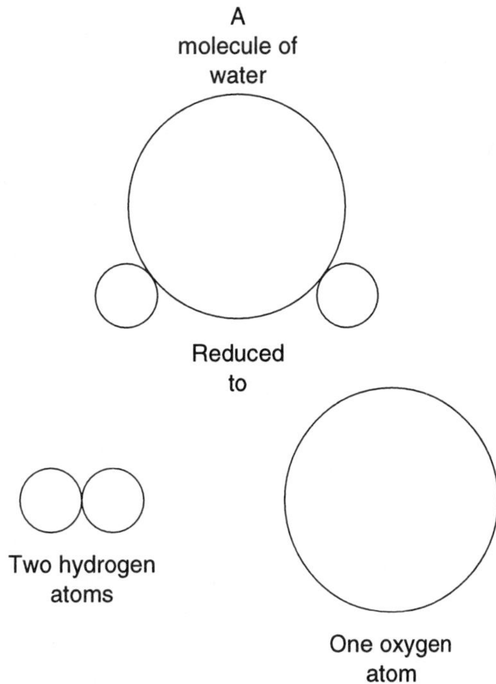

Figure 15–3 A molecule of water reduced

SUMMARY 15–1

- Matter can be defined as anything that occupies space and has mass.
- All matter, as indicated by the accepted atomic theory, has an electrical nature.
- Elements are considered to be the building blocks that make up all matter.
- An element is a substance containing only one kind of atom.
- The molecule is the smallest particle into which a compound can be reduced before being broken down into its basic elements.
- The atom is the smallest particle into which an element can be reduced and still retain its original properties.

REVIEW QUESTIONS 15–1

1. Name the three different characteristics of matter.
2. How many kinds of atoms do elements contain?
3. What are compounds?
4. What is the smallest particle of a compound?
5. What is the smallest particle of an element?

15–2 STRUCTURE OF AN ATOM

The word *atom* comes from the Greek meaning indivisible. Until only recently, the atom was considered to be indivisible. However, with new technology and knowledge, the atom can now be broken down. When the atom is broken down, that element no longer exists. This is because the smaller particles that result are just different elements. It is this difference in the number of smaller particles that causes an atom of one element to be different from another.

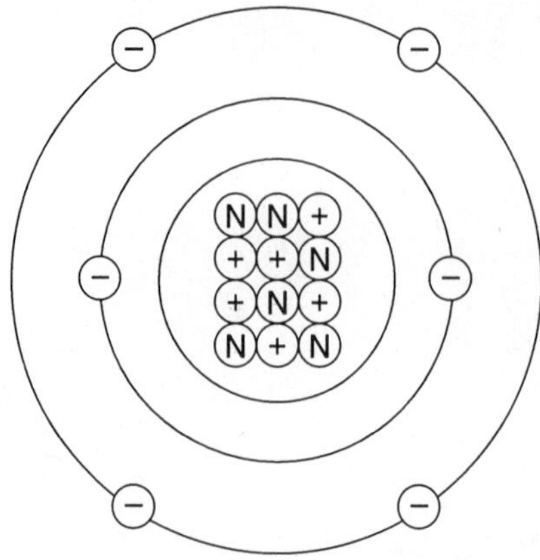

The nucleus contains 6 protons and 6 neutrons.
Six electrons orbit the nucleus.

Figure 15–4 A carbon atom

There are basically three types of these subatomic particles involved in the study of electricity. They are electrons, protons, and neutrons. The electrons are free-moving particles that revolve around the nucleus of an atom. The protons and neutrons are located in the nucleus, or center, of the atom. See Figure 15–4.

The Nucleus

The nucleus is located in the center of an atom. It is made up of protons, neutrons, and other subatomic particles. The number of protons contained in the nucleus is what causes the atoms of different substances to be different from each other. As an example, a hydrogen atom contains 1 proton; helium contains 2 protons and 2 neutrons; and copper contains 29 protons and 35 neutrons. See Figure 15–5.

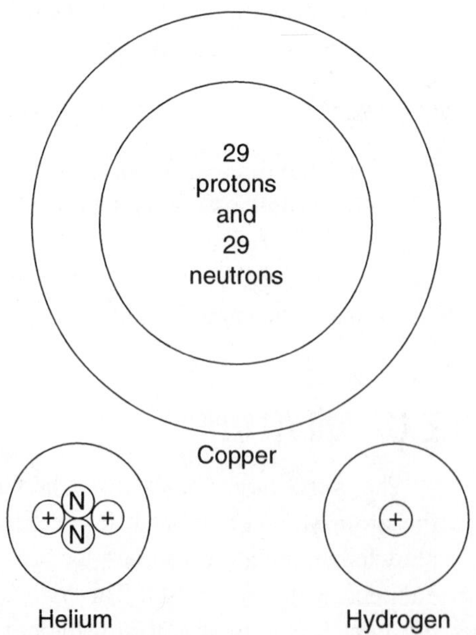

Figure 15–5 Nuclei of some typical atoms

The nucleus of an atom contains neutrons, which are neutral, and protons, which have a positive charge. The nucleus of any atom is, therefore, positive.
N
+
+
N
+
Helium

Figure 15–6 Protons in the nucleus

Also, it is the number of protons in the nucleus that determines the atomic number of the different elements. See Table 15–1.

The Neutron. This is actually a separate particle, but it is generally taken to be a combination of both the electron and the proton. They have a neutral electrical charge and are not really important in electrical theories.

The Proton. Protons are located in the nucleus of an atom. They contain a positive electrical charge. These are very small particles and are estimated to be 0.07 trillionth of an inch in diameter. Protons have exactly the same, but opposite, amounts of electrical charge as the electrons rotating in the orbit around the nucleus. They have a diameter measuring about one-third of the electron, but its mass is about 1,840 times that of the electron. See Figure 15–6.

Due to its weight, the proton is extremely difficult to dislodge from the nucleus. Therefore, in electrical theories, they are considered to be a permanent part of the nucleus. They actually make no contribution to the flow of electrical energy.

Because of their positive electrical charge, the lines of force are projected straight out in all directions from the proton. See Figure 15–7.

The Electron. The electron rapidly revolves around the nucleus in orbit. Electrons have a negative electrical charge. The diameter of the electron is about three times

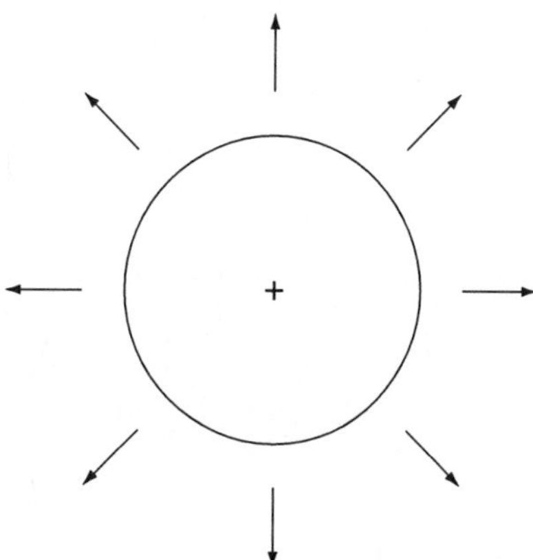
+

Figure 15–7 Electrical charge force lines from a proton

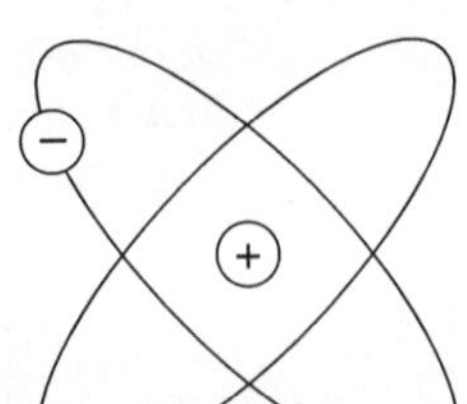

Figure 15–8 Electrons orbiting a nucleus

that of the proton. Electrons are easily moved from their nucleus and, therefore, are the particles that cause electricity to flow through a circuit. See Figure 15–8.

The lines of electrical force of an electron are straight into the electron from all directions. See Figure 15–9.

SUMMARY 15–2

- There are three types of subatomic particles involved in the study of electricity. They are electrons, protons, and neutrons.
- The protons and neutrons are located in the nucleus, or center, of the atom.
- The number of protons contained in the nucleus is what causes the atoms of different elements to be different from each other.
- It is the number of protons in the nucleus that determines the atomic number of the different elements.
- Protons have a positive electrical charge.
- Protons have exactly the same, but opposite, amounts of electrical charge as the electrons rotating in the orbits around the nucleus.
- Electrons have a negative electrical charge.
- Electrons are the particles that cause electricity to flow through a circuit.

REVIEW QUESTIONS 15–2

1. When an atom is broken down why does that element no longer exist?
2. Name the three types of subatomic particles.

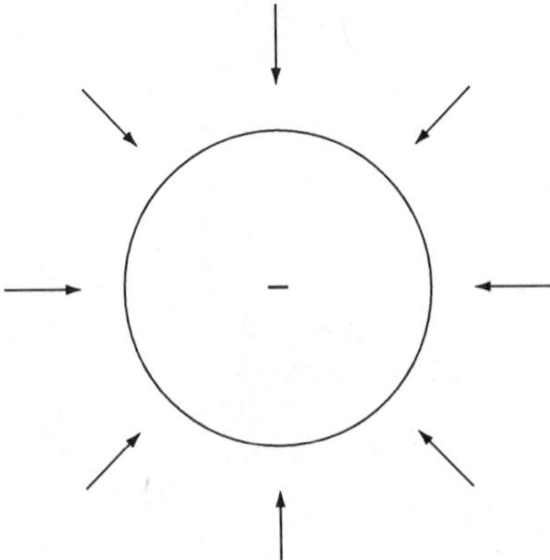

Figure 15–9 Electrical charge force lines into an electron

3. Why are some substances different from each other?
4. What electrical charge does the proton have?
5. What electrical charge does the electron have?

15–3 LAW OF CHARGES—STATIC ELECTRICITY

When electricity is at rest it is termed *static* electricity. There are several ways to generate static electricity. All of them involve the moving of electrons or protons from one substance to another.

Protons and electrons hold charges known as electrostatic charges. The positive charge on the proton is exactly the same as the negative charge on the electron; that is, the charges are equal but opposite in polarity. The electrical lines of force related to each particle produce electrostatic fields. These fields, because they are opposite in polarity, react in a certain manner to each other. Charged particles can either attract or repel other charged particles, depending on whether they have a positive or a negative charge.

This introduces one of the fundamental laws of electricity known as the Law of Charges. This law states that like charges repel each other and unlike charges attract each other. Thus:

- A positively charged proton (+) will repel another positively charged proton (+).
- A negatively charged electron (–) will repel another negatively charged electron (–).
- A positively charged proton (+) will attract a negatively charged electron (–).

See Figure 15–10.

Because of the weight of protons, they exert very little repulsive force on each other inside the nucleus.

Ions. In nature an atom is electrically stable because it has the same number of protons and electrons. It is, therefore, neutral in electrical charge. This is possible

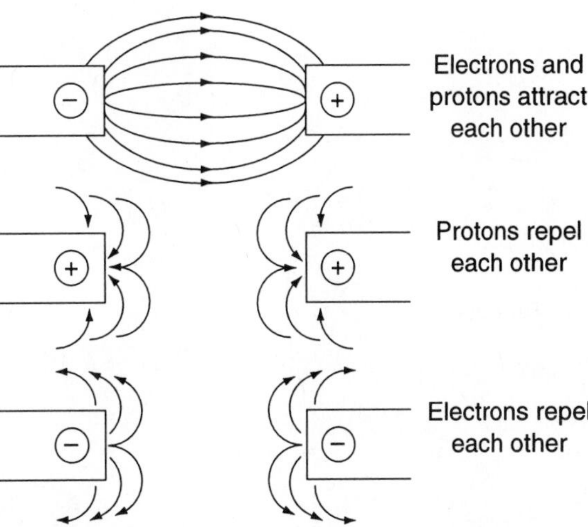

Figure 15–10 Actions of electrical lines of force

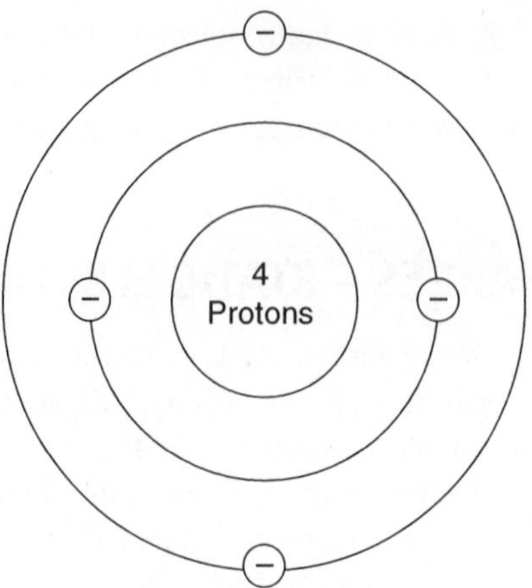

Figure 15–11 A neutral ion

because the positive and negative charges are exactly equal and cancel each other out. See Figure 15–11.

However, there are certain conditions under which electrons can leave the atom for a short period of time. Also, under certain conditions, an atom can accept more electrons for a short period of time. However, it is the number of protons in the nucleus of an atom that gives an element its properties. Only the electron is capable of moving without changing the properties of the atom.

When an atom collects more electrons than it normally has, or more electrons than protons, the atom will have a negative electrical charge. See Figure 15–12.

When the atom has more protons than electrons it will have a positive electrical charge. These atoms are called ions. See Figure 15–13.

SUMMARY 15–3

- When electricity is at rest it is termed "static" electricity.
- Protons and electrons hold charges known as electrostatic charges.
- The fundamental law of charges states that like charges repel each other and unlike charges attract each other.

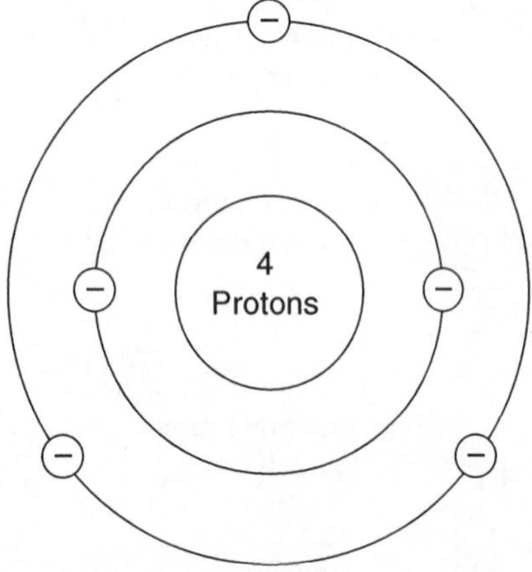

Figure 15–12 A negatively charged ion

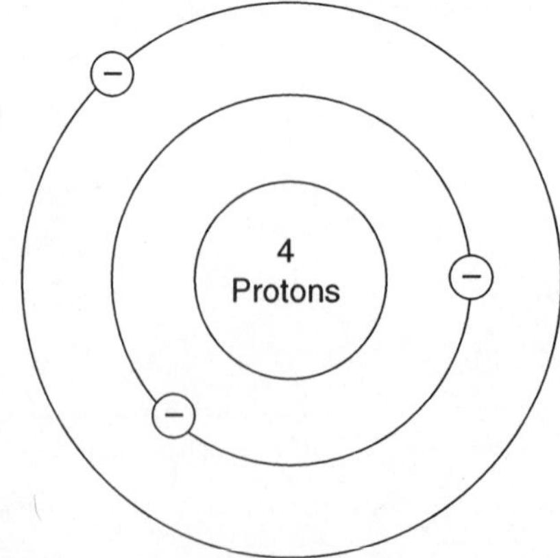

Figure 15–13 A positively charged ion

- In nature an atom is electrically stable because it has the same number of protons and electrons.
- Only the electron is capable of moving without causing a change in properties of the atom.

REVIEW QUESTIONS 15–3

1. What is done to generate an electrical charge?
2. What is required for one charged particle to attract another charged particle?
3. What makes an ion electrically stable in nature?
4. What particle can be made to move around without affecting the properties of the atom?

15–4 ELECTROSTATIC CHARGES

Atoms having either an excess or a deficiency of electrons are unstable. This instability occurs because the number of protons and electrons are unbalanced. The atom will rapidly and automatically balance itself again and become neutral in charge. This is an important point to remember. If we think of things that are large enough to be seen with the naked eye, the electron theory will be easier to understand. The things that we can see can be charged in several different ways and, because we can see the objects, the characteristics can be more easily seen.

Charging by Friction

Ancient Greeks are credited with discovering the principle of charging by friction. When a glass rod is rubbed with a piece of silk material, electrons will move from the glass rod to the silk. After this process, the glass rod will have a positive charge and the silk will have a negative charge. See Figure 15–14.

Charging by Contact

When a rubber rod is rubbed with a piece of fur, the rod will gather electrons from the fur and become negatively charged. If the rubber rod is now touched to another

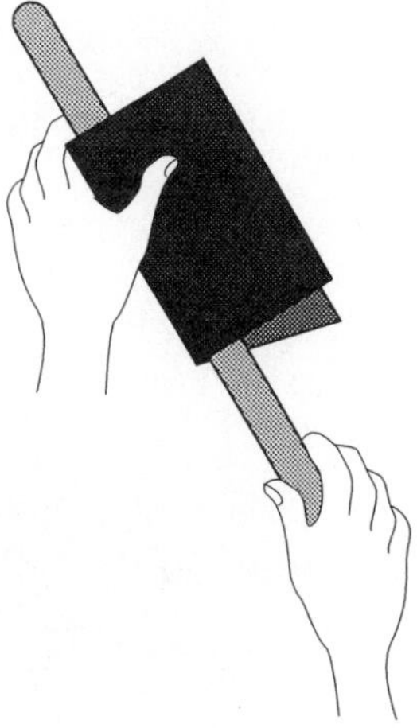

Figure 15–14 Making electricity by friction

object, some of the electrons will move from the rod to the other object. This transfer will continue until there is the same number of electrons on both objects.

If a copper rod was suspended with a string, see Figure 15–15, then touched with the negatively charged rubber rod, the copper rod would gather electrons and become negatively charged. See Figure 15–16. The transferred electrons would collect on the surface of the copper rod. See Figure 15–17. If the positively charged glass rod had been used, the electrons would have been transferred to the glass rod from the surface of the copper rod. The copper rod would have become positively charged.

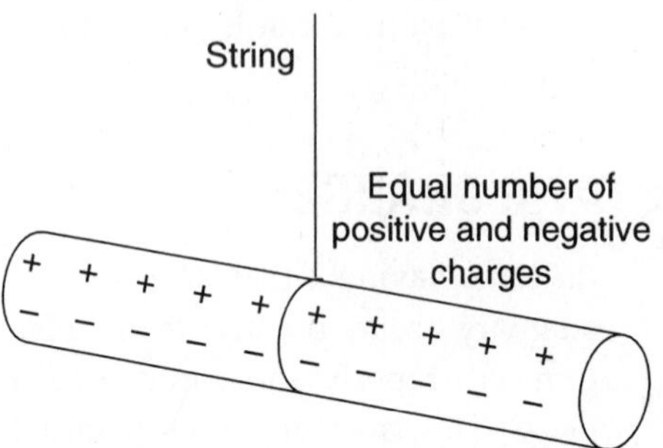

Figure 15–15 Neutral copper rod

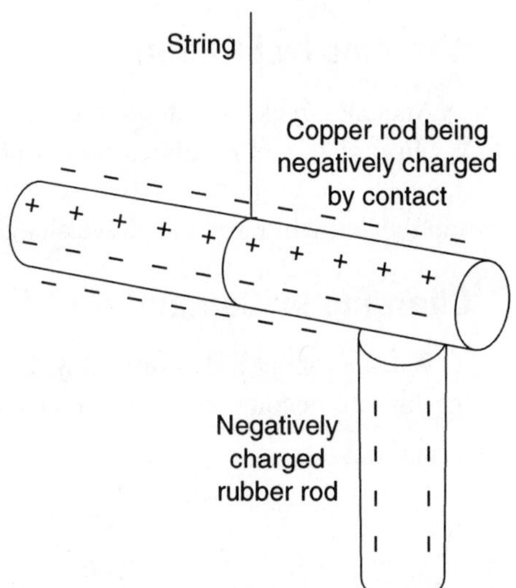

Figure 15–16 Touching rubber rod to copper rod

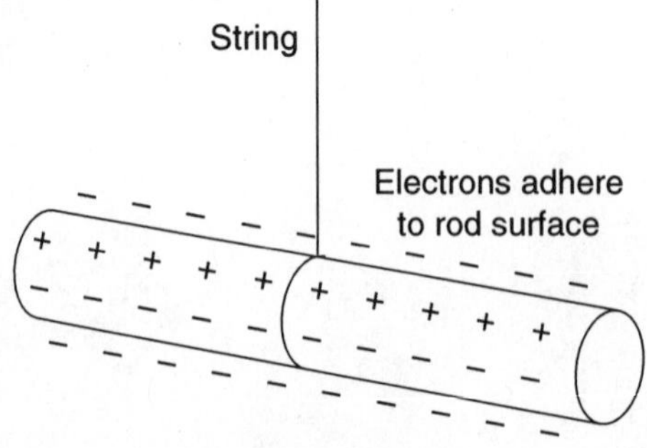

Figure 15–17 Electrons collected on a copper rod surface

Neutralizing an Electrical Charge

Before, we removed electrons from one object to another to cause the object to be charged. To neutralize an electrical charge, the number of electrons on each piece of material must be equal. To allow the electrons to equalize, the two objects must be brought together again without movement. That is, the glass rod and the piece of silk, or the rubber rod and the piece of fur, must touch each other. The attraction of the opposite electrical charges will cause the balance to occur; that is, the positively charged glass rod and the negatively charged piece of silk will cause the electrons to move from the piece of silk to the glass rod until the same number of electrons are on both pieces of material. Both pieces of material will then be electrically neutral.

This electrical charge can be transferred by some means other than directly touching the two materials together. For example, if we placed a conductor (wire) between the two objects the electrons would flow through the conductor just as if the two objects were touched directly. The electrons will flow until there is the same number on both materials.

SUMMARY 15–4

- Atoms having either an excess or a deficiency of electrons are unstable.
- Substances can be electrically charged by using either friction or touching the substances together.
- To neutralize an electrical charge, the number of electrons on each piece of material must be equal.

REVIEW QUESTIONS 15–4

1. What causes an atom to become unstable?
2. When objects of two different amounts of charge are touched, what happens?
3. What is done to neutralize an electrical charge between two objects?

15–5 ELECTROSTATIC FIELDS

An electrostatic field is the force that surrounds a charged body. The electrostatic lines of force between two charged bodies cause the repelling and attracting forces between them. These electrostatic lines of force surround all charged objects.

The lines of force surrounding a negatively charged object come straight into the object from all directions. The lines of force surrounding a positively charged object go straight out from the object in all directions. See Figure 15–18.

Depending on the charge, these lines of force will either strengthen or weaken the electrostatic field. There are three things that increase or decrease the electrostatic field between two objects. They are (1) when they are opposite in charge they will strengthen the field. When they have the same charge, they will weaken the field, (2)

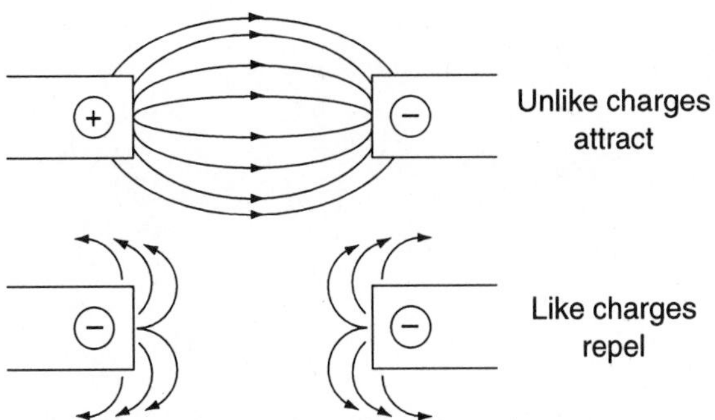

Figure 15–18 Electrostatic lines of force

the strength of the charge on each object, and (3) the distance between the two objects. A large electrical charge on each of the objects will cause more lines of force between them than a smaller charge. When the objects are close to each other, the lines of force will be more concentrated and will cause a stronger field between them.

SUMMARY 15–5

- An electrostatic field is a force surrounding a charged body.
- The lines of force surrounding a negatively charged object come straight into the object from all directions.
- The lines of force surrounding a positively charged object will go straight out from the object in all directions.
- Depending on the charge, these lines of force will either strengthen or weaken the electrostatic field.

REVIEW QUESTIONS 15–5

1. What is an electrostatic field?
2. Name the three things that will strengthen an electrostatic field.
3. If two objects were moved further apart, what happens to the electrostatic field between them?

15–6 ELECTRON ORBITS

Electricity is produced when the electrons leave their atoms because they then become charged ions. There are several methods used to cause these electrons to leave their orbits. To understand how this happens we must know more about the nature of the electron orbits surrounding the nucleus.

Electrons rotate around the nucleus within their orbits at high rates of speed. The centrifugal force caused by the high rate of speed has a tendency to cause the electron to fly away from the nucleus, or move into an outer orbit. However, the attraction of the positively charged nucleus tries to prevent the electron from leaving its shell. See Figure 15–19.

When enough force from the outside of the atom is applied, the electron could be forced out of its orbit. It then becomes a free electron. See Figure 15–20.

Orbital Shells

It is in the atomic shells that surround the atom of each element that the electrons are located. Depending on the nature of the element, it can have up to seven shells. Each material has a different number of shells and a different number of electrons. See Table 15–2.

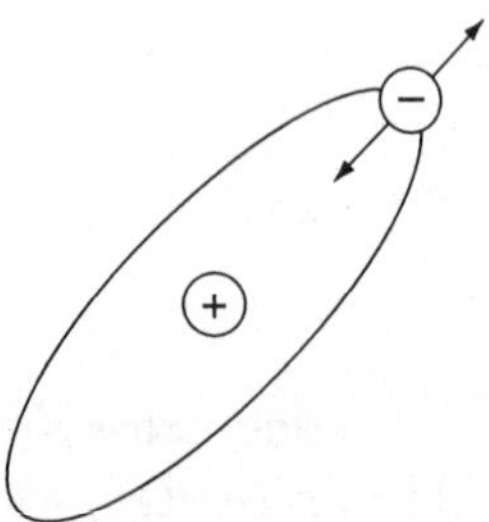

Figure 15–19 Forces on a revolving electron

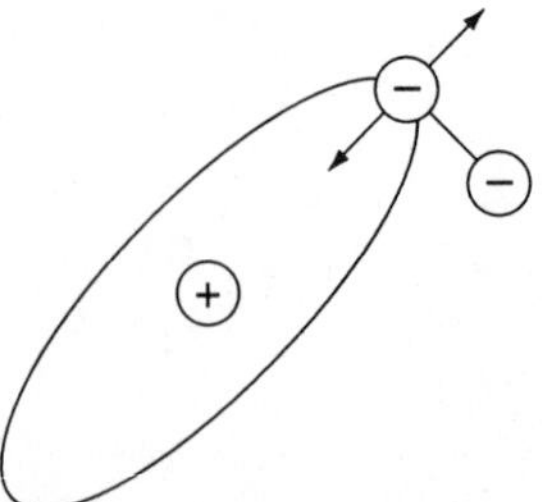

Figure 15–20 External force causes an electron to leave orbit

ATOMIC NO.	ELEMENT	ELECTRONS PER SHELL 1	2	3	4	5	6	7
1	Hydrogen, H	1						
2	Helium, He	2						
3	Lithium, Li	2	1					
4	Beryllium, Be	2	2					
5	Boron, B	2	3					
6	Carbon, C	2	4					
7	Nitrogen, N	2	5					
8	Oxygen, O	2	6					
9	Fluorine, F	2	7					
10	Neon, Ne	2	8					
11	Sodium, Na	2	8	1				
12	Magnesium, Mg	2	8	2				
13	Aluminum, Al	2	8	3				
14	Silicon, Si	2	8	4				
15	Phosphorus, P	2	8	5				
16	Sulfur, S	2	8	6				
17	Chlorine, Cl	2	8	7				
18	Argon, A	2	8	8				
19	Potassium, K	2	8	8	1			
20	Calcium, Ca	2	8	8	2			
21	Scandium, Sc	2	8	9	2			
22	Titanium, Ti	2	8	10	2			
23	Vanadium, V	2	8	11	2			
24	Chromium, Cr	2	8	13	1			
25	Manganese, Mn	2	8	13	2			
26	Iron, Fe	2	8	14	2			
27	Cobalt, Co	2	8	15	2			
28	Nickel, Ni	2	8	16	2			
29	Copper, Cu	2	8	18	1			
30	Zinc, Zn	2	8	18	2			
31	Gallium, Ga	2	8	18	3			
32	Germanium, Ge	2	8	18	4			
33	Arsenic, As	2	8	18	5			
34	Selenium, Se	2	8	18	6			
35	Bromine, Br	2	8	18	7			
36	Krypton, Kr	2	8	18	8			
37	Rubidium, Rb	2	8	18	8	1		
38	Strontium, Sr	2	8	18	8	2		
39	Yttrium, Y	2	8	18	9	2		
40	Zirconium, Zr	2	8	18	10	2		
41	Niobium, Nb	2	8	18	12	1		
42	Molybdenum, Mo	2	8	18	13	1		
43	Technetium, Tc	2	8	18	14	1		
44	Ruthenium, Ru	2	8	18	15	1		
45	Rhodium, Rh	2	8	18	16	1		
46	Palladium, Pd	2	8	18	18	0		
47	Silver, Ag	2	8	18	18	1		
48	Cadmium, Cd	2	8	18	18	2		
49	Indium, In	2	8	18	18	3		
50	Tin, Sn	2	8	18	18	4		
51	Antimony, Sb	2	8	18	18	5		
52	Tellurium, Te	2	8	18	18	6		
53	Iodine, I	2	8	18	18	7		
54	Xenon, Xe	2	8	18	18	8		
55	Cesium, Cs	2	8	18	18	8	1	
56	Barium, Ba	2	8	18	18	8	2	
57	Lanthanum, La	2	8	18	18	9	2	
58	Cerium, Ce	2	8	18	19	9	2	
59	Praseodymium, Pr	2	8	18	20	9	2	
60	Neodymium, Nd	2	8	18	21	9	2	
61	Promethium, Pm	2	8	18	22	9	2	
62	Samarium, Sm	2	8	18	23	9	2	
63	Europium, Eu	2	8	183	24	9	2	
64	Gadolinium, Gd	2	8	18	25	9	2	
65	Terbium, Tb	2	8	18	26	9	2	
66	Dysprosium, Dy	2	8	18	27	9	2	
67	Holmium, Ho	2	8	18	28	9	2	
68	Erbium, Er	2	8	18	29	9	2	
69	Thulium, Tm	2	8	18	30	9	2	
70	Ytterbium, Yb	2	8	18	31	9	2	
71	Lutetium, Lu	2	8	18	32	9	2	
72	Hafnium, Hf	2	8	18	32	10	2	
73	Tantalum, Ta	2	8	18	32	11	2	
74	Tungsten, W	2	8	18	32	12	2	
75	Rhenium, Re	2	8	18	32	13	2	
76	Osmium, Os	2	8	18	32	14	2	
77	Iridium, Ir	2	8	18	32	15	2	
78	Platinum, Pt	2	8	18	32	16	2	
79	Gold, Au	2	8	18	32	18	1	
80	Mercury, Hg	2	8	18	32	18	2	
81	Thallium, Tl	2	8	18	32	18	3	
82	Lead, Pb	2	8	18	32	18	4	
83	Bismuth, Bi	2	8	18	32	18	5	
84	Polonium, Po	2	8	18	32	18	6	
85	Astatine, At	2	8	18	32	18	7	
86	Radon, Rn	2	8	18	32	18	8	
87	Francium, Fr	2	8	18	32	18	8	1
88	Radium, Ra	2	8	18	32	18	8	2
89	Actinium, Ac	2	8	18	32	18	9	2
90	Thorium, Th	2	8	18	32	19	9	2
91	Protactinium, Pa	2	8	18	32	20	9	2
92	Uranium, U	2	8	18	32	21	9	2
93	Neptunium, Np	2	8	18	32	22	9	2
94	Plutonium, Pu	2	8	18	32	23	9	2
95	Americium, Am	2	8	18	32	24	9	2
96	Curium, Cm	2	8	18	32	25	9	2
97	Berkelium, Bk	2	8	18	32	26	9	2
98	Californium, Cf	2	8	18	32	27	9	2
99	Einsteinium, E	2	8	18	32	28	9	2
100	Fermium, Fm	2	8	18	32	29	9	2
101	Mendelevium, Mv	2	8	18	32	30	9	2
102	Nobelium, No	2	8	18	32	31	9	2
103	Lawrencium, Lw	2	8	18	32	32	9	2

Table 15–2 Electron shells

The closer the electron is to the nucleus, the more difficult it is to move it from its shell. This is because of the strong attraction of the positively charged nucleus and the shorter distance between the electron and the proton. When the electron gains more energy it will move to a shell further from the nucleus, lessening their attraction to each other because of the distance. It should be noted that the more electrons in an atom the more shells the atom will have.

Shell Capacity. For any given element, the orbital shell of an atom can hold only a certain number of electrons. In Table 15–2 notice that shell number 1 is the closest to the nucleus and can hold a maximum of 2 electrons; shell number 2 can hold only 8 electrons; shell number 3 can hold only 18 electrons; shell number 4 can hold only 32 electrons. See Figure 15–21.

The Valence Shell. The outermost shell is called the valence shell of the atom. As a result, the electrons that rotate in this shell are called valence electrons. See Figure 15–22.

The number of electrons in the valence shell is important because the smaller the number, the easier it is to free an electron.

Note that no shell will completely fill with electrons before electrons start filling the next shell. In Table 15–2 note that the third shell can hold a maximum of 18 electrons, but it does not hold more than 8 until the fourth shell has started gaining electrons. Each shell will not gather more than 8 electrons before the next shell is started, regardless of the number it can hold. From this we can see that the outer shell of an atom will never have more than 8 electrons.

Electron Energy. All electrons have some charge but not all of them have the same amount of electrical energy. The electrons located in shell number 1 have less energy than those in shells located further from the nucleus. The electrons in the

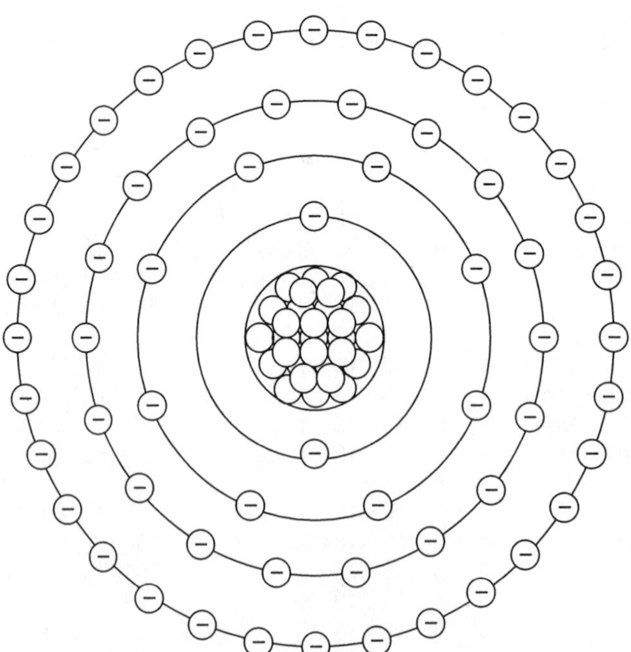

Figure 15–21 Maximum number of electrons contained in a shell

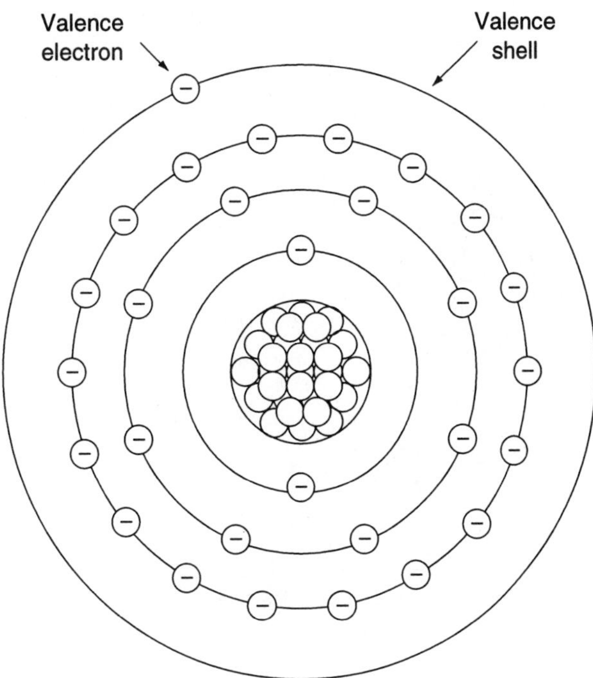

Figure 15–22 Valence electrons

shell the farthest from the nucleus will have the greatest amount of energy. See Figure 15–23.

As electrons gain strength they will move to the next higher shell. After a sufficient amount of energy is gained, the electron will eventually move out of the atom because there is no other shell for it to enter.

Producing Electricity. Electricity is produced when a valence electron leaves its atom. When the atom gains sufficient force or energy, the valence electrons will be

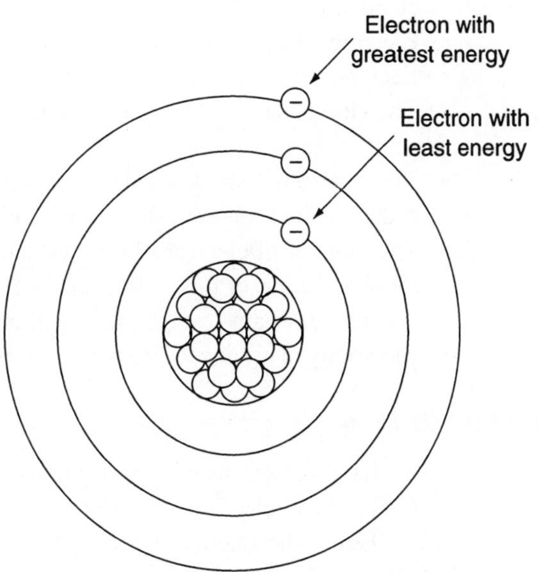

Figure 15–23 Energy in a valence electron

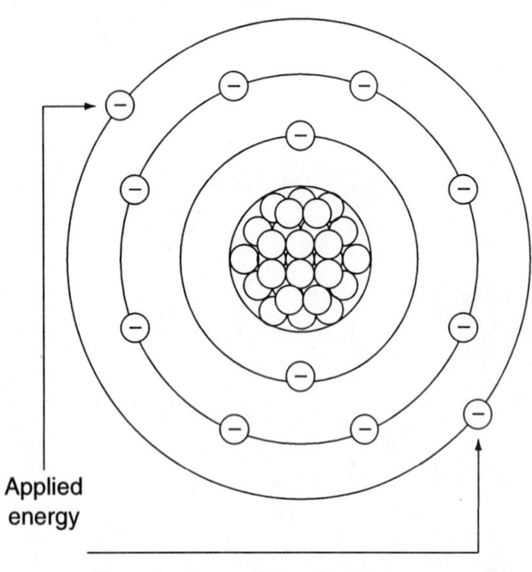
Applied
energy
Two electrons share the energy equally

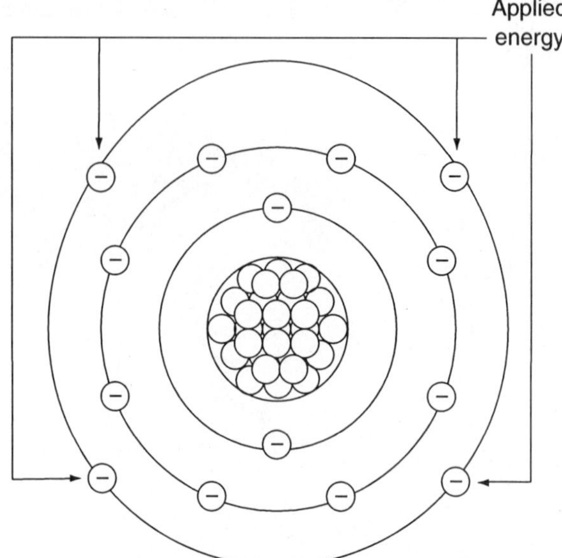
Applied
energy
Four electrons share the energy equally,
a smaller amount for each one

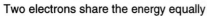

Figure 15–24 Energy applied to the valence electrons

set free. The energy applied to the valence shell is equally distributed among the valence electrons. Because of this, a material having fewer electrons in the valence shell will be a better conductor of electricity. See Figure 15–24.

SUMMARY 15–6

- Electricity is produced when electrons leave their atoms, because they then become charged ions.
- Electrons that leave their atom are called free electrons.
- Each material has a different number of shells and a different number of electrons.
- The closer the electron is to the nucleus the more difficult it is to move it from its shell.
- The outermost shell is called the valence shell of the atom. The electrons that rotate in this shell are called valence electrons.
- The number of electrons in the valence shell is important because the smaller the number the easier it is to free an electron.
- As electrons gain strength, they will move to the next higher shell.
- Electricity is produced when a valence electron leaves its atom.

REVIEW QUESTIONS 15–6

1. What do electrons become when they leave the atom?
2. When is an electron difficult to remove from its atom?
3. Where is the valence shell located?
4. Which electrons have the least amount of energy?
5. What causes an electron to leave its atom?

15–7 ELECTRICAL CONDUCTORS

Atoms with fewer valence electrons make the best conductors of electricity because less energy is required to free them from their atom. They are easier to free because any energy that enters that shell is shared by all of the electrons. Fewer electrons mean more energy for each one. Materials that make the best electrical conductors only have one or two valence electrons. The most popular electrical conductor is copper, with silver and gold closely following. See Figure 15–25.

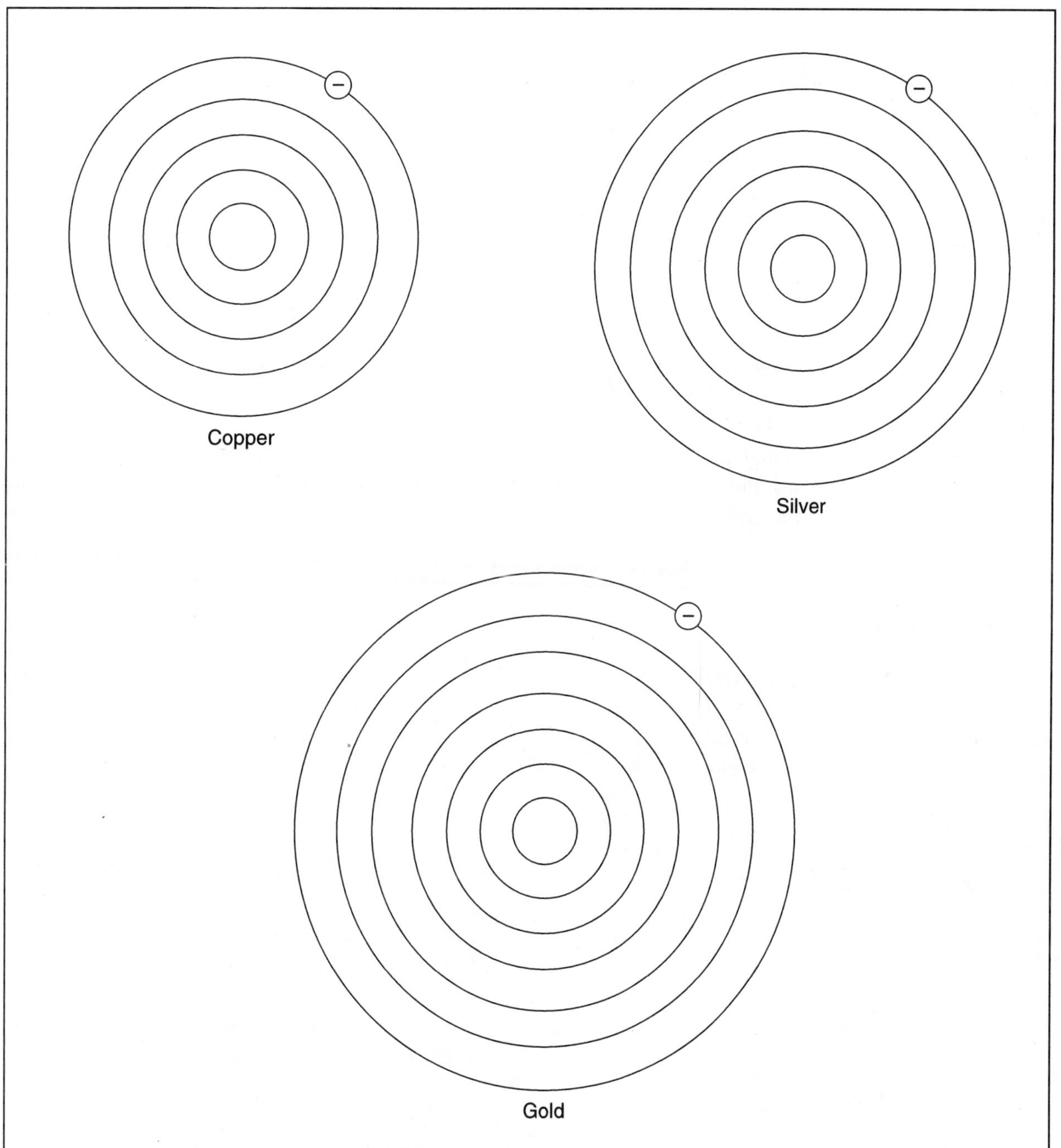

Figure 15–25 Atoms having only one valence electron

SUMMARY 15–7

- Atoms with fewer valence electrons make the best conductors of electricity because less energy is required to free them from their atom.
- Fewer electrons mean more energy for each one.

REVIEW QUESTIONS 15–7

1. Why do atoms with a small number of electrons in the valence shell make good conductors?
2. What type of material makes the best conductor?

15–8 ELECTRICAL INSULATORS

Materials that make good electrical insulators are poor electrical conductors. These materials have 5 or more valence electrons. The reason for this is because the energy is divided equally among all the valence electrons, thus leaving less energy for each one. Also, atoms having 8 valence electrons are very stable chemically and it is difficult to dislodge their valence electrons. Thus, if no electrons are flowing, there is no flow of electricity. See Figure 15–26.

These stable atoms resist any type of change. They will not combine with other atoms to form a compound. In all there are 6 atoms that are naturally stable. They are argon, helium, krypton, neon, radon, and xenon. All of these elements are known as inert gases.

Any time an atom has fewer than 8 valence electrons, it will attempt to gain enough electrons to become stable. Likewise, when a shell becomes less than half full, the atom will try to release the electrons from the unstable shell. Also, atoms that are more than half full will try to gain sufficient electrons to fill the shell and become stable. Materials whose atoms have 7 valence electrons easily attract more electrons and are generally used as insulators.

SUMMARY 15–8

- Materials that make good electrical insulators are poor electrical conductors. These materials have 5 or more valence electrons.
- Atoms that have 8 valence electrons are very stable chemically and it is difficult to dislodge their valence electrons.

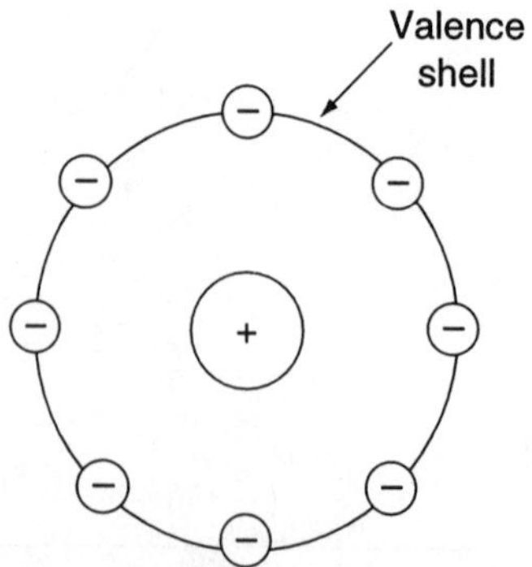

Figure 15–26 An atom with eight valence electrons is very stable chemically

- These stable atoms resist any type of change.
- Any time an atom has fewer than 8 valence electrons, it will attempt to gain enough electrons to become stable.
- When a shell becomes less than half full, the atom will try to release the electrons from the unstable shell.

REVIEW QUESTIONS 15–8

1. How many valence electrons does a good insulator have?
2. To become stable chemically, how many electrons are required?
3. What will happen when a valence shell has less than 8 electrons?
4. What types of material make good insulators?

15–9 ELECTRIC CURRENT

Actually, electricity is the flow of free electrons. In order to get electricity to flow, some type of force must be applied to the atoms to make the electrons move. Before the electrons start moving, it is called static electricity. The electrons must flow before any work can be accomplished.

In reality, the flow of electric current is many billions of free electrons moving through a conductor because of the repelling force caused by like charges. See Figure 15–27.

This repelling force is caused by the negative charge of the electrons in the conductor. It can be seen that the flow of electricity is from the negative to the positive terminals of the source of power. Remember that any given electron does not move completely from one end of the conductor to the other. The movement of electrons can be compared to a row of pool balls that are struck by a single ball at one end. See Figure 15–28.

This motion is called *impulse transfer of energy*. As a single ball hits the one on the end of the row, each ball down the line will bump the next ball in line. In this manner, the last ball moves almost at the same time that the first one hit on the other end of the line.

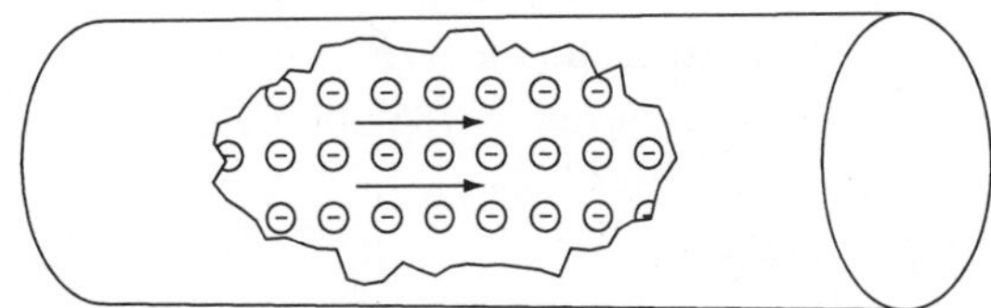

Figure 15–27 Free electrons moving through a conductor

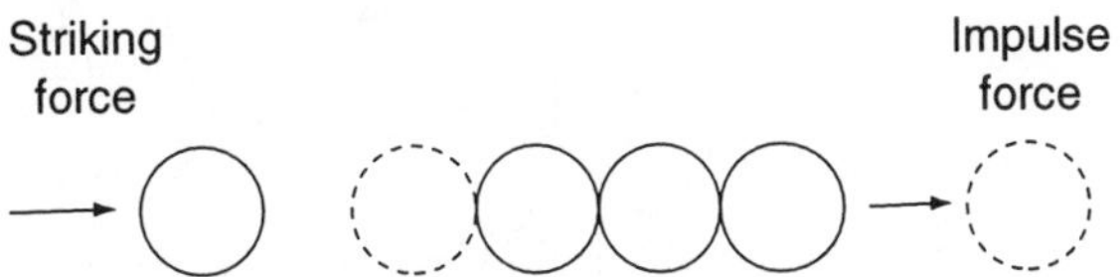

Figure 15–28 Impulse force of electron flow

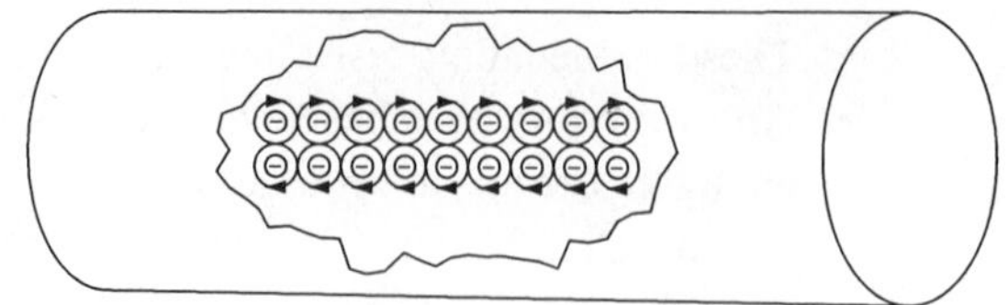

Figure 15–29 Electrons flow at random in a copper conductor

Free Electrons

Free electrons are those that have gained enough energy to move from their atom. This is possible because the valence electrons are held loosely in their orbit. In copper, the atoms are so close that their outer shells overlap each other. See Figure 15–29.

When the electron is revolving around in its orbit, it can easily be transferred to another atom during the overlap period. This shifting of electrons causes a chain reaction by causing an electron from that atom to be transferred to another atom. This random movement of electrons has no effect on the flow of electricity because valence electrons are not actually associated with any one atom. They are shared by all the atoms in that piece of material.

Electron Movement

To produce a flow of electrons through a conductor, the free electrons in that conductor must be forced to move in the same direction. If one end of the conductor was connected to a positive electrical charge and the other end connected to a negative electrical charge, such as some type of battery, the electrons would be forced to move in the same direction, from negative to positive, on the battery. See Figure 15–30.

This flow is possible because the negative electrons are repelled by the negative charge of the battery terminal. At the same time, they are attracted by the positive charge on the battery positive terminal. It is this attraction and repulsion by an external source that actually causes the electrons to flow. They will move in only one direction—from negative to positive. These electrons do not follow a straight path through the conductor, but move from atom to atom through the process of randomly entering another atom's valence shell. As the strength of the applied charge is increased, the electrons tend to flow faster and in more direct paths. The speed of electric current is 186,000 miles per second—the speed of light.

SUMMARY 15–9

- Electricity is the flow of electrons. In order to get electricity to flow, some type of force may be applied to the atoms to make the electrons move.
- The flow of electricity is from the negative to the positive terminals of the source of power.

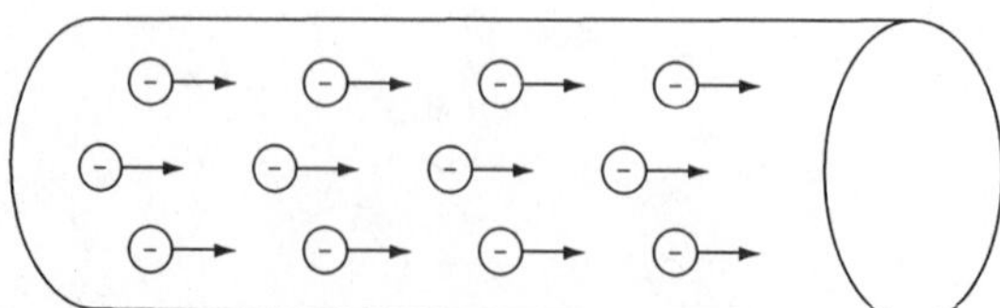

Figure 15–30 Forced flow of electrons

- Free electrons are those that have gained enough energy to move from their atom.
- When the electron is revolving around in its orbit, it can easily be transferred to another atom during the overlap period.
- To produce a flow of electrons through a conductor, the free electrons in that conductor must be forced to move in the same direction.
- This flow is possible because the negative electrons are repelled by the negative charge of the energy source. At the same time, they are attracted by the positive terminal of the source of power.

REVIEW QUESTIONS 15–9

1. At what point can electrical work be done?
2. What two forces cause electrons to flow through a circuit?
3. What are the electrons that move from their atom known as?
4. What does the movement of an electron from one atom to another cause?
5. What is required to make electricity flow through a conductor?

15–10 VOLTAGE (EMF—ELECTROMOTIVE FORCE)

Voltage is the electrical pressure or force that pushes the electrons through the conductor. It may be defined as the difference in electrical potential between two points. This pressure is measured in voltage and is designated by either the letter V or E. The terms used when discussing this area of electricity are potential, voltage, or EMF (electromotive force). These terms are used interchangeably. When a potential difference forces one coulomb of current to do one joule of work, the EMF is one volt. See Table 15–3.

A charge of 1 coulomb = 6.28×10^{18} electrons
An emf of 1 volt (V) = 1 coulomb doing 1 joule of work
1 microvolt (μV) = 1/1,000,000 volt
1 millivolt (mV) = 1/1,000 volt
1 kilovolt (kV) = 1,000 volts
1 megavolt (MV) = 1,000,000 volts
Conversion of units
volts (V) × 1000 = millivolts (mV)
volts (V) × 1,000,000 = microvolts (μV)
millivolts (mV) × 1,000 = microvolts (μV)
volts (V) ÷ 1,000 = kilovolts (kV)
volts (V) ÷ 1,000,000 = megavolts (MV)
megavolts (MV) × 1,000 = kilovolts (kV)
millivolts (mV) ÷ 1,000 = volts (V)
microvolts (μV) ÷ 1,000,000 = volts (V)
microvolts (μV) ÷ 1,000 = millvolts (mV)
kilovolts (kV) × 1,000 = volts (V)
megavolts × 1,000,000 = volts (V)
kilovolts ÷ 1,000 = megavolts
1 V = 1,000 mV = 1,000,000 μV = 0.001 kV = 0.000001 MV

Table 15–3 Units of voltage

SUMMARY 15–10

- Voltage is the electrical pressure or force that pushes the electrons through a conductor.
- Voltage may be defined as the difference in electrical potential between two points.
- The terms used when discussing this area of electricity are potential, voltage, or EMF.

REVIEW QUESTIONS 15–10

1. What is the pressure that causes electricity to flow known as?
2. What are the three names that voltage is often called?

15–11 CURRENT (AMPERAGE)

Electrical current is the movement of electrons through a circuit. A. M. Ampere (for whom the term *ampere* was named), a French physicist (1775–1836), worked with an electrical charge of 6.28×10^{18} displaced electrons. This amount of electrical charge was named the coulomb in honor of C. A. Coulomb. When one coulomb of current passes a point in one second, this indicates that one ampere of electrical current is flowing past that point. See Figure 15–31.

Electrical current is measured in several different ways. See Table 15–4.

The symbol indicating current is the letter I.

SUMMARY 15–11

- Electrical current is the movement of electrons through a circuit.
- When one coulomb of current passes a point in one second, this indicates that one ampere of electrical current is flowing past that point.

REVIEW QUESTIONS 15–11

1. How is the flow of electricity measured?
2. What is an electrical charge of 6.28×10^{18} moving electrons known as?

15–12 RESISTANCE

The opposition to current flow is termed *resistance*. All paths offer some resistance to the flow of electrons. Resistance is measured in ohms and is designated by the symbol Ω (omega). The ohm is named after the German scientist, George Simon

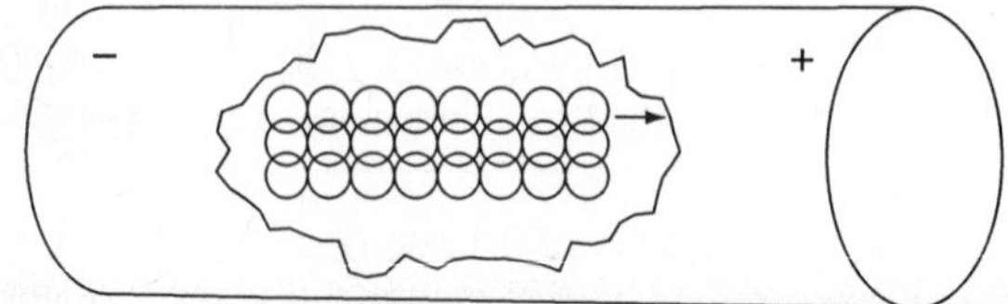

Figure 15–31 Flow of current through a conductor

1 ampere (A) = 1 coulomb/sec
1 milliampere (mA) = 1/1,000 ampere
1 microampere (μA) = 1/1,000,000 ampere
Conversion of units
amperes (A) × 1,000 = milliamperes (mA)
amperes (A) × 1,000,000 = microamperes (μA)
milliamperes (mA) × 1,000 = microamperes (μA)
milliamperes (mA) ÷ 1,000 = amperes (A)
microamperes (μA) ÷ 1,000,000 = amperes (A)
microamperes (μA) ÷ 1,000 = milliamperes (mA)
0.5 A = 500 mA = 500,000 μA

Table 15–4 Units of current

Ohm (1787–1854), who discovered, through scientific experiments, some very important electrical facts. See Table 15–5.

When an electrical circuit allows an EMF of one volt to cause a current of one ampere to flow, it has a resistance of one ohm.

SUMMARY 15–12

- The opposition to current flow is termed resistance. All paths offer some resistance to the flow of electrons.
- Resistance is measured in ohms and is designated by the symbol Ω (omega).
- When an electrical circuit allows an EMF of one volt to cause a current of one ampere to flow, it has a resistance of one ohm.

REVIEW QUESTIONS 15–12

1. What is the resistance to the flow of electricity known as?
2. What do all components in an electrical circuit cause?

15–13 OHM'S LAW

G. S. Ohm established the relationship between voltage (E), current (I), and resistance (R) in a closed circuit.

ohms ÷ 1,000 = kilohms (kΩ)
ohms ÷ 1,000,000 = megohms (MΩ)
kilohms (kΩ) ÷ 1,000 = megohms (MΩ)
kilohms × 1,000 = ohms (Ω)
megohms × 1,000,000 = ohms (Ω)
megohms × 1,000 = kilohms (kΩ)
500,000 ohms = 500 kilohms = 0.5 megohm or 500,000 Ω = 500 kΩ = 0.5 MΩ

Table 15–5 Units of resistance

He determined this relationship to be stated as: Current is directly proportional to voltage and inversely proportional to resistance. This can also be stated as:

1. If E is raised, I will also be raised;
2. If E is lowered, I will also be lowered;
3. If R is raised, I will be lowered;
4. If R is lowered, I will go up.

This can be stated in a mathematical formula as follows:

$$\text{Current in amperes (I)} = \frac{\text{Voltage in volts (E)}}{\text{Resistance in ohms (R)}}$$

This basic formula can be transposed to determine the value of any missing component as long as the other two are known. When transposed, the formula would become:

$$I = \frac{E}{R}\text{ R, } E = IR\text{, and } R = \frac{E}{I}$$

A simple way of remembering these formulas is to place them in a triangle. See Figure 15–32.

To use the triangle, simply cover the letter for the component that is missing. The remaining portion of the triangle will indicate the proper equation to use.

SUMMARY 15–13

- G. S. Ohm established the relationship and stated it as: Current is directly proportional to voltage and inversely proportional to resistance.
- The relationship between voltage, current, and resistance can be stated mathematically as: Current in amperes (I) = Voltage in volts (E) ÷ Resistance in ohms (R).

REVIEW QUESTIONS 15–13

1. Write the symbols used to indicate voltage, amperage, and resistance.
2. How is Ohm's Law triangle used?

15–14 ELECTRICAL CIRCUIT

The path through which the electrical current flows is known as the circuit. Its components are a source of electrical power, conductors, the load or loads, and some means of controlling the flow of electricity.

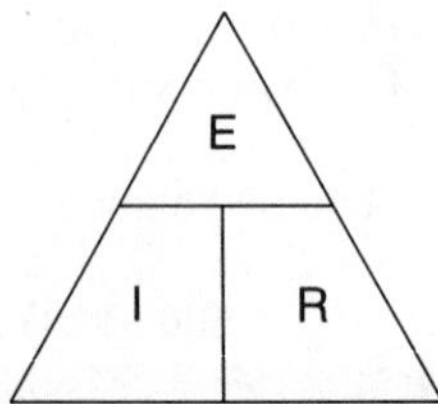

Figure 15–32 Ohm's Law triangle

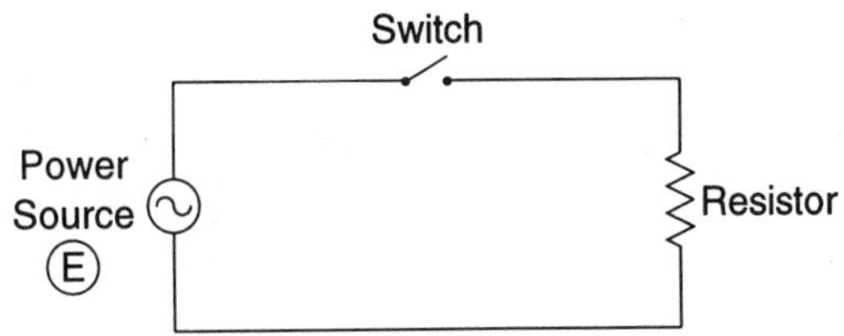

Figure 15–33 Basic electrical schematic diagram

The voltage source may be a battery, an alternating current generator, or any other source of electricity. The conductors provide the path from the source to the load and back to the source. In air conditioning and refrigeration, the load can be any or many of the components used to make a complete system. These loads may be an electric motor, a solenoid coil, a starter holding coil, or anything else that may require electricity to operate. To control the flow, a switch is usually placed in series with the load. A schematic diagram shows the various components required to make a complete electrical circuit. See Figure 15–33.

There are several basic facts that must be understood about electrical circuits because they sometimes appear quite complicated. Even though a circuit may appear to be complicated, it can only be one of three general types: series, parallel, or series-parallel.

Series Circuit

In a series circuit there is only one path for the current to flow through. The current flows through every component in that circuit on its way back to the source. A series circuit can be identified in two ways; there will always be only one conductor connected to one terminal, and there is only one path for the current to flow through from the power source through the load(s) and back to the power source. See Figure 15–34.

Resistance in a Series Circuit. In a series circuit all of the loads are connected into the circuit, one after the other. The total resistance is the sum of the total resistances. See Figure 15–35.

The formula used to calculate the total resistance in a series circuit is:

$$R_t = R_1 + R_2 + R_3 \ldots$$

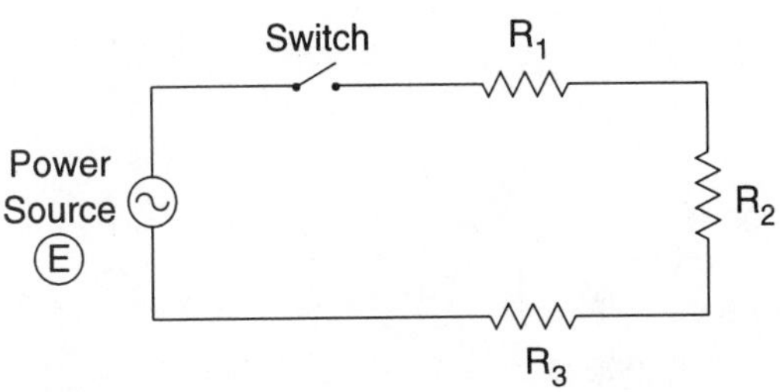

Figure 15–34 A simple series circuit

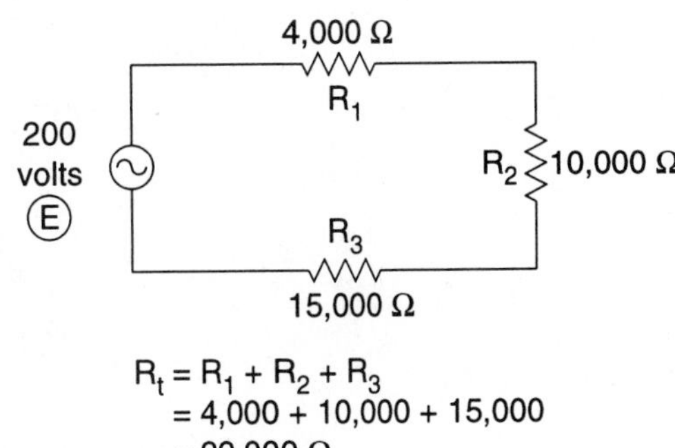

Figure 15–35 Example of the resistance in a series circuit

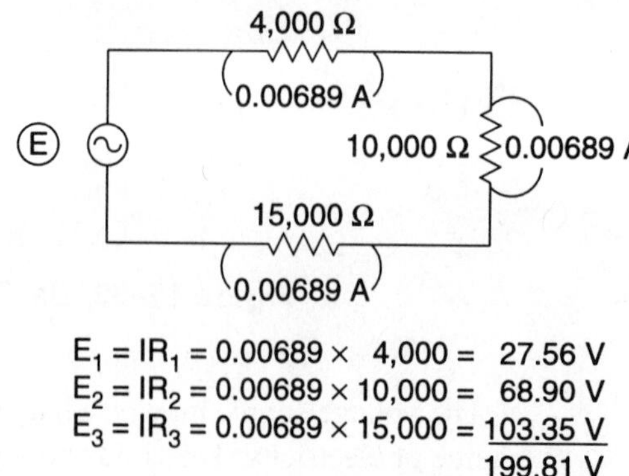

Figure 15–36 Current flow in a series circuit

Current in a Series Circuit. In a series circuit, Ohm's Law applies to the complete circuit as well as to any part in it. The current is the same throughout the complete circuit because there is only one path for it to follow. See Figure 15–36.

The current flow in this example is:

$$I = \frac{E}{R}$$
$$= \frac{200}{29{,}000}$$
$$= 0.00689 \text{ amp}$$

This is the amount of current flow that can be found at any point in the circuit.

Voltage Drop in a Series Circuit. As the current flows through a series circuit, the voltage drops at each resistive component. The sum of these voltage drops will always be equal to the source voltage. The amount of voltage drop through each load can be determined by using the formula E = IR for each resistance. See Figure 15–37.

The following will be the calculated voltage drops through the circuit:

$$E_1 = IR_1 = 0.00689 \times 4{,}000 = 27.56 \text{ V}$$
$$E_2 = IR_2 \times 0.00689 \times 10{,}000 = 68.90 \text{ V}$$
$$E_3 = IR_3 \times 0.00689 \times 15{,}000 = 103.35 \text{ V}$$
$$E_t = 199.81 \text{ V}$$

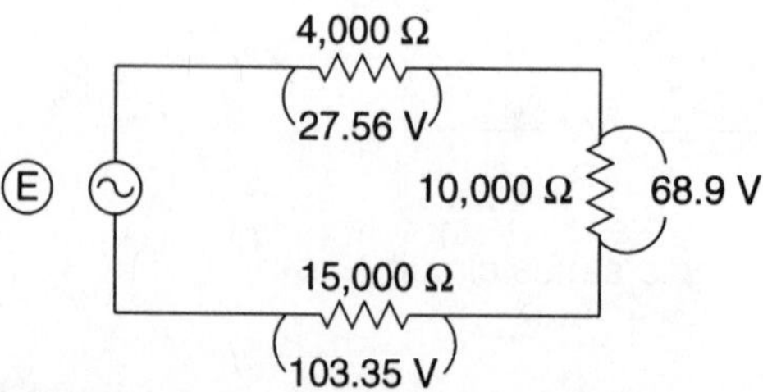

Figure 15–37 Voltage drops in a series circuit

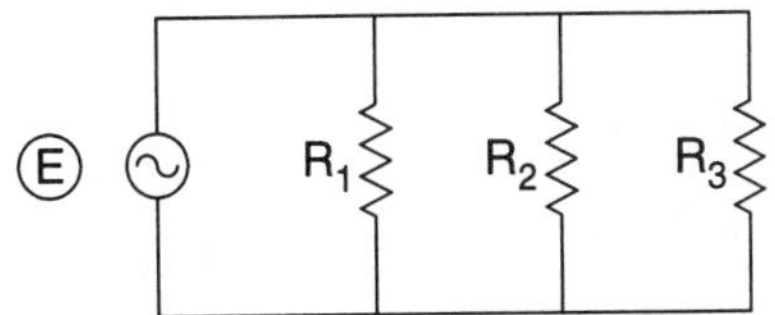

Figure 15–38 Simple parallel circuit

The total would have been more accurate if the calculations had been carried to more decimal places.

Kirchoff's Law states the following facts about series circuits:

1. The sum of the voltage drops around a series circuit will always equal the source voltage.
2. The current is the same when measured at any point in a series circuit.

Parallel Circuit

When two or more resistances are connected to a single power source, and different paths are provided for the current to flow, the resistances are said to be in parallel. See Figure 15–38.

When a circuit is connected in parallel, the total circuit resistance is reduced with the addition of each component.

Parallel circuits have been compared to water pipes. Two pipes can carry more water than 1 and 3 pipes can carry more water than 2. Each added pipe lowers the resistance to the flow of water. The same is true with parallel circuits; each time another path is added, the total circuit resistance drops.

Resistance in Parallel Circuits. In a parallel circuit, the total circuit resistance will always be smaller than the smallest resistance in the circuit. This is because the total circuit current is always greater than the current through any individual resistance. The total resistance in a parallel circuit is calculated by the formula:

$$Rt = \frac{1}{\frac{1}{R_1} + \frac{1}{R_2} + \frac{1}{R_3} + \cdots}$$

This formula can be used for any number of resistances in parallel. There is a simpler one that can be used when only two resistances are connected in parallel. It is known as the product over sum method:

$$R_t = \frac{(R_1 \times R_2)}{(R_1 + R_2)}$$

For an example see Figure 15–39.

When resistances of equal value are used, the following formula can be used:

$$R_t = \frac{\text{R (value of one resistor)}}{\text{N (number of resistors)}}$$

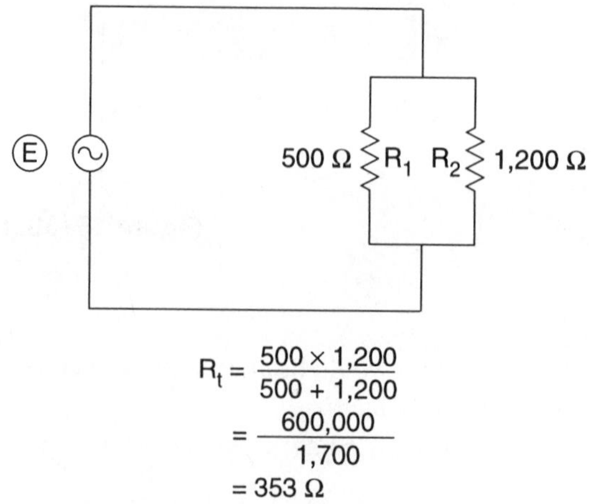

$$R_t = \frac{500 \times 1{,}200}{500 + 1{,}200} = \frac{600{,}000}{1{,}700} = 353\ \Omega$$

Figure 15–39 Calculating resistances in a parallel circuit

For an example see Figure 15–40.

Current in a Parallel Circuit. In a parallel circuit, the total current flowing through the complete circuit will be the sum of the individual branch currents. The current can be calculated with the following formula:

$$I_t = I_{R1} + I_{R2} + I_{R3}$$

For an example see Figure 15–41.

The laws governing parallel circuits are as follows:

1. The voltage across all branches of a parallel circuit is the same.
2. The total current is equal to the sum of the individual branch currents.

Compare these laws and those concerning series circuits.

Voltage Drops in Parallel Circuits. The voltage across each branch of a parallel circuit will be the same as the supply voltage. This is because each branch gets its voltage from the same source. See Figure 15–40. In those branches that have more than one resistance, the voltage will drop across each one of them. When a branch of a parallel circuit has more than one resistance, the voltage drops would be calculated just like those in a series circuit. Use the formula:

$$E = IR$$

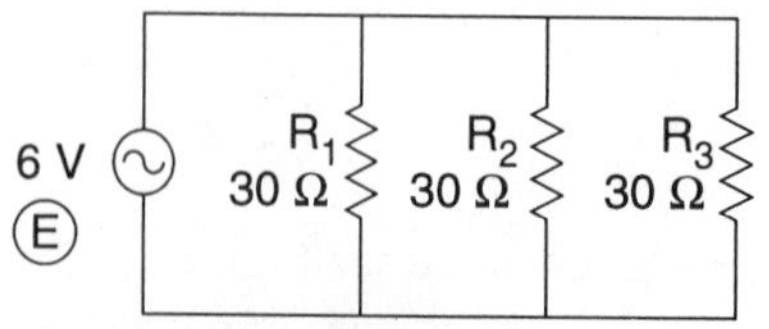

6 V E — R_1 30 Ω, R_2 30 Ω, R_3 30 Ω

$$R_t = \frac{30}{3} = 10\ \Omega$$

Figure 15–40 Using unequal resistance method

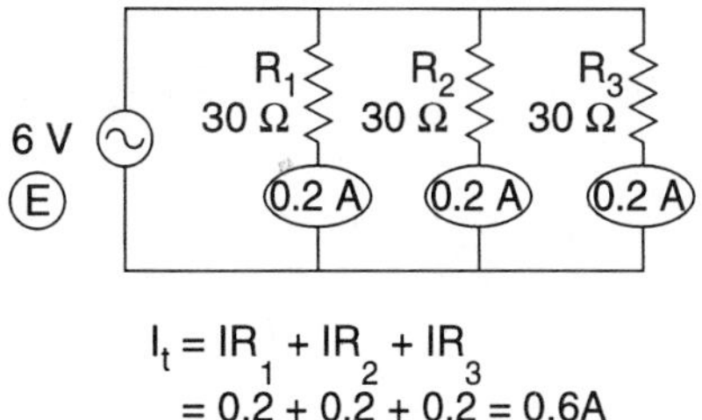

Figure 15–41 Parallel circuit current calculation

For an example see Figure 15–42.

Series-Parallel Circuits

These types of circuits are a combination of both series and parallel circuits. Oftentimes they are quite simple. However, they can be quite complicated, depending on the purpose of the circuit. They are used very often in the wiring of air conditioning and refrigeration systems.

There are certain basic factors that must be considered in any type of circuit. From the discussion in the previous paragraphs concerning series circuits and parallel circuits, we learned that the basic factors are: (1) the total current from the power source and the current in each part of the circuit, (2) the source voltage and the voltage drops across each component of the circuit, and (3) the total resistance and the resistance of each component of the circuit. When these factors are known, the others can be easily calculated.

When making calculations concerning any part of a circuit, it must be determined if it is in series or parallel. When this is known, use only the formulas that apply to that type of circuit. When calculating the electrical characteristics of a series-parallel circuit, some resistances are wired in series and some are wired in parallel. Thus, in series-parallel circuits, those components that are wired in series will be calculated with the series circuit formulas. Those that are wired in parallel will be calculated with the parallel circuit formulas.

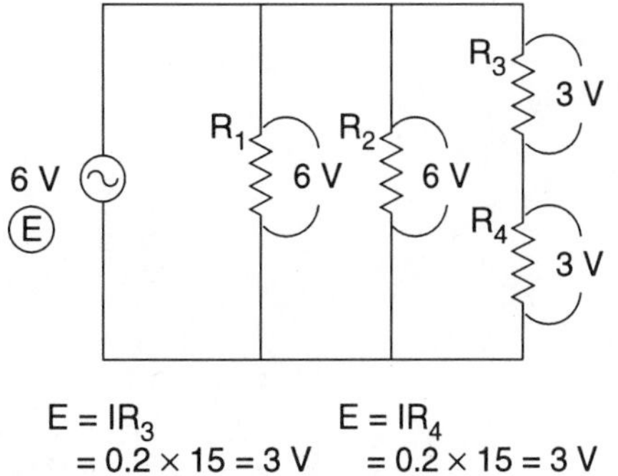

$E = IR_3$
$= 0.2 \times 15 = 3$ V

$E = IR_4$
$= 0.2 \times 15 = 3$ V

Figure 15–42 Calculating voltage drops in a parallel circuit

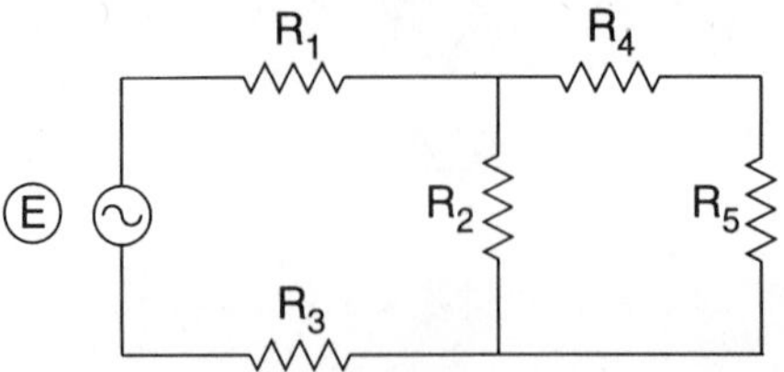

The circuit above can be redrawn to the circuit below

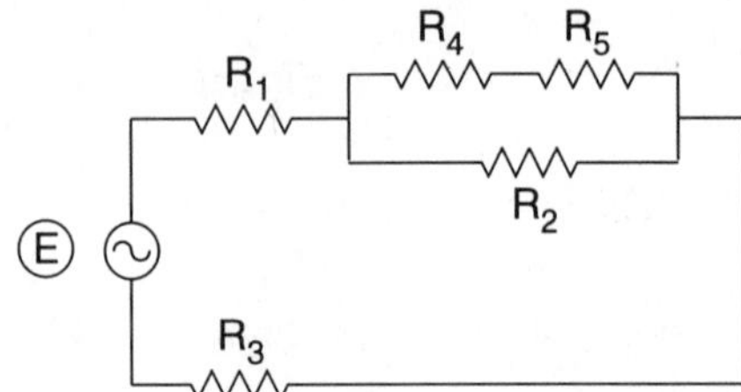

Figure 15–43 Redrawing a series-parallel circuit

Thus, it must be determined whether the system is series, parallel, or series-parallel before any problem solving can be accomplished. In simple circuits this is not too difficult. But, there are circuits in which making this determination is quite difficult. Sometimes it is easier to redraw the circuit in a simpler form so that the different circuits can be determined. See Figure 15–43.

The following steps are used to simplify series-parallel circuits:

Step 1. Determining the different types of circuits is the first step in calculating the resistances of a series-parallel circuit.

Step 2. This step is used to combine the series branch circuits. See Figure 15–44.

Step 3. In this step is the total resistance of the parallel circuit containing R_4, R_5, and R_2. See Figure 15–45.

Use the following formula:

$$Rt = \frac{1}{\frac{1}{R_1} + \frac{1}{R_2} + \frac{1}{R_3} + \cdots}$$

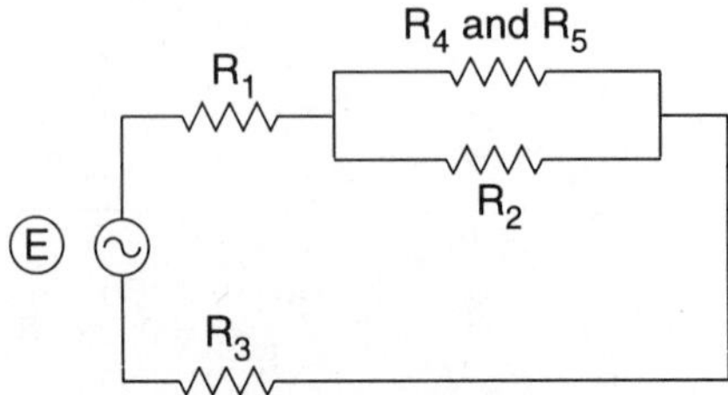

Figure 15–44 Combining series branch circuits

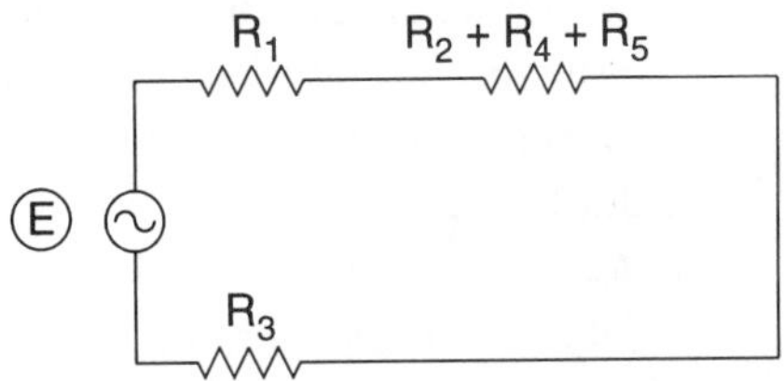

Figure 15–45 Calculating the total resistance of R_2, R_5, and R_4

The circuit can now be redrawn to make it simpler. See Figure 15–45.

Step 4. Compute the series circuit using the formulas that apply to that type of circuit. Use the formula:

$$R_t = R_1 + R_2 + R_3 \ldots$$

The circuit can now be simplified. See Figure 15–46.

It should be realized from this explanation that the total currents, the total voltage drops, or the total resistance cannot be calculated in a series-parallel circuit by using only the total current and the applied voltage. Each circuit must be calculated individually, determining the voltage across and the current through each load before starting on another branch. To accurately calculate series-parallel circuits takes experience and patience, after which you will start to develop your own methods and shortcuts for series-parallel circuits.

SUMMARY 15–14

- The path through which electrical current flows is known as the circuit.
- The components of the circuit are a source of electrical power, conductors, the load or loads, and some means of controlling the flow of electricity.
- In a series circuit there is only one path for the current to flow through.
- In a series circuit, all of the loads are connected into the circuit one after the other.
- In a series circuit, Ohm's Law applies to the complete circuit as well as to any part in it.
- As the current flows through a series circuit, the voltage drops at each resistor component.
- When two or more resistances are connected to a single power source, and different paths are provided for the current to flow, the resistances are said to be in parallel.
- In a parallel circuit, the total circuit resistance will always be smaller than the smallest resistance in the circuit.

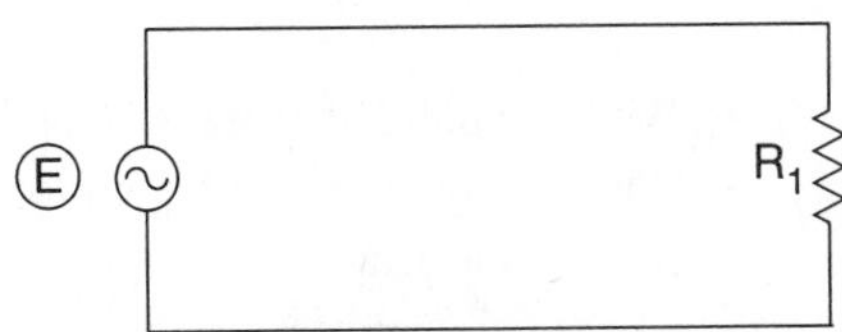

Figure 15–46 Calculating the total resistance

- In a parallel circuit, the total current flowing through the complete circuit will be the sum of the individual branch currents.
- The voltage across each branch of a parallel circuit will be the same as the supply voltage.
- Series-parallel circuits are a combination of both series and parallel circuits.
- When making calculations concerning any part of a circuit, it must be determined if it is in series or parallel.

REVIEW QUESTIONS 15–14

1. Name the components of an electrical circuit.
2. What does a series circuit provide?
3. How are the loads connected in a series circuit?
4. Does the current flow in a series circuit change with each resistance?
5. What will the sum of voltage drops across a circuit indicate?
6. Define a parallel circuit.
7. What will be indicated when the calculated resistance is greater than the smallest resistance in the circuit?
8. What will be the total current flowing through a parallel circuit?
9. To what is the voltage across all branches of a parallel circuit equal?
10. What must be determined before making calculations concerning a series-parallel circuit?

15–15 POWER AND ENERGY

Power is defined as the rate of doing work. In an electrical circuit the power is determined by multiplying the voltage times the amperage. The resulting power is in watts. To calculate power use the following formula:

$$P = E \times I$$

In Ohm's Law, equivalents are substituted for E and I, and these additional formulas are developed:

$$P = \frac{E^2}{R} \text{ and } P = I^2R$$

Electrical energy is the work done by expending electricity. It is measured in watt hours. This is how the power company calculates your electric bill each month. Use the formula:

$$W = PT$$

Where:
W = Electrical energy in watts
P = Electrical power in watts
T = Time in hours

SUMMARY 15–15

- Power is defined as the rate of doing work.
- Power is determined by multiplying the volts times the amperage. The resulting power is in watts.
- Electrical energy is the work done by expending electricity. It is measured in watt hours.

REVIEW QUESTIONS 15–15

1. What is power?
2. How is power indicated?

15–16 CONDUCTORS

Electrical conductors are the components of a circuit that provide the path for the electricity to flow through. The current-carrying capacity of a conductor, among other factors, is to a great extent dependent on the size of the conductor. A large conductor is capable of carrying more electrons than a smaller conductor. This is possible because there is less resistance in the larger conductor.

Electrical conductors are listed according to their size by the American Wire Gauge System. See Table 15–6. In this numbering system the larger the number the smaller the wire. Thus, a number 10 wire is larger than a number 22 wire.

The standard unit for measuring the cross-sectional area of a wire conductor is the circular mil. A mil is one thousandth of an inch. The circular mil is the cross-sectional area of a wire that is 1 mil in diameter.

It must be noted that the resistance offered to the flow of electricity by the conductor (wire) will vary inversely with its size or cross-sectional area.

The type of material will also make a difference in the amount of current that a conductor will carry. Some materials make good electrical conductors, while others make good electrical insulators.

There are several other factors that must be taken into consideration when making the conductor selection for a given application: the cost, its resistance, its weight, and its strength. A common method of comparing electrical conductors is by the amount of resistance offered by one circular mil-foot of the material being considered. See Figure 15–47.

Resistance in a Length of Wire

When selecting an electrical conductor, the length must be taken into consideration. Each foot of wire offers a given amount of resistance to the flow of electricity. When multiple feet of wire are used, the resistance will increase an equal amount. If, for example, a piece of wire 10 feet long has a certain amount of resistance, a piece of wire 100 feet long will have 10 times that amount of resistance. The total resistance is important because resistance in wires uses electrical power and causes electrical losses in the wire.

Conductor Temperature and Resistance

In most of the materials used in the manufacture of conductors, an increase in temperature will also result in an increase in the resistance of that conductor. When electrical circuits are to be enclosed, the circuit and enclosure are designed to dissipate the heat generated by the resistance of electron flow and help cool the circuit components. According to the National Electrical Code, any wires that are installed in a conduit or any other type of metal covering must be sized larger so that the amount of heat generated will be less and allow the desired amount of current to flow through the circuit.

B & S GAUGE NO.	DIAMETER OF BARE WIRE (INCHES)	OHMS PER 1,000 FT.		CURRENT CAPACITY (AMPERES)	
		70°F	167°F	RUBBER INSULATION	OTHER INSULATION
0000 (4/0)	0.460	0.050	0.060	160–248	193–510
000 (3/0)	0.410	0.062	0.075	138–215	166–429
00 (2/0)	0.365	0.080	0.095	120–185	145–372
0	0.325	0.100	0.119	105–160	127–325
1	0.289	0.127	0.150	91–136	110–280
2	0.258	0.159	0.190	80–118	96–241
3	0.229	0.202	0.240	69–101	83–211
4	0.204	0.254	0.302	60–87	72–180
5	0.182	0.319	0.381	52–76	63–158
6	0.162	0.403	0.480	45–65	54–134
7	0.144	0.510	0.606	45–65	54–134
8	0.128	0.645	0.764	35–48	41–100
9	0.114	0.813	0.963	35–48	41–100
10	0.102	1.02	1.216	25–35	31–75
11	0.091	1.29	1.532	25–35	31–75
12	0.081	1.62	1.931	20–26	23–57
13	0.072	2.04	2.436	20–26	23–57
14	0.064	2.57	3.071	15–20	18–43
15	0.057	3.24	3.873	15–20	18–43
16	0.051	4.10	4.884	6	10
17	0.045	5.15	6.158	6	10
18	0.040	6.51	7.765	3	6
19	0.036	8.21	9.792	3	6
20	0.032	10.3	12.35	–	–
21	0.028	13.0	15.57	–	–
22	0.025	16.5	19.63	–	–
23	0.024	20.7	24.76	–	–
24	0.020	26.2	31.22	–	–
25	0.018	33.0	39.36	–	–
26	0.016	41.8	49.64	–	–
27	0.014	52.4	62.59	–	–
28	0.013	66.6	78.93	–	–
29	0.011	82.8	99.52	–	–
30	0.010	106	125.50	–	–
31	0.009	134	158.20	–	–
32	0.008	165	199.50	–	–
33	0.007	210	251.60	–	–
34	0.006	266	317.30	–	–
35	0.005	337	400.00	–	–
36	0.005	423	504.50	–	–

Table 15–6 American standard wire gauges—dimensions and typical resistances of commercial copper wire

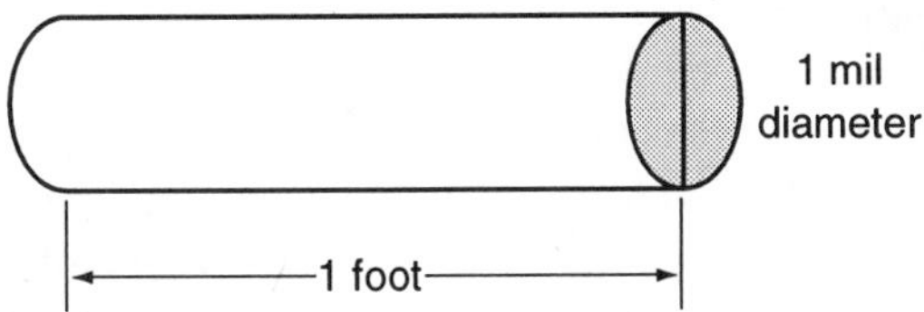

Figure 15–47 A representation of a circular mil

The resistance of wire varies with the following factors:

1. The size of the conductor
2. The length of the conductor
3. The kind of material that the conductor is made from
4. The temperature of the conductor

Stranded Conductors

At times it is desirable to use several small wires that have been twisted together rather than one solid wire to make an electrical cable. Stranded wire allows for greater flexibility and is much easier to install than a solid wire. When the installation is subject to vibration or some other type of movement, stranded wire offers more strength and less danger of breaking under stress. Stranded cables must not be confused with multiple wire conductors. Multi-conductor cables have different conductors that are insulated and color coded so that connections can be properly made.

SUMMARY 15–16

- Electrical conductors are the components of a circuit that provide the path for electricity to flow through.
- The current carrying capacity of a conductor, among other factors, is, to a great extent, dependent on the size of the conductor.
- When selecting an electrical conductor, the length must be taken into consideration. When multiple feet of wire are used, the resistance will increase an equal amount.
- In most of the materials used in the manufacture of conductors, an increase in temperature will also increase the resistance of that conductor.
- When the installation is subject to vibration or some other type of movement, stranded wire offers more strength and less danger of breaking under stress.

REVIEW QUESTIONS 15–16

1. Will the diameter of a wire affect its current carrying capacity?
2. Will a long conductor or a short conductor of the same material and diameter carry the most current?
3. What must be done to wires installed in a conduit?
4. What type of conductor is most desirable for use in a vibrating application?

15–17 RESISTORS

Remember that resistance is the opposition to electrical flow through a conductor. If it were not for resistance in an electrical circuit, there would be a lot of damage when it was connected to a source of power. The only way to prevent this type of damage is to place some form of resistance into the circuit. This resistance is usually in the form of resistors. Resistors are circuit components that are connected into the circuit to limit current flow. See Figure 15–48.

Modern day resistors are either made from powdered carbon or a high-resistance wire. Carbon is used because of its negative temperature coefficient (the resistance decreases with a rise in temperature). Carbon resistors can be built that have a very stable resistance over a wide range of temperatures. They can be made with a very precise ohmic value, but they are difficult to manufacture. The wire-type resistors are easier to manufacture than carbon resistors. They can be built to have a very precise ohmic value. However, they are less stable with varying temperatures.

Resistors are manufactured in a wide variety of shapes and sizes. It is the composition of the material from which they are made that allows their ohmic values to be accurately controlled. They are available in values ranging from one ohm resistance to several million ohms resistance. Resistors are rated by wattage and by their physical size. The larger the size of the resistor, the greater the amount of heat dissipated.

Variable Resistors

Sometimes it is desirable to be able to change the ohmic value of a resistor to suit certain applications, such as for modulating motors that control the valves and linkages for air dampers, and variable motor-speed control in air conditioning and refrigeration units. The variable resistor can be used to adjust the voltage to the desired amount to that specific device for the operation desired. Variable resistors are durable and can withstand frequent adjusting. Usually they are manufactured in a circular form, which may either be placed in some type of enclosure or left open. The element may be either wire wound, composition, or have a film on some type of sturdy backing. A movable contact is placed in contact with the resistance wire on

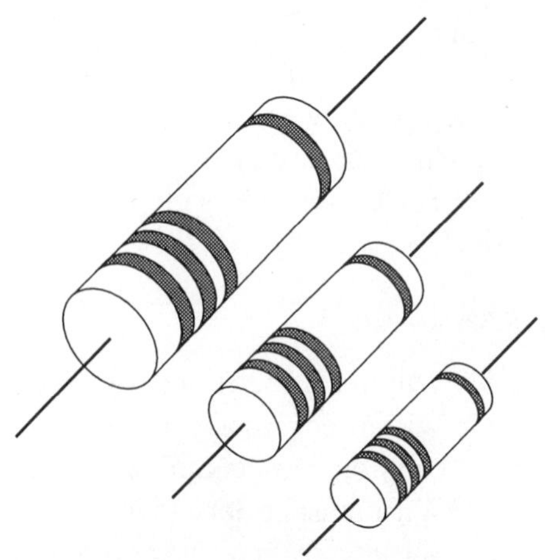

Figure 15–48 Resistors

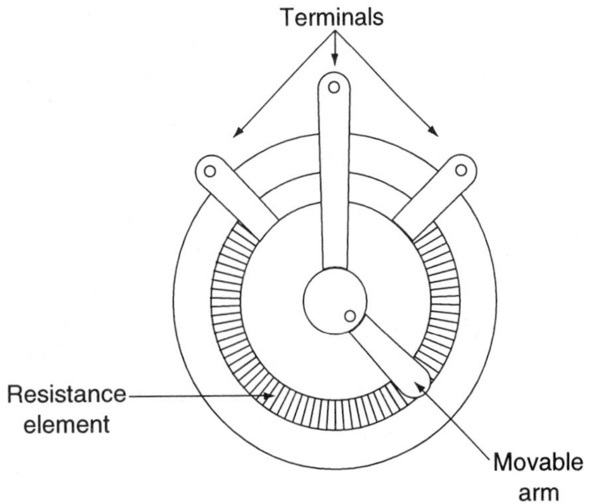

Figure 15–49 A variable resistor

one end and the circuit on the other. The movable contact can be moved along the wire to either increase or decrease the amount of resistance it introduces into the circuit. See Figure 15–49.

There are various means of moving the movable arm. It may be connected to the bellows of a pressure control, a thermostat, or just simply have a small handle that is used to make adjustments. Both ends of the resistance element and the movable contact are connected to external terminals on the housing. When all three terminals are wired into the circuit, the resistor is known as a potentiometer. When only the movable contact and one end of the resistance wire are connected to the outside, it is known as a rheostat.

SUMMARY 15–17

- Resistance is the opposition to electrical flow through a conductor.
- Modern day resistors are either made from powdered carbon or a high-resistance wire.
- The variable resistor can be used to adjust the voltage to the desired amount to a specific device for the operation desired.

REVIEW QUESTIONS 15–17

1. What would happen to a circuit connected to a power source without any resistance?
2. Which type of resistor is less stable for use with varying temperatures?
3. To what application is the variable type resistor suited?

15–18 MAGNETISM

The effects of magnetism are very important in the operation of refrigeration and air conditioning systems. Scientists discovered many years ago that magnetism had some of the same properties that electricity had. Later it was discovered that magnetic properties and electrical properties aided each other. In present day circuits, the two are so entwined that to separate them would be almost impossible. Magnetism is used in the production of electricity. When the electrons flow through a circuit, a magnetic field is generated outside the conductor and circuit components. Therefore, any knowledge of one only enhances the knowledge of the other.

Laws of Magnetism

A magnet will attract iron. If two bar magnets were hanging on a flexible wire or string so that they could rotate freely, it can be demonstrated how the poles of the magnets react to each other. When opposite poles come close to each other the bar magnets will be attracted to each other. See Figure 15–50.

Also, when like poles come close to each other there is a repulsive force exerted between them. This action proves the laws of magnetism that unlike poles attract and like poles repel each other.

A third law states that the forces between two magnetic fields will change when the distance between them is changed. When the magnetic fields are brought close together the field of force will be stronger, and when they are separated further the field of force will be weakened. This change in the field of force varies inversely with the square of the distance between the magnetic poles. As an example, when the distance between two magnetic poles is increased to two times the distance, the field of force will be reduced to one-fourth of the original strength.

Magnetic Fields

There is a magnetic field that exists around all magnets. This field is very much like an electrostatic field. A magnetic field is the area surrounding a magnetic pole and its force can be felt. This field of force has a definite direction and its strength can be measured.

If we place some iron filings on a piece of paper and place two magnets under the paper, the iron filings will align themselves with the lines of force within the field. See Figure 15–51.

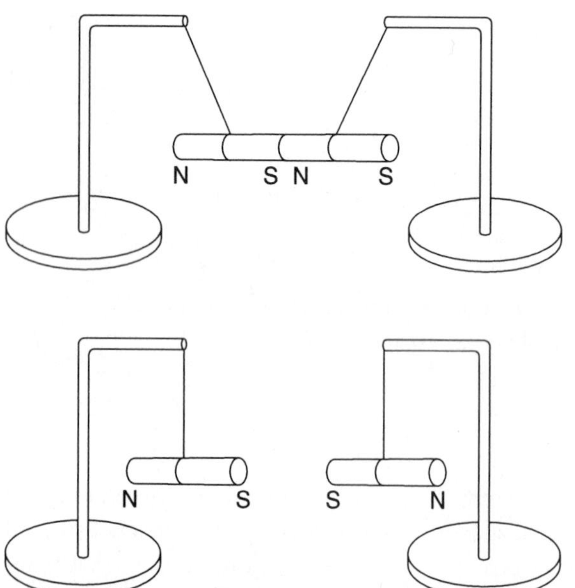

Figure 15–50 Laws of magnetism

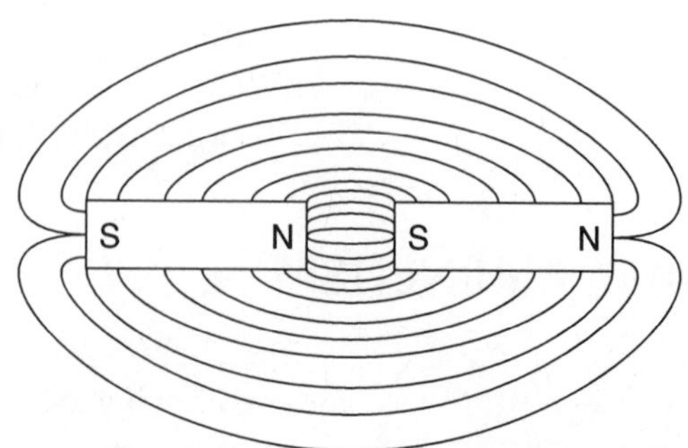

Figure 15–51 Magnetic lines of force between two attracting magnets

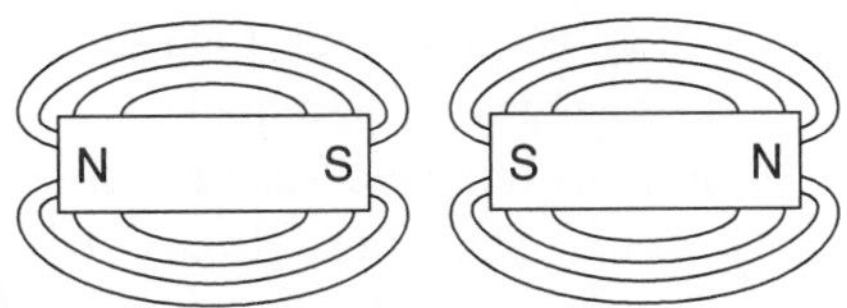

Figure 15–52 Magnetic lines of force between two repelling magnets

It should be noted that the strength of the field between the N and S poles of the magnets is very strong.

If the poles of the magnets were reversed so that like poles were close to each other, a lack of flux lines between the two magnets would be shown. See Figure 15–52.

It should be noted that the lines of force never cross each other regardless of the strength of the field or whether the field is attracting or opposing.

Magnetizing Materials

Basically there are only two ways to magnetize a piece of material: by stroking it with another magnet or using an electric current. The metal being magnetized must be ferromagnetic (ferrous means iron), otherwise it will not keep the magnetism.

When making a magnet by stroking a piece of ferromagnetic material, the stroking must always be in the same direction with the same pole of the magnet. This is so that the particles of the material can become aligned in the direction of the strokes. We know that metal particles already have magnetic polarities. However, the magnetic properties of the complete piece of metal are arranged in a random manner such that the entire piece is electrically neutral. See Figure 15–53.

When magnetizing a piece of metal with an electric current, an unmagnetized piece of the metal is placed within a coil of electric wire. The wire is then connected to a source of electrical energy, such as a battery. As the electric current flows through the wire, a magnetic field is built up around the wire that causes the atoms to align themselves, changing the piece of iron into a magnet. See Figure 15–54.

SUMMARY 15–18

- The effects of magnetism are very important in the operation of refrigeration and air conditioning systems.
- Magnetism is used in the production of electricity.
- Unlike magnetic poles attract and like magnetic poles repel each other.
- The forces between two magnetic fields will change when the distance between them is changed.

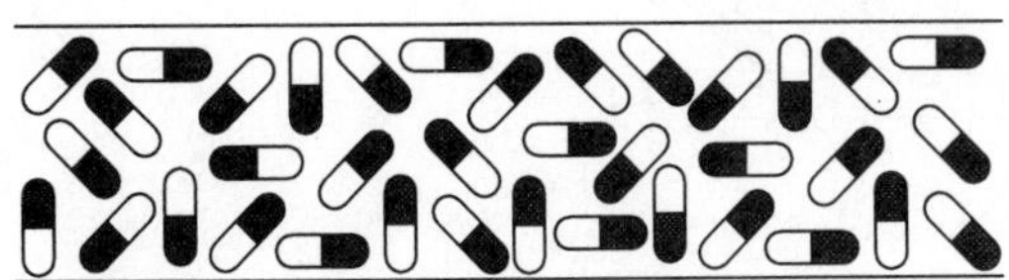

Figure 15–53 Non-magnetized metal

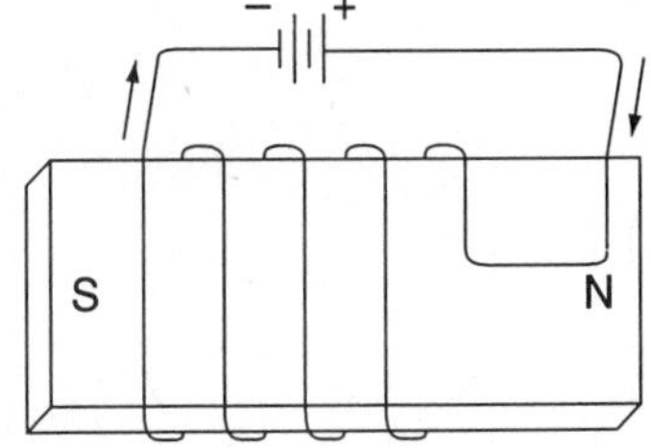

Figure 15–54 Using electricity to make a magnet

- There is a magnetic field that exists around all magnets. This field is very much like an electrostatic field.
- It should be noted that the strength of the field between the N and the S poles of the magnets is very strong.
- Basically there are two ways to magnetize a piece of material: by stroking it with another magnet or using an electric current.

REVIEW QUESTIONS 15–18

1. What do magnetic and electrical properties do to each other?
2. When two magnets are free to move, what happens when the N poles of both magnets are brought close together?
3. When two magnets are moved apart, what happens to the force between them?
4. What happens when the magnetic lines of force cross?

15–19 ELECTROMAGNETS

Because electrons spin in their orbit, they create their own magnetic field. There is no magnetic field around static electrons because their opposite spins tend to cancel out the magnetic field of the others. Therefore, static electricity produces no magnetic field.

When electrons are flowing through a wire and cannot pair off with those having opposite spins, they tend to produce a magnetic field around that wire. When they are flowing in the same direction, their magnetic fields tend to add to each other and increase the strength of the field.

When a wire carrying current is coiled into a loop, the direction of the magnetic lines of force are in the same direction through the center of the loop. See Figure 15–55.

If several loops of wire were wound around a magnetic core, the field will include the complete coil. See Figure 15–56.

The lines of force leave the right-hand end of the core, go around the outside of the core, and reenter on the left-hand end.

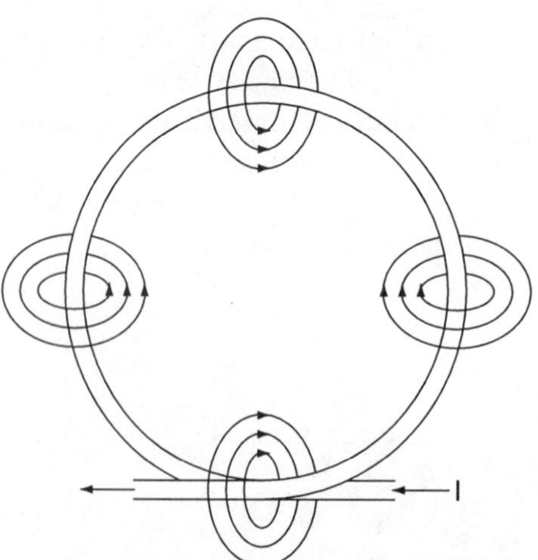

Figure 15–55 Electromagnetic field inside a coil of wire

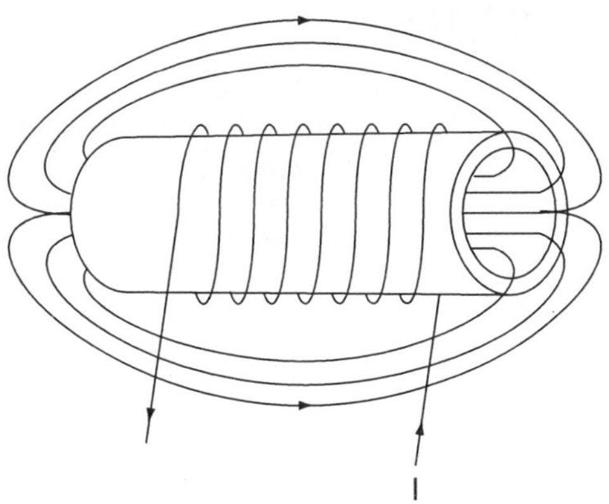

Figure 15–56 Electromagnetic field around a coil

The Solenoid

When a current-carrying conductor is wound into a larger coil, or solenoid, the individual electromagnetic lines of force surrounding the conductor tend to join together and increase the strength of the field. The coil will produce a magnet having the N pole and the S pole at opposite ends. See Figure 15–57.

Notice that the magnetic lines of force leave the N pole and enter the S pole of all magnets.

The strength of the electromagnetic field depends on two things: (1) the number of turns of wire in the coil, and (2) the current in amperes flowing through the wire. When the number of turns are multiplied by the current, the strength of the magnetic field can be calculated. This is the way that solenoid coils for use in electrical circuits are designed.

SUMMARY 15–19

- Because electrons spin in their orbit, they create their own field.
- When electrons are flowing through a wire and cannot pair off with those having opposite spins, they tend to produce a magnetic field around that wire.
- When a wire carrying current is coiled into a loop the direction of the magnetic lines of force are in the same direction through the center of the loop.
- The strength of the electromagnetic field depends on two things: (1) the number of turns of wire in the coil, and (2) the current in amperes flowing through the wire.

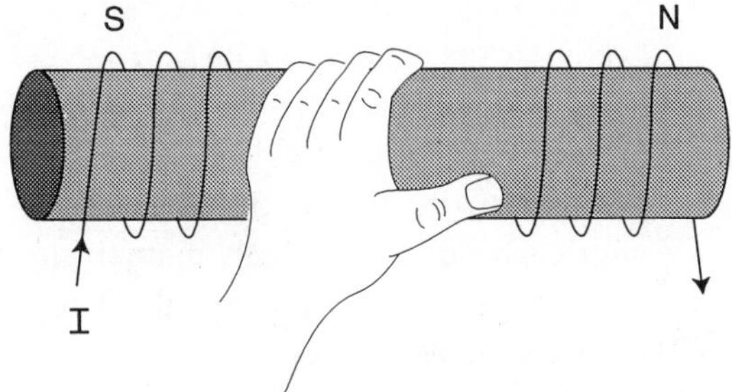

Figure 15–57 Magnetic polarity of a solenoid

REVIEW QUESTIONS 15–19

1. Why does static electricity not produce a magnetic field?
2. In what direction are the lines of magnetic force through the center of a coil?
3. Name the two things that determine the strength of a solenoid coil.

15–20 INDUCTORS

An inductor is defined as any coil of wire that has an electric current flowing through it. The major emphasis in electricity is the characteristics of the circuit that involve inductance, capacitance, resistance, and some combination of these characteristics.

Inductance

Inductance is defined as that property of an electric circuit that resists a change in the flow of current through the circuit. This resistance to the flow of current occurs because of the electrical energy that is stored in the magnetic field surrounding a coil. The strength of this field is measured in Henries. The letter L indicates inductance. Inductance in a coil is created by the counter EMF generated when current flows through the coil. All inductive voltages oppose the applied voltage. Because of this it is generally termed counterelectromotive force and is designated by (CEMF). Sometimes it is called back EMF.

The measurement Henrie is representative of the inductance of a coil when one volt of induced EMF is produced when the current changes at the rate of one ampere per second.

This can be more simply stated as: inductance is the counter-EMF that is induced back into a conductor every time a magnetic line of force cuts through that conductor. The time lag caused by the delay in cutting through the conductor causes the EMF to be 180° out of phase with the applied voltage. Thus, it is exactly opposite to it.

Self-Induction. The expanding and collapsing of the magnetic field as electric current flows through a conductor cuts across the conductor, creating a CEMF that is in direct opposition to the applied voltage and opposes a change in the current flow. This is known as self-induction. Self-induction is increased by adding to the number of turns in the coil or the wire, by the relationship between the length of the coil to its diameter, and the permeability of the core material.

Mutual Induction. Mutual induction generally occurs when two coils are in close proximity to each other. However, there is also mutual-induction between two conductors. This is a weak induction and is not usually usable. Mutual induction is defined as the ability of one electric circuit to induce a voltage into another when the two inductors are electrically separated. See Figure 15–58.

Mutual inductance is the transfer of electrical energy from one circuit to another by use of a magnetic field.

When two coils are placed close enough so that their magnetic fields will link with each other, they have mutual inductance. When placed in this position, the magnetic fields will cut across the wires of the other coil causing the induced voltage in that coil. When they are placed at some distance from each other, there will be very little linkage between the two coils.

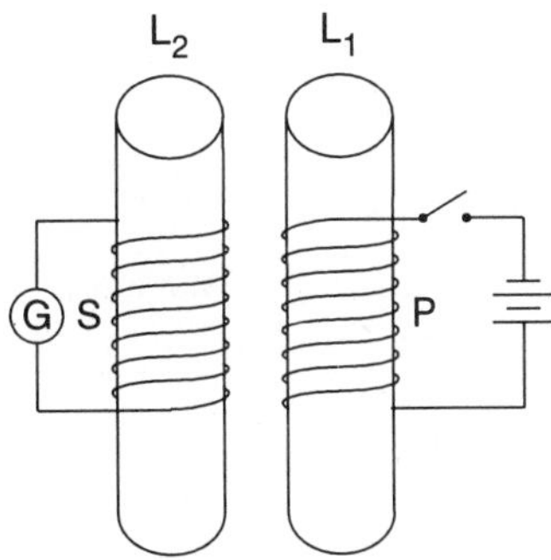

Figure 15–58 Mutual induction principle

When a common core is placed between the two coils, the mutual inductance between the two will be greatly increased. The amount of flux lines that interact with the other coil is called coupling. Unity coupling is reached when all the lines of force cut across all the wires in the other coil. Unity coupling is the greatest coupling possible.

Mutual induction must be given consideration when two inductors are placed into a single circuit.

The Transformer

The operation of a transformer is dependent on mutual induction between the two coils. Transformers are used to furnish power in the control circuits in air conditioning and refrigeration systems. Transformers may be defined as a device used to transfer electrical energy from one circuit into another by electromagnetic induction.

Transformers are made with two or more coils wound around a laminated iron core. When manufactured in this manner, unit coupling can almost be reached. Transformers have no moving parts and require very little maintenance. Basically, they are simple, rugged, and very efficient in operation.

The coil that is connected to the line voltage is known as the primary winding. The other, secondary winding is connected to the control circuit. See Figure 15–59.

The electrical energy in the secondary winding is caused by mutual induction between the two coils. The voltage supplied to the primary winding must be varying in order to have an expanding and collapsing field force to cut across the wires in the other winding. Because of this, transformers must have alternating current or pulsating DC current to operate. We are concerned mostly with step-down transformers that produce 24-V or low voltage for the control circuits used in air conditioning systems.

SUMMARY 15–20

- An inductor is defined as any coil of wire that has an electric current flowing through it.
- Inductance is defined as that property of an electric circuit that resists a change in the flow of current through the circuit.
- The expanding and collapsing of the magnetic field as electric current flows through a conductor cuts across the conductor, creating a CEMF that is in direct opposition to the applied voltage, and opposes a change in the current flow.

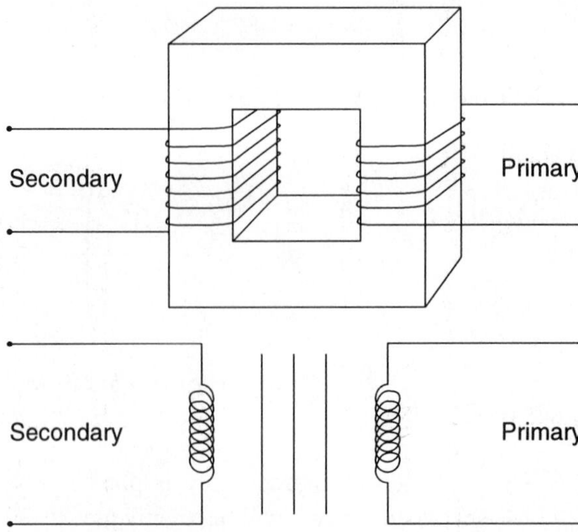

Figure 15–59 Construction of a transformer and its symbol

- Mutual induction is defined as the ability of one electric circuit to induce a voltage into another when the two inductors are electrically separated.
- When a common core is placed between the two coils, the mutual induction between the two will be greatly increased.
- The operation of a transformer is dependent on mutual induction between the two coils.
- Transformers are made with two or more coils wired around a laminated iron core.

REVIEW QUESTIONS 15–20

1. How is inductive resistance measured?
2. What is inductance?
3. What causes inductive CEMF?
4. What is mutual induction?
5. By what principle do transformers operate?

15–21 CAPACITANCE

Capacitors are used in air conditioning and refrigeration to give motors the extra torque required for starting heavy loads. These are relatively simple devices. Capacitance may be defined as that property of an electric circuit that opposes any change in voltage. Inductance is that property of a circuit that opposes any change in current.

The capacity of a capacitor is determined by the number of electrons that it can hold on its plates for each volt of applied electricity. Capacitance is measured in farads. A farad is a charge of one coulomb that raises the electrical potential one volt. Capacitance is calculated by the following formula:

$$C = \frac{Q}{E}$$

Where: C = Farads
Q = Coulombs
E = Volts

The ones used in air conditioning and refrigeration systems are measured in microfarads. Microfarads are represented by the Greek letter "μ," which means micro. The numeric value of the capacitor is then listed along with the μ.

The capacitance of a capacitor is determined by the following factors:

1. The surface area between the capacitor plates
2. The distance between the plates
3. The type of insulation (dielectric) between the plates

As the surface area of the plates increases, the capacity of the capacitor increases. When the dielectric (insulation) is increased, the capacity of the capacitor is also increased. As the distance between the plates is increased, the capacity of the capacitor is decreased.

Capacitors

A capacitor can be described as two plates of a conductive material separated by an insulator, or dielectric. See Figure 15–60.

It can be seen from the figure that a capacitor is a very simple device. It is merely two plates that are separated by an air insulator. If DC current were applied to this circuit it would appear to be an open circuit because of the separation between the plates. See Figure 15–61.

If the amperage was checked when the circuit was first energized, there would be only a momentary flow of current. The electrons flow from negative terminals on the DC power source to one of the capacitor plates. These electrons will repel the electrons that have collected on the other plate and they will be attracted to the positive terminal of the power source. The capacitor and the power source now have equal electrical potential. But these two charges have opposite polarities. When a capacitor is removed from a circuit it will maintain the charge on its plates because the charges are stored in an electric field.

It should be noted that in the DC circuit no electrons flowed through the circuit. From this we can say that a capacitor will block DC voltage. When the DC current was applied to the capacitor, one plate became positively charged and the other one became negatively charged. This separation caused a strong electric field between the two plates. The dielectric (insulating) materials used in the manufacture of capacitors will have a varying capacity to support an electric field. This is known as the dielectric constant of the insulator.

Capacitors in Series. Wiring the capacitors into the circuit in series has the same effect as separating the capacitor plates of a single capacitor more. The total capacitance in the circuit then becomes less than the capacitance of the smallest capacitor

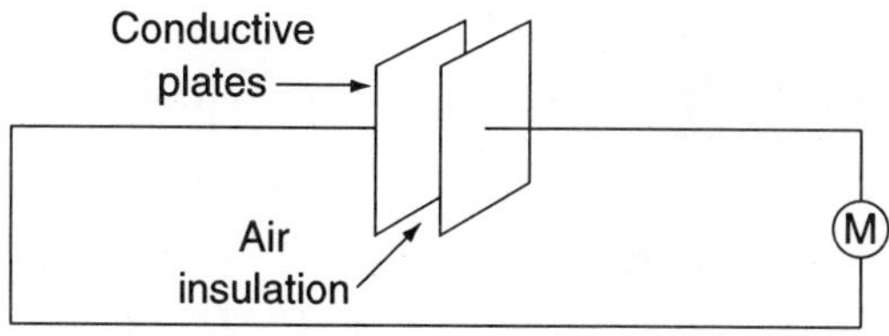

Figure 15–60 A simple capacitor

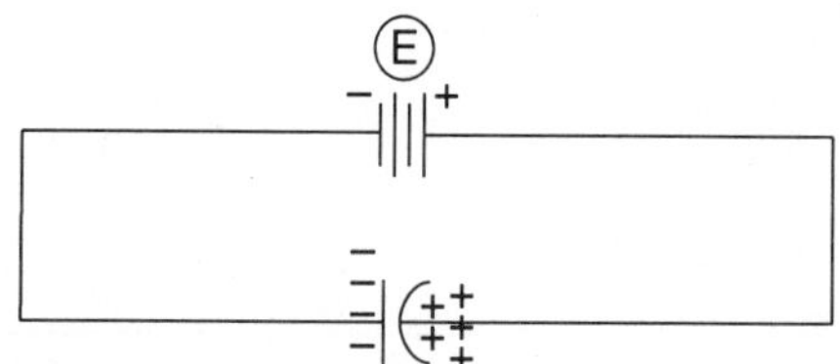

Figure 15–61 Operation of a simple capacitor

in the circuit. The capacity of capacitors in series can be calculated by using the following formulas:

For two capacitors of equal capacitance (see Figure 15–62) use:

$$C_t = \frac{(C_1 \times C_2)}{(C_1 + C_2)}$$

For capacitors that are unequal in capacitance (see Figure 15–63) use:

$$Ct = \frac{1}{\frac{1}{C_1} + \frac{1}{C_2} + \frac{1}{C_3} + \ldots}$$

Capacitors in Parallel. Connecting capacitors in parallel has the same effect as enlarging the plate area of one capacitor. Remember that the total plate area of each capacitor is exposed to the applied voltage. As a result, when capacitors are connected in parallel, the total capacitance value is increased. Thus, when capacitors are connected in parallel, their capacitance can be calculated by the following formula:

$$C_t = C_1 + C_2 + C_3 \ldots$$

To calculate the total capacitance of parallel capacitors, see Figure 16–64.

Capacitance in AC Circuits

When capacitors are subjected to an alternating current, they have a reactive quality. This is called capacitive reactance, which is an opposition to a change in the current flow in the circuit. Capacitive reactance is indicated by the symbol X_c. X_c is measured in ohms and is therefore a resistance in the circuit. The amount of capacitive reactance that a capacitor has is dependent on the capacitance and the phase angle of the applied voltage. Capacitive reactance is calculated by using the formula:

$$X_c = \frac{1}{(2 \pi f C)}$$

Where:
X_c = Capacitive reactance
2 = A constant
f = Voltage source frequency in hertz (cycles per second)
C = Capacitance of the capacitor

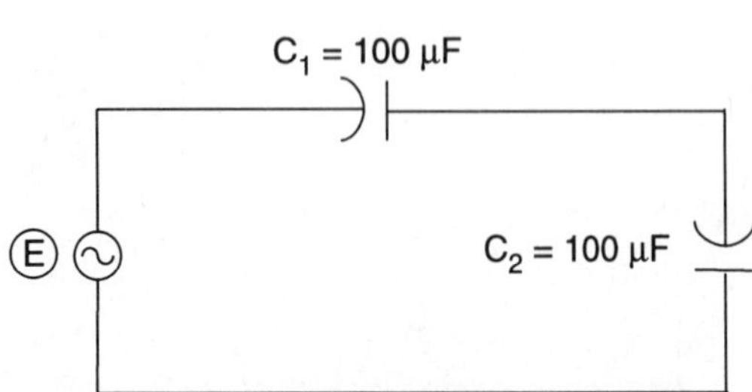

Figure 15–62 Calculating capacity of series capacitors

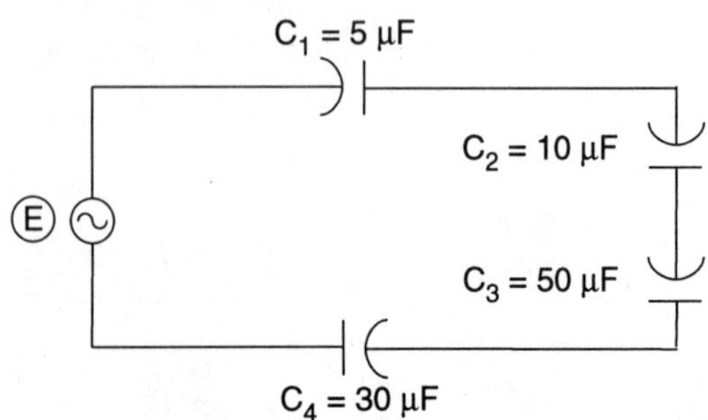

Figure 15–63 Calculating capacity of series capacitors of unequal value

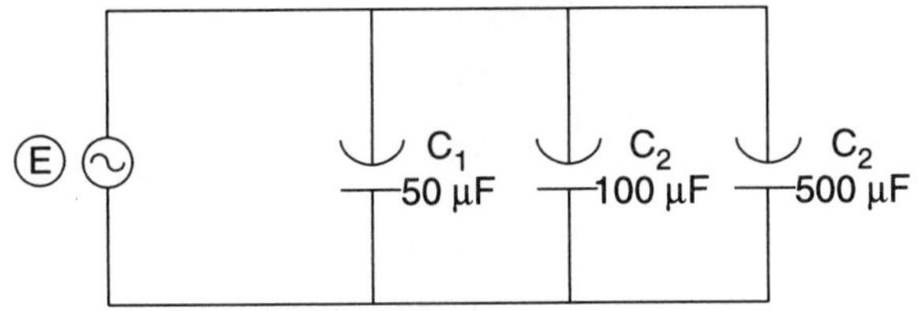

Figure 15–64 Calculating the capacity of capacitors in parallel

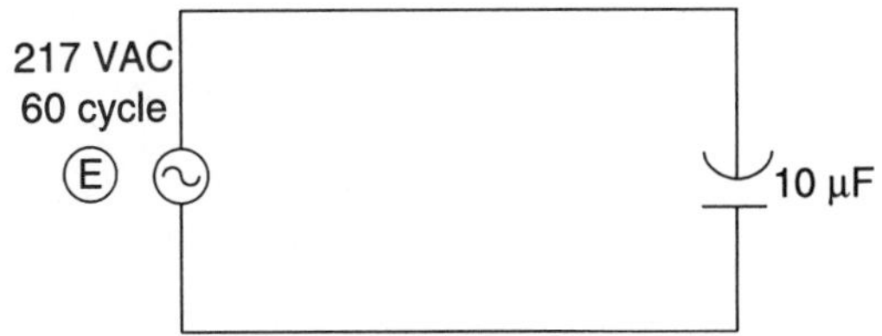

Figure 15–65 Calculating capacitive reactance

For an example refer to Figure 15–65. What is the capacitive reactance of a 15 µF capacitor operating in a circuit having a frequency of 60 hertz?

$$X_c = \frac{1}{(2\,\pi\, f\, C)}$$

In capacitive reactance circuits, the applied voltage and current are out of phase by 90° with the current leading the voltage. See Figure 15–66.

When the capacitive reactance has been determined, it can be treated just as any other resistance in the circuit.

When current is first applied to a capacitor, there is no opposition to current flow. It will immediately reach the peak flow for that circuit. Also, the electrons immediately begin to collect on the plates of the capacitor. One plate will become positive and the other one will become negatively charged. The polarity of the charge on the capacitor plates is exactly opposite to that of the applied voltage. As the charge increases on the plates, there is an opposition to current flow, which is also increased. With an increase in this opposition, the current flow decreases. The maximum flow of current reaches a zero value at exactly the same instant that the applied voltage reaches the peak value.

Reactive Power

When reactive power is considered it can be compared to an inductive circuit in that the stored power is returned to the circuit. This power is stored in the magnetic field between the capacitor plates. When the capacitor is discharged, the stored energy is returned to the circuit. Thus, the electrical energy is used only temporarily by the reactive circuit. This is termed wattless power. A power-wave form indicates that equal amounts of positive and negative power are used by the circuit. This results in zero power consumption by the circuit. Thus, the true power is zero. However, the apparent power is equal to the product of the effective voltage and the effective amperage.

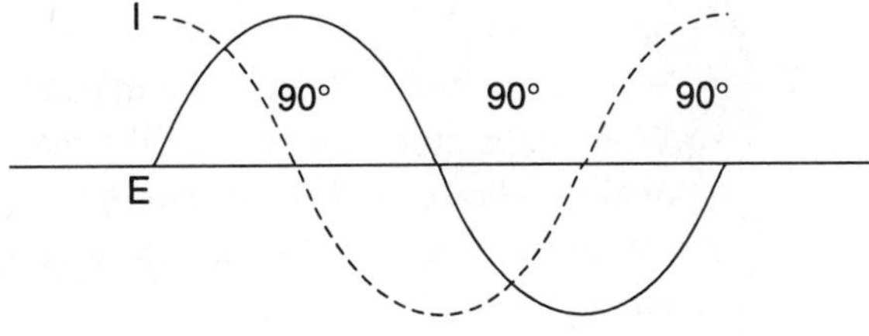

Figure 15–66 Phase relationship of current and voltage in a capacitive circuit

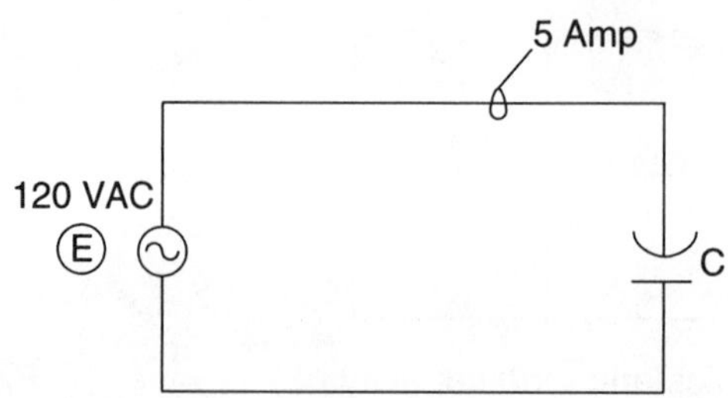

Figure 15–67 Calculating reactive power in a capacitive circuit

Example: If 120 Vac was applied to a capacitive circuit and there is a current flow of 5 amperes, what will the apparent power be? (120 × 5 = 600 volt-amperes.) See Figure 15–67.

Power factor is the relationship existing between the apparent power and the applied power in an AC circuit. To calculate the power factor, use the following formula:

$$PF = \frac{\text{True power}}{\text{Apparent power}}$$

In our previous example the power factor would be:

$$PF = \frac{0}{\text{600-volt-amps}} = 0$$

SUMMARY 15–21

- Capacitance may be defined as that property of an electric circuit that opposes any change in voltage.
- The capacity of a capacitor is determined by the number of electrons that it can hold on its plates for each volt of applied electricity.
- Capacitance is measured in farads.
- A capacitor can be described as two plates of a conductive material separated by an insulator, or dielectric.
- DC current will be blocked by a capacitor.
- Wiring capacitors into the circuit in series has the same effect as separating the capacitor plates of one capacitor more.
- The total capacitance when capacitors are connected in series becomes less than the capacitance of the smallest capacitor.
- Connecting capacitors in parallel has the same effect as enlarging the plate area of one capacitor.
- When capacitors are connected in parallel, the total capacitance value is increased.
- When capacitors are subjected to an alternating current, they have a reactive quality called capacitive reactance.
- Capacitive reactance (X_c) is measured in ohms and is therefore a resistance in the circuit.
- In capacitive reactance circuits, the applied voltage and current are out of phase by 90°, with the current leading the voltage.

REVIEW QUESTIONS 15–21

1. Does capacitance oppose a change in voltage or current?
2. Define farad.
3. Will capacitors function in a DC circuit?
4. When connecting capacitors in series, is the total capacity increased or decreased?
5. When connecting capacitors in parallel, is the capacitance increased or decreased?
6. What does capacitance in a circuit oppose?
7. Does capacitive reactance add resistance to a circuit?
8. In a circuit having capacitive reactance, how do the applied voltage and current relate to each other?
9. In a capacitive circuit, when does the maximum current flow reach zero?
10. How does a capacitor store electrical power?

CHAPTER

16

ELECTRIC MOTORS AND CONTROLS

OBJECTIVES

Upon completion of this chapter, you should be able to:

- Better understand the theory of electric motor operation
- Better understand how capacitors function
- Know more about the use of capacitors
- Know more about split-phase motor operation
- Know more about capacitor-start motors
- Know more about capacitor-start/capacitor-run motors
- Know more about shaded-pole motors
- Better understand the theory of two-speed motor operation
- Know how single-phase motor protectors operate
- Know how to test single-phase motors
- Know how to determine compressor terminal use
- Know more about single-phase motor starting devices

INTRODUCTION

In the study of electric motors and the controls that cause them to function as we desire, the laws of magnetism and magnetic induction are very important. It is magnetism that causes electric motors to operate, as well as many of the controls that are used in refrigeration and air conditioning systems. Because of this it is necessary that these theories be learned and understood.

16–1 ELECTRIC MOTOR THEORY

The basic components of an electric motor are the rotor, a permanent magnet mounted on a movable shaft, and two magnetic poles that are mounted on the outside motor shell. See Figure 16–1.

The magnetic poles are made of wire coils wound around a pole. Thus, they are stationary poles and are known as the stator. The electric current flowing through these coils produces a strong magnetic field in the area surrounding them.

The strength of the starting torque built into the motor starting circuit represents the major difference in the different single-phase motors being used in air conditioning and refrigeration systems. It is the method used to develop this torque that makes the difference between the motors.

The laws of magnetism say that like magnetic poles repel and unlike magnetic poles attract. In the example in Figure 16–1 the pole marked N (north) is producing a magnetic field that attracts the other pole marked S (south) on the rotor (moving part). Since the stator is stationary and cannot move, when the magnetic field becomes strong enough, the rotor (movable part) will start to move toward the stator.

In the United States, alternating current is used almost extensively. As the current alternates between positive and negative, the direction of current flow is alternately changed. It changes from a high negative to a high positive valve during every cycle. This change occurs 60 times every second, on 60-cycle electric systems. During these alternations, the polarity of each stator winding is reversed from a positive value to a negative value.

These alternations cause a push-pull action between the rotor and the stator. This rapid change in the electricity keeps the rotor turning and prevents it from lining up with unlike poles and stopping the motor. The momentum of the rotor carries it past the in-line position where it could possibly stall.

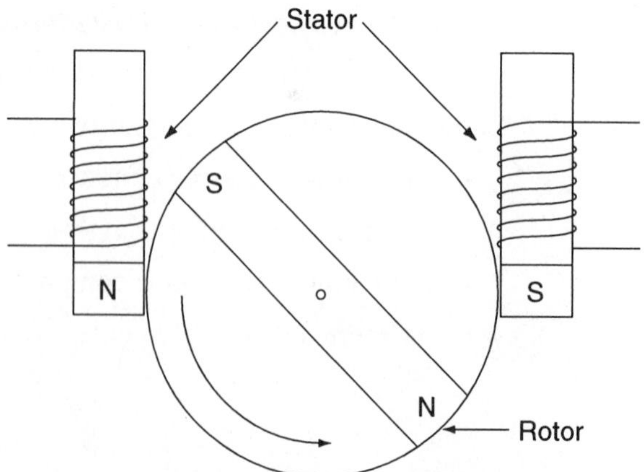

Figure 16–1 Basic components of an electric motor

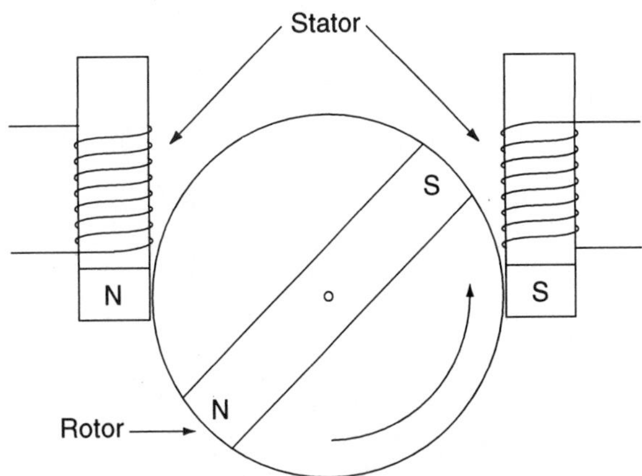

Figure 16–2 Magnetic repulsion and attraction of the rotor

When the electrical cycle changes, reversing the polarity of the stator poles, a magnetic field of the opposite polarity is created around each of the stator poles. This opposite magnetic field causes the magnetic field of the stator and that of the rotor to repel each other, causing a pushing force that pushes the rotor further through the rotation. See Figure 16–2.

The constant reversing of the magnetic fields and the rotation of the rotor keeps the motor turning and pulling the load.

During operation, the motor will automatically adjust its speed to correspond to the 60-cycle electricity. This is a two-pole motor that will theoretically turn at 3,600 revolutions per minute (rpm). For a two-pole motor, this is the synchronous speed. The synchronous speed of a motor can be determined by mathematical calculation as follows:

$$\text{Synchronous Speed} = 120 \times \frac{\text{Frequency}}{\text{Number of Poles}}$$

Where: 120 = the number of times the magnetic field builds and collapses each second (60 × 2 = 120)

Frequency = frequency of the electrical supply

Number of Poles = number of poles in the motor

Example: The synchronous speed of a two-pole motor is:

$$\text{Synchronous Speed} = 120 \times \frac{\text{Frequency}}{\text{Number of Poles}}$$

$$= 120 \times \frac{60}{2}$$

$$= 3{,}600 \text{ rpm}$$

Example: The synchronous speed of a four-pole motor is:

$$\text{Synchronous Speed} = 120 \times \frac{\text{Frequency}}{\text{Number of Poles}}$$

$$= 120 \times \frac{60}{4}$$

$$= 1{,}800 \text{ rpm}$$

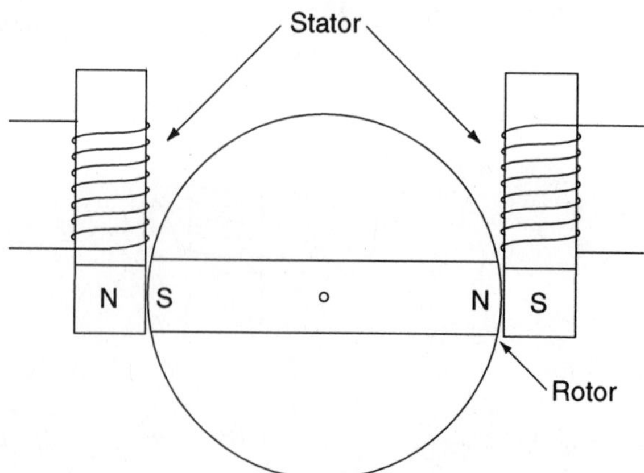

Figure 16–3 Rotor stopped with poles in line with stator poles

It can be seen from this explanation that as long as the electricity keeps alternating between positive and negative, the motor will continue to turn. The problem, however, lies not in running the motor but in starting it.

Should the motor stop with the stator poles in line with the rotor poles, the motor would probably not start again. See Figure 16–3.

Should the motor stop with the south pole next to the north pole of the stator, the motor would probably just simply hum and not turn because of the attracting magnetic fields. When the polarity of the poles changes, the motor would probably still not turn because the opposing magnetic forces would be at right angles to each other and cause the rotor to stall.

If, however, another pole was installed in the stator, the motor would start from any position. See Figure 16–4.

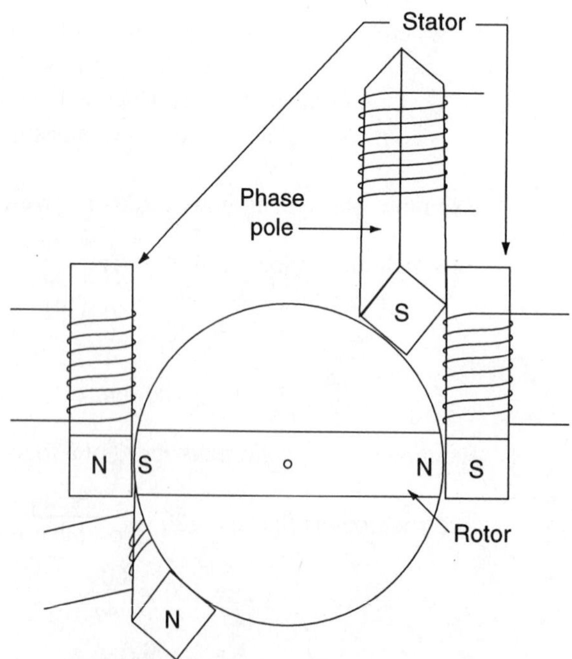

Figure 16–4 Stator with a phase pole

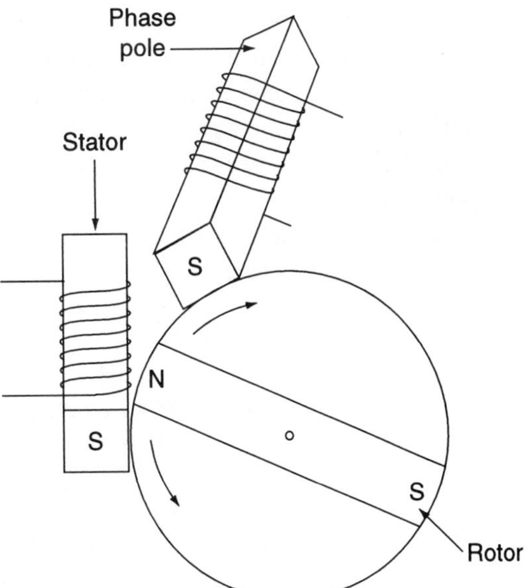

Figure 16–5 Magnetic field equal between two stator poles

The phase (start) pole, which has the same magnetic polarity as the main stator pole, would attract the rotor pole, causing it to move toward it, starting to turn the rotor.

Even with this additional pole, there could possibly still be a problem in starting the motor. Suppose that the rotor stopped with the pole in the mid-position between the two stator poles. The motor would probably still not start because the two magnetic fields would produce the same force strength to prevent the motor from starting. See Figure 16–5.

A possible solution would be to have one stator pole strong when the other one is weaker. In this motor circuit, the stator windings would be connected to the same source of electrical power. See Figure 16–6.

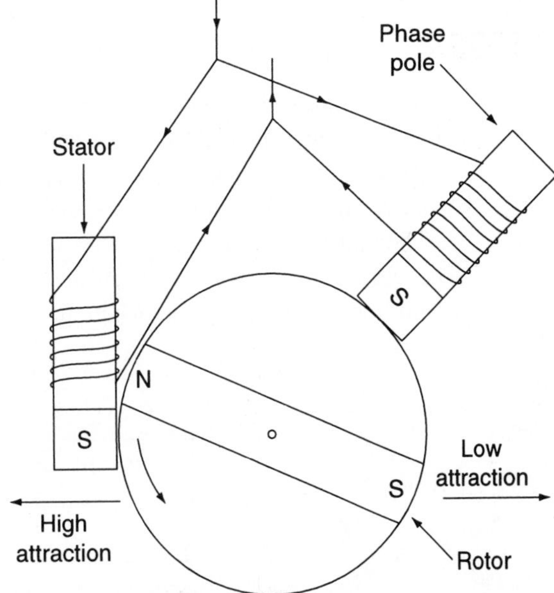

Figure 16–6 Low and high force magnetic fields attracting the rotor poles

In this figure, electricity enters the motor at point A. At this point the electricity is divided and part of it flows to the pole on the left and the remainder flows to the pole on the right. If some means were used to cause the electricity to flow to the left before it flows to the right, the left stator pole will become stronger before the right pole. The left pole, because it is strong first, would attract the rotor pole and start the motor turning.

The most common method of causing this current division is the capacitor. The capacitor actually causes a two-phase current to flow inside the motor windings.

SUMMARY 16–1

- The basic components of an electric motor are the rotor, a permanent magnet mounted on a movable shaft, and two magnetic poles mounted on the outside motor shell.
- The strength of the starting torque built into the motor starting circuit is the major difference in the different single-phase motors used in air conditioning and refrigeration systems.
- The laws of magnetism say that like magnetic poles repel and unlike magnetic poles attract.
- When the electrical cycle changes, reversing the polarity of the stator poles, a magnetic field of the opposite polarity is created around each of the stator poles.
- During operation, the motor will automatically adjust its speed to correspond to the frequency of the power source.

REVIEW QUESTIONS 16–1

1. Name the basic components of an electric motor.
2. What is the major difference between single-phase motors?
3. What are the laws of magnetism?
4. What is the synchronous speed of a four-pole motor?
5. What causes the current division in an electric motor?

16–2 CAPACITORS

To understand how a capacitor produces a second phase of electricity, a review of some basic facts about AC voltage and current is necessary.

As an illustration, see Figure 16–7, when the voltage is applied to points A and B and a wave form is produced, as shown in the right of the figure.

The current flow through a pure resistive circuit will be in phase with the voltage. The illustration shows that when the current (dashed line) is at a peak, the voltage (solid line) is also at a peak. When the current is at zero, the voltage is also at zero.

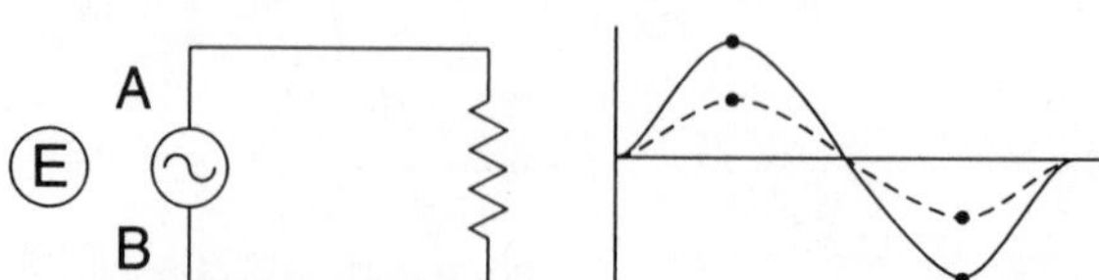

Figure 16–7 Current and voltage through a pure resistive circuit

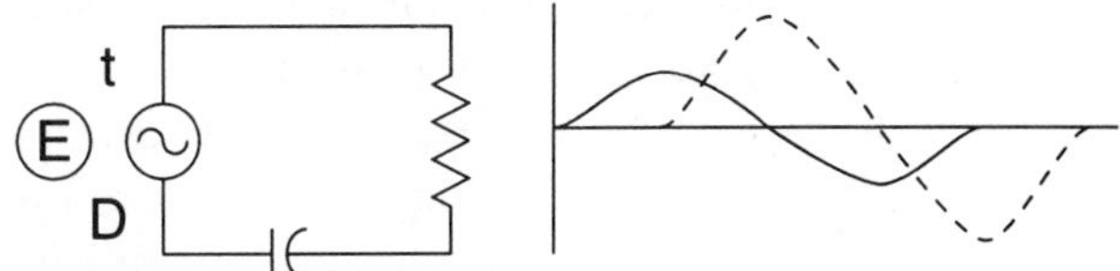

Figure 16–8 Current and voltage through a capacitive circuit

When a capacitor is installed in series with the circuit resistance, the current flow through the load will lead the applied voltage. See Figure 16–8.

The voltage is applied to points C and D and produces a sine wave identical to that shown in Figure 16–7. The current portion of the wave form now leads the applied voltage.

The capacitor is a device that can be added to an electrical circuit to provide the two-phase power necessary to start an electric motor. It is necessary that the basic characteristics of a capacitor be known so that their full use can be realized.

The measurement used to rate capacitors is the microfarad (mfd). A capacitor having a high mfd rating will contain either large metal plates or a lesser amount of insulation between them. A small mfd rating is possible through the use of small metal plates or more insulation between them. See Figure 16–9.

When a capacitor having a high mfd rating is placed between the resistive load and the applied voltage in an electric circuit, a large phase shift will be produced. It is this phase shift that provides the torque necessary to start the rotor turning in single-phase motors. A small mfd capacitor will produce a smaller phase shift, resulting in a lower starting torque motor.

In addition to the torque generated by a capacitor, the amount of electric current flowing through a series circuit when capacitors with different mfd ratings are used is important. Capacitors having a large mfd rating will allow a high current to flow through the series load. Thus, capacitors having small mfd ratings will allow low current to flow through the circuit.

In refrigeration and air conditioning equipment there are two types of capacitors used. The size of the capacitor case is in no way related to the mfd rating of the

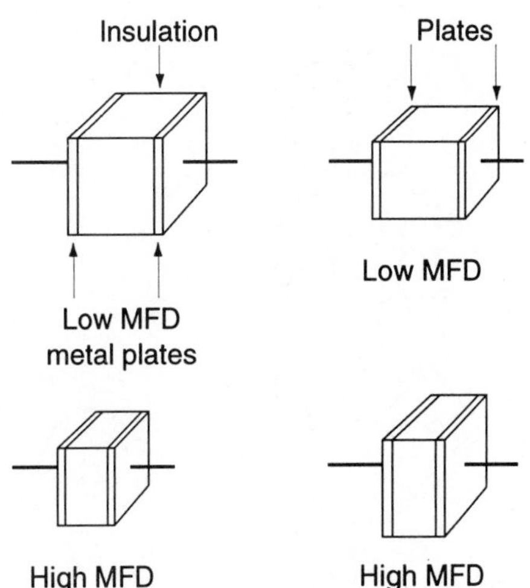

Figure 16–9 Effects of plate size and thickness of insulation on capacitor capacity

capacitor. Run capacitors are usually rated with a small mfd. However, the case is usually very large in physical size when compared to starting capacitors. This is because of the special insulation between the plates and the electrolyte in continuous use capacitors. The purpose of the extra insulation is to dissipate the heat generated by the current flowing through it during motor operation. Run capacitors are used for low starting torque applications. There is only a relatively small current flow through them when connected in series with the load. These capacitors are designed to remain in the circuit all the time the motor is running.

Starting capacitors usually have a high mfd rating, but they are smaller in physical size than running capacitors. These capacitors are not designed to pass high current flow for an extended period of time. If they remain in the circuit for more than a few seconds, permanent damage will likely be done to them. Thus, they are switched out of the circuit after the motor has reached approximately 75% of its running speed.

To see where a capacitor is installed in the electric circuit, we will consider the motor discussed earlier that had a phase (start) winding in the motor. See Figure 16–10.

Both of the pole windings are connected to the supply voltage. The capacitor is connected in series with the pole winding on the right. The rotor is stopped between the two poles, where it would be almost impossible to start without the capacitor. As stated before, when a capacitor is connected in series with the load, the current leads the voltage.

Now, the current flowing through the right stator pole winding leads the current flowing through the left stator pole winding. This is because of the action of the capacitor. The north stator pole on the right will become strong before the north stator pole on the left. Under these conditions, the starting torque is created and the south pole of the rotor will be deflected toward the right stator pole.

When the current reverses its polarity in the right stator pole winding, the polarity of the stator pole is reversed. See Figure 16–11.

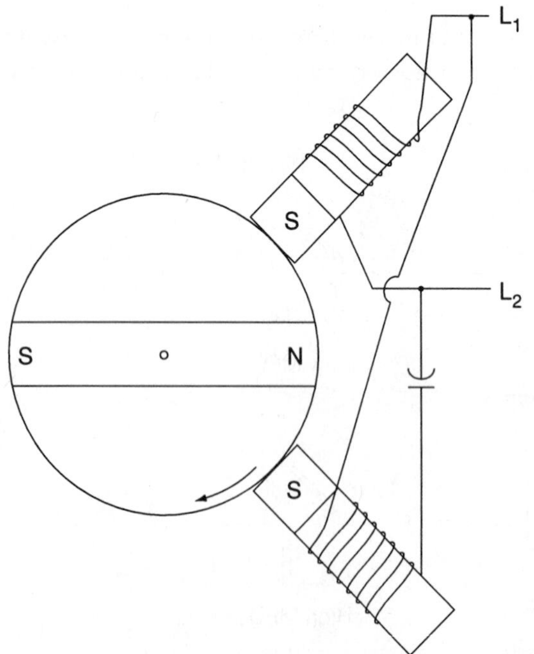

Figure 16–10 Capacitor location in motor circuit

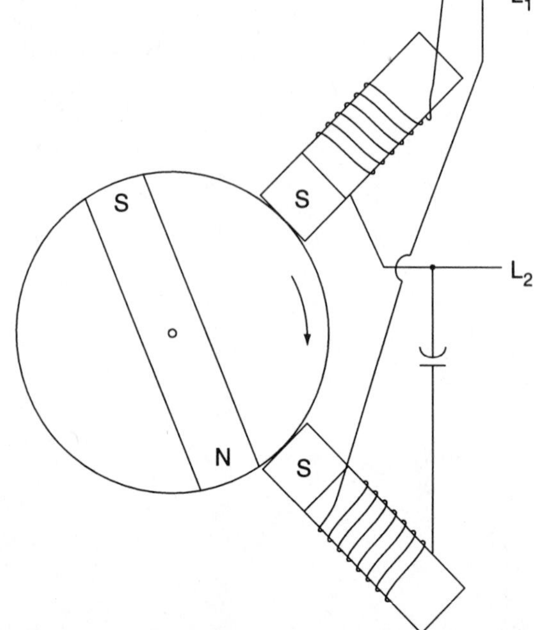

Figure 16–11 Magnetic field deflecting rotor

At this instant, the south stator pole is repelling the south pole of the rotor. The force of this repulsion adds turning force to the motor shaft. A higher efficiency motor is created when the capacitor remains in the circuit through the complete running cycle of the motor.

There are five types of electric motors used in air conditioning and refrigeration units. They are: (1) split-phase, (2) capacitor-start (CSR), (3) permanent-split capacitor (PSC), (4) capacitor-start/capacitor-run (CSCR), and (5) shaded-pole.

SUMMARY 16–2

- The capacitor is an electrical device that can be added to an electrical circuit to provide the two-phase power necessary for starting an electric motor.
- The measurement used to rate capacitors is the microfarad (mfd).
- Capacitors having a large mfd rating will allow a high current to flow through the series load.
- Starting capacitors are not designed to pass high current flow for an extended period of time.

REVIEW QUESTIONS 16–2

1. How does the current flow in relation to the voltage in a pure resistive circuit?
2. How is a capacitor connected into an electric circuit?
3. What will cause a large phase shift in an electric motor circuit?
4. Name the two types of capacitors used in refrigeration and air conditioning system motors.
5. At what point is a starting capacitor switched out of the circuit?
6. Of the capacitors used in refrigeration and air conditioning motors, which one is physically smaller?

16–3 SPLIT-PHASE MOTORS

Split-phase motors are popular for use when the required motor size falls between about 1/20 to 1/3 horsepower. They are popular on applications such as blowers, oil burner motors, and water pumps. Their popularity is because of their moderate starting torque. Moderate starting torque motors have just a little more starting torque than running torque.

These are single-phase induction motors that use a nonwound rotor. The stator windings are embedded into the insulated slots of a laminated steel core. In these motors, the stator windings are made up of two separate windings known as the main, or running, winding and the phase, or starting, winding. These two windings are wired in parallel in the motor circuit. See Figure 16–12.

The phase winding is placed physically in a magnetic position that is in a relationship with the main winding to produce the required two-phase electrical field. The two-phase electricity helps to start and bring the motor up to speed. When the motor has reached approximately 75% of its running speed, a centrifugal switch removes the starting winding from the circuit, allowing the motor to operate as a single-phase induction type motor. See Figure 16–13.

The phase (start) winding is made from wire much smaller than the main (run) winding. Also, there are more turns of wire in the start winding than the run winding. Due to these differences there are two factors that must be remembered: (1) the start winding has a higher resistance than the run winding, and (2) the start winding is not designed to carry high current draw for a long period of time.

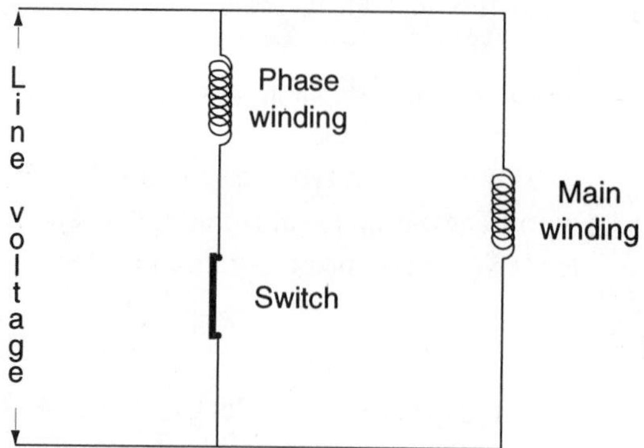

Figure 16–12 Split-phase motor connection diagram in the starting position

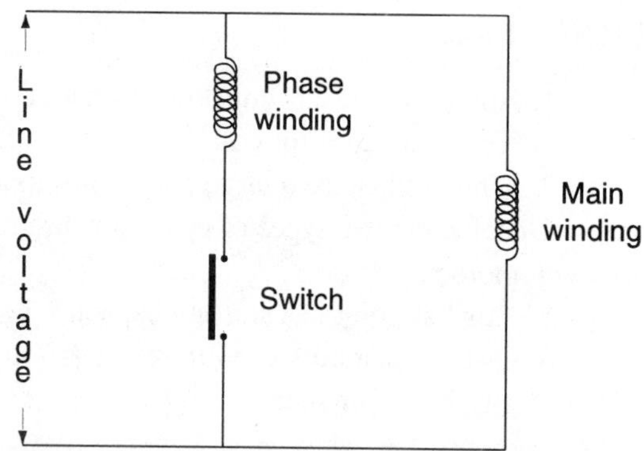

Figure 16–13 Split-phase motor diagram in the operating position

SUMMARY 16–3

- Split-phase motors are popular on applications such as blowers, oil burners, and water pumps.
- They have a moderate starting torque, which is just a little more starting torque than running torque.
- The run-and-start windings are wired in parallel in the motor circuit.
- When the motor has reached approximately 75% of its running speed, a centrifugal switch removes the starting winding from the circuit, allowing the motor to operate as a single-phase induction type motor.
- The phase winding is made from smaller wire than the run winding.

REVIEW QUESTIONS 16–3

1. Name the two windings used in single-phase motors.
2. Why is two-phase electricity used in single-phase motors?
3. Which winding in a single-phase motor is made from the largest wire?
4. Which winding cannot carry high current draw for an extended period of time?

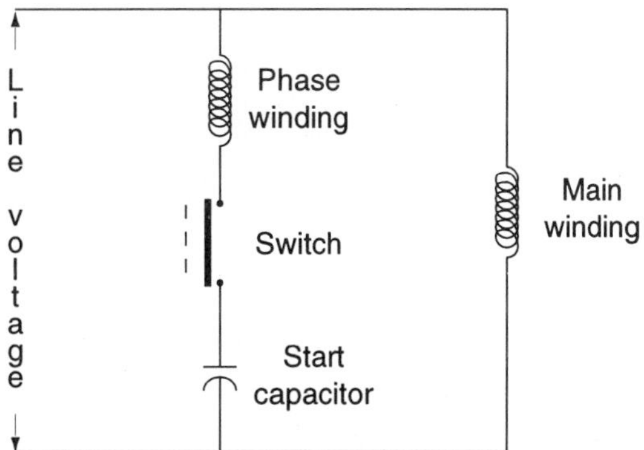

Figure 16–14 Capacitor-start motor connection diagram

16–4 CAPACITOR-START (CSR) MOTORS

Capacitor-start motors operate very similar to split-phase motors, the major exception being that a starting capacitor is connected into the motor starting circuit. See Figure 16–14.

The reason for the capacitor in the circuit is to increase the starting torque of the motor. After the motor has reached about 75% of its running speed, the capacitor is removed from the starting circuit by the starting switch.

When the capacitor is removed, the motor operates as a split-phase motor. It also has the same running torque characteristics of a split-phase motor of the same size.

These types of motors are popular in applications when the required starting torque is high but the required running torque is moderate. CSR motors are used as compressor motors on large fans and blowers and on water pumps.

SUMMARY 16–4

- The major difference between capacitor-start and split-phase motors is that a starting capacitor is connected into the motor starting circuit. The starting capacitor increases the starting torque of the motor.
- Capacitor-start motors are popular in applications requiring a high-starting torque and a moderate-running torque.

REVIEW QUESTIONS 16–4

1. What is connected into the starting circuit in a capacitor-start motor?
2. During the running mode, how does a capacitor-start motor operate?

16–5 PERMANENT-SPLIT CAPACITOR (PSC) MOTORS

Permanent-split capacitor (PSC) motors are very popular for use in refrigeration compressor motor applications. They are also used on fans and blowers. They have a medium starting torque and excellent running efficiencies and characteristics. When this type of motor is used in compressor applications, the refrigerant pressures should equalize during the OFF cycle. This is to allow the motor to start with as little load as possible.

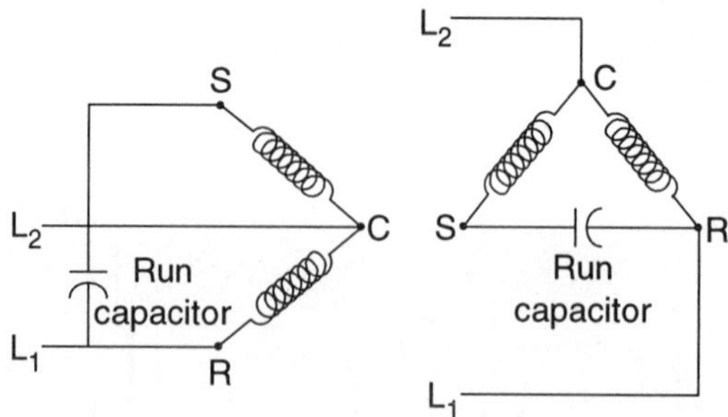

Figure 16–15 Permanent-split capacitor connection diagram

PSC motors include a running capacitor connected in electrical series with the phase (start) winding. See Figure 16–15.

There are two purposes for using the running capacitor in this application. One, the capacitor provides the split-phase electrical power necessary in starting the motor. Two, the capacitor is not removed from the circuit and provides split-phase power to the motor. This causes the motor to operate more efficiently.

Since the start winding is made from small wire, it cannot carry high current flow for long periods of time without burning out. Because of the resistance of the winding, the running capacitor must have a small mfd rating to limit the amount of current flowing through the start winding. Running capacitors having only a small mfd rating will produce a small phase angle in the winding. This small phase angle allows the motor to have only a small current flow, which will produce a moderate starting torque. Remember that the running capacitor is always connected between the start and run terminals of the motor.

SUMMARY 16–5

- Permanent-split capacitor (PSC) motors are very popular for use in refrigeration compressors.
- PSC motors have a medium starting torque and excellent running efficiencies and characteristics.
- PSC motors include a running capacitor connected in electrical series with the phase (start) winding.
- Because of the resistance of the phase winding, the running capacitor must have a small mfd rating to limit the amount of current flowing through the start winding.
- The run capacitor is connected between the start and the run terminals of the motor.

REVIEW QUESTIONS 16–5

1. What are the starting and running characteristics of PSC motors?
2. When a PSC motor-compressor is used, what is required before the compressor will start?
3. What do running capacitors cause in an electric motor?

16–6 CAPACITOR-START/CAPACITOR-RUN (CSCR) MOTORS

CSCR motors have the best characteristics of both the permanent-split capacitor (PSC) motor and the capacitor-start motor (CSR). Thus, the high starting torque of

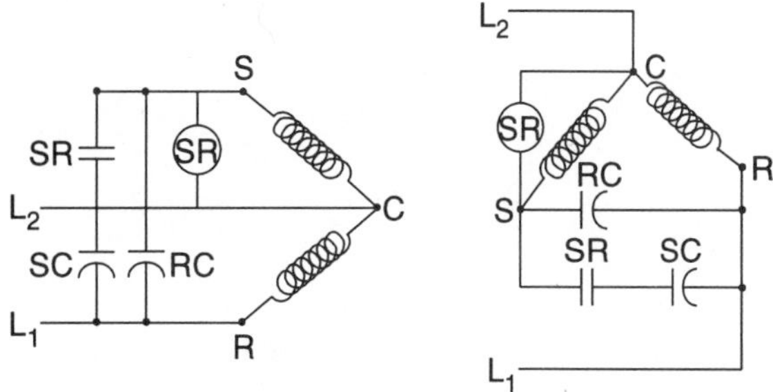

Figure 16–16 Capacitor-start capacitor-run wiring diagram

the CSR type motor and the higher running efficiency of the PSC type motor are combined into this type of motor.

The wiring connections are a combination of the two different types of motors. The running capacitor is connected the same as with the PSC motor. The CSR circuit uses a starting relay to remove the start capacitor from the circuit after the motor has reached about 75% of its rpm. The relay makes it possible to use this type of circuit on hermetic and semihermetic compressor motors. The centrifugal switch cannot be used in these types of installations because of the problems that may occur with it. If it was inside the compressor housing, it would be impossible to service without replacing the compressor. See Figure 16–16.

During a normal cycle, when the control circuit calls for the compressor to run, the compressor is started as both a capacitor-start and a permanent-split capacitor motor. When the motor has reached about 75% of its normal operating speed, the starting relay removes the start capacitor from the circuit. For the remainder of the cycle, the motor operates as a permanent-split capacitor (PSC) motor.

SUMMARY 16–6

- Capacitor-start/capacitor-run motors have a high starting torque and the higher running efficiency of the PSC motor.
- The running capacitor is connected in electrical series with the phase winding.
- There is a starting relay used on hermetic and semi-hermetic compressor motors to remove the starting capacitor when the motor has reached about 75% of its running speed.
- The CSR compressor starts as both a capacitor-start and a permanent-split capacitor motor.

REVIEW QUESTIONS 16–6

1. How does a CSR motor start?
2. Into what winding is the running capacitor connected?
3. On a hermetic motor-compressor, what takes the start capacitor from the circuit?

16–7 SHADED-POLE MOTORS

Shaded-pole motors operate with very low torque characteristics. They usually have a maximum size of about 1/2 horsepower. They are used mostly on light torque applications such as small fans, pumps, and timer motors. Shaded-pole motors are relatively inexpensive and have a great longevity.

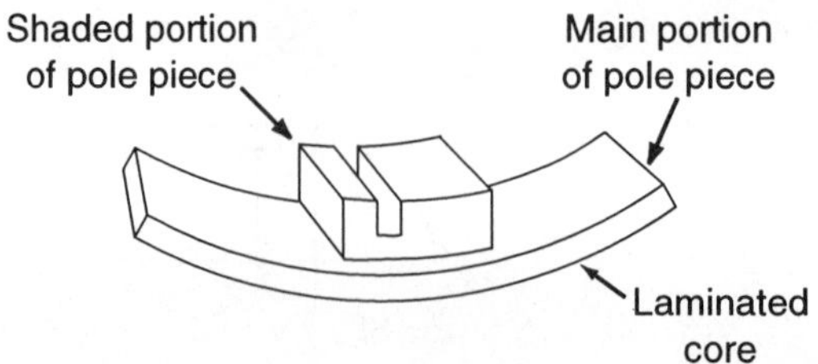

Figure 16–17 Shaded-pole motor pole piece

In this type of motor the pole pieces are quite different from those in other types of motors. Each pole is equipped with a slot cut into its face. See Figure 16–17.

The windings of a shaded-pole motor are distributed differently than the other types of motors. It is a type of induction motor. A shading coil is placed in the slot in the pole face. There are several different methods used to create the shading coil. Some manufacturers use a single piece of copper wrapped around the shading portion of the pole piece and others use stranded wire to make the shading coil. See Figure 16–18.

The shading coil forms a closed loop that is not connected to the motor main winding supply voltage. The remainder of the pole piece provides a place for the main windings to be installed.

In operation, when electricity is applied to the motor, the magnetic force around the wire is constantly changing direction, thus creating a magnetic force having constantly changing magnetic fields from N to S. The magnetic lines of force that are built up around the pole reach a maximum in one direction, then build up to a maximum in the other direction. The shading pole changes the speed of the building and collapsing of the magnetic fields.

This creates a situation where the run winding has a magnetic field that is building up and collapsing at a different rate than the shading portion of the pole piece. At some point in the cycle, one of the magnetic fields will momentarily get in front of the other. When this occurs, the magnetic strength is not the same in both windings and a torque is developed, causing the rotor to start turning. It should be noted that the motor will always turn toward the shading pole.

SUMMARY 16–7

- Shaded-pole motors operate with very low torque characteristics.
- In shaded-pole motors, the pole pieces are quite different from those in other types of motors. Each pole is equipped with a slot cut into its face.

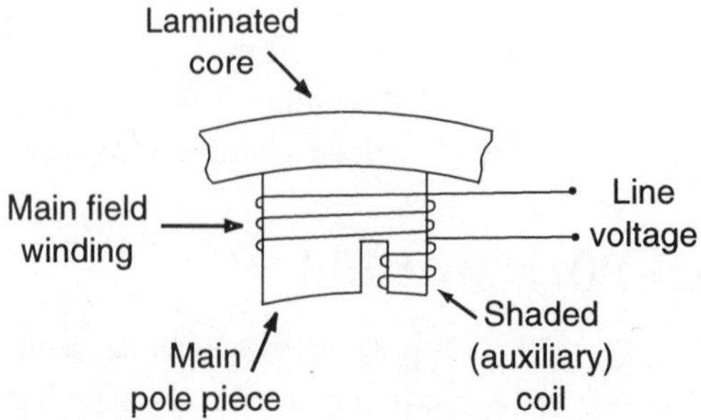

Figure 16–18 Shaded-pole motor windings

- The shading coil forms a closed loop that is not connected to the motor main winding supply voltage.
- The motor will always turn toward the shading pole.

REVIEW QUESTIONS 16–7

1. Where are shaded pole motors most popular?
2. What type of motors are shaded-pole motors?
3. In what direction will shaded-pole motors turn?

16–8 TWO-SPEED MOTORS

These types of motors are available in both the split-phase and capacitor-start types. There are three different windings used in two-speed motors. They are the phase-winding, the low-speed main winding, and the high-speed main winding. See Figure 16–19.

Both the high- and low-speed are connected through an external switch or relay that changes the motor in response to the control circuit. When the motor is operating in the high-speed mode, it functions just like a single-speed motor using only the phase winding and the high-speed winding. When they are operating in the low-speed mode, the phase winding and the high-speed windings are both used to start the motor. When the motor has started turning, the electric power is automatically switched to the low-speed main winding. See Figure 16–20.

Two-speed motors are used in applications where the fans operate at different speeds during different times in the cycle, or when changing from cooling to heating or heating to cooling. This speed change is needed so that the proper amount of air can be delivered to the conditioned space at the right time. They are also used when variable volumes of air are required that demand that the fan motor operate at different speeds.

SUMMARY 16–8

- Two-speed motors are available in both split-phase and capacitor-start types.
- There are three different windings used in two-speed motors: the phase-winding, the low-speed winding, and the high-speed winding.

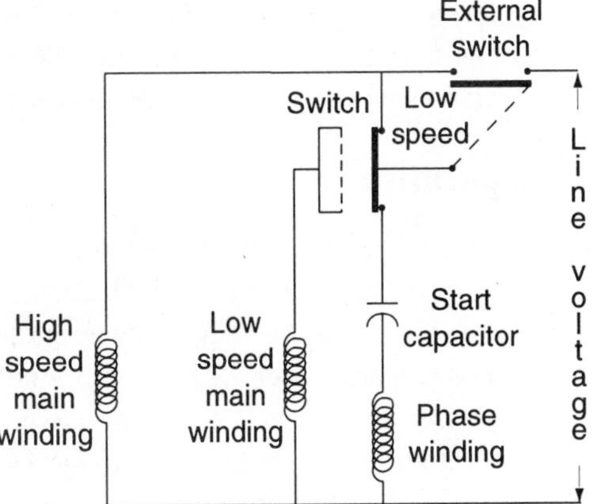

Figure 16–19 Two-speed motor wiring connections

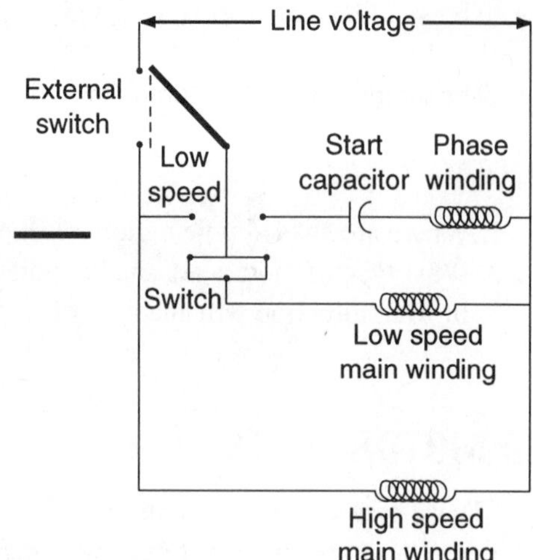

Figure 16–20 Low-speed connections of a two-speed motor

- Two-speed motors are used in applications where it is desired that the fan operate at different speeds during the different times in the cycle or when changing from cooling to heating or heating to cooling.

REVIEW QUESTIONS 16–8

1. Name the different windings used in two-speed motors.
2. In what mode does a two-speed motor start for low-speed operation?
3. In what application are two-speed motors popular?

16–9 CENTRIFUGAL SWITCH

The centrifugal switch is the device that automatically removes the start winding from the circuit after the motor has reached about 75% of its normal operating speed. The principal components of the switch are the moving contact arm, the stationary contact plate, the governor weight, and the weight spring. The contact arm pivots on the switch contacts. It is held in position with two pins or screws. There is a compression spring mounted on the top pin to hold the contacts open when the switch is in the normal operating position. The electrical line terminal studs, and the overload device when used, are mounted on the stationary contact plate assembly.

Operation

When the motor is not running, the riding edge of the governor weight pushing against the contact arm holds the contacts closed, completing the electrical circuit to the starting winding. When power is supplied to the motor it starts turning. When the motor reaches about 75% of its normal running speed, the centrifugal forces of the governor weight overcomes the governor-weight spring and snaps outward on the pin, moving the contact arm toward the rotor. This action opens the contact and removes the phase-winding from the motor circuit. The motor will operate in this condition until it stops and the contacts are closed again, ready for the next start-up.

SUMMARY 16–9

- The centrifugal switch is a device that automatically removes the start winding from the circuit after the motor has reached about 75% of its normal operating speed.
- At about 75% of the motor operating speed, the centrifugal switch forces of the governor weight overcomes the governor-weight spring and snaps outward on the pin, moving the contact arm toward the rotor.

REVIEW QUESTIONS 16–9

1. What is the purpose of the centrifugal switch in an electric motor?
2. What causes the centrifugal switch to operate?

16–10 SINGLE-PHASE MOTOR PROTECTORS

The purpose of single-phase motor protectors is to protect the motor from damage by conditions such as overheating, overcurrent, or both. Motor protectors may be mounted either internally or externally, depending on the design of the equipment. The contacts in these controls are normally closed and open in response to an overload condition.

Internal Overload

There are two types of internal overloads: the thermostatic type and the line-break type. The thermostatic type is usually wired into the control circuit to interrupt the power to the motor starting device. The line-break type is usually wired directly into the power voltage to the motor winding to interrupt the power to the winding. Both protectors are located precisely in the calculated center of the heat-sink portion of the windings to protect the motor from current draw and high temperatures. See Figure 16–21.

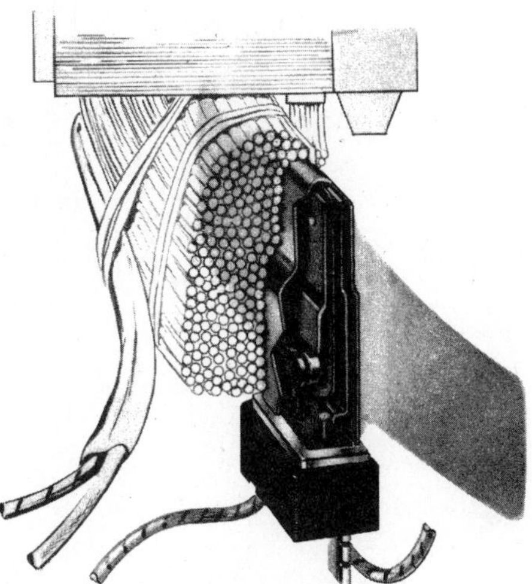

Figure 16–21 Internal overload location (Courtesy of Tecumseh Products Company)

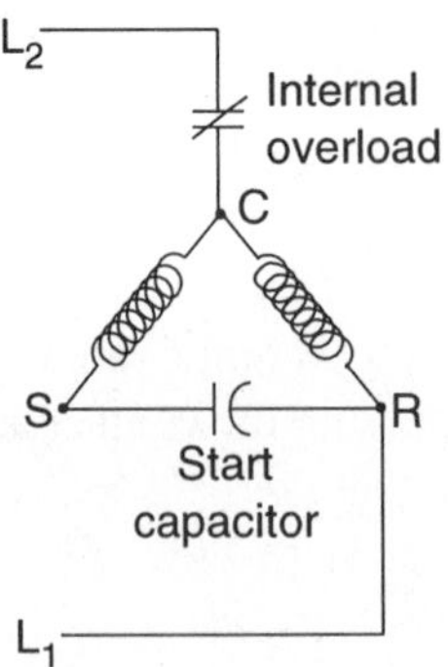

Figure 16–22 Wiring diagram for a line-break internal overload

The line-break overload is wired into the common terminal of the motor winding so the electrical circuit to both windings will be broken to stop the motor in case of overload. See Figure 16–22.

The internal thermostat is also used to sense over-current and over-temperature conditions. It will stop the motor if either or both conditions exist. The thermostat is wired into the control circuit shown in Figure 16–23.

If it is determined that either of these controls has caused the motor to stop, the motor must be cooled down before it will rest and resume operation. The motor can sometimes be cooled much faster if a slow stream of water is allowed to trickle over the motor housing. Use caution to prevent water from entering the terminal box and causing an electrical short. Perhaps a safer method is to allow the motor to sit idle for several hours to cool enough for the overload to reset. When the internal thermostat will not reset, the wiring can be modified to permit operation to determine exactly what the problem is. If the internal line-break overload will not reset when cooled sufficiently, the motor must be replaced.

Figure 16–23 Internal thermostat overload (Courtesy of Tecumseh Products Company)

Figure 16–24 Typical external overload location (Courtesy of Tecumseh Products Company)

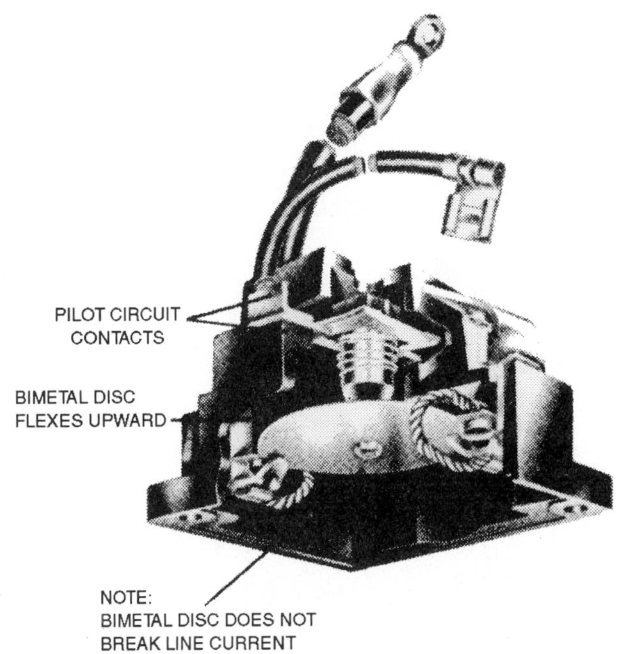

Figure 16–25 External line break overload (Courtesy of Tecumseh Products Company)

External Overload

The purpose of the external overload is to protect the motor from high current draw, high temperatures, or both, depending on the manufacturer's design. These types of protectors are mounted on the outside of the motor and can be replaced if found to be faulty. This type of overload is mounted in direct contact with the hottest area of the motor housing or the hermetic compressor housing to sense the temperature at that point.

External overloads may be of the line-break or the thermostatic type. They must be installed in the place designated by the motor manufacturer. See Figure 16–24.

When this control is found to be faulty, it must be replaced with an exact replacement to maintain the proper motor protection.

The operating part of the line-break overload is a bimetal disk. See Figure 16–25.

The contacts are normally closed and open when an overload condition exists. See Figure 16–26.

The contacts are wired in the circuit to the common terminal of the motor winding. See Figure 16–27.

When the resistance heater in the external line-break overload heats sufficiently to indicate an overloaded condition, the bimetal disk warps, separating the contacts to interrupt the electrical power to the motor winding. The motor will not restart until the overload has cooled enough for the bimetal to warp back into position and close the contacts.

The thermostatic type of external overload is mounted in exactly the same way that the line-break control is. The difference is that the contacts are wired into the control circuit rather than in the motor common terminal wire. See Figure 16–28.

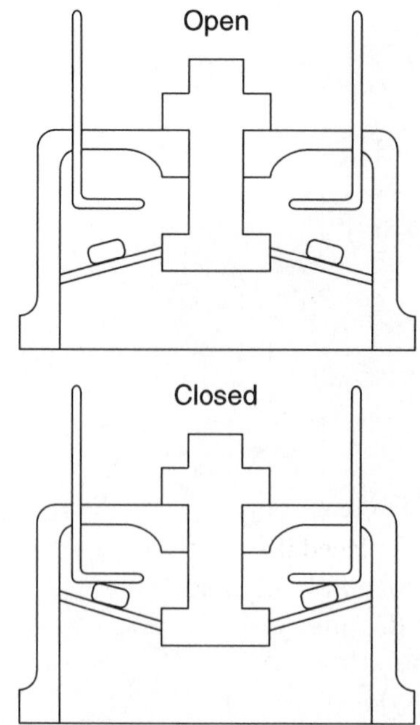

Figure 16–26 Operation of external overload contacts

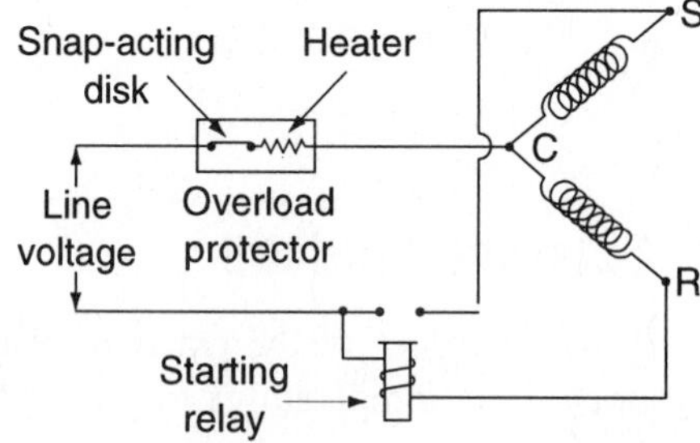

Figure 16–27 Wiring diagram for external overload

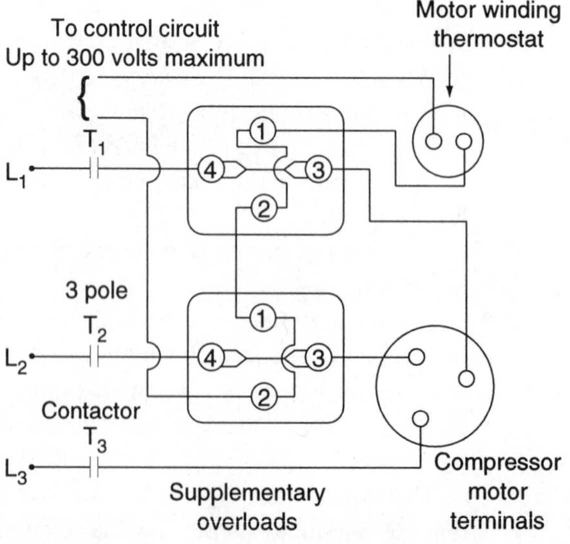

Figure 16–28 Three-phase wiring diagram for external thermostatic overload

The control circuit is interrupted to stop the motor during an overload condition. This causes the motor starter or contactor coil to be de-energized while the overload condition exists.

SUMMARY 16–10

- The purpose of single-phase motor protectors is to protect the motor from damage caused by conditions such as overheating, overcurrent, or both.
- There are two types of internal overloads: thermostat and the line-break type.
- Both protectors are located precisely in the calculated center of the heat-sink portion of the windings to protect the motor from high current draw or high temperature or both.
- External motor protectors are mounted on the outside of the motor and can be replaced if found to be faulty.
- External overloads may be of the line-break type or of the thermostatic type.
- External overloads must be replaced with an exact replacement.

REVIEW QUESTIONS 16–10

1. What do overload protectors sense?
2. Where are internal overloads located?
3. How is the line-break internal overload wired into the system?
4. Where are external overloads mounted?
5. When a motor has tripped the overload, when will it start again?

16–11 TESTING SINGLE-PHASE MOTORS

Troubleshooting electric motors is not complicated when the basic theory is understood. Probably the most difficult type to troubleshoot is the capacitor-start/capacitor-run. However, a systematic approach and proper troubleshooting techniques make the job much easier.

When troubleshooting electric motors, all safety precautions should be followed to prevent damage to the equipment and personal injury. The following is a list of the most basic safety precautions:

- Make certain that the electrical system is disconnected from the electrical supply.
- Be sure to discharge all capacitors before handling them. Use a 20,000 ohm resistor to touch both capacitor terminals and bleed off any electrical charge stored in it. Do not short the terminals with a screwdriver or other nonresistor material. To do so could cause the capacitor to explode.
- When checking the voltage of a running motor, be sure to set the meter on the highest possible scale to prevent damage to the meter and possible personal injury. This is because the EMF developed in the windings can be much higher than the applied voltage.

Locating Compressor Terminals

It is not necessary to have the manufacturer's data to determine the common, start, and run terminals of a compressor. This task is easily done with a good ohmmeter. The first step is to draw a diagram showing exactly where the terminals are located and their relation to each other. Then number the terminals on the diagram to make it easier to keep them straight. See Figure 16–29.

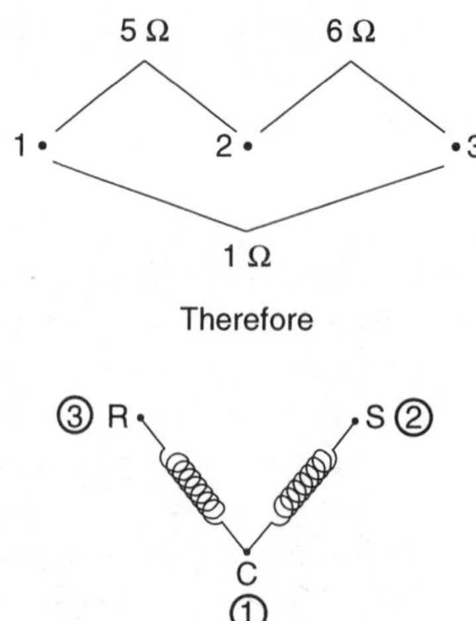

Figure 16–29 Locating compressor terminals

Next, use the ohmmeter to measure the resistance between each of the terminals and mark them on the diagram. Apply the following rule: The greatest amount of resistance will be between the start and run terminals. The medium amount of resistance will be between the start and common terminals. The least amount of resistance will be between the run and common terminals.

When the resistance is determined, draw a line between the two terminals where the measurement was taken and mark the reading on the diagram. In Figure 16–29, the greatest amount of resistance is between terminals 1 and 2; the medium amount of resistance is between terminals 2 and 3. The least amount of resistance is between terminals 1 and 3. In our example, terminal 3 is common (C), terminal 2 is start (S), and terminal 3 is run (R). It should be noted that the sum of the resistances between the common terminal and the start terminal and the common terminal and the run terminal should equal the amount of resistance between the run-to-start terminals.

Grounded Windings

To check for grounded motor windings, the ohmmeter should be set on the highest possible scale, R × 10,000 if possible. Be sure that the ohmmeter is properly zeroed. Touch one of the meter leads to the motor housing. Then touch the other meter lead to each of the motor terminals in turn. See Figure 16–30.

As a general rule of thumb, hermetic motors of 1 horsepower or less should have a minimum of one million ohms between the motor winding and the motor housing. Motors larger than one horsepower should have a minimum resistance between the winding and the housing of 1,000 ohms per volt. Always use the manufacturer's specifications when available to determine the exact readings for each winding. Resistance readings will be more accurate if the motor is warm. Also, a warm motor winding is more likely than a cold winding to show a weakened condition.

Open or Shorted Windings

An open winding occurs when the motor winding has separated, interrupting the path for current flow. When a motor winding is open the motor will not operate. In fact, it will not even try. The ohmmeter will indicate an infinity reading when testing an open circuit.

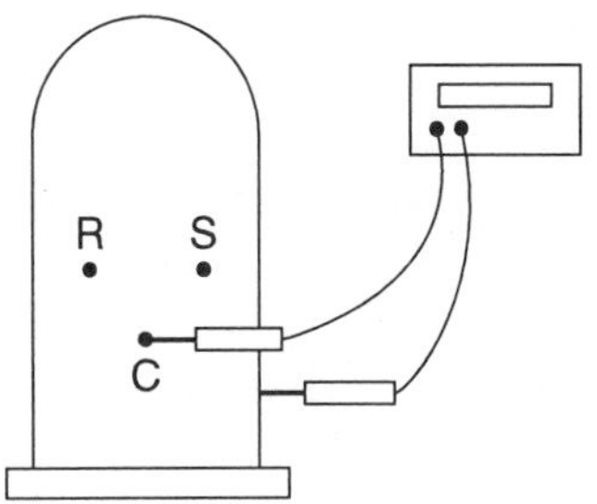

Figure 16–30 Checking for grounded motor windings

A shorted winding occurs when the winding insulation breaks down and allows the current to flow past a part of the winding. The resistance will be less than normal. This is probably the most difficult condition to confirm.

When testing for open or shorted windings, set the ohmmeter on the R × 1 scale. It is a good idea to check the manufacturer's specifications when checking the motor winding resistance. Be sure to zero the ohmmeter before taking any measurements and when the scale is changed. Clean the motor terminals so that a good reading can be obtained. Then touch one of the meter leads to one of the terminals, alternately touching the other lead to the other terminals in turn. Be sure that the meter leads make good contact with the terminals. See Figure 16–31.

SUMMARY 16–11

- The only tool needed to determine the common, start, and run terminals of a compressor motor is an ohmmeter.
- The first step is to draw a diagram showing exactly where the terminals are located and their relation to each other.
- As a general rule of thumb, hermetic motors of one horsepower or less should have a minimum of one million ohms between the motor winding and the motor housing.
- Motors larger than one horsepower should have a minimum resistance between the winding and the housing of 1,000 ohms per volt.
- An open winding occurs when the motor winding has separated, interrupting the path for current flow.
- A shorted winding occurs when the insulation breaks down and allows the current to flow past a part of the winding.

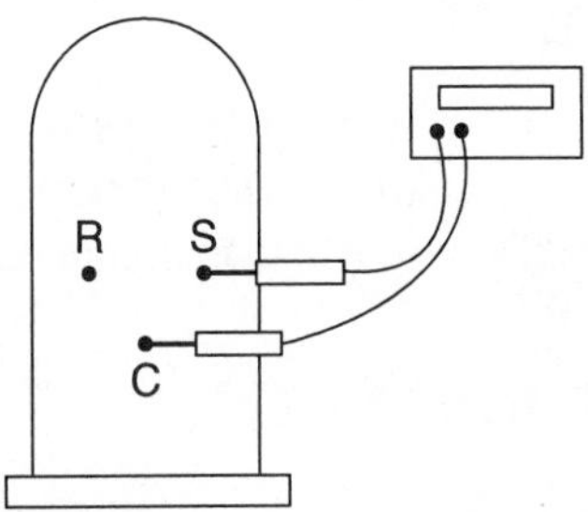

Figure 16–31 Checking for open or shorted motor windings

REVIEW QUESTIONS 16–11

1. What is required to make troubleshooting electric motors easier?
2. What is the first safety precaution to take when troubleshooting electric motors?
3. What should first be done when determining compressor-motor terminals?
4. To what should the sum of the resistances between the common terminal and the start terminal, and between the common terminal and the run terminal equal?
5. When is a motor winding most likely to show a weakened condition?
6. What will be the resistance of a shorted motor winding?

16–12 REPLACING CAPACITORS

It is not always possible to have the exact capacitor replacement on the service truck. Therefore, the technician must know how to determine just what capacitors to use to obtain the capacity required by the motor. Following is a list of general rules that can be used in capacitor replacement:

1. The voltage rating of any capacitor must be equal to or greater than the capacitor being replaced.
2. When replacing starting capacitors, the capacitance must be equal to but not greater than 20% more than that of the capacitor being replaced.
3. When replacing a running capacitor, the capacitance rating may vary no more than ± 10% of the capacitor being replaced.

Rules for capacitors in parallel:

1. All capacitors must have a voltage rating equal to or greater than the capacitor being replaced.
2. The total mfd rating is the sum of the mfd rating of each capacitor used.

Rules for capacitors in series:

1. The sum of the voltages of the replacement capacitors must be equal to or greater than that of the capacitor being replaced.
2. The total capacitance of capacitors in series can be determined by using the formula:

$$C_t = \frac{C_1 C_2}{C_1 + C_2}$$

SUMMARY 16–12

- It is not always possible to have the exact capacitor replacement on the service truck. Therefore, the technician must know how to determine just what capacitors to use to maintain the capacity required by the motor.

REVIEW QUESTIONS 16–12

1. When replacing a capacitor, how should the voltage rating be determined?
2. When capacitors are connected in parallel, how is the mfd rating of the capacitors determined?

16–13 STARTING RELAYS

The purpose of a starting relay is to remove the starting components from the circuit when a single-phase motor has reached about 75% of its normal operating speed. It is the replacement for the centrifugal switch used on open-type motors. There are several types of starting relays. However, we will discuss only the most popular types here, which are the amperage (current) relay, the potential (voltage) starting relay, and the solid-state starting relay. The type of relay used is generally determined by the horsepower and the design of the equipment manufacturer.

Amperage (Current) Starting Relay

The amperage relay is normally used on motors of 1/2 horsepower and smaller. It has an electromagnetic coil wired in series with the motor run winding. It is an electromagnetic relay. Because the coil is connected in series with the motor run winding, the wire must be large enough to carry the running current of the motor. See Figure 16–32.

These types of starting relays are positional-type. That is, they must be mounted in the correct position or they will not operate properly. A weighted armature that depends on gravity to open the contacts is used. The contacts on this relay are normally open.

During operation, when the electrical circuit to the motor is closed, the heaviest amount of current is flowing through both the circuit and the relay coil. The heavy current flow generates a magnetic field around the relay coil, which pulls the armature in and closes the relay contacts. See Figure 16–33.

When the contacts close, a circuit is completed through the start winding to provide the necessary phase shift to start the motor turning. When the motor reaches about 75% of its normal operating speed, the amperage draw through the relay coil and the run winding is somewhat less than the starting current. This drop in current flow is due to the build up of counter emf in the motor windings. This counter emf opposes the applied voltage, causing the amperage to drop. The reduced amperage flow causes a reduction in the strength of the electromagnetic field in the relay coil, allowing the

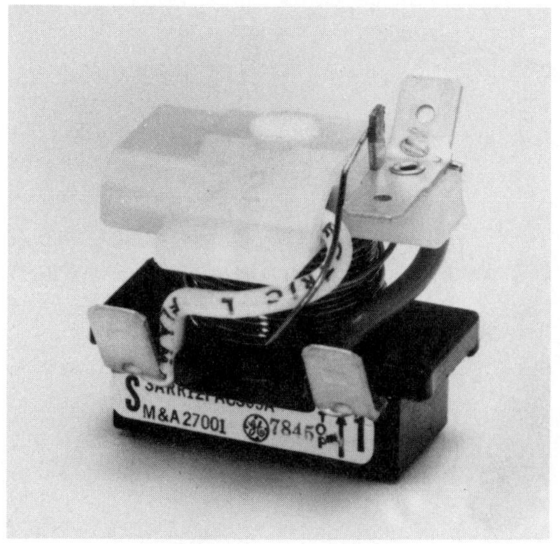

Figure 16–32 Amperage relay (Courtesy of Motors and Armatures, Inc.)

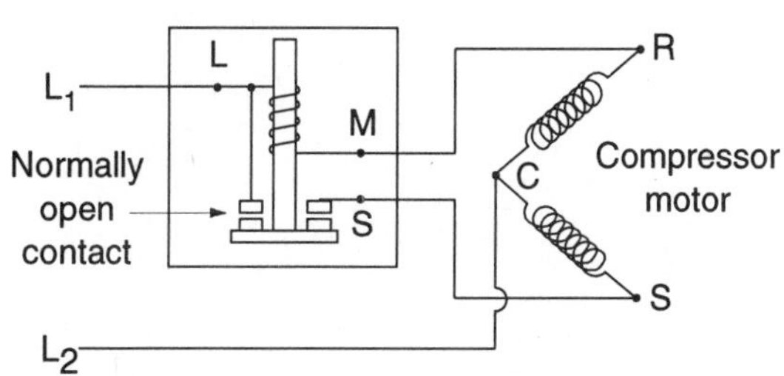

Figure 16–33 Wiring diagram for an amperage relay

gravity to overcome the magnetic field and open the relay contacts. When the contacts open, the starting winding is removed from the circuit. The motor continues to run on the run winding until the motor control is satisfied and interrupts the electricity to the motor, causing it to stop. The contacts in the relay will remain open until the motor circuit is again energized, causing the required electromagnetic field.

These types of starting relays must be sized according to the horsepower and the amperage draw of the motor. When a relay that is too large for the application is used, the relay contacts may not close to energize the starting circuit and the motor may not start. When the relay is too small for the application, the current draw of the run winding may keep the contacts closed all the time, causing the motor to draw too much current. This condition could cause permanent damage to the motor windings. A motor protector must be used with an amperage starting relay.

Potential (Voltage) Starting Relay

These types of starting relays depend on an electromagnetic field for their operation. They have a coil of very fine wire that is wound around a core piece. Potential relays are designed for use on almost all sizes of motors. However, they must be sized for the motor used to prevent damage to the motor windings. Potential relays have normally closed contacts that open when the armature in the relay coil is pulled in. There are three connections to the inside of the relay. They are numbered 1, 2, and 5. The terminals numbered 4 and 6 are auxiliary terminals used for external wiring connections. See Figure 16–34.

The relay is wired into the circuit with the number 5 terminal on the relay connected to both the electrical line to the motor and the common motor terminal. Terminal number 2 on the relay is connected to the start winding terminal. Terminal number 1 on the relay is connected to one terminal on the starting capacitor. See Figure 16–35.

During operation, when the electrical circuit is completed to the motor windings, electrical power is supplied to the start winding through the relay contacts located

Figure 16–34 Potential starting relay (Courtesy of Motors and Armatures, Inc.)

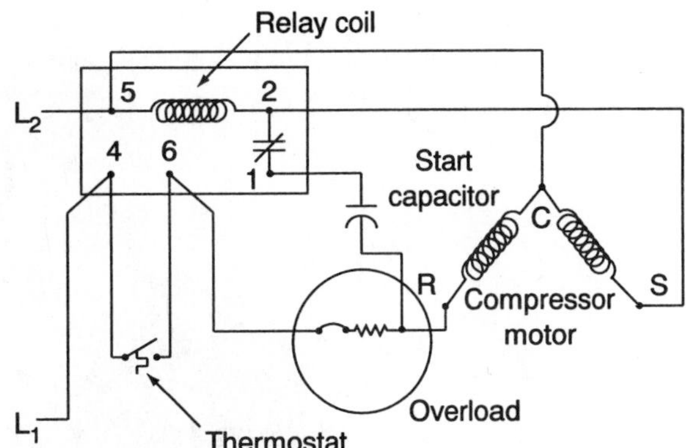

Figure 16–35 Potential starting relay wiring diagram

between terminals 1 and 2. When the motor reaches about 75% of its normal operating speed, the counter emf in the start winding has increased enough to build the electromagnetic field in the relay coil to the point that it will "pick up" the armature, opening the relay contacts. When the relay contacts open, the starting circuit is opened, removing the starting components from the circuit. All of the circuits will remain in this condition until the electrical power to the motor is interrupted. As the motor slows down, the counter emf in the start winding is reduced, weakening the strength of the electromagnetism in the relay coil. When a given voltage is reached, the relay coil will be de-energized, closing the contacts. This voltage is known as the "drop-out voltage." The relay remains in this position until electricity is again supplied to the motor circuit.

These are nonpositional relays. That is, they may be mounted in any position as long as they are not mounted on a vibrating surface that will cause the contacts to chatter and burn much faster than normal. The sizing of these relays is not as critical as the current relay. Check with the manufacturer for the recommended potential starting relay for a given motor. If this is not available, start the motor and check the voltage between the start and common terminals on the motor when it is operating at full speed. Multiply this voltage by 0.75. This will be the pick-up voltage required for the relay.

Solid-State Starting Relay

Solid-state starting relays use a self-regulating ceramic whose electrical resistance increases as its temperature increases. This allows the starting components to be removed from the circuit and reduces the current draw by the starting winding to a milliamp level. It takes approximately 0.35 seconds for the material to reduce the current flow to this level. This rapid current draw reduction allows this type of relay to be used on domestic refrigerator and freezer compressors without being sized for each and every model compressor. They can be used on virtually any 120-volt AC compressor up to about 1/3 horsepower. They fit directly on the compressor terminals just as the amperage relay does. See Figure 16–36.

Solid-state starting relays are wired into the motor circuit, with the ceramic material between the line and the compressor starting terminal. See Figure 16–37.

In this manner the relay is wired in electrical series with the start winding. When the relay is energized, the ceramic material heats up and increases in resistance to the

Figure 16–36 Solid-state motor start relay (Courtesy of Klixon Controls Division, Texas Instruments, Inc.)

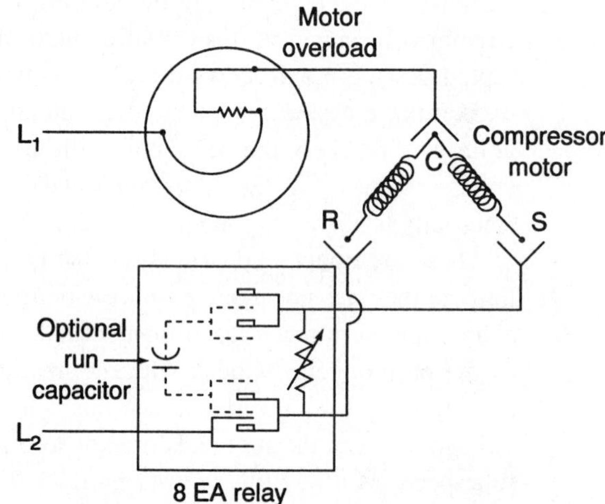

Figure 16–37 Solid-state starting relay wiring diagram

point that virtually all current flow is stopped in 0.35 seconds. The ceramic remains at this temperature until the power is turned off. When the power is turned off, the ceramic requires a few seconds to cool down before it is ready for another start up.

Solid-State Hard Start Kit

This kit is designed to provide the required starting torque for PSC motors that have difficulty starting because of low line voltage or other problems. The phase-change is caused by the use of a positive temperature coefficient ceramic material (ptc). The ptc material increases in resistance with an increase in temperature. When the material reaches its anomaly temperature, the resistance increases at a very rapid rate. See Figure 16–38.

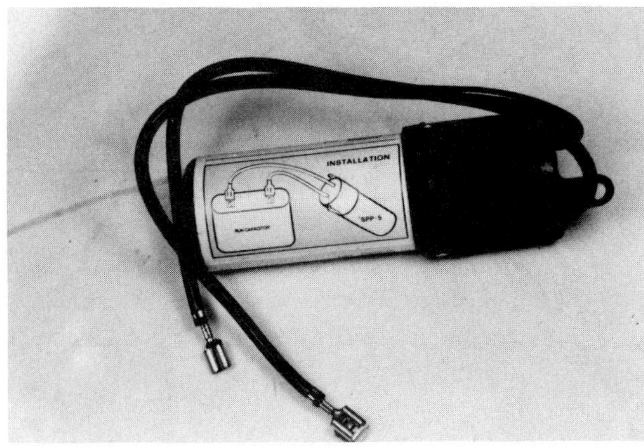

Figure 16–38 Solid-state hard start kit (Photo by Billy C. Langley)

The purpose in using this type of kit is to provide the surge of current needed to start the motor. When the ceramic material has heated to its anomaly temperature, the starting current is then reduced, allowing the motor to operate as a normal PSC motor. When the electrical supply is turned on to the PSC motor, current flows through the start winding and through the parallel combination of the run capacitor and the low resistance of the ptc. See Figure 16–39.

The low resistance in the ptc material does two things. First the starting current is increased, and second, the angular displacement between the starting current and the run winding current is reduced. This is a big advantage because this angle displacement is usually greater than 90%.

The surge current on starting increases the motor starting torque and helps to heat the ptc material to the anomaly temperature. The time required for the ptc to heat to its anomaly temperature is dependent on the amount of material being heated, the anomaly temperature required, resistance, and the applied voltage to the ptc. The amount of time required is not dependent on when the motor starts but on the above factors. When a 240-volt motor equipped with a 9 EA start assist is initially energized with the proper voltage, the relay switching time is 16 electrical cycles. However, when the same motor is started with 25% less voltage, the switching time is changed to 32 electrical cycles. This way, hard-starting assistance is provided when low voltage or other hard starting conditions are present.

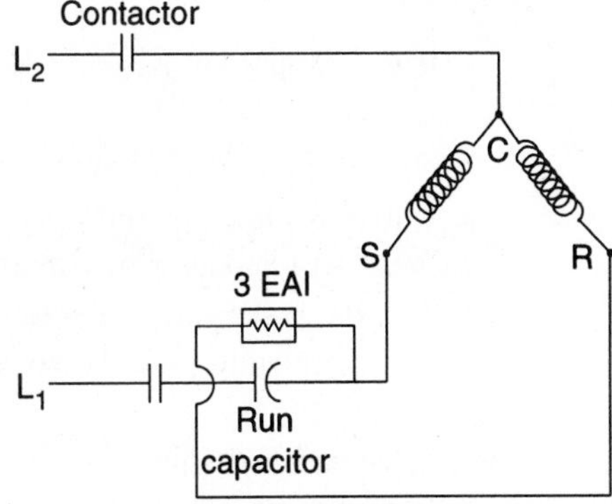

Figure 16–39 Solid-state hard start kit wiring diagram

When the ptc has heated up to its anomaly temperature, its resistance increases to about 80,000 ohms. This high resistance effectively switches it out of the circuit without the use of an electro-mechanical relay. During normal running conditions, the relay draws only about 6 milliamps of current. This low current draw does not affect the normal motor operation.

When the electrical current is interrupted to the unit the motor stops and the ptc starts cooling down. Should power be restored before the ptc has cooled below its anomaly temperature, the motor will attempt to start in the normal PSC mode. The ptc should be allowed about one minute to properly cool down below its anomaly temperature. This will assure that the motor will have the start assist on the next start cycle.

SUMMARY 16–13

- The purpose of the starting relay is to remove the starting components from the circuit when a single-phase motor has reached about 75% of its normal operating speed.
- An amperage relay has an electromagnetic coil wired in series with the motor run winding.
- Amperage relays are positional-type relays.
- Amperage relays must be sized according to the horsepower and amperage draw of the motor.
- Potential relays must be sized for the motor used to prevent damage to the motor windings.
- Potential relays have normally closed contacts that open when the armature in the relay is pulled in.
- Potential relays are nonpositional relays. However, they must not be mounted on a vibrating surface.
- Solid-state relays use a self-regulating ceramic whose electrical resistance increases as its temperature increases.
- Solid-state relays can be used on domestic refrigerators and freezer applications without being sized for each and every compressor model.
- Solid-state hard-start kits are designed to provide the required starting torque for PSC motors that have difficulty starting because of low-line voltage or other problems.
- When the ptc has heated to its anomaly temperature, its resistance increases to about 80,000 ohms.

REVIEW QUESTIONS 16–13

1. What device removes the starting components from the circuit to an electric motor?
2. How is the coil of an amperage starting relay wired to the motor?
3. Why are amperage relays of the positional type?
4. What will happen when an amperage relay is sized too large for the motor?
5. To what is the coil of a potential starting relay wired?
6. In what position are the contacts of a potential relay when no energy is applied to the coil?
7. What causes a solid-state starting relay to operate?
8. For what reasons are hard-start kits used?

16–14 STARTERS AND CONTACTORS

Refrigeration and air conditioning systems use several size motors, but the compressor motor represents the greatest switching load for the control circuit. When considering starters and contactors for a given installation, the fan motors, pump motors, and any other equipment that may be used will also draw current through the starter or contactor contacts. Therefore, the starter or contactor must be sized to adequately handle the complete electrical load passing through it.

Contactor

The definition of a contactor is a device for repeatedly establishing and interrupting an electric power circuit. Each contactor has features that are common to all of them such as an electromagnetic coil, switching contacts, and stationary contacts. The contacts are used to make the electrical circuit when closed by the electromagnetic coil, when it is energized. Contactors are used for switching heavy current, high voltage, or both. One single contactor may be used to switch more than one circuit. See Figure 16–40.

During operation, when the coil is energized, an electromagnetic field is built up around it, causing the armature to be pulled into its center. The movable contacts are mounted on the armature and are pulled against the stationary contacts to complete the electrical circuit to the motor or other load. When the electric circuit to the contactor coil is interrupted, the electromagnetic field collapses and allows the armature to return to its de-energized position. When the armature drops, it takes the movable contacts away from the stationary contacts, opening the power circuit to the compressor or load and is ready for the next start cycle.

Starter

A motor starter is a contactor with additional components to satisfy some system requirement. The additional components are generally overload relays, holding

Figure 16–40 Two-pole contactor (Photo by Billy C. Langley)

Figure 16–41 Motor starter (Photo by Billy C. Langley)

contacts, step resistors, disconnects, reactors, or other hardware that may be necessary to complete the required starter package. See Figure 16–41.

Starters operate almost exactly as the motor contactor. The major difference is in the changes that the additional components make.

SUMMARY 16–14

- The starter or contactor must be sized to adequately handle the complete electrical load passing through it.
- A contactor is a device for repeatedly establishing and interrupting an electric power circuit.
- A motor starter is a contactor with additional components to satisfy some system requirement.

REVIEW QUESTIONS 16–14

1. What component represents the greatest electrical load in refrigeration and air conditioning systems?
2. What are contactors in refrigeration and air conditioning systems used for?
3. What is the difference in operation between a starter and a contactor?

GLOSSARY

A

Absolute humidity: The weight of water vapor in grains of moisture that are actually contained in one ft^3 of the air and moisture mixture.

Absolute pressure: The sum of the atmospheric pressure (14.7 psi) and gauge pressure.

Absolute temperature: At this temperature the substance theoretically contains no heat and all molecular motion stops.

Absorbent: The ability of one substance to absorb another into its composition.

Accessible hermetic: The component that contains both the motor and compressor and may be serviced in the field.

Accumulator: A component that is placed in the suction line to prevent liquid refrigerant from entering the compressor crankcase. The liquid is vaporized before it enters the compressor.

Acid condition: A condition in which the refrigerant and/or oil contains another fluid that is acidic in nature.

ACR tubing: The type of tubing that is used for refrigeration lines to connect the various components of the system.

Activated alumina: A material that is made up of aluminum oxide and is used in refrigerant driers.

Activated carbon: A carbon that has been specially treated and is used for cleaning the air in a building.

Adiabatic compression: During this process, a vapor is compressed without the addition or removal of heat.

Air: A mechanical mixture of oxygen and nitrogen with traces of other gases and moisture, which is known as humidity. The weight of dry air is 0.075 lb/ft^3. It takes one Btu to raise fifty-five cubic feet of air one degree F. Air is also an elastic gas.

Alternating current: This is electrical current that changes direction periodically. In the United States it changes 60 times per second. It is also known as 60-cycle current.

Altitude correction: Atmospheric pressure changes in direct relation to a change in altitude. The vapor bellows used in some refrigeration and air conditioning controls is affected by an altitude change. The effects of altitude are greater than those caused by barometric changes in pressure and they are more permanent. Because of this, any adjustments made to controls at a given altitude will remain set for that operating pressure.

Ambient temperature: This is the temperature that surrounds all objects and is measured with a dry bulb thermometer.

Ammeter: This is an instrument used to measure the current flow through an AC electrical circuit.

Amperage: This is the unit of electrical measure that is equivalent to the flow of one coulomb per second past a given point.

Analyzer: An instrument that is used to determine the condition of electrical components.

Anneal: The process of heat-treating metal so that it will take on the desired softness and ductility.

Anode: This is the positive terminal of an electric cell or battery.

Armature: The part of an electric motor or generator that is caused to move by magnetism.

ASTM standards: The standards that are set by the American Society of Testing Materials.

Atmospheric pressure: The pressure that is exerted in all directions by the atmosphere. It is measured with a barometer. Atmospheric pressure at sea level is considered to be 14.7 lb/in^2.

Atom: The smallest part of a substance that can exist either alone or in some combination with other elements.

Atomize: To cause a liquid to change into a fine spray or small particles.

Automatic control: To cause a piece of equipment to operate in its various modes without it being manually adjusted to do so.

Automatic expansion valve (AXV): A pressure-operated refrigerant control that reduces the pressure and permits it to pass into the evaporator at the evaporation rate of the refrigerant inside.

B

Back pressure: This refers to the refrigerant pressure in the low side of the system; from the outlet of the flow control device to the suction of the compressor.

Back-seating: The position of a service valve that allows full refrigerant flow through it. In this position, a gauge port is usually closed off. This is the normal operating position of the service valve.

Barometer: One of the instruments used to measure atmospheric pressure.

Battery: A device that uses the interaction of chemicals and metals to produce electricity.

Bearing: A device that is used for aligning and maintaining the alignment of moving components. It is a low-friction device.

Bellows: A corrugated metal cylinder used to respond to changes in pressure and provide a seal during the movement of the parts.

Bellows seal: The seal between moving parts that prevents leakage between them. The seal expands and contracts with a change in pressure.

Bleed: The process of slowly releasing pressure from a system by slightly opening a bleed valve.

Bleed valve: A valve that is used to control the flow rate of pressure that is being released. It is usually opened to permit flow and closed to stop the flow.

Boiling point: The temperature at which a liquid will boil when under atmospheric pressure.

Bourdon tube: A thin-walled, elastic metal tube that is used in pressure gauges. It is made in a circular shape that will tend to become straight when the pressure inside it increases.

Boyle's Law: A physics law concerning the volume of gas as its pressure is changed. With a constant temperature, the volume will vary. On an increase in pressure, its volume is decreased. A reduction in pressure allows an increase in the volume.

Braze: The process of using a nonferrous metal as a filler to join two other pieces of metal together. A temperature between 800°F and the melting temperature of the base metal.

British thermal unit (Btu): The amount of heat required to raise the temperature of one pound of water 1°F. This is the generally accepted definition for use in refrigeration and air conditioning work.

Bypass: The pipe or duct that is used to allow a fluid to flow around a component or object. The flow is usually controlled by a valve or a damper.

C

Calibrate: To adjust the pointer of a control to a given position.

Calorie: A unit of heat that is usually used in engineering work. The amount of heat required to raise 1 kilogram of water 1°C.

Capacitance: The property of an electric current that allows electrical energy to be stored in an electrostatic field and released at some later time as required by the circuit.

Capacitor: A device that is used to store electrical energy.

Capacitor-start motor: An electric motor that uses a capacitor to provide the required starting torque and runs as an induction motor.

Capacity: In refrigeration and air conditioning work this is the capacity of the system to absorb a certain amount of energy per unit of time. It is usually measured in Btu per hour.

Capillary tube: A copper tube with a small inside diameter that is used to control the flow of refrigerant into the evaporator.

Centigrade: A thermometer that has 100 degrees between 0 and 100° on its scale. It is a metric temperature measurement.

Centrifugal compressor: A rotating compressor that uses centrifugal force to compress the refrigerant vapor.

Change of state: The process of making a substance change its state, such as from a liquid to a vapor and from a solid to a liquid.

Charge: In refrigeration and air conditioning work, this is the amount of refrigerant that is put into the system. It also refers to putting refrigerant into the system.

Charles' Law: A law of physics that states that the volume of a gas at a constant pressure will vary according to any temperature change of the gas.

Check valve: A system component that will allow a flow in one direction while stopping it in the other direction.

Chlorodifluoromethane: A type of refrigerant that is commonly called R–22.

Circuit: The tubing, piping, or electrical wiring used to allow energy to flow from the source through the circuit and back to the source.

Circuit breaker: An electrical safety switch used to protect an electrical circuit when an overloaded condition exists.

Circuit, parallel: Arranging of electrical devices so that the source of power is the same to all circuits.

Circuit, series: Arranging electrical devices so that all of the electrical current passes through them one after the other.

Clearance pocket: The space in a cylinder above the piston that is not swept clear when the piston is at the top of its compression stroke.

Closed circuit: An electrical circuit that is complete, allowing the electricity to flow through it.

Coefficient of heat transmission (U): The amount of heat that is transmitted from air to air in 1 hr/ft^2 of the partition for every 1°F temperature difference between the two sources of air.

Coefficient of performance (COP): The ratio of the amount of work performed to the amount of energy consumed in the process.

Cold: This is a relative term that refers to the absence of heat.

Cold storage: The enclosure where perishables are preserved on a large scale by the use of refrigeration.

Commutator: That part of the rotor in an electric motor that transfers the electrical energy to the motor windings.

Compound gauge: A gauge that will measure pressure both above and below atmospheric pressure. They are usually used to determine the pressure in the low-side of a refrigeration system.

Compression ratio: The ratio of the clearance volume to the total volume of the compressor cylinder. It is also the ratio of the absolute suction and discharge pressures.

Compression system: A refrigeration system that uses a compression device (a compressor) to cause the refrigerant to flow through the system.

Compressor: A device used for increasing the pressure on refrigerant inside a system and causing it to flow through the system.

Compressor displacement: The volume in inches represented by the area of the top of the piston multiplied by the stroke of the piston in inches.

Compressor seal: The component on open compressors that prevents the leakage of refrigerant and oil between the crankshaft and the compressor body.

Condensate: Moisture that occurs because heat has been removed from a vapor, lowering its temperature below the dew-point temperature.

Condenser: That part of the refrigeration system where the refrigerant vapor is cooled and liquified by the removal of heat.

Condenser fan: The fan that causes the cooling air to flow over the condenser coil.

Condensing pressure: The pressure inside the condenser where the vapor has given up enough latent heat of condensation to be liquefied. This pressure will change with the temperature of the cooling medium.

Condensing temperature: The temperature inside the condenser where the refrigerant vapor has given up enough latent heat of condensation to become a liquid. This temperature will change with a change in refrigerant pressure.

Condensing unit: Sometimes referred to as the high side. It consists of the compressor, condenser, receiver, fan, fan motor, all placed on a designed frame, and the required accessories.

Conduction: The transfer of heat through and by matter.

Conductivity: The amount of heat that can be transferred through a homogenous material 1 inch thick in 1 hour for each degree of temperature difference between the two surfaces of the material.

Conductor: A substance that is capable of conducting electrical or heat energy.

Contaminant: A foreign matter such as dirt, moisture, or some other foreign material that is in the refrigerant and/or the oil of a refrigeration system.

Control: A device that is used for the regulation of a unit, either manual or automatic. It may respond either to pressure or temperature, but it will not usually respond to both at the same time.

Control system: This is an electric circuit that is made up of the devices needed to provide automatic control of a given process.

Control valve: A valve that regulates the flow of a medium in order to cause some effects on a controlled process. It is controlled by a remote signal from some other device that may be either pneumatically, electrically, or electrohydraulically operated.

Convection: The transfer of heat with the circulation of a fluid, such as air. When convection occurs naturally, it is caused by the difference in weight between the cooler and the hotter of the fluids.

Convector: A surface that is designed to transfer heat to the surrounding air through the convection process.

Cooler: A heat exchanger that is used to transfer heat from one substance to another.

Cooling tower: A device that is used to cool water down to the wet-bulb temperature by evaporating some of the water.

Copper plating: A condition that exists inside a refrigeration system when moisture is present. It occurs because some of the copper from the tubing is being electrolytically deposited on the steel components inside the system.

Corrosion: The deterioration of metal that is caused by chemical action.

Coulomb: The amount of electrical energy that flows past a given point in a circuit when electricity is flowing at a rate of one ampere per second.

Counter EMF: Electrical energy that is the result of electricity reversing its direction of flow as the magnetic field around the conductor changes.

Counterflow: Fluids that flow in an opposite direction to each other when the coldest portion of one meets the warmest portion of the other.

Coupling: In piping work it is the mechanical device used to join two pipes. In electrical work it occurs when two electromagnetic fields interact with each other.

Crank throw: The distance measured from the center line of the main bearing journal to the center of the crankpin or eccentric.

Crisper: The drawer or compartment of a refrigerator where vegetables are stored. It helps to keep the vegetables at the desired humidity level for storage without spoiling.

Critical pressure: The condition of a compressed refrigerant at which the vapor and liquid have the same properties.

Critical temperature: Above this temperature a vapor cannot be liquified regardless of the pressure applied to it.

Cross-charged: The combination of two fluids that will cause a desired pressure-temperature curve.

Current: The flow of electrical energy through a circuit. It occurs when electrons change positions.

Current relay: A component that is used for starting an electric motor-compressor. It operates in response to a change in the flow of electric current in the circuit.

Cut-in: The temperature or pressure at which a set of contacts close to complete an electric circuit.

Cut-out: The temperature or pressure at which a set of contacts open to stop the flow of electrical current.

Cycle: The operation of a refrigeration system through its complete course of operations, including the four major functions of compression, condensation, expansion, and evaporation.

Cylinder: The round passage in a compressor through which the piston travels when compressing refrigerant vapor.

Cylinder head: A cap or plate that covers the open end of the cylinder.

D

Dalton's Law: A law in physics that states that a vapor pressure exerted in a container by a mixture of gases is equal to the sum of each individual gas pressure in the mixture.

Damper: A mechanical blade that is placed in the duct system to control the flow of air through the duct.

De-aeration: The separating of air from some substance.

Decibel: A unit of measurement used to measure the noise level in a given place.

Decomposition: The spoiling and decaying of a perishable substance.

Defrost: To remove any accumulated frost from a cooling coil.

Defrost cycle: The cycle of a refrigeration system in which the refrigerant flows in the opposite direction to heat a cooling coil and remove any accumulation of frost on it.

Defrost timer: An electrical device wired into the electrical circuit to start the defrost cycle and keep it operating until all the frost is melted from the cooling coil.

Degree: A unit of measure on a thermometer. It represents temperature.

Degree day: A measurement that represents a difference of one degree in the temperature between the inside and the outside temperature and the average outdoor air temperature for a single day. This measurement is usually used in estimating energy requirements of a particular installation.

Degree of superheat: The difference between the actual boiling temperature of a liquid and the temperature of the vapor above the boiling temperature.

Dehumidifier: A unit that is used to lower the humidity within a given space.

Dehumidify: The process of removing water or moisture from the air; to remove water vapor or moisture from a material.

Dehydrated oil: A lubricant that has had the moisture removed from it to some acceptable level of dryness.

De-ice control: An electrical component that is used to control the compressor for the proper removal of ice from the coiling coil.

Density: The weight per unit volume of a substance.

Deodorizer: A device that contains a substance such as activated charcoal to absorb odors.

Desiccant: A material that is used in refrigeration system driers to absorb moisture from the refrigerant.

Design pressure: The highest pressure that the system is expected to reach during normal operation.

Dew point temperature: The temperature at which a vapor starts to condense. In the atmosphere this is usually 100% relative humidity.

Diaphragm: A device made from either metal or rubber and placed between two chambers to keep them separated.

Dichlorodifluoromethane: A member of the halocarbon refrigerant family popularly known as R–12.

Differential: The difference between the cut-in and the cut-out points of a control. It may control either pressure or temperature.

Direct-expansion evaporator: An evaporator that uses either an automatic or a thermostatic expansion valve. A capillary tube may also be used.

Double pole: The nomenclature that is used to designate the operation of a set of contacts that includes two separate forms of contacts; i.e., two single-pole contact assemblies.

Double throw: A term that is applied to a contact arrangement that denotes that each contact included is a make-break; i.e., one contact opens its connection to another contact and then completes its connection to a third set of contacts.

Drier: A chamber containing a desiccant used to remove moisture from a refrigeration system.

Drip pan: A pan that is placed below a cooling coil to collect any condensate leaving the evaporator surface.

Dry bulb thermometer: A thermometer used to measure the temperature of the ambient air.

Dry bulb temperature: The actual temperature of the ambient air. The temperature of the air indicated by a thermometer that is not affected by the moisture content of the air.

E

Eccentric: A disc that is mounted off center on a straight shaft and is used to transfer a reciprocating motion to the piston.

Eddy currents: The induced electrical currents that flow in a laminated core.

Electric defrosting: The defrosting of a cooling coil by using electrically heated elements.

Electrolysis: The chemical reaction between two substances because of the current flow through them.

Electrolytic capacitor: A plate or surface that is capable of storing small electrical charges.

Electromagnet: A coil of wire wrapped around a soft iron shaft or core. When electric current flows through the wire, an electromagnet results.

Electromotive force (EMF): The electrical force that makes the free electrons flow through a conductor or circuit.

Electron: The basic component of an atom. It has a negative electrical charge.

Electronic leak detector: An electronic instrument used to locate refrigerant vapor in the air around a pipe or component. It may indicate leaks by one of several means—a light, a buzzer, or a beeping noise.

End bell: The end of an electric motor that holds the bearings and also keeps the shaft in the center of the motor housing.

End play: The movement of the shaft in an electric motor, usually parallel with the shaft.

Energy: The ability to do work.

Enthalpy: The total amount of heat contained in a substance. In refrigeration work it is calculated from a base of –40°F.

Entropy: The amount of energy in a system. It is usually used in engineering calculations.

Environment: The surroundings and their condition.

Evacuation: The mechanical removal of air, moisture, and refrigerant vapor from the inside of a refrigeration system.

Evaporation: The changing of a liquid to a gas. This is the point during this process where the greatest amount of heat is transferred.

Evaporative condenser: A condenser especially designed to remove the heat from a refrigerant gas by the evaporation of water.

Evaporator: The part of the system that actually does the cooling by evaporating refrigerant liquid.

Evaporator fan: The fan used to force air through the evaporator.

Exhaust valve: The port through which any condensed refrigerant is allowed to leave the cylinder. It is commonly called the discharge valve.

Expansion valve: The flow control device that feeds refrigerant into the evaporator as it is needed. It causes a reduction in the pressure of the refrigerant so that it will evaporate.

External equalizer: The tube that connects the low-pressure side of the expansion valve diaphragm to the low-pressure line at the outlet of the evaporator. It allows the expansion valve to sense the suction pressure at this point to help in controlling operation of the valve.

F

Fahrenheit scale: This is the scale on thermometers used in the United States. It has a freezing temperature of 32°F and a boiling temperature of 212°F.

Fail-safe control: A control that is designed to open an electric circuit when abnormal conditions exist.

Fan: A mechanical air-moving device that uses an enclosed propeller to move the air. In refrigeration and air conditioning work it is used to designate any device used to move air.

Farad: The unit used to rate the capacity of capacitors.

Female thread: The internal thread used on valves, fittings, and pipe.

Field pole: A part of an electric motor stator that concentrates the magnetic field in the field winding.

Filter: A component used to remove solid particles from air and fluids, such as refrigerant.

Fin: The metal extension on the tubing of coils. Its purpose is to increase the efficiency and capacity of the coil.

Finned tube: A tube that has built-up extensions in the form of fins.

Flammability: The ability of a substance to sustain burning.

Flammable liquids: Liquids that have a flash point below 140°F and have a vapor pressure at 100°F less than 40 psi.

Flapper valve: The thin metal piece used as suction and discharge valves in a refrigeration compressor.

Flare: The enlargement of the end of copper tube. It is used when joining two pieces of copper tube together. It usually has a 45° angle for the seat.

Flare fitting: A type of soft copper tubing connector that requires the tubing to be flared to accommodate the fitting. The flare is used to make a mechanical seal between the two pieces.

Flare nut: A screw type fitting that is placed over a flared piece of tubing and screwed onto the female part of the fitting to make a seal between the two pieces.

Flash gas: Gas produced by the evaporation of some liquid refrigerant in the flow-control device to help cool the remaining liquid down to evaporator temperature.

Flash point: The temperature at which the vapor from a combustible material will ignite, but the vapor will not support combustion.

Float valve: A valve that uses a float in a chamber to control the flow of refrigerant into the evaporator.

Flood: A condition when liquid refrigerant enters the low side of the system or the compressor.

Flooded system: A system that allows liquid refrigerant to collect in the evaporator in a puddle. The evaporator will have a puddle of liquid in it during a normal operating cycle.

Flow meter: An instrument used for measuring the velocity or flow of a fluid.

Fluid: Commonly defined as matter in any state that will take the shape of its container. A gas or a liquid.

Flush: To force foreign particles or fluids from a refrigeration system by admitting other types of refrigerant or fluids into the system.

Flux: A paste or liquid that is applied to a tubing joint to be soldered or silver soldered. It keeps the cleaned joint from oxidizing during the heating process.

Foaming: The bubbling of the oil in the crankcase because of liquid refrigerant in the oil. The refrigerant rapidly boils off. Foaming is more likely to occur on compressor start-up.

Foam leak detector: Soap bubbles or any of the plastic liquid leak detectors that are spread over the suspected joint to locate leaks.

Force: The accumulated pressure expressed in pounds.

Forced convection: Fluid being moved by force as with a pump or a fan.

Forced-feed oiling: A type of compressor lubrication that uses a pump to force the oil through the moving parts.

Freezer: A unit that is used to freeze perishable products for storage.

Freezer alarm: Any type of indicator that lets the user or operator know that the unit is not operating properly. The indicator may be a bell or a light.

Freeze-up: When ice forms in the flow-control device because of excessive moisture in the system. Frost formation on the outside of the evaporator surface that will stop the circulation of air over the coil.

Freezing: The changing from a liquid to a solid.

Freezing point: The temperature at which a liquid will change into a solid on the removal of enough heat. The temperature at which a given substance will freeze.

Frostback: When liquid refrigerant leaves the evaporator and enters the suction line. It may or may not be accompanied by frost.

Frost-free refrigerator: A refrigeration unit that operates with a predetermined defrost period, preventing a large accumulation of frost on the evaporator.

Frozen: In mechanics it refers to a bearing frozen because of a lack of lubrication. It also is used to refer to a matter that has changed from a liquid to a solid.

Fuse: An electrical safety device that has a strip of metal through which the current flows. When the current flow exceeds the rating of the strip it will melt and burn apart, stopping the flow of current through it.

Fusible: Something that can be melted under certain circumstances.

Fusible plug: A safety plug that is placed in refrigerant cylinders to prevent them from rupturing from excessive pressure inside.

G

Gas: A substance in the vaporous state.

Gasket: The flexible material placed between mating surfaces to prevent leaking at that point.

Gauge: An instrument used to measure pressures both above and below atmospheric in a refrigeration system.

Gauge manifold: A series of valves and ports on which the compound and pressure gauges are installed so that certain service operations can be more easily done. They also have hose connections to connect the service hoses.

Gauge port: The opening or port provided on service valves so that the refrigeration gauges can be connected to the system.

Gauge pressure: The measure of pressure taken with a gauge. It is pressure that is measured from atmospheric pressure rather than absolute pressure.

Ground wire: A wire in the electrical system that will safely conduct electricity from a structure or a piece of equipment to the ground should an electrical short occur.

H

Halide refrigerants: A family of synthetic refrigerants containing halogen chemicals.

Halide torch: A type of leak detector that uses an open flame to locate refrigerant leaks. It draws the refrigerant into a hose connected at the base of the flame. When the refrigerant is burned, the flame changes to a bright blue-green color.

Halogens: Substances that contain fluorine, chlorine, bromine, and iodine.

Hanger: A support placed in the desired place to support long refrigerant lines.

Head: Pressure that is usually expressed in feet of water.

Header: A piece of pipe that is large enough to carry the total volume of fluid that may flow through it from several different pipes. It may also be used to carry fluids to a given point.

Head pressure: The pressure caused by the condenser against which the compressor must pump the vapor refrigerant.

Head-pressure control: A control that is operated by pressure. It opens the electrical circuit to stop the compressor if high head pressure is experienced.

Heat: A form of energy produced by the expenditure of some other form of energy.

Heat of compression: The heat developed by the compressor when compressing the vapor refrigerant.

Heat content: The amount of heat expressed in Btu per pound absorbed by a refrigerant when raising its temperature from some predetermined point to some final condition and temperature. When a change of state is experienced, the latent heat of the substance will be required for this change to occur.

Heat exchanger: A device used to transfer heat from one substance into another.

Heat of fusion: The amount of heat required to change a solid to a liquid or a liquid to a solid with no change in its temperature. Also, the latent heat of fusion of a substance.

Heat intensity: The concentration of heat in a substance indicated by the temperature of that substance. Heat that can be measured with a thermometer.

Heat lag: The amount of time required for heat to travel through a substance when only one side is heated.

Heat leakage: The heat that flows through a substance having a temperature difference on each side.

Heat of the liquid: The amount of heat required to raise the temperature of a liquid from a predetermined point to some final temperature. The heat content of the liquid.

Heat load: The amount of heat in Btu required to change the temperature of the inside of an enclosure.

Heat sink: A surface that is relatively cold and will absorb heat. A heat sink is usually used as a place to locate heat sensing controls.

Heat transfer: The movement of heat from one point to another. It may be transferred by radiation, convection, or conduction.

Heat unit: This usually refers to the Btu.

Heat of the vapor: The heat content of a gas. The heat necessary to raise the temperature of a liquid from a predetermined point to the boiling temperature, plus the heat of vaporization required to change the liquid to a gas.

Hermetic compressor: A unit that has both the compressor and motor in one single housing. The motor is continuously surrounded by refrigerant vapor, which helps to keep it cool.

Hertz (Hz): Used to designate the frequency of an electrical distribution system.

Hg (mercury): At room temperature it is a liquid metal. It is used in many thermostats to complete the electrical circuit when it moves to cover two contacts in a mercury bulb.

High-pressure cut-out: A switch used to sense the pressure in the high side of a refrigeration system and stop the compressor when it reaches a predetermined pressure.

High side: That part of a refrigeration system that, during operation, is subjected to the discharge pressure of the compressor. Sometimes it is used when making reference to the condensing unit.

High-side charging: The introduction of liquid refrigerant into the high side of a refrigeration system. This procedure must be used on some types of equipment.

High-vacuum pump: A vacuum pump that has the capability to pump a vacuum in the range of 1,000 to 1 microns.

Holding charge: The partial charge placed in the system after it has been evacuated. It is usually used for shipping purposes.

Holding coil: The electrical coil used in a relay, starter, contactor, or other electromagnetically operated control that makes the control function when it is energized.

Horsepower: A unit of power. The amount of energy that must be expended to raise 33,000 lb through a distance of 1 ft in one minute.

Hot gas: This is the refrigerant discharged from the compressor.

Hot-gas bypass: Sometimes used as a capacity control device. It consists of a connection between the suction and discharge of the compressor and is automatically controlled by a valve.

Hot-gas defrost: A method used for defrosting an evaporator coil. It uses the hot gas discharged from the compressor to melt the frost or ice.

Hot-gas line: The line that connects the compressor to the condenser and provides the path for the refrigerant to get into the condenser.

Hot wire: This is a wire made from a high resistance material. It is used in some starting relays and in other heating applications.

Humidity: The amount of moisture in the air.

Humidity, relative: The amount of moisture that is actually in a sample of air compared to the amount it could hold at that temperature. It is expressed as a percentage.

Hunting: The erratic operation of a control that is attempting to establish some system operating equilibrium under adverse conditions.

Hydrocarbon: An organic compound that contains only hydrogen and carbon atoms in various combinations.

Hydrometer: An instrument used for measuring the specific gravity of a liquid. It operates by floats that indicate the specific gravity of the liquid.

I

ICC (Interstate Commerce Commission): A U.S. government agency that oversees the design, construction, and shipping of pressure containers.

Ice cream cabinet: A commercial refrigeration cabinet that operates with a temperature of about 0°F inside. It is used to store ice cream until it is sold.

Impedance: This is an opposing force in an electrical circuit. It resists the flow of an alternating current much like the resistance in a circuit.

Impeller: The part of a pump that actually pushes the water through the pipes.

Induced draft cooling tower: A cooling tower that uses one or more fans to blow air through the cabinet to remove the heat and saturated air.

Induced magnetism: The magnetism in a piece of metal caused by the magnetic induction of an electric current.

Inductive reactance: The property of an electric circuit that creates a CEMF in the circuit as the current changes. Inductive reactance is in direct opposition to the applied current.

Inhibitor: A substance that prevents the chemical reaction such as corrosion and oxidation of a piece of metal.

Instrument: The term used to indicate a tool used for measuring, recording, indicating, and controlling a unit.

Insulation, electrical: A material that has practically no free electrons that is placed on electric conductors to prevent a short.

Insulation, thermal: A material having a high resistance to the flow of heat. It is placed around something to reduce the amount of heat lost from it.

Interlock: The part of a control that prevents something else from operating until a required function has occurred or is occurring.

Ion: This is a group of atoms that has either a negative or a positive electrical charge.

IR drop: This is a term used to indicate the voltage drop in an electric circuit. It is found by multiplying amps times resistance (I×R).

Isothermal: A term used to describe some change in volume or pressure of a substance that is under constant temperature conditions.

Isothermal expansion and contraction: The expansion and contraction of a material that has no change in temperature.

J

Joint: The connection between two surfaces, such as two pipes.

Journal, crankshaft: The bearing surface on the crankshaft of a compressor. Its purpose is to provide the smooth surface of the rod bearing to run on.

Junction box: An electrical box where electrical connections are made.

K

Kilowatt (kW): The kilowatt represents 1,000 watts of electrical power.

Kilowatt hour (kWh): This is 1,000 watt hours of electrical energy.

King valve: A service valve installed on the outlet of the receiver tank, when used.

L

Lag: This term represents the delay in the response to a demand from another control.

Latent heat: Latent heat is the amount of heat added to a substance to change its state without a change in its temperature.

Latent heat of condensation: The amount of heat that must be removed from a substance to make it change from a vapor to a liquid and result in no change in its temperature.

Leak detector: A tool used to locate leaks in a refrigeration system. It may be a halide torch, an electronic leak detector, soap bubbles, an ultrasonic leak detector, or one that uses a dye to locate the leak.

Liquid: A substance that has free-moving molecules that are closer together than those in a vapor or gas.

Liquid charge: A term used to designate the charge in the power element of a thermostatic expansion valve and some temperature controls.

Liquid filter: A strainer made of very fine material and used for removing foreign particles from the liquid refrigerant.

Liquid indicator: An accessory placed in the liquid line to indicate if a solid column of liquid is flowing past that point. It has a glass eye to allow observation of the refrigerant as it passes through the part.

Liquid line: The line that connects the condenser or receiver to the evaporator coil. During operation, it contains liquid refrigerant.

Liquid receiver: A tank that is connected to the outlet of the condenser and is used to store liquid refrigerant during the normal operating cycle. It is sometimes used to store refrigerant during system repairs.

Liquid-receiver service valve: Also known as a king valve. It may be either a two- or a three-way manually operated valve located on the outlet of the receiver. Its main purpose is for use during service procedures.

Liquid sight glass: This is a glass eye that is placed in the liquid line so the technician can determine if liquid refrigerant is flowing past that point. It is also known as a liquid indicator.

Liquid stop valve: This is an electromagnetically operated valve that is placed in the liquid line to control the flow of refrigerant to the evaporator. It operates in response to some other control such as a temperature control. It is sometimes referred to as a solenoid valve.

Liquid strainer: See liquid filter.

Liquid-vapor valve: This is a dual type hand valve used on refrigerant cylinders to permit the removal of either liquid or gaseous refrigerant as desired.

Load: The required removal of heat from a substance or material. The amount of heat removed in a certain amount of time places a load on the system for a particular application.

Locked rotor amps (LRA): The amount of current that flows to an electric motor when it starts, or when there is some type of malfunction preventing the movement of the motor shaft. It is usually as much as six times the full load amperage of the motor.

Low-pressure control: This is a pressure operated switch that senses the system low-side pressure. When the low-side pressure drops to a predetermined point, the control contacts open to stop the compressor to prevent damage to it. It is also used to control the temperature of the storage compartment.

Low-side charging: The procedure used when charging refrigerant into the low side of the system. It is usually used when only a small amount of refrigerant is needed and can be safely charged into the compressor suction.

Low-side pressure: The refrigerant pressure in the low side of a refrigeration system.

M

Machine: This refers usually to a complete refrigerating unit.

Machine room: A room where the refrigeration and other equipment is installed. The evaporator is usually installed inside the cabinet.

Magnetic across-the-line starter: A control used to start and stop a motor or a compressor. Full line voltage can be applied to the contacts of this control.

Magnetic field: The field of force caused around a conductor because of current flowing through that conductor.

Magnetic gasket: The rubber strip placed around refrigerator doors to prevent the loss of heat through the crack around the door.

Magnetism: The magnetic attraction of magnetic materials that are made of iron or some other magnetic material.

Male thread: The thread on the outside of a piece of pipe, fitting, or a valve that permits making a screwed connection.

Manifold: The place in a refrigerant line where several branch lines are connected together. It is sometimes only one piece of pipe that allows refrigerant from more than one pipe to flow through it.

Manifold, discharge: This is a manifold used to collect the discharge gas from more than one compressor cylinder or several compressors.

Manifold, service: A manifold equipped with gauge and hose connections and is used by the technician to check the refrigerant pressures in the system and to perform various service operations.

Manual shutoff valve: A hand-operated valve used to manually control the flow of refrigerant through a pipe.

Manual starter: A motor switch that is operated manually to start and stop an electric motor. It is usually equipped with an overload mechanism.

Mass: A quantity of matter that forms one body.

Master switch: A large electrical switch that is manually operated to control the starting and stopping of a complete system.

Mechanical efficiency: The amount of work done by a unit compared to the amount of energy used to do the work.

Megohm: This is a measure of electrical resistance. One megohm is equal to one million ohms resistance.

Megohmmeter: This is an instrument used for measuring very high electrical resistances.

Melt: The changing of a substance from a solid to a liquid.

Melting point: The temperature at which a substance will melt when exposed to atmospheric pressure.

Mercury bulb: A small glass tube that contains mercury and is used in controls to either make or break an electrical circuit in response to some other function.

Meter: An instrument used to take measurements of various functions.

Micro: This is equal to $^1/_{1,000,000}$ of a specified unit.

Microfarad: This is an electrical unit used to express the capacity of a capacitor. It is equal to $^1/_{1,000,000}$ of a farad.

Micrometer: A very precise measuring instrument used to measure the diameter of bearing journals and other fittings that require very close tolerances.

Micron: A unit of length used in the metric system. It is equal to $^1/_{1,000}$ of an inch. In refrigeration work, it is used to indicate the amount of vacuum inside the system.

Micron gauge: A very accurate instrument used to measure the vacuum inside a refrigeration system.

Milli: This is a measurement that is equal to $^1/_{1,000}$ of a specified unit.

Miscibility: The characteristics of several substances that allow them to mix.

Modulating: A type of control that changes its demands in very small increments rather than being either ON or OFF.

Modulating control: A type of control system that uses modulating controls to regulate the flow of air, refrigerants, or some other fluid through a system.

Molecular: This refers to a material that consists of molecules.

Molecular weight: The weight of an average molecule of a substance.

Molecule: The smallest particle of a substance that can exist alone.

Molliers diagram: A graph or diagram that indicates refrigerant pressure, heat, and temperature. It is used sometimes to determine the operating efficiency of a unit.

Motor: A device that changes electrical energy into mechanical energy.

Motor burnout: This condition occurs when the winding in a motor has shorted and allows the current to bypass a part of the winding.

Motor, capacitor: A type of motor that uses a capacitor to start, then it operates as an induction-type motor. It has a higher starting torque than the single-phase motor.

Motor, capacitor start and run: An electric motor that uses one capacitor to start and one to run. Both the starting and running efficiencies are improved. The run capacitor is designed to remain in the circuit while the motor is operating. The starting capacitor is switched out after the motor has started.

Motor control: An electrical device used to start and stop a motor in response to another function.

Motor, shaded pole: These are small induction type motors. They use a shading pole to start. They have a very small amount of starting torque.

Motor starter: A device that contains more than one set of contacts that are opened or closed at the same time. They are electromagnetically operated by a control circuit.

Movable contact: The part of a set of contacts designed to move. They are moved by an actuating system inside the switch.

Muffler: A noise-limiting device placed in the discharge line of the compressor. It reduces or eliminates the pulsating noise generated by the compressor.

Mullion: A stationary part of the frame located between two doors on a refrigerator cabinet.

Mullion heater: An electrical resistance heating tape that is installed in the mullion to prevent it from sweating.

Multiple system: A type of refrigeration system that has more than one evaporator connected to a single condensing unit.

N

Natural convection: The natural movement of air resulting from only a temperature difference. This is the method used to cool some domestic refrigerator condensers and for air circulation inside the cabinet.

Natural-draft cooling tower: A type of cooling tower that depends on the natural circulation of air through it to cool the water to the desired temperature.

Neoprene: This is a type of synthetic rubber used to make gaskets for use in refrigeration systems.

No-frost refrigerator: A type of refrigeration cabinet in which there is no frost accumulation on the evaporator or any of the materials stored inside it.

Nominal-size tubing: A type of tubing that has the same inside diameter as a piece of iron pipe the same size.

Noncondensable gas: Any type of foreign gas that may enter a refrigeration system. They cannot be condensed at the pressures and temperatures normally encountered in a refrigeration system.

Nonferrous: A type of metal alloy that has no iron in its makeup.

Nonfrosting evaporator: A type of evaporator that never has a collection of frost or ice on its surface.

Normal charge: A refrigerant charge that is part liquid and part vapor under all operating conditions.

Normally closed contacts (NC): A set of contacts that are closed when the relay is in the de-energized condition.

Normally open contacts (NO): A set of contacts that are open when the device is de-energized.

North pole, magnetic: The end of a magnet from which the lines of force flow.

Nucleus: This is the center of an atom. It is electrically neutral. It has the same number of electrons and protons.

O

Odor: Contaminants that affect the sense of smell.

OFF cycle: The time when a refrigeration system is not operating.

Ohm (R): This is a unit of electrical resistance. One ohm of resistance exists when one volt causes one ampere of current to flow in an electrical circuit.

Ohmmeter: A test instrument used to measure the resistance of some component.

Ohm's Law: A relationship between voltage, current, and resistance. It has mathematical representation by the letters E, I, R. It may be expressed in the formula: Voltage (E) = Amperes (I) × Ohms (R).

Oil binding: A condition in which a film of oil over the refrigerant will prevent it from evaporating at its normal operating pressure and temperature.

Oil check valve: A type of check valve installed between the compressor suction manifold and the crankcase to allow oil to return to the crankcase but prevent oil from leaving the crankcase on start-up.

Oil, compressor lubricating: A special type of lubricant designed specifically for use in refrigeration compressors.

Oil, entrained: The droplets of oil that are carried out into the system by the high velocity refrigerant gas.

Oil equalizer: A small pipe that is installed between two or more compressors connecting their crankcases together. Its purpose is to maintain the desired oil level in all the crankcases.

Oil filter: A type of filter that removes any foreign particles from the oil before they can reach the bearing surfaces.

Oil level: The lubricant level in a compressor crankcase. It is the amount required to properly lubricate the compressor.

Oil loop: A loop placed in the piping at the bottom of a riser that causes the refrigerant to force the oil up the riser.

Oil-pressure-failure control: A control that senses the compressor oil pressure and will stop the compressor if the desired pressure is not reached within a specified amount of time.

Oil-pressure gauge: A gauge that indicates the amount of pressure developed by the oil pump inside the compressor.

Oil pump: A gear-type pump that is installed on the compressor crankshaft. Its purpose is to pump the oil to the compressor parts under pressure to provide proper lubrication for them.

Oil-return line: The small line between the oil separator and the compressor crankcase. Its purpose is to allow the oil collected by the separator to re-enter the compressor crankcase.

Oil separator: A shell-like device that is installed in the compressor discharge line. Its purpose is to separate the oil from the refrigerant and return the oil to the compressor crankcase.

Oil sight glass: A glass eye placed in the compressor body at the desired oil level. It will indicate the amount of oil in the compressor crankcase.

Oil sludge: A thick, slushy substance that indicates that the oil is contaminated.

Oil trap: A low place or sag in a refrigerant line, or any place where oil will collect.

ON cycle: The period when the unit is energized and is operating.

Open circuit: An electrical circuit that has been interrupted and the flow of electricity stopped.

Open compressor: A compressor whose drive motor is located on the outside of the compressor body.

Open display case: A commercial refrigeration type case that is designed to keep its contents at the desired temperature, even though the cabinet is not enclosed.

Open-type system: A type of refrigeration system that uses an external source of power to turn the compressor.

Operating cycle: The period when the unit is operating automatically to maintain the desired conditions inside the space.

Operating pressure: The actual pressure that exists when the unit is operating.

Output: The amount of energy that the system is delivering in a given period of time.

Overload: This condition occurs when the unit is experiencing a load greater than that for which it was designed.

Overload protector: A control that is designed to stop operation of the motor if an overload condition occurs.

Overload relay: This is a thermal device that will interrupt an electrical circuit when an excessive amount of current flows through a heater coil.

Oxidize: A material that has deteriorated from corroding, rusting, or slow burning is said to be oxidized.

Oxygen: That part of the atmosphere that is essential for animal life.

P

Package units: This is a complete unit that is usually located inside the refrigerated space. All the components needed are placed in a single cabinet.

Packing: A type of resilient material that is placed around the stems of some valves and the shafts of some types of pumps. Its purpose is to prevent leakage of refrigerant or water at this point.

Packless valve: A valve that has no packing around its stem.

Partial pressure: This is a condition that occurs when two or more gases occupy the same space and each gas causes a part of the total pressure.

Pascal's Law: This law states that a pressure on a fluid is transmitted equally in all directions.

Performance: This is a term often used to indicate the efficiency of a unit.

Performance factor: The ratio of the amount of energy delivered by a refrigeration system as compared to the amount of energy consumed for operation.

Permanent magnet: A type of material that has all its molecules aligned in one direction and can create a magnetic field around itself. Also, a piece of material that has been magnetized.

Phase: A definite part of an operating cycle of a refrigeration system.

Photoelectricity: A condition in which a flow of electricity is caused by light waves.

Pilot control: A secondary valve that is designed to sense the refrigerant pressure inside the line and control the main valve in response to this pressure.

Piston: The bucket-type cylinder placed inside the compressor cylinder, connected to the crankshaft by a connecting rod, and used to increase the pressure on the refrigerant and force it through the system.

Piston displacement: The volume in a cylinder that is swept by the piston as it travels through its stroke.

Pitch: The amount of slope of a line used to help in draining the oil to some preferred place in the system. Usually expressed in inches per foot.

Polystyrene: An insulation used in refrigeration cabinets that is made from plastic.

Potential, electrical: The electric force that makes the electrons flow through a conductor or some resistance.

Potential relay: A type of starting relay that opens its contacts when the proper voltage is sensed by its coil. Its contacts close when the electrical supply drops to a predetermined value.

Potentiometer: This is a wire-wound coil that is used to measure or control an operation by sensing very small changes in electrical resistances.

Power: A measure of the amount of time and energy required to do a job.

Power element: The sensing element of a temperature-operated control. They are also used on thermostatic expansion valves to sense the temperature of the suction line.

Power factor: This is the correction factor used on alternating electricity because the voltage and current are constantly changing.

Ppm (parts per million): A measurement of the concentration of a contaminant in a solution.

Precooler: A type of cooler used to cool a product before it is shipped, stored, or processed.

Pressure: The amount of force exerted on an object. It is measured in pounds per square inch (psig) and absolute pressure (psia).

Pressure, condensing: See pressure, discharge.

Pressure, crankcase: The amount of pressure present in the compressor crankcase.

Pressure, discharge: The amount of pressure that the compressor must pump against while operating.

Pressure drop: The loss of pressure due to some restriction or lift.

Pressure gauge: A tool used to determine the pressure in a given vessel. It is read in psig (pounds per square inch gauge).

Pressure, gauge: The pressure measured above atmospheric pressure. Thus, the reading in a gauge is 14.7 psi lower than the absolute pressure at that time.

Pressure, heat diagram (P-H chart): A graph that represents the refrigerant pressure, heat, and temperature characteristics. It is also sometimes referred to as a Mollier diagram.

Pressure limiter: A type of control that remains closed until a predetermined pressure is reached, then it opens to release pressure to some other part of the system or it may open or close a set of electrical contacts.

Pressure motor control: A type of control that opens or closes a set of electrical contacts in response to the changes in system pressure.

Pressure-operated altitude valve (POA): A suction pressure regulating valve that maintains a constant low-side pressure when operating regardless of the altitude.

Pressure-regulating valve: A type of valve that maintains a constant outlet pressure at its outlet regardless of the amount of pressure variation on its inlet side.

Pressure regulator, evaporator: A type of valve that is installed in the suction line as it leaves the evaporator to maintain some predetermined pressure inside.

Pressure, saturation: The pressure of a vapor in contact with its liquid. The pressure exerted by the refrigerant inside a cylinder.

Pressure, suction: The pressure of the refrigerant leaving the evaporator and entering the compressor suction port.

Pressure switch: A type of electrical switch operated by either a rise or fall in pressure.

Pressure tube: The small line that carries the pressure to the sensing element of a pressure control.

Pressure water valve: A valve used to control the amount of water flowing through a water-cooled condenser. It acts in response to the discharge pressure.

Process tube: A short length of copper tubing connected to the dome of a hermetic compressor. It is used to perform service procedures on the system.

Protector, circuit: A type of heat-sensing electrical switch that will open its contacts when a high temperature is sensed.

Proton: The particle of an atom having a positive electrical charge.

Pump: A device used to deliver a fluid to some component or process. Their main purpose is to force fluids through a pipe or a series of piping.

Pump, centrifugal: A pump that moves fluids by centrifugal force.

Pumpdown: A procedure in which the pressure in a part of a refrigeration system is reduced to near atmospheric pressure.

Purge: The release of pressure from a system. It may be used to remove noncondensables or impurities from a system.

Q

Quench: To submerge a hot, solid material in some type of cooling fluid like oil or water.

Quick-connect coupling: A type of coupling that allows the connection of system components together by using compression-type fittings.

R

Range: The change between the operating limits of a control.

Rating: This indicates the capacity of a unit.

Reactance: The part of impedance in an alternating current circuit because of capacitance, inductance, or both.

Receiver, auxiliary: An extra receiver tank that increases the capacity of the original. It is useful when the system must be pumped down for service operations on the system.

Reciprocating motion: This describes the back-and-forth motion in a straight line used in reciprocating compressors.

Recirculated air: Air that is brought back to the unit for reconditioning and is returned to the conditioned space.

Reclaim: A term used in refrigeration meaning to bring the refrigerant back to new standards. Reclaim requires that a chemical analysis be performed that certifies that the refrigerant meets new refrigerant requirements.

Recording ammeter: An instrument used to read and make a recording of the amount of current flowing to a given piece of equipment.

Recording thermometer: A temperature-sensing instrument that senses the temperature and makes a recording of the temperature at given times during the cycle.

Recover: The recovery of refrigerant is to remove refrigerant in any condition from a system and store it in an external container without necessarily processing or testing it in any way.

Recycle: This is the process of removing the refrigerant from a system and cleaning it to some degree. The refrigerant is not necessarily brought to new refrigerant standards.

Reed valve: A very thin piece of flat, tempered steel that is placed on valve plates inside a compressor and regulates the direction of refrigerant flow through the compressor.

Refrigerant: A fluid that absorbs heat as it expands or evaporates. It is generally considered to be any substance used as a medium to remove heat from another body and move it to another place.

Refrigerant control: A type of control designed to meter the amount of refrigerant that flows into the evaporator. It is a dividing point between the high and low side of the refrigeration system. In some instances it may designate any type of control that meters the flow of refrigerant, not necessarily into the evaporator.

Refrigerant tables: These are tables that list the various properties of a refrigerant at its saturation pressure and temperature.

Refrigerant velocity: The movement of the refrigerant through the system. It may be at different levels between the high side and the low side of the system.

Refrigerating capacity: The rate at which a refrigeration system can remove heat from a space. It is usually expressed in Btu/h.

Refrigerating effect: The amount of heat that a pound of refrigerant will absorb when changing from a liquid to a vapor at a given pressure and temperature.

Refrigeration: The process of removing heat from a space or material and maintaining that space or material at that temperature.

Refrigeration cycle: This designates the complete sequence of operation used in providing refrigeration.

Relay: An electromagnetic device that senses some demand from another control and either opens or closes a set or sets of electrical contacts to energize or deenergize another circuit.

Relay, control: An electromagnetic device that either opens or closes a set or sets of contacts on demand from another control.

Relay, thermal overload: A temperature sensing control that has a set of normally closed contacts that will open and interrupt the controlled circuit when the temperature rises too high for safe operation.

Relief valve: A type of valve that is designed to open and release excessive pressure from a refrigerant cylinder or tank.

Reluctance: Magnetic lines of force that act as a resistance in an electric circuit.

Remote bulb: A remote bulb senses the temperature or pressure at a given location and transfers that reading to another part of the control, which will react in accordance to the demand from the remote bulb.

Remote-bulb thermostat: A type of temperature control that senses a change in temperature and relays that change to the system controls. In effect, this is a thermostat.

Remote system: A refrigeration unit that has the evaporator and fan located at a different location than the compressor, condenser, and fan.

Resistance: The opposition to the flow of electrical current. It is measured in ohms.

Resistor: A device that provides electrical resistance in an electrical circuit to protect a component or to control something.

Restrictor: A cross section of a pipe that is smaller than the rest of the pipe. It reduces the amount of pressure passing through it.

Reverse-cycle defrost: A defrost method that reverses the direction of refrigerant flow through the system. The hot gas is directed to the evaporator to melt the ice or frost from it.

Reversing valve: A type of valve that changes the direction of refrigerant flow through a system. It directs the hot gas to the evaporator first.

Riser: Refrigerant piping that carries refrigerant from a low point to a high point in the system.

Riser valve: A type of valve used to control the refrigerant flow in a vertical pipe. It may either be manual or automatic.

Rotary-blade compressor: A type of rotary compressor that pumps the refrigerant with moving blades used in place of cylinders.

Rotary compressor: A type of compressor that pumps the refrigerant through the use of a rotating motion.

Rotor: The part of an electric motor or generator that turns.

Running time: The amount of time that the unit operates during each hour of the day.

Running winding: The winding in an electric motor that allows current to flow through it any time the motor is energized. The larger winding in an electric motor.

S

Saddle valve: A type of valve that can be placed over the line and a hole punched in the tubing so that the pressure can be measured. Also a tap-a-line valve.

Safety control: This is a type of control that signals the unit to stop operating when unsafe pressures or temperatures are sensed.

Safety factor: The amount of extra capacity added to heat load calculations.

Safety plug: A type of pressure safety that releases refrigerant pressure from a container when high pressures are sensed.

Saturation: A condition that exists when liquid refrigerant is in contact with the vapor. It is the pressure measured in a refrigerant cylinder.

Schrader valve: A spring-loaded type of valve used to measure the pressures in a refrigeration system. It has a center pin that is depressed when the service hose is connected to it.

Screw pump: A type of compressor that compresses the refrigerant between the lobes of the screws in the compressor.

Seat: The machined surface on which a companion component will touch and stop the flow of refrigerant through that component.

Seat, front: A condition that occurs when the valve stem is screwed all the way in. The flow through the valve is stopped completely.

Second Law of Thermodynamics: This law states that heat will always flow from an object that has a higher temperature to an object that has a lower temperature.

Self-contained unit: A unit that has all of the components placed in one cabinet.

Self-inductance: The magnetic field that is introduced into a conductor carrying an electric current.

Semiconductor: A type of material that has the ability to conduct electrical current. It is neither a good conductor nor a good insulator.

Semi-hermetic compressor: A type of compressor that has the motor and compressor in one housing but it can be serviced in the field.

Sensible heat: The heat that causes a change in the temperature of an object but does not cause a change of state.

Sensor: An electronic control that goes through a physical change or a characteristic change as the surroundings change.

Sequence controls: A type of control that makes a change in small increments in a series or a timed order.

Serpentining: A type of tube arrangement that provides circuits of some desired length. The design is to keep the pressure and velocity drop to a minimum.

Service valve: A manually operated valve placed in the refrigeration system so that service operations can be more easily completed.

Servo: This is a low-power device. It may be either electrical, hydraulic, or pneumatically operated to control a more complex or a more powerful system.

Shell and coil: A type of heat exchanger used mostly in chillers and water-cooled condensers. It consists of a shell with a coil inside.

Shell and tube: A type of heat exchanger that has a tube placed inside a shell. It is usually used in water-cooled condensers and chillers.

Short-circuit: An electrical condition that occurs when a part of the circuit has been shorted out of the main circuit. It allows the current to take an undesired path.

Short cycle: The frequent starting and stopping of the unit. This is usually undesirable because it makes extra wear and tear on the equipment and causes an increased electrical bill.

Shroud: A housing placed over the fan blades of a condenser or evaporator fan to increase the amount of air blowing through the coil.

Silica gel: An absorbent type of material used in liquid and suction-line driers that are used in refrigeration systems.

Silver brazing: The process in which the brazing alloy contains some silver, making it more ductile.

Sine wave: A wave form of a single-frequency alternating current.

Single-phase motor: A type of electric motor that operates on single-phase electricity.

Single-pole, double-throw switch (SPDT): An electric switch that has two sets of contact points and only one blade. When the switch changes position, one set of contacts is always made and one set is always open.

Single-pole, single-throw switch (SPST): An electrical switch that has only one set of contacts. When the blade changes positions the contacts are either made or broken.

Skin condenser: A type of condenser that uses the skin of the appliance as the condenser. Most popular in domestic refrigerators and freezers.

Slugging: A condition that occurs when there is liquid refrigerant or oil passing through the compressor cylinder. It will cause a hammering noise and can damage the compressor.

Solder: The process of joining two metals together by the adhesion process. It usually is done at a temperature below 800°F.

Solenoid valve: An electromagnetically operated stop valve placed in refrigerant lines to stop the flow of refrigerant. It consists of a moving core and an electric coil.

South pole, magnetic: The end of a magnet that the lines of force are always entering; the magnet pole.

Specific gravity: This is the weight of a liquid when compared to water. Water has the specific gravity of one.

Specific heat: The amount of heat required to raise the temperature of one pound of a substance one degree Fahrenheit.

Specific volume: The volume per unit of mass. It is usually expressed as cubic feet per pound.

Splash system, lubrication: A method of lubricating moving parts by agitation of the oil or splashing the oil onto the moving parts.

Split-phase motor: A motor that operates on single-phase power. This type of motor has two windings. Both windings are used to start the motor rotating. Then one is removed from the circuit and the motor operates on only one winding, the main or run winding.

Split system: A unit that has the evaporator located in one place and the condensing unit located in another place. They are connected by the wiring and refrigerant piping.

Squirrel cage: A type of centrifugal fan used in many air-moving applications.

Standard atmosphere: A condition that exists when the air is at 14.7 psia, 68°F temperature, and 36% relative humidity.

Standard conditions: The conditions used to rate refrigeration systems. They are a temperature of 68°F, a pressure of 29.92 hG, and a relative humidity of 30%.

Starting relay: An electromagnetic relay used to remove the starting winding from operation after the motor is running at about 75% of its normal operating speed.

Start winding: The winding in a motor that is made of smaller wire. It is used to either help start the motor or to increase its efficiency.

Static head: The weight of a fluid that results in a pressure being imposed on the remainder of the column. It is also considered to be the resistance that results from lifting the fluid.

Strainer: A device that has a very small screen wire through which the refrigerant must flow. Its purpose is to remove foreign particles from the refrigerant before they can enter a system component and cause damage.

Subcooling: The process of cooling a refrigerant below its condensing temperature.

Subcooling coil: An auxiliary coil installed in the liquid line between the condenser and evaporator. Its purpose is to cool the liquid refrigerant below its condensing temperature.

Sublimation: This occurs when a substance passes from the solid state to the gaseous state with no apparent liquid state.

Suction line: The refrigerant pipe that connects the outlet of the evaporator to the suction port of the compressor.

Suction pressure: The pressure of the refrigerant in the evaporator, the suction line, and the compressor crankcase.

Suction riser: A vertical tube that carries the suction gas from an evaporator that is located on a lower level to the compressor located on a higher level.

Suction service valve: A manually operated valve located on the compressor inlet. It may be used as either a shutoff valve, a service gauge connection, or closed to both ports for normal operation. It is used when performing service operations on the system.

Suction side: The low-pressure side of the system that extends from the evaporator to the compressor suction port.

Suction temperature: The temperature of the refrigerant in the evaporator and the suction line. It is close to the evaporating temperature of the refrigerant in the evaporator.

Sump: A reservoir in which a fluid is kept until needed for some operation.

Superheat: The temperature of a gas above the saturation temperature. The amount of temperature above the boiling point is the amount of superheat.

Superheated gas: A gas that is heated above its boiling temperature. The number of degrees above boiling is the amount of superheat.

Swage: The process of enlarging one end of a tube so that the end of another tube can be inserted to make a leaktight joint.

Switch, disconnect: A switch that controls the electric power to the complete unit. It may or may not have fuses inside it. When this switch is opened, all the electricity to the unit is turned off.

T

Tail pipe: The outlet tube extending from the evaporator.

Temperature: The measure of heat or intensity.

Temperature, discharge: The temperature of the gas as it leaves the compressor cylinder.

Temperature, dry-bulb: The temperature of the air measured with an ordinary thermometer. It is sensible heat.

Temperature, entering: The temperature of a substance as it enters a unit or piece of equipment.

Temperature, evaporating: The temperature of the refrigerant that is boiling under the existing pressure.

Temperature, final: The temperature of the substance as it leaves the process.

Temperature, saturation: The boiling temperature of a refrigerant when subjected to a given pressure. Generally, it is considered to be the evaporator temperature in a refrigeration system.

Temperature, suction: See suction temperature.

Temperature, wet-bulb: The temperature of air when measured with a wet-bulb thermometer.

Test charge: A small amount of refrigerant placed in a system to test for leaks or for shipping.

Test light: A light that has test leads to be used when testing electrical circuits.

Therm: A unit of heat that represents 100,000 Btu.

Thermal overload element: A relay that is equipped with an alloy to sense heat. The element melts when its temperature reaches a predetermined temperature.

Thermal relay: A heat-operated starting relay that opens or closes the starting circuit, depending on whether the motor is starting or stopping, during normal operation.

Thermistor: This is basically a semiconductor whose electrical resistance will vary in response to its temperature.

Thermodisk defrost control: An electrical control that uses a bimetal disk to open and close the switch contacts in response to temperature changes.

Thermometer: An instrument used to measure sensible temperatures.

Thermometer well: A small pocket into which a thermometer is placed for measuring the temperature of a fluid inside the pipe.

Thermostat: An electrical heat-sensing control used to control the operation of a piece of equipment. It starts or stops the unit in response to temperature changes.

Thermostatic expansion valve: A flow-control device operated by both temperature and pressure in the outlet of the evaporator.

Three-phase: A term that designates the type of electricity in a given circuit. The three current circuits differ in phase by one-third of a cycle.

Time-delay relay: A type of relay that is actuated after a given period of time has passed.

Timer: A clock-operated control used to open or close an electric circuit at some predetermined time.

Ton of refrigeration: The capacity of a refrigeration unit that will remove 200 Btu per minute, 12,000 Btu per hour, or 288,000 Btu per day.

Torque: The turning force created by an electric motor.

Torque, starting: The amount of force available to start a load and bring it up to speed.

Total heat: The sum of both the sensible and latent heats of a substance.

Transformer: An electrical device that changes one value of voltage to another value. In refrigeration work, step-down transformers are generally used.

Triple point: A temperature-pressure condition in which the substance is in equilibrium. It can occur in solids, liquids, and gases.

Tube-within-a-tube: A type of heat exchanger used in a water-cooled condenser. It is made from two pieces of tubing, one placed inside the other.

Tubing: A thin-walled type of pipe.

Two-temperature valve: A pressure-operated type valve that is installed in the suction line on multiple refrigeration systems to maintain the evaporators at different temperatures.

U

Ultraviolet: Invisible radiation waves that are shorter than X-ray waves, which are visible.

Urethane foam: A type of insulation placed between the shell and liner of refrigeration cabinets.

Useful oil pressure: This is represented by the difference in pressure between the discharge and suction ports of the oil pump.

V

Vacuum: A vacuum is produced by lowering the pressure inside an enclosed system below the surrounding atmospheric pressure. It is usually measured in microns or sometimes inches of mercury (hG).

Vacuum pump: A pump used to pump a vacuum on a refrigeration system. They are available in single-stage and two-stage.

Valve: A device used to control or to stop the flow of a fluid.

Valve, cap seal: A valve cap that is placed over the valve stem when not in use to protect the stem from damage and to keep dirt from entering the valve parts.

Valve, charging: A charging valve is located on the liquid line, usually near the outlet of the condenser or the receiver. It is used to charge liquid refrigerant into the system.

Valve, condenser shutoff: A shut-off valve installed in the discharge line before the condenser inlet.

Valve, cylinder discharge: A thin strip of special steel located on the compressor valve plate that allows the refrigerant to flow in only one direction—out of the compressor.

Valve, cylinder suction: A thin strip of special steel located on the compressor valve plate that allows the refrigerant to flow in only one direction—into the compressor.

Valve, expansion: A refrigerant flow-control device that controls the flow of refrigerant into the evaporator on system demand. It maintains a constant pressure and temperature in the evaporator.

Valve plate: This is the part that is placed over the compressor cylinder to enclose it for compression purposes. It is located between the compressor body and the cylinder head.

Valve port: The passage in a valve that opens and closes to control the flow of fluid in response to the valve position or the valve button to the valve seat.

Valve, purge: A valve used to slowly remove noncondensables from a refrigeration system.

Vapor: A fluid in its gaseous form. It is formed by the evaporation of a liquid.

Vaporization: The process of changing a liquid to a vapor.

Vapor pressure: The pressure caused by a vapor.

Vapor, saturated: The vapor that is almost at the condensation temperature. If the temperature is lowered slightly, droplets of the liquid will be formed.

Velocity: The speed or rapidly changing position of a body.

Vent: A port used to relieve pressure from an enclosed container.

Voltmeter: An instrument used to measure the voltage in an electrical circuit.

Volumetric efficiency: The relationship between the actual pumping capacity of a compressor and the calculated capacity. It is based on compressor displacement.

W

Walk-in cooler: A commercial refrigeration unit that is large enough for a person to walk in. It is usually used in supermarkets and other places that have a large amount of goods to cool.

Water-cooled condenser: A condenser that uses water as its source of cooling to condense the refrigerant.

Water treatment: The treatment of water so that it will not deposit solid contaminants on the heat-transfer surfaces.

Watt: A unit of electrical power.

INDEX

Note: Page numbers in bold type reference non-text material.

Absolute
  humidity, 28
  pressure, 46
AC circuits, capacitance in, 340–41
Accumulators, 218–19
Air-cooled
  compressors, 144
  condensers, 150–51
    fans for, 153
    maintenance of, 153–54
    See also Condensers
Alternating current circuits, capacitance in, 340–41
Aluminum tubing, 98
Ammeters, 88–89
Atmospheric pressure, 44
Atoms, 296–97
  structure of, 297–300
Automatic expansion valve (AXV), 64–65, 187–91

Bending, tubing, 111–12
Boyle's law, 49, 50
Brazing
  described, 84
  procedures, 83–84
  silver, 84
British Thermal Unit (Btu), 13
Brushes, fitting, 79
Btu (British Thermal Unit), 13

Capacitance, 338–42
  AC circuits and, 340–41
Capacitor analyzers, 89
Capacitors, 339–40
  motors and, 350–53
  replacing, 368
Capacitor-start (CSR) motors, 355
Capacitor-start/capacitor run (CSCR) motors, 356–57
Capillary tube
  refrigeration systems, 62–63
  selection of, 256
Centrifugal
  compressors, 129–30
  switch, 360–61
Change of state, described, 9
Charging
  by contact, 303–4
  cylinders, 78
  by friction, 303
  hoses, 76–77
Charles' law, 50–51
Check valves, 236–37
Chemical compounds, described, 6
Chlorodifluoromethane, 251–52
Circuits
  AC, capacitance in, 340–41
  circulation, 3–4
  parallel, 321–22
  series, 319–21
  series-parallel, 323–25
Circulation circuit, refrigerant and, 3–4
Clean Air Act, 280–81
  Section-608
    compliance dates, 287
    contractor self-certification, 285
    evacuation standards, 285–86
    leak repair requirements, 286
    recovery-recycle unit, 287–88
    technician certification, 284–85
Clearance volume, compressors, 143
Closed refrigeration circuit, described, 3
Cloth, sand, 115
Compliant scroll compressors, 126–28
Compound
  gauge, 74
  retard gauge, 74
Compounds, 284
Compression
  fittings, 109–10
  heat of, 21–22
  ratio, compressors, 141–42
  systems, principles of, 57–59
Compressor valves
    checking condition of, 137–38
    discharge, 136–37
    service, 235–36
    suction, 134–36
Compressors
  centrifugal, 129–30
  clearance volume, 143
  compliant scroll, 126–28
  compression ratio of, 141–42
  cooling, 144–45
  design
    hermetic, 132–33
    open-type, 131–32
    semi-hermetic, 132
  lubrication methods, 139–41
  output, factors controlling, 146–47
  reciprocating, 118–20, 121
  rotary, 122–24
  screw, 124–25
Compressor terminals, locating, 365–66
Condensation
  latent heat of, 19
  water-cooled,
    shell-and-tube, 155–57
    tube-in-tube, 157–58

Condensers
  air-cooled, 151–54
  capacity of, 161
  cleaning, 164–65
  evaporative, 159
  location of, 165
  non condensable gases and, 163–64
  purpose of, 150–51
  temperature and, 162–63
  valves, water-flow control, 166–67
  water, counter flow of in, 160–61
  water-cooled, 154–58
Condensing pressure, refrigerants, 243, 246
Conduction, heat, 14
Conductors
  electrical, 311–12, 327–29
    stranded, 329
Contactors, motors and, 375
Control valves, water-flow, 166–67
Convection, heat, 15
Cooling
  defined, 25
  evaporation and, 25–26
  expansion and, 26–27
  subcooling, 27
  water, condensers and, 160–61
Copper tubing
  cutting, 98–100
  seemless, 96–98
Crankcase heaters, 233–35
Critical
  pressure, 46
  temperature, 38
CSR (Capacitor-start) motors, 355
Current, 313–14, 316
  parallel circuits and, 322
  series circuit and, 320
Cylinders
  charging, 78
  refrigerant, handling of, 276

Dalton's law of partial pressure, 51
Defrosting, evaporators, 179–80
Dehydration, described, 29–30
Density, defined, 33
Dew-point temperature, 41–42
Dichlorodifluoromethane, 250, 251
Digital gauges, 76
Direct-expansion (DX) evaporators, 172–73
Discharge
  mufflers, 232–33
  valves, 136–37
    checking condition of, 138
Driers, filter
  liquid line, 220–21
  suction-line, 222–24
Dry
  bulb temperature, 39
  evaporators, 172–73
  refrigeration systems, 63–65
DX (Direct-expansion) evaporators, 172–73

Electric defrost, evaporators, 179–80
Electricity
  capacitance and, 338–42
  charge, neutralizing, 305
  circuit,
    described, 318–19
    parallel, 321–22
    series, 319–21
    series-parallel, 323–25
  conductors of, 311–12, 327–29
  current, 313–14, 316
  inductors and, 336–37
  insulators, 312
  magnetism and, 331–34
  Ohm's law, 317–18
  resistance, 316–17
  resistors and, 330–31
  static, 301–2
Electric motors
  capacitors and, 350–53
    replacing, 368
  capacitor-start, 355
  capacitor-start/capacitor run (CSCR), 356–57
  centrifugal switch and, 360–61
  contactors and, 375
  hard start kit, 372–74
  permanent-split capacitor (PSC), 355–56
  protectors, single-phase, 361–65
  shaded-pole, 357–58
  single-phase, testing, 365–67
  split-phase, 353–54
  starters and, 375–76
  starting relays
    amperage (current), 369–70
    potential (voltage), 370–71
    solid-state, 371–72
  theory of, 346–50
  two-speed, 359
Electromagnets, 334–35
Electromotive force (EMF), 315
Electron, 299–300
  energy, 308–9
  free, 314
  movement, 314
  orbits, 306–10
Electronic
  leak detectors, 93–94
  temperature testers, 89–90
Electrostatic
  charges, 303–5
  field, 305–6
Elements, 294, 295
EMF (Electromotive force), 315
Energy
  defined, 31
  described, 326
Enthalpy, 24–25
Evaporation
  cooling, 25–26
  refrigeration by, 56–57
Evaporative condensers, 159

Evaporators
  defrosting, 179–80
  design factors of, 177–78
  dry, 172–73
  flat-plate type, 174
  flooded, 172
  gravity type, 174
  heat transfer and, 175–76
    calculation of, 176–77
  oil circulation in, 180–81
  styles of, 173–74
  temperature difference and humidification and, 178–79
Expansion
  cooling and, 26–27
  tube, selection, 256
External overload, single-phase motor protectors, 363–65

Fans, air-cooled condensers and, 153
Filter-driers, 220–24
Fitting brushes, 79
Fittings
  compression, 109–10
  flare, 107
    joining, 107–9
  hose, 110
  low-pass, 77
  "O"-rings, 110–11
  sweat, 100–101
    joining, 102–6
Flare fittings, 107
  joining, 107–9
Flaring tools, 69–70
Flat-plate type evaporators, 174
Flooded
  evaporators, 172
  refrigeration systems, 62–63
Flow-control devices
  theory of operation, 185–87
  types of, 184–87
Flux, solder and, 115
Foods, specific heat of, 17
Force, defined, 31
Forced lubrication, compressors, 139–41
Free electrons, 314
Fusion, latent heat of, 19

Gases, described, 8–9
Gauges
  compound, 74
  compound retard, 74
  digital, 76
  manifolds, 73
  pressure, 45–46, 75
General gas law of pressure, 52
Gravity type evaporators, 174
Grounded windings, 366

Hand tools
  fitting brushes, 79
  flaring, 69–70
  pinch-off, 72
  reversible ratchets, 72–73
  swaging, 70–71
  tubing
    benders, 72
    cutters, 68–69
    reamers, 71–72
Hard start kit, 372–74
HCF-134a, 252, 255
  saturation/temperature table, 261
HCFC-123, 252, 253
HCFC-124, 252, 254
Heat
  British Thermal Unit (Btu), 13
  calculation, 23–24
  compression and, 21–22
  conduction, 14
  convection, 15
  enthalpy, 24–25
  flow, described, 14
  latent, 18–20
  mechanical equivalent of, 25
  movement, described, 12–13
  radiation, 15–17
  sensible, 18
  specific, 17–18
  temperature and, 20–21
Heaters, crankcase, 233–35
Heat transfer
  evaporators and, 175–76
    calculation of, 176–77
Hermetic analyzers, 91
Hermetic compressors, 132–33
High (discharge) pressure, 47
Horsepower, defined, 32
Hoses
  charging, 76–77
  fittings for, 110
Hot-gas defrost, 180
Humidification, effects of, 28–29
Humidity
  absolute, 28
  defined, 27
  relative, 28

Inductance, described, 336
Inductors, 336–37
Insulators, electrical, 312
Internal overload, single-phase motor protectors, 361–62

Latent heat, 18–20
Leak detectors
  electronic, 93–94
  ultraviolet, 94
Liquid
  described, 8
  line driers, 220
  receivers, 167–69
  receiver safety devices and, 168–69
Low (suction) side pressure, 47

Low-pass fittings, 77
Lubrication, compressors, 139–41

Magnetic fields, 332–33
Magnetism, 331–34
  electromagnets and, 334–35
  laws of, 332
Manifolds
  gauge, 73
  valve, 73–74
Materials, magnetizing, 333
Matter
  atoms, 296–97
    structure of, 297–300
  change of state, 9
  compounds, 294
  elements, 294, 295
  forms of, 8–9
  molecule, 295
Micron meters, 90
Millivolt meters, 90
Moisture-liquid indicators, 226–27
Molecules, 295
  described, 6
  motion of, 6–7
Motion, defined, 31
Motors
  capacitors and, 350–53
    replacing, 368
  capacitor-start, 355
  capacitor-start/capacitor run (CSCR), 356–57
  centrifugal switch and, 360–61
  contactors and, 375
  electric, theory of, 346–50
  hard start kit, 372–74
  permanent-split capacitor (PSC), 355–56
  protectors, single-phase, 361–65
  shaded-pole, 357–58
  single-phase, 365–67
  split-phase, 353–54
  starters and, 375–76
  starting relays
    amperage (current), 369–70
    potential (voltage), 370–71
    solid-state, 371–72
  two-speed, 359
Mufflers, discharge, 232–33
Multimeters, 86
Mutual induction, 336–37

Neutrons, 299
Noncondensable gases, condensers and, 163–64
Noncondensables, 30

Ohmmeters, 87–88
Ohm's law, 317–18
Oil
  circulation, evaporators, 180–81
  cooled compressors, 145
  refrigerants and, 259–60
  separators, 227–30
  functions of, 229
  purpose of, 228
  sizing of, 231
Open-type compressors, 131–32
Open windings, 366–67
Orbital shells, electron, 306–10
"O"-rings, fittings for, 110–11
Ozone, stratospheric, 278–79

Parallel circuits, 321–22
Pascal's law, 52
Permanent-split capacitor (PSC) motors, 355–56
Pinch-off tools, 72
Pipe. *See* Tubing
Pocket thermometers, 77–78
Potential relay analyzers, 91
Power
  defined, 32
  described, 326
Pressure
  absolute, 46
  atmospheric, 44
  Boyle's law, 49, 50
  Charles' law, 50–51
  critical, 46
  Dalton's law of partial, 51
  effect of on boiling point, 241–42
  evaluation of, 47
  gauge, 45–46
  gauges, 75
  general gas law, 52
  high (discharge), 47
  low (suction) side, 47
  measurement, 47–49
  Pascal's law, 52
  temperature and, 53–54
Protons, 299
PSC (Permanent-split capacitor) motors, 355–56
Psychrometer, sling, 92

R-11 Trichlorofluoromethane, 249–50
R-12 Dichlorodifluoromethane, 250, 251
R-22 Chlorodifluoromethane, 251–52
Radiation, heat, 15–17
Ratchets, reversible, 72–73
Reactive power, described, 341–42
Receivers, liquid, 167–69
Reciprocating compressors, 118–20, 121
  condensing pressure, chart, 263–64
  two-stage, 121–22
Refrigerant-cooled compressors, 144–45
Refrigerants
  boiling point, effects of pressure on, 241–42, 244–45
  capillary tube selection and, 256
  characteristics of, 240
  charging, 257–58
  circulation circuit and, 3–4
  Clean Air Act and, 280–81
    Section-608, 284–88
  condensing pressure, 243, 246
  critical temperature of, 242–43

cylinders, handling of, 276
equipment, recovery, 289
expansion tube selection and, 256
federal tax and, 283
latent heat of vaporization of, 247, 249
oil and, 259–60
P-H diagram, 262, 265–75
actual cycle, 272–75
cycle performance, 271–72
saturation/subcooling/superheat, 265–70
quantity of, 257
recovery methods, 289–91
recycling/emission-reduction program, 282
return gas/discharge temperature and, 256
standard conditions of, 243
tables, 260
pocket temperature, 262
temperature/pressure chart, 248
types of
HCF-134a, 252, 255, (261)
HCFC-123, 252, 253
HCFC-124, 252, 254
R-11 Trichlorofluoromethane, 249–50
R-12 Dichlorodifluoromethane, 250, 251
R-22 Chlorodifluoromethane, 251–52
R-500, 257
R-502, 257–58
vaporizing pressure, 246–47
Refrigeration
by evaporation, 56–57
compression systems principles and, 57–59
cycle, 59–62
defined, 3
dry systems, 63–65
flooded systems, 62–63
history of, 2–3
Relative humidity, 28
Relay analyzers, potential, 91
Relays, starting
amperage (current), 369–70
potential (Voltage), 370–71
solid-state, 371–72
Resistance
conductor temperature and, 327, 329
electrical, 316–17
parallel circuits and, 321
series circuits and, 319, 320
wire lengths and, 327
Resistors
described, 330
variable, 330–31
Restrictor refrigeration systems, 62–63
Retard gauge, compound, 74
Return gas/discharge temperature, refrigerants and, 256
Reversible ratchets, 72–73
Rotary compressors, 122–24

Safety, welding unit and, 81–82
Sand cloth, 115
Saturation, temperature, 38–39
Scales, temperature, 36–38
Screw compressors, 124–25
Scroll compressors, compliant, 126–28
Seemless
copper tubing, 96–98
steel tubing, 98
Self-induction, 336
Semi-hermetic compressors, 132
Sensible heat, 18
Series circuits, 319–21
Series-parallel circuits, 323–25
Service valves, compressor, 235–36
Shaded-pole motors, 357–58
Shell-and-tube condensers, 155–57
Shorted windings, 366–67
Silver
brazing, 84
solder, 114
Single-phase motor
protectors, 361–65
external overload, 363–65
internal overload, 361–62
testing, 365–67
Sling psychrometer, 92
Soft solder, described, 113–14
Soft soldering, 84
Solder
flux for, 115
silver, 114
soft, described, 113–14
Soldering, soft, 84
Solenoids, 335
Solids, described, 8
Solid-state
starting relays, 371–72
hard start kit, 372–74
Specific heat, 17–18
Splash lubrication, compressors, 139
Split-phase motors, 353–54
Starter, motors and, 375–76
Starting relays
amperage (current), 369–70
potential (voltage), 370–71
solid start, hard start kit, 372–74
solid-state, 371–72
Static electricity, 301–2
Steel
tubing, seemless, 98
welding, 83
Strainers, 225–26
Stranded conductors, 329
Stratospheric ozone, 278–79
Subcooling, 27
Sublimation, latent heat of, 19
Suction
line drivers, 222–24
valves, 134–36
checking condition of, 137–38
Superheat, 23
Swaging tools, 70–71
Sweat fittings, 100–101
joining, 102–6

Temperature
    condensers and, 162–63
    critical, 38
    dew point, 41–42
    dry-bulb, 39
    heat and, 20–21
    pressure and, 53–54
    saturation, 38–39
    scales, 36–38
    wet-bulb, 39–41
Temperature testers, electronic, 89–90
Test equipment
    ammeters, 88–89
    capacitor analyzers, 89
    care of, 85–86
    charging
        cylinders and, 78
        hoses, 76–77
        hoses and, 76–77
    compound
        gauge and, 74
        retard gauge and, 74
    digital gauges and, 76
    gauge manifolds and, 73
    hermetic analyzers, 91
    leak detectors, electronic, 93–94
    low-pass fittings and, 77
    micron meters, 90
    millivolt meters, 90
    multimeters, 86
    ohmmeters, 87–88
    potential relay analyzers, 91
    pressure gauges and, 75
    repair of, 85
    selection of, 85
    sling psychrometer, 92
    temperature testers, electronic, 89–90
    thermometers, pocket and, 77–78
    ultrasonic, 94
    valve manifold and, 73–74
    voltmeters, 86
    wattmeters, 89
    welding unit, 80–83
Thermometers, pocket, 77–78
Thermostatic expansion valve (TXV), 65
Tools, hand. *See* Hand tools
Transformers, 337
Trichlorofluoromethane, 249–50
Tube-in-tube condensers, 157–58
Tubing
    aluminum, 98
    bending, 111–12
    changing size of, 112
    copper, cutting, 98–100
    cutters, 68–69
    equivalent lengths of, 112–13
    reamers, 71–72
    seemless
        copper, 96–98
        steel, 98
Two-speed motors, 359
Two-stage reciprocating compressors, 121–22
TXV (Thermostatic expansion valve), 65

Ultrasonic test equipment, 94
Ultraviolet leak detectors, 94

Valence shell, 308
Valve manifold, 73–74
Valves
    check, 236–37
    compressor service, 235–36
    discharge, 136–37
        checking condition of, 138
    suction, 134–36
        checking condition of, 137–38
    water-flow control, 166–67
    water-regulating, 237
Vaporization, latent heat of, 19, 247, 249
Vaporizing pressure, refrigerants, 246–47
Variable resistors, 330–31
Vibration eliminators, 230–32
Voltage, 315
    drop
        parallel circuits and, 322
        series circuit and, 320–21
Voltmeters, 86

Water
    cooled
        compressors, 144
        condensers, 154–58
            shell-and-tube, 155–57
            tube-in-tube, 157–58
        See also Condensers
    counterflow of, condensers and, 160–61
    flow control valves, 166–67
    regulating valves, 237
Wattmeters, 89
Welding
    procedure, 83–84
    steel, 83
    unit, 80–83
Wet-bulb
    depression, 41
    temperature, 39–41
Windings
    grounded, 366
    open/shorted, 366–67
Wire, resistance in length of, 327, 328
Work, defined, 31–32